Asteroid and Space Debris Manipulation: Advances from the Stardust Research Network

Asteroid and Space Debris Manipulation: Advances from the Stardust Research Network

EDITED BY

Massimiliano Vasile
University of Strathclyde, Glasgow, United Kingdom

Edmondo Minisci
University of Strathclyde, Glasgow, United Kingdom

Volume 250
Progress in Astronautics and Aeronautics

Timothy C. Lieuwen, Editor-in-Chief
Georgia Institute of Technology
Atlanta, Georgia

Published by
American Institute of Aeronautics and Astronautics, Inc.

American Institute of Aeronautics and Astronautics, Inc.
12700 Sunrise Valley Drive, Suite 200, Reston, VA 20191-5807

1 2 3 4 5

ISBN 978-1-62410-323-0

TABLE OF CONTENTS

PREFACE

Asteroids and space debris represent a major hazard for both space and terrestrial assets, not to mention life on the planet. Asteroids, by general definition, are considered as celestial objects orbiting the Sun that do not possess a disc indicative of a planet and are not observed to have the characteristics of an active comet. In March 2016, the IAU Minor Planet Center (the official worldwide organization tasked with collecting observational data for minor planets) counted more than 1.3 million objects in the inner and outer solar system. Out of these objects, only 750,000 are sufficiently well characterised to have a numbered designation. It is estimated that in the asteroid belt alone there are up to 2 million objects larger than a kilometre in diameter, and over 25 million objects with diameters of 100 m. This means that only a small percentage of asteroids, and their orbits, are known and for the ones that are known there is substantial uncertainty on ephemerides, composition, shape and mass. A subset of these objects, together with some comets, are classified as Near Earth Objects (NEOs) if their closest approach to the Sun is less than 1.3 Astronomical Units. As of March 2016, the MPC is tracking 14229 NEOs, with 1690 listed as Potentially Hazardous Asteroids. This further sub-classification is for those asteroids or minor planets with the greatest potential for close approaches to the Earth. Looking at the damage, from 1994 to 2013, there were over 850 recorded impacts with objects ranging from 1 m up to 20 m in diameter.

Looking instead at manmade objects, in 2013 NASA estimated that over 500,000 objects, classified as space debris, are orbiting the Earth. The growth of space debris population represents a collision threat for satellite and manned spacecraft in orbit around the Earth. In recent years it has become clear that the ever-increasing population of space debris could lead to catastrophic consequences in the near term. The Kessler syndrome, where the density of objects in orbit is high enough that a collision could set off a cascade, is more realistic now than when it was first proposed in 1978.

Asteroids and space debris share a number of commonalities:

- both are uncontrolled objects whose orbit is deeply affected by a number of perturbations,
- both have an irregular shape and an uncertain attitude motion,
- both are made of inhomogeneous materials that can respond unexpectedly to a deflection action,
- for both, accurate orbit determination is required, and
- both need to be removed before they impact with something valuable for us.

This list of commonalities demonstrates that a number of underpinning technologies needs to be developed to mitigate the risk posed by both types of objects: observation, orbit determination, and state estimation techniques are required to discover, identify, track and monitor both debris and asteroids. Impact prediction is required to inform decision makers, and derive actions and requirements. Forms of manipulation, removal or deflection technologies/strategies need to be developed and implemented, and finally one can think of exploiting asteroids and the consequences of disposing of space debris (like the re-entry of large objects). The observation, manipulation and disposal of space debris represent one of the most challenging goals for modern space technology. It represents a key scientific and commercial venture for the future in order to protect the space environment for essential telecommunication, navigation and Earth observation services.

In order to address these issues, many countries have created dedicated programs, such as the Space Situational Awareness or the Clean Space programmes of the European Space Agency, or have invested significant funding to start new medium to long term research projects. Among projects supported by the European Commission, Stardust is a four-year EU-wide programme, funded by the FP7 Marie Curie Initial Training Networks (ITN) scheme. One of the key goals of Stardust is to train the next generation of engineers and scientists to turn the threat represented by asteroids and space debris in an opportunity and mitigate, if not remove, the threat of an impact. Stardust integrates multiple disciplines, from robotics to applied mathematics, from computational intelligence to astrodynamics, to find practical and effective solutions to the asteroid and space debris issue. Stardust recruited 15 research fellows who developed new ideas and methods, exploring advanced concepts and solutions in the three main areas covered by the Stardust research programme: modelling and simulation, orbit and attitude determination and prediction, and active removal and deflection of uncooperative targets. Stardust put together two different sectors, academia and industry, two communities, minor planets and space debris, and people with different expertise and background, such as planetologists, astronomers, mathematicians, aerospace engineers, physicists, experts in robotics and computer science in order to push the boundaries of space research to save our future.

This book represents a collection of lectures delivered during the opening training school of the Stardust network in November 2013. The chapters were edited by the research fellows and contain a mix of original material developed by the lecturers and material taken from the relevant literature on the subject. Each lecture was written up, transcribing the material presented into a book chapter, in most cases by a pair of research fellows, with each research fellow responsible for two chapters. This interesting approach allowed the young researchers a chance to delve much deeper into the material, and gave everyone a first chance to work together, both with the other research fellows and with experts in the field. As Benjamin Franklin famously said, "tell me and I forget, teach me and I may remember, involve me and I learn."

Chapter 1 was edited by Helene Ma and contains the material delivered by Dr Alessandro Rossi on the Dynamics and Long-Term Evolution of the Space Debris. Chapter 2 was edited by Ioannis Gkolias and is partially based on a lectured delivered by Prof Colin McInnes on Analytic Modeling of the Long-Term

Evolution of Orbital Debris. Chapter 3 was edited by Davide Amato and is based on a lecture delivered by Dr Claudio Bombardelli on Advanced Orbit Propagation Methods and Application to Space Debris Collision Avoidance. Chapter 4 was edited by Stefano Maro' and is based on a lecture delivered by Dr Ettore Perozzi on The Accessibility of the Near-Earth Asteroids. Chapter 5 was edited by Georgios Tsirvoulis and is based on a lecture delivered by Dr Patrick Michel on Physical Properties of Near-Earth Asteroids. Chapter 6 was edited by Georgios Tsirvoulis and is based on the lecture delivered by Dr Bojan Novakovic on Linking Near-Earth Asteroids to Their Main-Belt Source Regions. Chapter 7 was edited by Fabien Gachet and based on a lecture delivered by Prof Alessandra Celletti on Regular and Chaotic Motions in Dynamical Systems with Applications to Asteroid and Debris Dynamics. Chapter 8 was edited by Davide Amato and is based on a lecture delivered by Dr Juan Luis Cano on Orbital Dynamics About Small Bodies. Chapter 9 was edited by Clemens Rumpf and is based on a lecture delivered by Dr Edmondo Minisci on Numerical Methods for Entry Flow Simulations. Chapter 10 was edited by Stefano Maro' and is based on a lecture delivered by Dr Giovanni F. Gronchi on Classical Methods of Orbit Determination. Chapter 11 was edited by Chiara Tardioli and is based on a lecture delivered by Dr Francesco Topputo and Dr Pierluigi Di Lizia on Optimal Control Problem and Space Trajectory Optimization. Chapter 12 was edited by Clemens Rumpf and Helene Ma and is based on a lecture delivered by Dr Seishiro Kibe on JAXA's Approach to the Space Debris Situation. Chapter 13 was edited by Marko Jankovic and Nicolas Thiry and is based on a lecture delivered by Dr Pierre Bourdon on Lasers for Asteroid and Debris Deflection. Chapter 14 was edited by Marko Jankovic and is based on a lecture delivered by Dr Jan Paul on Robotic Active Debris Removal and On-Orbit Servicing. Chapter 15 was edited by Fabien Gachet and Ioannis Gkolias and is based on a lecture delivered by Prof Hiroshi Yamakawa on Asteroid and Space Debris Mitigation Techniques. Chapter 16 was edited by Natalia Ortiz Gómez and is based on a lecture delivered by Dr Scott J. Walker on Rotational Dynamics and Attitude Control for Space Debris Rendezvous and Capture. Chapter 17 was edited by Nicolas Thiry and is based on a lecture delivered by Prof Massimiliano Vasile on Methods and Techniques for Asteroid Deflection.

These chapters cover the range of topics and disciplines developed within Stardust and provide a mixture of fundamental textbook material, practical applications and examples of key enabling technologies for the future.

Massimiliano Vasile

Edmondo Minisci

August 2016

CHAPTER 1

Dynamics and Long-Term Evolution of the Space Debris

Helene Ma* and Alessandro Rossi*
University of Pisa, Pisa, Italy

I. INTRODUCTION

At the present time, several radar and optical observation facilities are tracking space debris orbiting around the Earth. Notwithstanding the untraceable ones, several investigations since the last few decades have indicated that all these objects constitute a growing threat for space operations.

Moreover, associated with the ongoing space activities, this evolution of population would lead to a collision cascade effect in the near-Earth environment, and thus to a significant growth of the debris population in the future, as exposed in the "*Kessler Syndrome*" scenario.

It became necessary to implement such complex models and numerical codes, whose purpose is to predict the future global evolution of Earth-orbiting bodies, in order to acquire insight into the mechanisms which determine its pace. Through the outcomes of the simulations, international space agencies could point out possible mitigation measures to curb the debris proliferation and to prevent the possible onset of a collisional "*reaction*," representing a serious and long-term hazard to all space activities.

The reason of this trend of evolution will be explained in this chapter. In fact, satellites launched into space are continuously exposed to various perturbations depending on their positions from the Earth. The magnitude of the perturbing effects on a satellite also changes after a few weeks, years, or even centuries and depends on the physical characteristics of the spacecraft. Furthermore, the result of the debris evolution takes into account some hazardous phenomena which are predicted, such as the breakups of spacecraft.

All these factors give this peculiar distribution of these bodies, which will be described in the first section. The main long-term perturbations which induce such distribution will be also defined in this chapter.

*Department of Mathematics.

Besides, we take the opportunity to underline a peculiar problem by taking the example of the so-called "*lunisolar resonances*" in MEO† region, which provide an additional effect on the evolution.

Since the end of the 1980s, several generations of simulation codes have been developed, aimed at modeling the global evolution of the debris population over long time spans. The first simple model will be exposed in the last two sections, with a brief description of the recent versions of the model, and thus, a conclusion of the future work will be expected for the next few years.

The contents of this chapter refer to the lecture given by Alessandro Rossi who gave the essential information about the dynamics and the long-term evolution of the Earth-orbiting population during the Opening Training School in Glasgow in November 2013.

II. ORBITAL DISTRIBUTION

The following populations can be distinguished in the distribution of the space debris around the Earth (see also Fig. 1):

1. Starting from low altitudes until 2,000 km above the Earth's surface, we find *LEO‡ objects*, where the region is the most-crowded zone around the Earth. They represent satellites and upper stages associated to civil and military missions for Earth observation, surveillance, and telecommunication, and

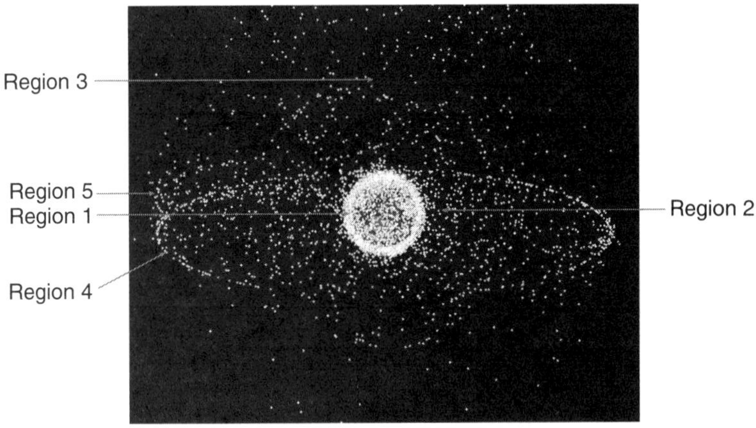

Fig. 1 Overview of space debris distribution around the Earth-localization features. (Image from NASA Orbital Debris Program Office Website.)

† Medium Earth orbits.

‡ Low Earth orbits.

also the fragments from a significant number of fragmentations happening in these orbits.

2. At altitudes between 10,000 and 35,000 km, there are the *MEO objects*, some of them forming a kind of sphere over the LEO region and corresponding to navigation constellations.
3. A ring profile, at about 36,000 km of altitude, corresponds to the region populated by the *GEO*[§] *objects*, which initially remain almost fixed with respect to a ground station.
4. From about 530 km of perigee altitude to 40,000 km of apogee distance, with 63.4 deg of inclination, we find the so-called *Molniya orbits*, where the Russian telecommunication satellites are located. These orbits are exploited since countries located at high latitudes have difficulties to detect satellites positioned in the region described in 3.
5. Other satellites concentrated at the same altitudes as for the GEO region, with higher inclinations, which will be defined in Sec. 2.

In the next few sections, the reason for this specific distribution in circumterrestrial space will be explained, as well as the problems that space debris can cause to active satellites.

III. ORBITAL PERTURBATIONS

In Fig. 2, we plot the magnitude of acceleration for a given perturbation that a satellite undergoes, as a function of its distance from the center of the Earth. The vertical black lines correspond to the following thresholds: the upper limit of LEO at about 8,400 km of distance from the center of the Earth, the GPS orbits at around 26,000 km, and the GEO region orbits at around 42,000 km.

It can be noticed how the magnitude and the relative importance of the perturbations significantly depend on the altitude. The main perturbations are as follows:

1. The main effect is due to the monopole term of the Earth gravity field, expressed as *GM*.
2. The air drag perturbation, which is extremely important for LEO region.
3. The J_2, J_{22}, and J_{88} terms are the most significant coefficients of the expansion of the Earth gravity field in spherical harmonics, due to the nonspherical shape of the Earth (see Sec. A.1. Earth Gravity Field).
4. The gravitational attraction of other planets, mainly the third-body attraction of the Moon and the Sun.
5. The SRP[¶] which is a nongravitational perturbation.

[§] Geostationary Earth orbits.
[¶] Solar radiation pressure (SRP).

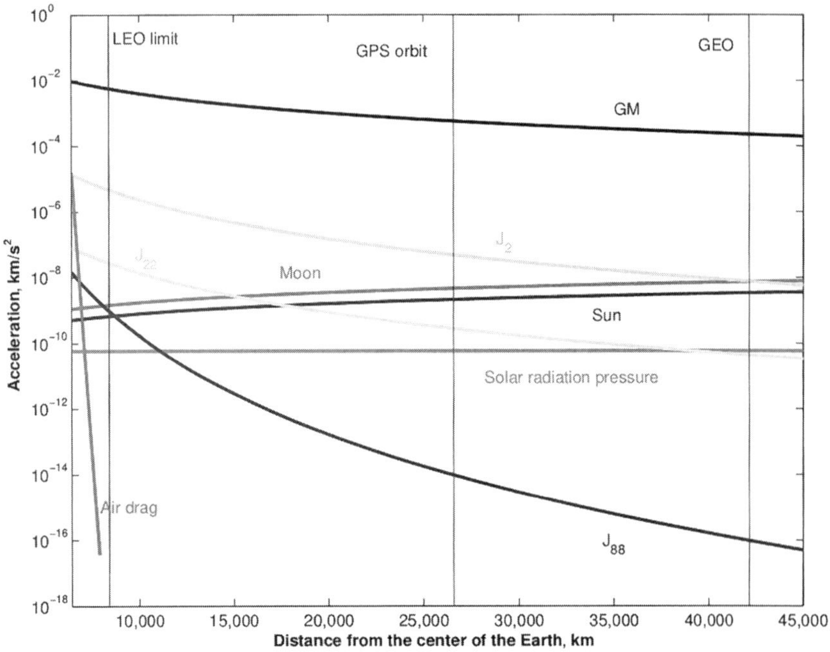

Fig. 2 Orbital perturbations acting on a satellite: we plot the acceleration (km/s^2) as a function of the distance from the center of the Earth (km). The spacecraft is assumed as a spherical shape with an area-to-mass ratio of 0.1 m^2/kg.

We can see from Fig. 2 that different perturbations are more relevant for objects in different regions of space: for example, the J_{22} effect is more significant than the attraction of the Moon for distances less than 15,000 km, and is less important at greater distances.

A. GRAVITATIONAL PERTURBATIONS

1. EARTH GRAVITY FIELD

The expansion of the gravitational potential of the Earth in spherical harmonics is given by

$$V(r,\theta,\lambda) = \frac{GM}{r}\left[1+\sum_{l=2}^{\infty}\sum_{m=0}^{l}\left(\frac{R_e}{r}\right)^l P_{lm}(\sin\theta)\left(C_{lm}\cos(m\lambda)+S_{lm}\sin(m\lambda)\right)\right] \quad (1)$$

r, θ, and λ are respectively the radius, the latitude, and the longitude in a coordinate system whose origin is the center of mass of the body. R_e represents the characteristic physical dimension (e.g., the larger semiaxis of the body, so the equatorial radius of the Earth).

The spherical harmonics expansion is characterized by the Legendre-associated functions P_{lm} and the coefficients of the expansion C_{lm} and S_{lm}, with respect to the degree l and order m.

The gravitational potential of a sphere is equal to GM/r, with G the universal constant of gravitation and M the mass of the body. If we assume a planet as a perfect sphere, all the coefficients of the spherical harmonics expansion [Eq. (1)] are zero. On the other hand, for a body like the Earth, which is not a perfect sphere, some of these terms must be accounted.

For example, J_2 term is the *quadrupole term*, related to the polar flattening of the Earth and represents the most important effect after the monopole term GM/r.

The higher-gravity terms of the satellite are inversely proportional to the distance: the more the altitude h increases, the less it is subjected to these effects, whose magnitude is reduced by a factor

$$\left[\frac{a}{(a+h)}\right]^{(l+1)} \tag{2}$$

with l the harmonic degree of the potential and a the semimajor axis of the satellite's orbit. n and l are the same, i.e. they represent the degree subscript of the spherical harmonics.

These perturbations influence the orbits of the satellites: generally, they tend to distort the shape of the orbits and change the orientation of the orbital plane. Indeed, one of their effects is to mainly change the angular arguments of the satellite orbit: for example, the main effects of J_2 are on Ω and ω, causing the orbital plane to precess.**

However, these perturbations have no impact on the energy of the orbits. Therefore, there is no secular accumulated effect on the semimajor axis, which is important for the long-term evolution of the orbits of space debris.

2. THIRD-BODY PERTURBATION FROM THE MOON AND THE SUN

The gravitational effect on the satellite due to the presence of other celestial bodies, like the Moon and the Sun, is not negligible. The perturbation is caused by the "*tidal terms*" of the gravitational attraction, that is, by the difference between the force on the Earth and the force on the satellite. The perturbation is proportional

**There is a secular regression of the orbital node and a precession of the argument of perigee.

to the distance of the satellite and the perturbing body and in case of the Moon, it is of the order of

$$\left(\sim \frac{r}{r_{\text{Moon}}^3}\right) \tag{3}$$

where r, r_{Moon} are respectively the distance of the satellite and the Moon from the center of the Earth.

Looking at the curves representing the effects of the Moon and the Sun in Fig. 2, we can see that these perturbations are less effective in the LEO region.

If we go further from the surface of the Earth, the more the altitude increases, the closer the satellite will be to these perturbing bodies, and the lunisolar effects will become more significant.

We will see that for high Earth orbits, for example, for satellites with an orbital period equal or longer than 12 h, this kind of effect should be taken into account.

We will stress how the lunisolar perturbation influences the space debris population through Figs. 3 and 4.

The evolution of inclination for cataloged objects in the geosynchronous region is plotted in Fig. 3 as a function of their launch dates.

Two populations can be identified: there is a population of objects which are basically on the equator (zero inclination), representing the real active GEO control satellites launched from 1990 to 2000. Another population represents the

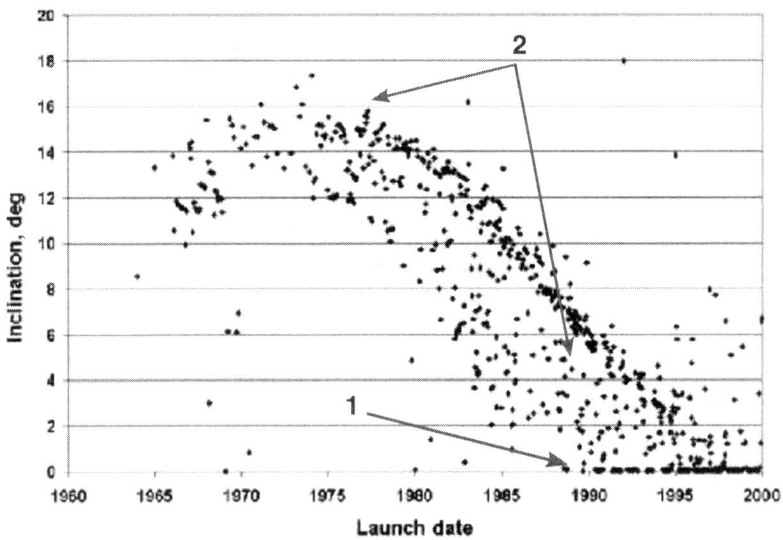

Fig. 3 Lunisolar effects on space debris according to the launch date (in years) and the inclination of the orbits (in degrees).

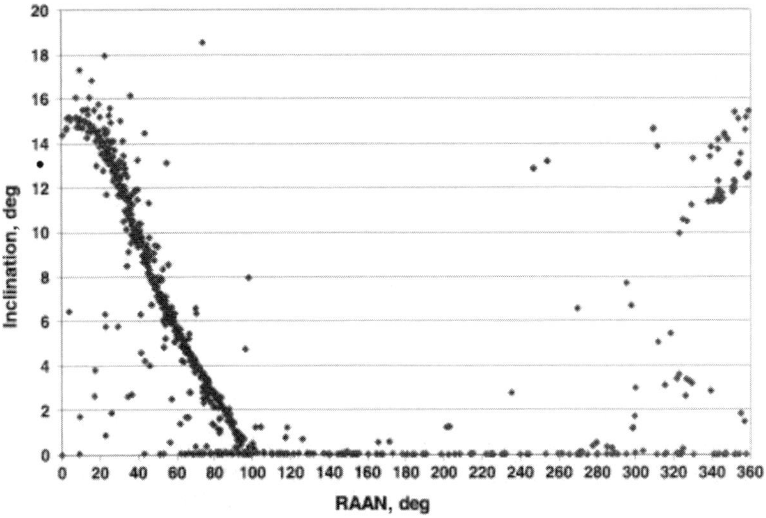

Fig. 4 Lunisolar effects on space debris according to the RAAN and the inclination of the orbit (in degrees).

objects launched from 1965 to 2000, whose inclination shows a long-period evolution, with oscillations between 0 and 15 deg around the stable Laplace plane at an inclination of 7.5 deg.

A similar view is given in Fig. 4, where we plot the inclination of the orbital plane and the RAAN[††] of the same population orbits. Again, a clear correlation between the orbital elements and two different populations can be spotted. Indeed, the population of objects located at zero inclination corresponds to the active GEO satellites mentioned previously, and the population of objects with an RAAN lower than ~100 deg or higher than ~280 deg has the same variations of inclination as seen before.

In fact, the objects with larger inclinations represent the satellites which are abandoned in GEO region. Initially, they orbited close to the equatorial plane with ~0 deg of inclination. Once abandoned in the geostationary ring at the end of life, the perturbation effects due to the Moon and the Sun generated a precession of the orbit pole which in turn causes the observed change in inclination and RAAN.

In other words, the orbit pole precesses around the pole of the ecliptic due to the solar attraction, and also around the pole of the Moon's orbit from the lunar attraction: the result is a kind of a gyroscopic effect, where the mass of the Moon tends to rotate the orbital plane so that it contains the pole of the Moon, and similarly for the Sun.

[††] Right ascension of the ascending node.

Because of this oscillation of the inclination of the orbital plane, the abandoned spacecraft is crossing the GEO ring with a high relative velocity‡‡ and increased collision probability against the controlled operational satellites. This abandoned population refers to population 5 in Fig. 1.

Consequently, in order to avoid the collisions between these objects, basically when a geostationary satellite arrives at the end of its operations, its orbital altitude shall be raised above the operational GEO ring and the plane has to be reoriented (with respect to the Sun and the Moon) so that the satellite will be put into an orbit, which is less affected by the lunisolar perturbations: this orbit is in fact the Laplace plane, which was proved as a robust long-term disposal orbit for GEO satellites in Ref. [1].

3. RESONANCES

Different kinds of resonances, between the motion of the satellite, the rotation of the Earth, and the motion of the perturbing bodies, can significantly alter the behavior of an orbiting object producing long-term and even secular perturbations.

In Ref. [2], a survey of the orbital dynamics in the MEO region is performed in order to understand the phenomenon related to the MEOs instability. The behavior of these objects is well known and is due to the resonance between the orbital motion of the satellite and the rotation of the Earth.

Besides, a so-called *mean motion resonance* appears when the period of revolution of a satellite is in phase with the Earth rotation: this satellite passes over the same region of the Earth at the same time, and therefore undergoes the same gravity anomaly.

In the mean motion resonances, the exact condition for commensurability is that the satellite performs β nodal periods while the Earth rotates α times relative to the precessing satellite orbital plane.§§ After this interval, the path of the satellite relative to the Earth repeats exactly, leading to the resonance. For a long time scale, this phenomenon is so called the "*deep resonance.*"

A typical example of deep resonance happens close to the GPS region. Normally, the overall shape of the constellation is kept under control to assure the global coverage of the Earth. In this respect, the global relative precession of the orbital planes is taken into account and kept under control along with the orbital eccentricities. However, the deep mean motion resonance in this region can produce important changes in the orbital parameters (mainly the eccentricity).

Similarly, the resonance with the motion of the Moon and the Sun greatly enhances the eccentricity growth, giving way to the so-called *lunisolar resonance.* It appears when the secular motions of the lines of apsides and nodes become

‡‡Whereas the controlled GEO satellites at nearly zero inclination have a very low relative velocity.

§§α and β are mutually prime integers derived from the resonance argument expressed in Ref. [2]. The 2:1 mean motion resonance involves the MEOs with α equal to 1 and β equal to 2.

commensurable with the mean motion of the two bodies. We will detail this resonance in Sec. 3.a below.

Concerning the resonances issues in MEO regions, a navigation spacecraft located in MEOs is subjected to the 2:1 mean motion resonance, with a period of half a sidereal day (i.e., $\approx$12 h). This resonance causes long-period variations in the orbital eccentricity and can, for example, modify the configuration of the navigation constellations.

Moreover, the interaction between the third-body and the geopotential perturbations can cause a more complex resonance which induces a very long term and nearly secular modification of the eccentricity (see e.g., Ref. [2]). As we will see in the following section, this can represent a critical hazard for the long-term disposal of the spent satellites and upper stages of rockets in the MEO region.

a. Resonance Onset

Figures 5–7 show the eccentricity evolution over 100 years, for a GPS-like orbit propagated with increasing sophisticated dynamical models, from a simple case with only the gravity monopole term plus lunisolar perturbations (Fig. 5), to a

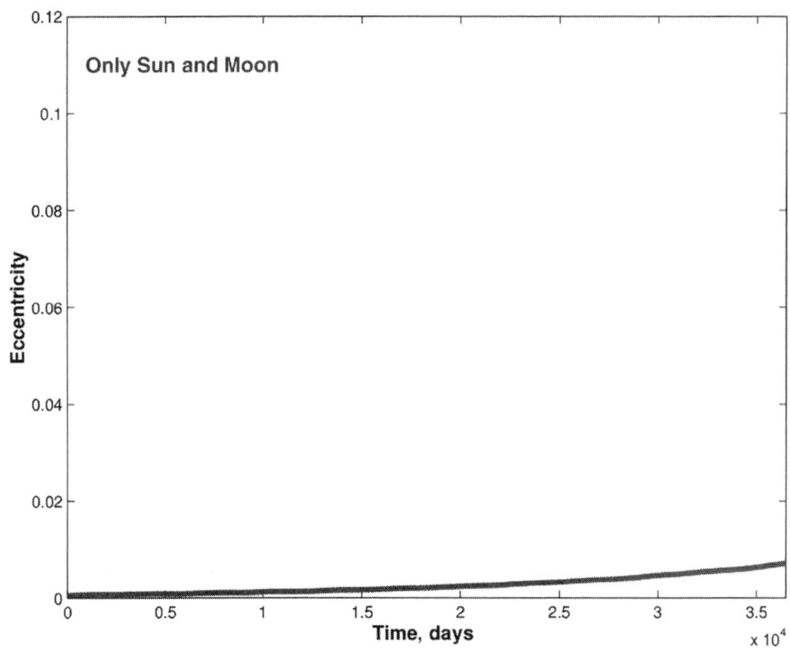

Fig. 5 Resonance onset: Time evolution of the eccentricity considering only the gravity monopole term with the lunisolar perturbations.

full model with Earth gravity harmonics up to degree and order 10 and lunisolar perturbations (Fig. 7).

The first long-term effects of the 2:1 resonance appear when the harmonics up to $\ell = 3$ and $m = 3$ are included. The main point to stress is that the real striking change in the pace of the eccentricity growth happens when a full model including Earth gravity harmonics and lunisolar perturbation is assumed. The eccentricity undergoes a very long-term (quasi-secular) perturbation that, at the end of the 100-year time span, leads to a value more than 3 times larger than in the case without lunisolar perturbations. This is due to a *lunisolar resonance*.

These figures illustrate the problem related to the presence of the resonances: since the growth of the eccentricity will not only distort the shape of the orbit and the constellation with the need of frequent orbital corrections, it will also strongly affect the long-term evolution of disposed satellites at their end of life, as described in the next section.

b. Etalon 2-Orbit Evolution (200 Y)

To understand the dynamics in the MEO region, two bodies were identified as references for the simulation of the orbital evolution in MEOs: the satellites Etalon

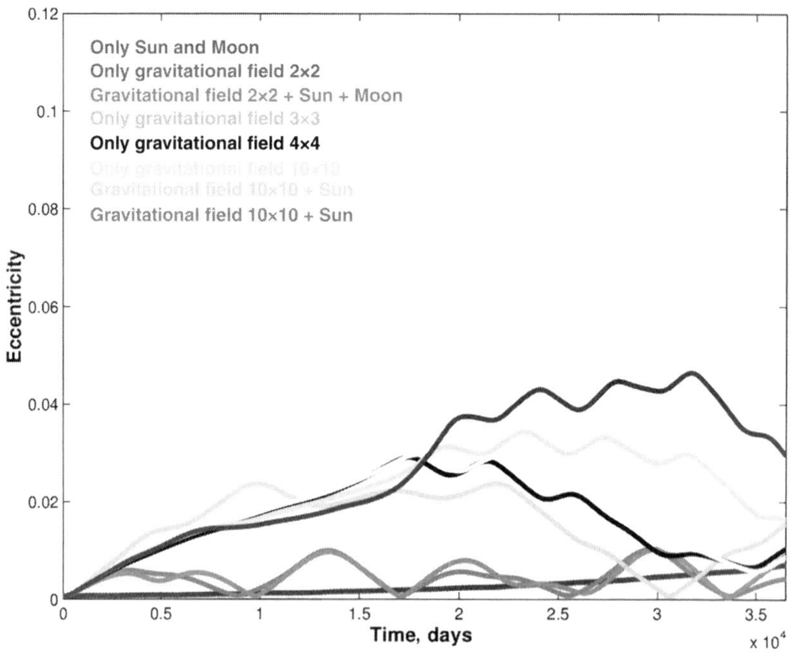

Fig. 6 Resonance onset: Impact of the different perturbations on the eccentricity as a function of time. In all the cases, we have periodic perturbations of the eccentricity, with an increasing period.

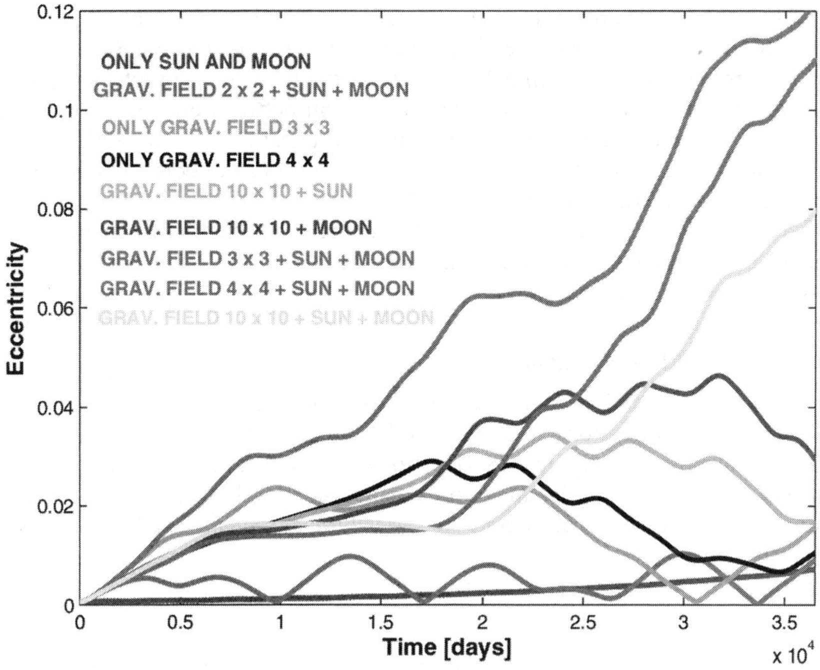

Fig. 7 Resonance onset: Impact of the different perturbations on the eccentricity as a function of time. The full dynamical model is represented by the gravity harmonics up to degree $l = 10$ and order $m = 10$ with the lunisolar perturbations: it gives the most realistic evolution of the eccentricity.

1 and 2. Thanks to their spherical shape and their very low A/M[¶¶] ratio, the model was easy to perform since the effects of the nongravitational perturbations were particularly low.

In Fig. 8, the very long-term evolution of the apogee (top) and perigee (bottom) of Etalon 2 was simulated and plotted. Etalon 2 was initially placed at a circular orbit. However, due to the resonances met in this orbit, the eccentricity growth will cause the satellite to cross the GPS nominal altitude in about 150 years.[***]

Typically, there is no problem for the active satellites of the GNSS constellations to perform maneuvers in order to keep the eccentricity close to zero during their operational mission. However, at the end of their lifetime, the satellites have to be disposed outside the operational region of the respective constellations.

¶¶Area-to-Mass.

***The features of the satellites Etalon 1 and 2 minimize the effects of the nongravitational perturbations. As a consequence, the largest perturbations seen in the figure are gravitational perturbations.

The prevention policy adopted for the GPS satellites is to move them into a circular orbit at least 500 km (depending on the constellation) above the operational orbit. This procedure is intended to minimize the effect of the perturbing resonances so that it prevents any accidental collision between operational and old satellites. Unfortunately, it was shown that the instability of the GPS disposal orbit leads to an increasing of its eccentricity, and thus a periodic and dangerous crossing of the operational orbits.

As a consequence, it is important that the study of the dynamics has to be carried out carefully in this case, with the best possible disposal solutions for the GNSS constellations.

c. Spacecraft and Debris Around GNSSs

Figure 9 shows the distribution of the perigee and apogee for the current population in the navigation constellations. The solid horizontal lines define the operational altitude of GLONASS, GPS, and GALILEO constellations.

The dashed lines limit an altitude band of ±500 km around the operational orbits, which is considered as a "*safety operational band*" around each constellation.

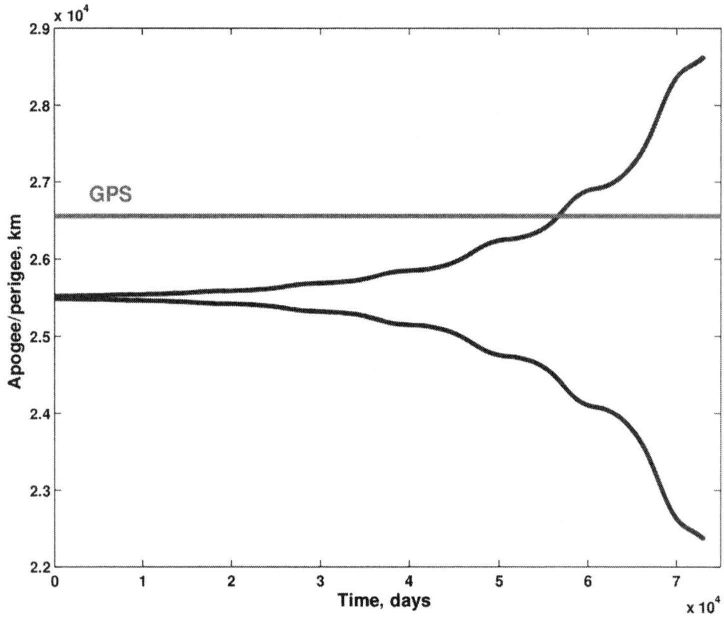

Fig. 8 Evolution of the apogee/perigee of the satellite Etalon 2, propagated for 200 years. The GPS nominal altitude, at about 26,500 km, is shown by the horizontal grey line.

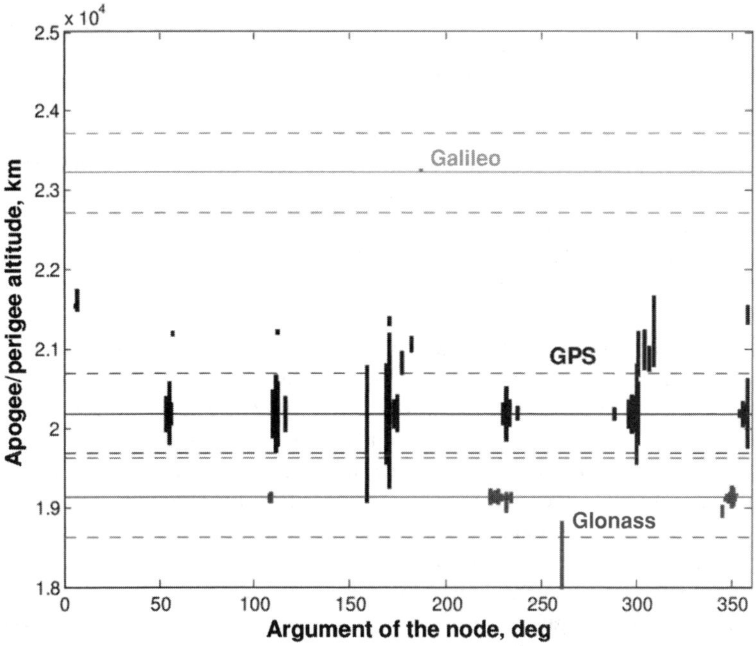

Fig. 9 Apogee/perigee altitude (in km) of spacecraft and debris in the GPS (black) and GLONASS (dark grey) constellations as a function of the argument of the node (in degrees). Forty four are GPS-related objects, while 115 are GLONASS-related objects.

As it can be noticed, while the GLONASS-related objects are significantly numerous, they tend to remain confined within the GLONASS band.

On the other hand, in the GPS constellation, which is more subjected to resonances, we find some objects with considerable eccentricities, which overtake the constellation safety band and moreover, even interconstellation intrusions take place with the GLONASS constellation.

This picture again indicates the effect of the orbital resonances and the need for a thorough global study of the dynamics in the MEO region to avoid the onset of the debris problems already seen in the LEO region.

B. NONGRAVITATIONAL PERTURBATIONS

1. AIR DRAG ACCELERATION

The air drag a_D is a nonconservative acceleration causing a secular effect on the semimajor axis (i.e., a decay of the satellite into the atmosphere). Its expression is the following:

$$a_D = -\frac{1}{2} C_D \frac{A}{M} \rho V_r^2 \tag{4}$$

C_D is a dimensionless coefficient describing the interaction of the atmosphere with the surface materials of the satellite. V_r is the velocity of the satellite with respect to the atmosphere whose density is defined by ρ. With the cross-sectional area A of the satellite and its mass M, the A/M ratio is a parameter which weighs the effect of the perturbation: a satellite with a large area and a small mass will be subjected to a higher effect of the air drag than another with a high mass and a small surface.

The accuracy of the air drag acceleration model has to be studied carefully, seeing as, when we want to propagate the satellite orbit into the atmosphere, the two main uncertain parameters seen in Eq. (4), which are the cross-sectional area of the spacecraft and the atmospheric density, do not remain constant.

Indeed, the atmospheric density $\rho(h, t)$ decreases exponentially with respect to the altitude and moreover, it depends on time, since it is altered by the incoming solar radiations which inflate and deflate the atmosphere.

Therefore, the air drag perturbation is modulated by the solar activity whose period is 11 years: when the Sun activity is at its maximum, the atmosphere is more heated, inflated, and then more expanded. There is also a shorter modulation due to the day and night sides of the atmosphere which becomes respectively warmer and cooler.

Likewise, the cross-sectional area of the spacecraft $A(t)$ might be changing with time. Since the satellite moves and rotates along its orbit, a mean (spherical) cross section is usually taken averaging over the original dimensions of the spacecraft. For very peculiar cases, as will be shown in the following paragraph, this approximation can not be applied.

Figure 10 shows the spacecraft GOCE which ended its mission and performed its reentry in the Earth atmosphere at the end of 2013.

Typically, to represent a spacecraft in the reentry phase simulations, an average cross section is adopted. In the GOCE case, this approximation is particularly not accurate: the spacecraft had a peculiar shape and even wings for the stabilization since it was orbiting at very low altitudes to study the Earth gravity field. Hence, in the crucial phases of the atmospheric reentry, the behavior of GOCE was not easily predictable by the standard models.

The normalized lifetime of a satellite, with a typical $A/M \sim 0.1\ \mathrm{m^2/kg}$, as a function of its altitude is shown in Fig. 11. The index $F10.7$ is a measure of the solar activity and is variable with a 11-year cycle.

For instance, if a satellite is located at 600 km of altitude when the solar activity is low,[†††] it may need more than 1000 days to perform its reentry. Generally, the

[†††]"Quiet atmosphere" is defined by F10.7 equal to 75 and specifies the atmosphere with low solar activity.

Fig. 10 The Gravity Field and Steady-State Ocean Circulation Explorer (GOCE) satellite.

duration for the reentry increases when the level of the solar activity is lower, as well as if the satellite is located at higher altitudes.

The air drag is a nonconservative force, and thus it represents a mean to clean up the space below a certain altitude. Nonetheless, it must be emphasized that although it takes the energy away from the orbit, the lifetime of a satellite in an orbit with just a few hundreds of kilometers of altitude can be very long. In particular, for the most-crowded regions of the LEO zone (between 800 and 1,000 km of altitude), the atmospheric effect is negligible.

2. SOLAR RADIATION PRESSURE

Another important nongravitational perturbation is due to the SRP.

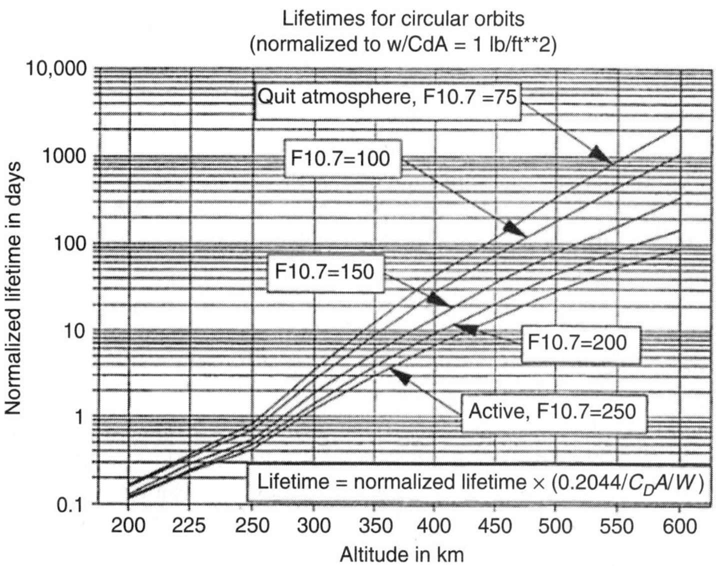

Fig. 11 Air drag acceleration: estimated lifetime (in days) as a function of the altitude (in km). (Source: From Chobotov, *Orbital Mechanics* [8]; reprinted with permission.)

A satellite exposed to solar radiations is subjected to a small force coming from the absorption or reflection of photons from the Sun.

The main effect of the SRP on the orbit is to change the eccentricity. This perturbation does not impact significantly on the total energy of the orbit: the other orbital elements, for example, the semimajor axis, the inclination, and the mean motion, are not affected.

The SRP acceleration is given by the following expression:

$$\ddot{r} = -\frac{A}{M}\frac{\Phi_{\odot}}{c}C_R\left[\frac{a_{\odot}}{\|r_{\odot}\|}\right]^2\frac{r_{\odot}}{\|r_{\odot}\|} \tag{5}$$

The radiation pressure coefficient C_R expresses how the Sun is reflected on the body. $\Phi_{\odot}$ defines the solar flux, c defines the velocity of light, and the satellite is featured by its position r and its A/M ratio. Generally, GEO satellites have large solar panels and antennas and their typical A/M ratio is about 0.1 m^2/kg.

3. SRP: THE CASE OF HIGH A/M OBJECTS

An extreme case of the SRP perturbation effect will be briefly described in this section.

Several years ago, the European Space Agency's 1-m telescope in Tenerife observed and identified a population of faint objects in a peculiar orbit near GEO region (see Ref. [3]): these objects orbited with mean motions of about 1 rev/day, with the same semimajor axis of the GEO satellites, and with orbital eccentricities as high as 0.6.

Note that nothing is launched close to those particular orbits. A study provided by the NASA Orbital Debris Program Office was completed to explain the existence of this specific population: the results from the simulations indicated that the SRP could cause a considerable increase of the eccentricity for objects with very high A/M ratios.

Equation 6 estimates the long-periodic effects in eccentricity and semimajor axis, coming from the solar radiation perturbation on the spacecraft body

$$(\dot{a}\dot{e})_{lp} \simeq \frac{1}{2\Pi}\left(\frac{A\phi_{\odot}}{Mc}\right)P \tag{6}$$

where P is the orbital period of the satellite and $(A\phi_{\odot}/Mc)$ gives the order of magnitude of the total solar radiation force per unit mass of the satellite.

This effect appears as an oscillation of the satellite distance over one orbital period: the amplitude of this oscillation will change roughly with the rate expressed by Eq. (6) and a yearly periodicity [4].

The evolution of the mean eccentricity, as a function of different values of the A/M ratio, is plotted in Fig. 12a. As can be seen, objects with extremely high A/M ratios can actually reach those highly eccentric orbits under the effect of the SRP.

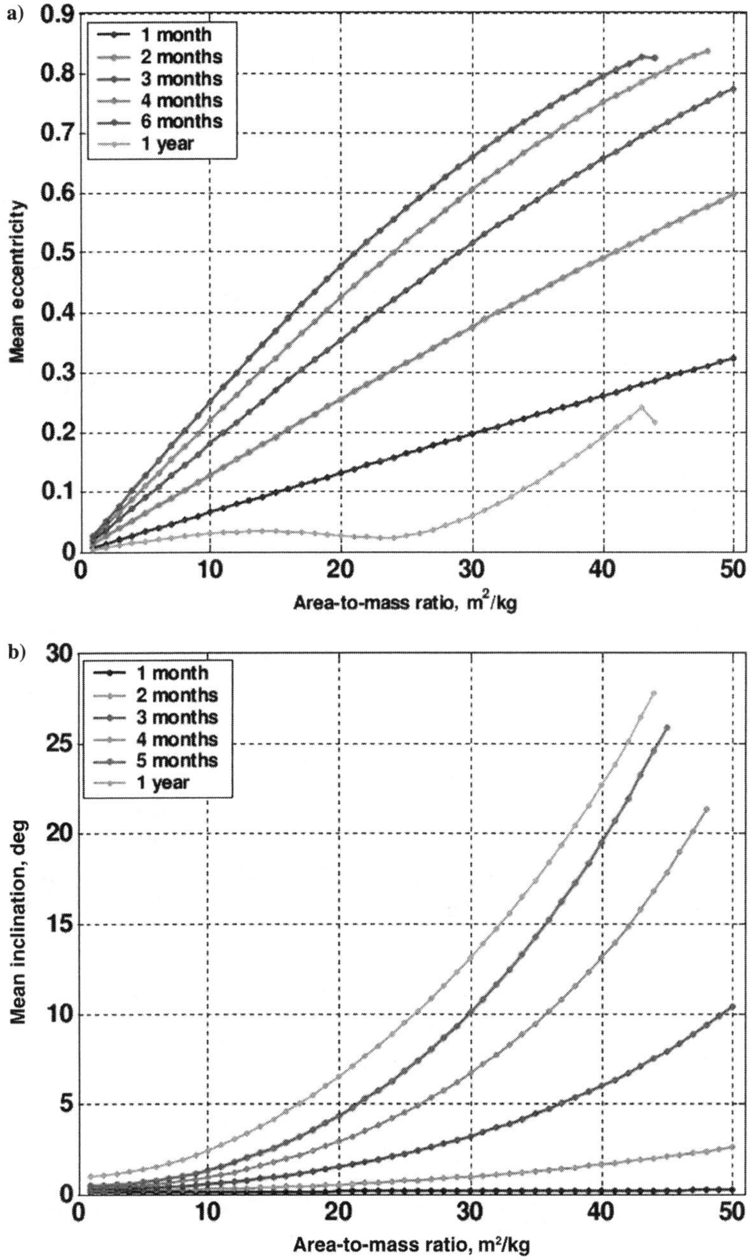

Fig. 12 Evolution of a) the mean eccentricity and b) inclination as a function of the *A*/*M* ratio (m^2/kg) for different durations of SRP perturbation on the satellite.

In fact, these objects are commonly dubbed *as high area-to-mass ratio (HAMR) objects.*

For example, the objects of 30–40 m^2/kg of ratio will reach a mean eccentricity of about 0.7–0.8 after being exposed to the solar radiations for six months.

Figure 12b also demonstrates the modification of the orbital inclination of the HAMR objects, strongly affected by the SRP.

Possible sources of these HAMR objects detected in GEO are for instance the thermal blankets or MLI.[‡‡‡] The MLI pieces are generated by two possible mechanisms: the breakup of satellites or upper stages, and the degradation of the satellite or upper-stage surface due to impacts by small meteoroids or orbital debris [3].

Note that the curve in Fig. 2, representing the magnitude of acceleration for the SRP, was computed assuming a typical $A/M \sim 0.1\,m^2/kg$: these "*normal*" objects could never reach such extreme values of the orbital eccentricity under the effect of the SRP.

The dynamical problem faced by the HAMR objects is different and is sometimes called the "*photo-gravitational problem*," where the solar radiation perturbation becomes comparable or even more important than the monopole effect.

The peculiar nature of the HAMR objects can also be seen from Fig. 13. The picture shows the normalized light-curve obtained observing an HAMR object from the Loiano telescope (located in central Italy), with different filters according to the observation sequence (see Ref. [5] for details).

The normalized light-curve basically allows a comprehensive physical characterization of the observed object, by giving the amount of reflected light coming from this object, as a function of time, its geometry, and rotation.

As an example, a sinusoidal and periodic light-curve would be expected with an ellipsoidal object rotating around its principal axis.

Looking at this figure, from the fast and nonperiodic variation of the light coming from the object, this extremely high variability of the object reflectance can deduce to the complex (definitely non-spherical) shape of this object, which changes very quickly. This is compatible with the hypothesis of a foil of an insulating material which folds and unfolds rapidly, changing its aspect angle with the Sun.

Consequently, a long time propagation of this kind of objects is not easy to be performed with an average A/M ratio: we need to make accurate computations in order to track these objects, and some special catalogs are maintained by different observatories to monitor this category of population.

IV. MODELS OF THE FUTURE EVOLUTION

To simulate the future proliferation of the overall population, since the late 1980s, a generation of mathematical models which try to accommodate the dynamical features mentioned in this chapter, started to be developed.

---

‡‡‡ Multilayer insulation.

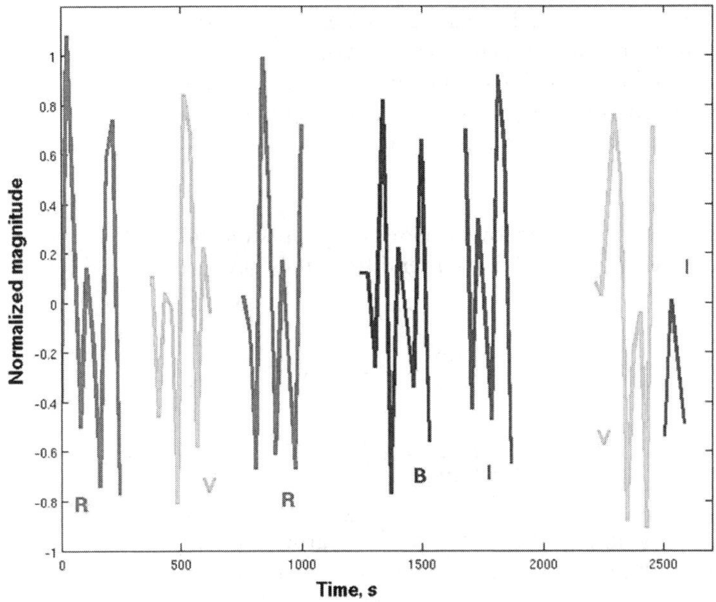

Fig. 13 Normalized light-curve for the HAMR object 84980 (time in seconds) observed from the Loiano telescope [5].

A. A SIMPLE VOLTERRA-LIKE MODEL

In 1991, the first model of the evolution of space debris in LEO, similar to the Volterra model, was developed in Pisa by P. Farinella and A. Cordelli [6].

The Volterra model is a simple mathematical model of the long-term evolution of the space debris population. The so-called "*prey–predator model*" is featured by a system of two differential equations representing two populations: the preys and the predators. Its purpose is to predict the evolution of these populations as a function of time.

From the early 1990s, mathematical models with increasing levels of complexity were developed in order to study the evolution of the debris population by the group in Pisa for the European Space Agency. In the same years, similar models were developed by NASA (see Ref. [9]) and in the UK (see Ref. [10]).

In the case of the LEO space debris model from P. Farinella and A. Cordelli, the two populations of orbiting bodies are defined as the following (the values of the parameters, which are cited in the next two paragraphs, refer to the "*standard*" model made in 1991 by P. Farinella and A. Cordelli described in Ref. [6]):

n is the number of fragments referring to the projectiles (i.e., the predators) and represents the number of small bodies capable to cause a catastrophic breakup when impacting a satellite (these fragments correspond to typical sizes of the order of 1 cm and masses of a few grams).

N describes the population of satellites, which are the targets (i.e., the preys), and are assigned as for large objects (of cross sections on the order of a few square meters and of masses of hundreds of kilograms).

Therefore, the interaction of these two populations in LEO region is studied by the following system of first-order differential equations:

$$\begin{cases} \dfrac{dN}{dt} = A - xnN \\ \dfrac{dn}{dt} = \beta A + \alpha xnN \end{cases} \tag{7}$$

The term A is interpreted as the satellites growth rate, which is the difference between the number of satellites launched and the number of objects reentered in the atmosphere ($A = 100$ expresses the order of magnitude of the rate of "*big*" objects inserted into the orbit at the epoch of simulation).

x is a constant coefficient which characterizes the number of collisions per unit cross section and per year between two particles. It is proportional to the ratio between the average collision velocity V and the volume W of the circumterrestrial shell containing the population of orbiting bodies[§§§] (with $V \sim 10\,\text{km/s}$ and $W \sim 10^{12}\,\text{km}^3$, the ratio $V/W \sim 10^{-11}\,\text{km /s}$, i.e., $x \sim 3 \cdot 10^{-10}\,\text{m/year}$).

Hence, with the coefficient x, the part xnN represents the number of collisions between projectiles n and targets N: it defines the number of satellites N destroyed by collisions.

α is the number of fragments generated from a typical collision ($\alpha \sim 10^4$). Thus, the expression αxnN gives the number of objects produced in xnN collisions.

Assuming that every year, A new satellites enter into LEO, consequently, at the same time, new "*primary*" and "*collisional*" fragments are mostly created during or after launches, involving rockets or second/third stages. The parameter β defines this mean number of primary fragments ($\beta \approx 70$).

To solve this system, we need suitable initial conditions that are the number of projectiles n_0 and the number of satellites N_0 at the beginning of the simulation [e.g., $N_0 = N(0) = 2 \cdot 10^3$ and $n_0 = n(0) = 5 \cdot 10^4$].

The results of the system [Eq. (7)] with the initial conditions and values of the parameters described previously are plotted in Fig. 14: this is a typical example of

[§§§] By analogy, the satellites are assumed to move like perfect gases in a box.

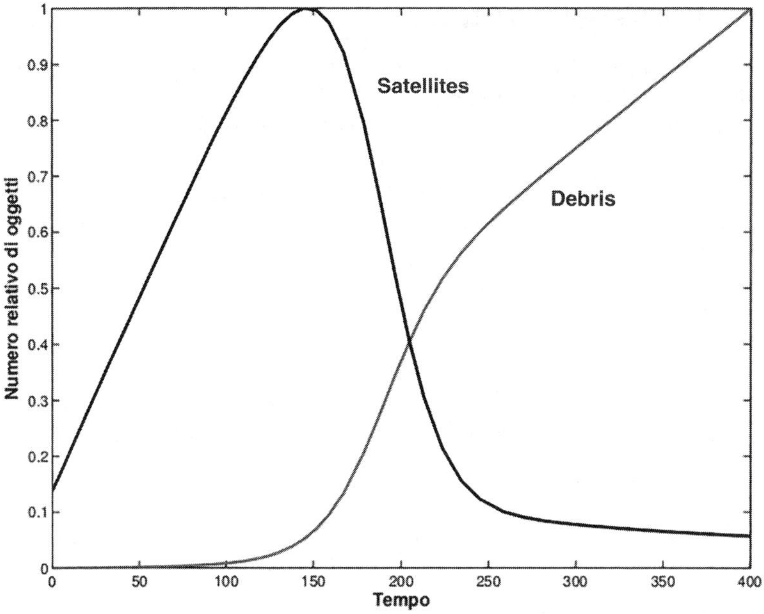

Fig. 14 The number of LEO satellites (black line) and debris (grey line) predicted by the mathematical model of P. Farinella and A. Cordelli over a time span of 400 years in the future. Note that each population is normalized to 1 in order to keep both curves in the same plot since here we are only interested in the relative growth of the two populations.

the evolution of the both populations obtained with this simple model, over a time span of 400 years in the future.

At the beginning of the simulation, the number of satellites is higher than the number of fragments. Therefore, since the collisions are not important, both populations grow almost linearly with time.

After about 150 years, the number of fragments starts to increase suddenly. The number of satellites reaches its maximum, and thereafter declines abruptly since they are destroyed by the large number of debris already present in space. At the same time, the new collisional fragments continue to be created until about 200 years.

And afterward, at about 300 years, it ensues a quasi-steady abundance of the satellites, while the fragment proliferation continues to grow considerably.

This simple model demonstrates that there is an intrinsic limit in the number of satellites in LEO. Because of collisions, the number of fragments is expected to grow significantly only after a few decades in the future. This model could be very useful to obtain some main physical features of the Earth-orbiting bodies for the long-term evolution of these objects, as we will see in the next section.

B. TOWARD INCREASED COMPLEXITY......

A. Rossi, A. Cordelli, P. Farinella, and L. Anselmo have extended the model described in Ref. [6] to a more complex one [7].

The Earth-orbiting objects can be subjected to hypervelocity mutual collisions; thus, bins of mass are considered for the objects.

The space around the Earth is split into a number of discrete-altitude shells, with collision probabilities at each altitude computed from a set of actual orbiting objects.

The evolution process was assumed to be caused by a number of source and sink mechanisms, such as launches, explosions, mutual collisions, and mitigation measures.

Therefore, the future evolution model is represented by the following set of coupled, first-order differential equations for $N(m_i, h_j, t)$, which is the number of objects at time t in the bin centered at mass m_i and in the shell centered at altitude h_j:

$$\frac{\mathrm{d}N(m_i, h_j, t)}{\mathrm{d}t} = \beta(m_i, h_j) - \frac{N(m_i, h_j)}{\tau(m_i, h_j)} + \frac{N(m_i, h_{j+1})}{\tau(m_i, h_{j+1})} + \sum_{k,l} f(m_k, m_l, m_i) p(h_j) \sigma(m_k, m_l) N(m_k, h_j) N(m_l, h_j) \quad (8)$$

In Eq. (8), $\beta(m_i, h_j)$ is the matrix that accounts for the launches, the release of small objects, and the explosions.

Given the mass and altitude of an object, the second and third terms express the drag-induced orbit decay, assuming a characteristic residence time τ related to the re-entry in the atmosphere.

The fourth term accounts for the collisions. $f(m_k, m_l, m_i)$ describes the number of objects of mass m_i that are produced or destroyed by a collision between two bodies of mass m_k and m_l.

$p(h_j)$ is the collision probability for objects as a function of altitude h_j.

$\sigma(m_k, m_l)$ is the squared sum of the radii of the same two bodies with respective mass m_k and m_l.

The different cases were simulated by varying both the parameters related to the physical properties of the existing bodies, and the possible future policy decisions on the launch and removal rates of objects. Their simulations and results are detailed in Ref. [7].

The outcomes of this mathematical model showed the same trend for the growth of collisional fragments as found in the simple model described in Sec. A.

For the nominal case defined in Ref. [7], depending on the altitude and after a certain period of steady population growth, the small-sized population increases exponentially. That is explained by the collisional fragments which are swiftly generated in contrast to the intrusion of non collisional debris into space. Thus, the occurrence of breakup events happens more frequently, leading to a chain reaction in certain altitude bands.

This runaway growth still occurs even changing some of the parameters mentioned previously. For instance, the case with possible active removal of small debris using high-power laser sweepers was tested. More precisely, 1,000 objects smaller than 13.5 g and located between 700 and 900 km in altitude were assumed to be removed per year, considering no more explosions: it turns out that this action does not prevent this exponential growth of the fragment population.

C. THE LAST MODELS: SDM

In reality, detailed studies are required to evaluate the best-mitigate measures able to limit the growth of space debris population. This goal requires more complex models capable to accurately propagate single objects with detailed dynamical models, including the main perturbations detailed in the previous sections.

Since the early 90's, the research group at the CNUCE Institute in Pisa, Italy started to implement one of these models called SDM,[¶¶¶] which was dedicated to study the long-term evolution of orbital debris and to evaluate the effectiveness of mitigation measures.

With a semianalytic propagator, this model propagates the orbits of all the larger objects individually, with a dynamical model including the geopotential harmonics, the third-body perturbations, the SRP (including shadows), and the atmospheric drag. In addition, it checks the probability of collisions between these objects by measuring their orbital crossings in space.

Various source and sink terms are included: launches, explosions, collisions, RORSATs-like events, solid rocket motors exhausts (slag), mitigation measures, collision avoidance, and active debris removal (ADR).

Each single-fragmentation event has to be modeled with an ad hoc physical model. It must be stressed that when two objects collide in LEO with an average velocity around 9–10 km/s, we are dealing with a so-called *hypervelocity impact*, where the speed of the projectiles is higher than the sound speed in the target.

These kinds of events are very difficult to simulate on the ground, thus making the modeling of the resulting debris cloud quite complex and subject to significant uncertainties in terms of mass, shape, and orbital distribution of the fragments.

D. SDM: NUMBER OF OBJECTS LARGER THAN 10 cm

To derive common guidelines for the future space activities to minimize the growth of the debris population, different international space agencies try to

[¶¶¶] Space Debris Mitigation Model.

compare the projection of the future environment drawn from the models similar to SDM.

The main committee working on this subject is the IADC (Inter-Agency Space Debris Coordination Committee), whose members are space agencies from different nations, taking care of all the space debris issues. Within IADC, the agencies are trying to reach a kind of consensus on the future evolution of the orbital population.

Figure 15 gathers the models of the future LEO evolution over 200 years, coming from six space agencies.****

This representation compares the different models for objects larger than 10 cm of size, assuming that the launch traffic and space management operations of the next 200 years will mimic the current ones.

Generally, the conclusions of these models are very consistent and all of them point to the steady growth of the population in the future (with a 11-year modulation due to the change in the atmospheric density). This will happen notwithstanding the application of the proposed mitigation measures, such as the "*25-year*" rule prescribing the deorbiting of the spacecraft at its end of life into an

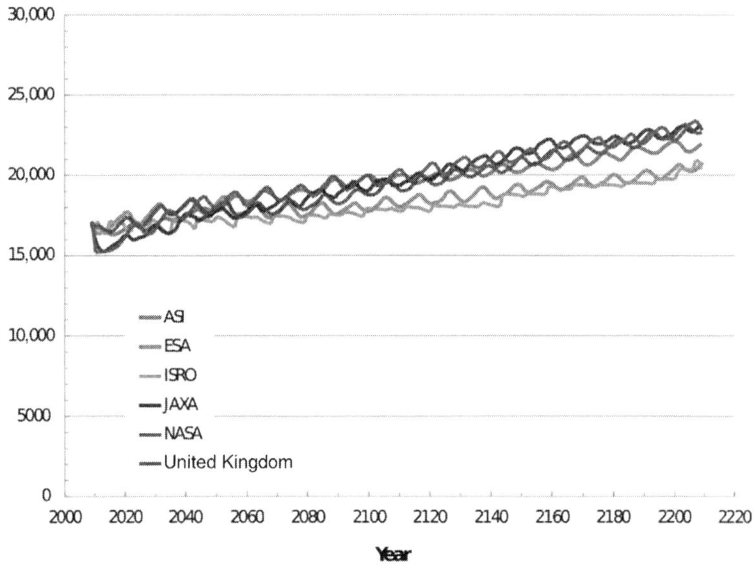

Fig. 15 Projection of LEO populations over 200 years obtained by six different long-term evolution models.

****ASI, ESA, ISRO, JAXA, NASA, and UK Space Agency (initially BNSC).

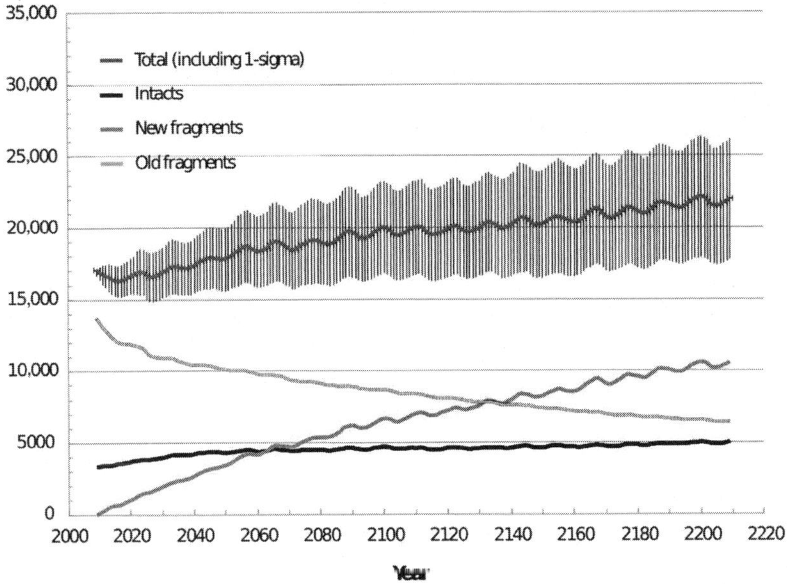

Fig. 16 Breakdown of ASI's projection of the LEO populations over 200 years.

orbit with a residual lifetime of 25 years (actually, a 90% success rate of this kind of disposal maneuver is assumed in the simulation for the sake of realism).

The reason for this trend of evolution is explained in Fig. 16. Starting from the top of the figure, the first curve represents the evolution of the total LEO populations with its standard deviation: SDM is a statistical model which performs multiple times in a Monte Carlo fashion and all the results are averages over a given number of Monte Carlo runs, typically of the order of 50.

The descending curve in light grey colour shows the number of old fragments, existing in our initial population. The evolution of the intact objects injected in space is illustrated by the black curve ("stable" evolution in y-axis) and these objects are mainly satellites and rocket bodies. Then, the significant ascending curve in dark grey colour illustrates the evolution of the number of new fragments, which are generated by collisions happening after the beginning of the simulation.

We can underline the nearly linear growth of the intact objects launched in space, and the significant increase of the number of fragments coming from future collisions for the next 200 years. Therefore, the number of future collisions explains the evolution of the LEO populations.

The simulation in Fig. 16 does not include the ADR or the collision avoidance activities in space. These results are used by the IADC in order to push forward the need of these two actions in order to reverse the trend of this evolution.

V. CONCLUSION

We have seen that all six IADC members yielded very similar results and confirmed the instability of the current LEO population. However, their simulations do not consider the mitigation measures. Many actions remain to undertake in order to model the long-term evolution of space debris.

Starting from the model in Fig. 16, we have to assess the actual need of the ADR activities, that is, to find a common agreement of their performances and to confirm their exact level of activity.

Indeed, it is essential for the ADR process to consider the decision of the debris to be removed based on their degree of danger and their location,[††††] and at the same time, the choice of the number of objects that have to be removed in order to curb the growth of the space debris population.

Moreover, we have to study effective disposal measures for high Earth orbits. From the dynamics point of view, mapping the stability and the instability of the Earth orbital regions has to be done in order to find the best mitigation practices for the MEO and GEO satellite disposal.

Another action to consider is to share and spread the awareness of space debris issues: there are a lot of operators who still do not deorbit their spacecraft correctly after its operational life.

This is explained in the last two figures where we talk about 90% of PMD[‡‡‡‡] success which is very optimistic: we assume that in the next 100 years, 90% of the objects will be, at the end of their lifetime, disposed out of their operational orbits, and will be brought into orbits that have a residual lifetime under the cumulative effects of the Earth under 25 years. Today, the actual percentage of PMD is more or less close to 50%.

ACKNOWLEDGMENTS

The author and coauthor thank G.F. Gronchi and D. Serra from the University of Pisa, for contributing to review this chapter.

REFERENCES

[1] Rosengren, A. J., Scheeres, D. J., and McMahon, J. W., "The Classical Laplace Plane as a Stable Disposal Orbit for Geostationary Satellites," *Advances in Space Research*, Vol. 53, No. 8, April 2014, pp. 1219–1228.

[2] Rossi, A., "Resonant Dynamics of Medium Earth Orbits: Space Debris Issues," *Celestial Mechanics and Dynamical Astronomy*, Vol. 100, No. 4, April 2008, pp. 267–286.

†††† Ranking and identification of the best candidates, order of removal.

‡‡‡‡ Postmission disposal.

[3] Liou, J.-C., and Weaver, J. K., "Orbital Dynamics of High Area-to-Ratio Debris and Their Distribution in the Geosynchronous Region," *Proceedings of the Fourth European Conference on Space Debris (ESA SP-587)*, edited by D. Danesy, ESA/ESOC, Darmstadt, Germany, April 2005. ESA Publications Division, Noordwijk, The Netherlands, pp. 285–290.

[4] Milani, A., Nobili, A. M., and Farinella, P., *Non-Gravitational Perturbations and Satellite Geodesy*, Dipartimento di Matematica, Universit di Pisa, A. Hilger, Bristol, 1987.

[5] Rossi, A., Marinoni, S., Cardona, T., Dotto, E., Santoni, F., and Piergentili, F., "The Loiano Campaigns for Photometry and Spectroscopy of Geosynchronous Objects," Paper IAC-12,A6,1,3,x13587, *63rd International Astronautical Conference*, Vol. 3, Naples, Italy, Oct. 2012, p. 339.

[6] Farinella, P., and Cordelli, A., "The Proliferation of Orbiting Fragments: A Simple Mathematical Model," *Science Global Security*, Vol. 2, March 1991, pp. 365–378.

[7] Rossi, A., Cordelli, A., Farinella, P., and Anselmo, L., "Collisional Evolution of the Earth's Orbital Debris Cloud," *Journal of Geophysical Research: Planets (1991–2012)*, Vol. 99, No. E11, Nov. 1994, pp. 23195–23210.

[8] Chobotov, V.A. (ed.), *Orbital Mechanics*, 3rd ed., AIAA Education Series, AIAA, Reston, VA, 2002.

[9] Liou, J.-C., and Johnson, N. L., "Risks in Space from Orbiting Debris," *Science*, Vol. 311, No. 5759, 2006, pp. 340–341.

[10] Lewis, H. G., Swinerd, G.G., Williams, N., and Gittins, G., "DAMAGE: A Dedicated GEO Debris Model Framework," *Proceedings of the Third European Conference on Space Debris*, The European Space Agency (ESA), 2001, pp. 373–378.

CHAPTER 2

Review of Analytic Modeling of the Long-Term Evolution of Orbital Debris

Ioannis Gkolias*
University of Rome "Tor Vergata," Rome, Italy

I. INTRODUCTION

It was on 4th of October 1957 when Earth's first-ever artificial satellite, Sputnik I, was launched. It was a spherical object with size similar to a beach ball. Ninety minutes after its launch, it was confirmed that Sputnik made it through the atmosphere and reached its final elliptical orbit around the Earth. However, it was not the only satellite that got in orbit. The final stage of the launcher, 30 m long and weighting more than 4 tons, as well as the protective shroud, which helped the satellite ascent through the atmosphere, also made it to orbit. These were the first ever space debris we deposited in the near-Earth space environment and at that time nobody would imagine that a few years later we would have to deal with pollution problems in space.

Nowadays, almost 60 years later, more than 500,000 artificial objects with size larger than 1 cm are orbiting the Earth. The most populated region is the low-earth orbit (LEO), which is the region of space within 2000 km from the Earth's surface (Fig. 1). Almost 95% of the objects in LEO are nonfunctional space debris. This dense population is threatening not only operational satellites but also upcoming space missions. For instance, it is reported that the International Space Station is forced to make, on average, one collision-avoid maneuver per year because of space debris. In the last few years the problem became more evident and arouse a global concern about our local space environment. The major space agencies decided to take action, forming the Inter-Agency Space Debris Coordination Committee (IADC). According to IADC the following definition holds for space debris [1]:

Definition—Space debris: Space debris, also known as orbital debris, are all man-made objects, including fragments and elements thereof, in Earth orbit or reentering the atmosphere, that are nonfunctional.

This chapter is partly based on a lecture delivered by Colin McInnes for the Stardust network.
*Department of Mathematics.

Fig. 1 View of the population of space debris in low-Earth orbit. (Image courtesy of NASA.)

But, where do all these objects come from? Like in the case of Sputnik, in order for a satellite to successfully get on orbit, several other pieces of equipment make it to the orbit too. Upper stages of launch vehicles, pyrotechnic devices, sensor covers, and yo-yo masses are examples of operational debris. Also, the orbital lifetime of satellites in LEO is multiple times longer than their operational lifetime. Because, they spend a good amount of time rotating Earth as space debris, before they reenter the atmosphere due to air drag and burn.

However, the deposit of new objects in LEO is not the only reason for a constant increase in the space debris population. Other very important effects are intentional and unintentional collision as well as orbit explosions. Explosions may occur to fuel tanks, which are in orbit and have still an amount of stored propellant or other high-pressure fluid. This was the main source of new space debris until 2007. Then two catastrophic collisions dramatically changed the evolution of the space debris population (Fig. 2). In 2007, China intentionally broke up the weather satellite Fengyun-1C during an antisatellite missile test. Two years later, Cosmos-2251, a Russian satellite which was no more actively controlled, run into Iridium-33, a U.S. operational communications satellite. The two satellites collided with a speed of almost 12 km/s, producing a huge amount of smaller orbital debris.

The importance of collisions on the space debris population evolution had been pointed out many years earlier from the scientific community. In fact, as the LEO environment becomes more and more populated, the probability of on-orbit collisions increases dramatically. The possibility for a cascade of catastrophic collisions that will produce a huge amount of space debris and will render the near-Earth space environment totally unusable, was first proposed in 1978 by Kessler [2]. This scenario, also known as "*the Kessler Syndrome*," is nowadays more frightening than ever.

Space agencies are really concerned about the future LEO environment, because it is the main field of their activities. Consequently, they have developed

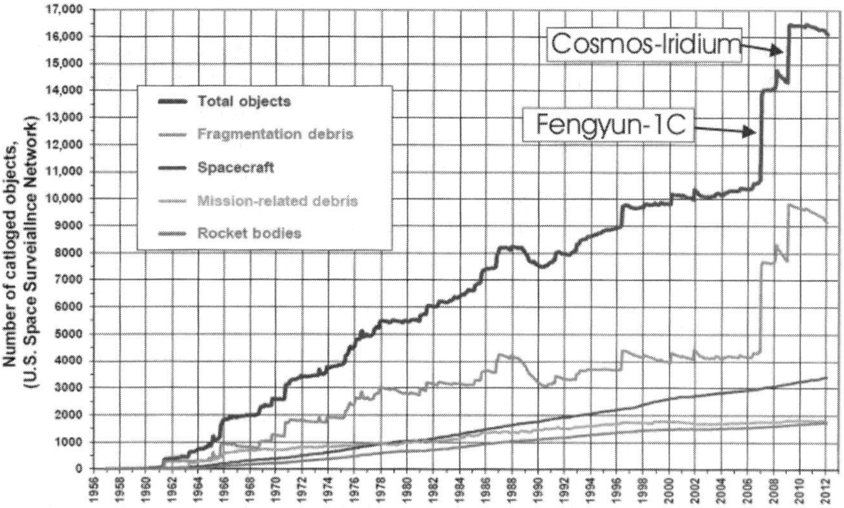

Fig. 2 The growth of the cataloged space debris population the last 56 years. The intentional break-up of Fengyun-1C and the catastrophic collision of Cosmos-2251 and Iridium-33 dramatically changed the number of space debris the last 5 years. (Image courtesy of IADC.)

various tools for simulating the long-term evolution of space debris and trying to predict the future population (Table 1). Based on state-of-the-art modeling of orbit propagation and collisions between debris, these numerical codes are used as high-fidelity operational tools.

TABLE 1 OPERATIONAL TOOLS FOR PREDICTING THE FUTURE SPACE DEBRIS ENVIRONMENT

Space Agency	Simulation Tool
ASI	Space debris mitigation long-term analysis program (SDM)
ESA	Debris environment long-term analysis model (DELTA)
ISRO	KS canonical propagation model (KSCPROP)
JAXA	LEO debris evolutionary model (LEODEEM)
NASA	LEO-to-GEO environment debris model (LEGEND)
UKSA	Debris analysis and monitoring architecture for the geosynchronous environment (DAMAGE)

One of the main goals of these tools is to determine whether the future LEO environment will be stable or not. Using as initial conditions, a baseline environment of debris provided by the MASTER model of ESA for the year 2009, IADC asked all the agencies to predict the future population of space debris with size 10 cm or larger. Even in the best case scenario, when a future 90% post mission disposal rate for both satellites and launch vehicles is assumed, the results are still not satisfying. In Fig. 3 the evolution of the future space debris population, produced by different models, is plotted. Ignoring the fluctuations in the curves, which are mainly due to the solar activity circle, it is evident that all models predict an increasing trend for the LEO population.

Although computer simulations are very accurate and produce a lot of useful information, they lack the simplicity and the elegance of an analytic approach. In this chapter, we try to tackle the problem of the long-term evolution of space debris with the means of pencil and paper mathematics. Mathematical modeling will not give us better or more accurate solutions than scientific computing. However it will provide us with the tools to get a different insight into the problem.

In the preceding sections, we present two very basic approaches for the evolution of the space debris population. The first one is based on ordinary differential equations (ODEs), whereas the second one on partial differential equations (PDEs). In both cases, simple equations, rather than sophisticated algorithms, will be used in order to predict the future of near-Earth space environment.

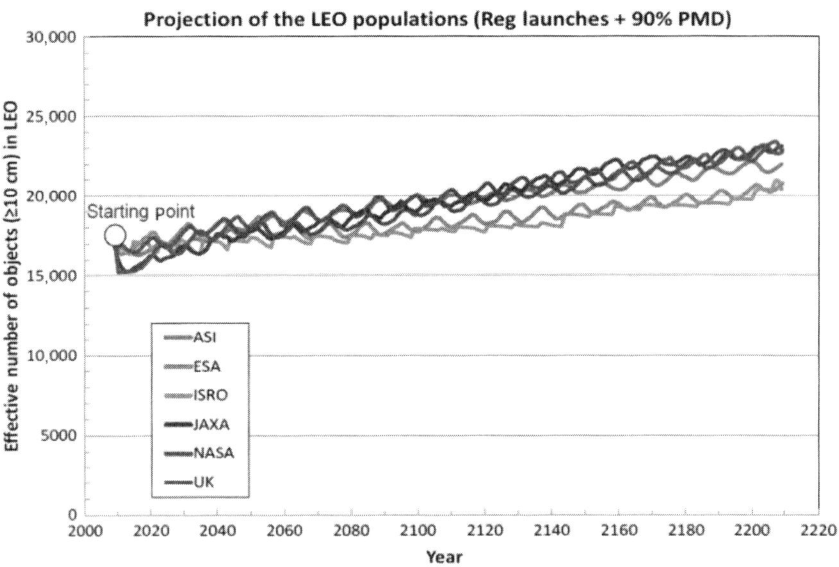

Fig. 3 Numerical tools for space debris population models.

II. MODELING WITH ODEs

In this section, the evolution of the population of orbital debris will be modeled with the simple means of ODEs. In this context, two different approaches will be presented. The first one follows the "*particles in a box*" approximation and the problem is modeled with a single differential equation. The second one is inspired by the models used in ecology to describe the interaction between species, also known as "*Lotka-Volterra*" models. In this case, a set of two coupled differential equations is used to describe the system.

A. PARTICLES IN A BOX MODEL

The particles in a box model (PIB), introduced by Talent in 1992 [3], is probably the simplest mathematical tool used to explore the dynamics of the space debris population. A first-order differential equation is used to describe the time evolution of the system. The state variable of the system is chosen to be the total number N of objects in LEO. Then, considering the orbital debris as identical particles in the same environmental box the equation for the rate of change of N can be derived:

$$\frac{\mathrm{d}N}{\mathrm{d}t} = A + BN + CN^2 \tag{1}$$

where in Eq. (1) the coefficients are

1. A the deposition coefficient
2. B the removal coefficient
3. C the collision coefficient

Those three coefficients describe all the physical mechanisms that affect the population. Coefficient A is used to model the deposition rate of new objects in LEO. It is related to the number of launches per year, as well as the average number of pieces that get in orbit for each launch. It also models the increase in the population due to on-orbit explosions. Coefficient B models the rate of removal of objects from the environment. The main natural mechanism that sweeps the LEO environment is the air drag. Because of the air drag, the orbital debris spiral inward until they reenter the atmosphere and burn in the top layers of it. Finally, coefficient C models the growth of the population due to the collisions. C is calculated under the particles in a box assumption, using already known formulas from the kinetic theory of gases. Determining the values of A, B, and C will allow us to have a better insight on the LEO environment.

But even without assigning values to the coefficients, an analytic treatment of the problem can be done under some simple considerations. As a first step, one can notice that the equilibrium solutions of the Eq. (1), namely those for

$\dot{N} = 0$, are given by a quadratic equation. The solution of the equation yields:

$$\frac{dN}{dt} = 0 \Rightarrow N_{1,2} = \frac{-B \pm \sqrt{B^2 - 4AC}}{2C} \tag{2}$$

At this point, the role of the discriminant of the quadratic equation $A + BN + CN^2$ should be discussed. It is given by the formula

$$q = \underbrace{B^2}_{\text{Sinks}} - \underbrace{4AC}_{\text{Sources}} \tag{3}$$

and it represents the difference between sources and sink terms. Depending on the value of q, three possible scenarios for the evolution of the population are available

1. $q > 0 \Rightarrow$ *Sinks* $>$ *Sources* $\Rightarrow$ *Conditionally stable*
2. $q < 0 \Rightarrow$ *Sinks* $<$ *Sources* $\Rightarrow$ *Unstable*
3. $q = 0 \Rightarrow$ *Sinks* $=$ *Sources* $\Rightarrow$ *Threshold*

In the case $q > 0$ the sinks are more than the sources and the system relaxes into a stable end state. On the other hand, when $q < 0$ the source term is dominating leading to an explosion in the debris population, a scenario similar to

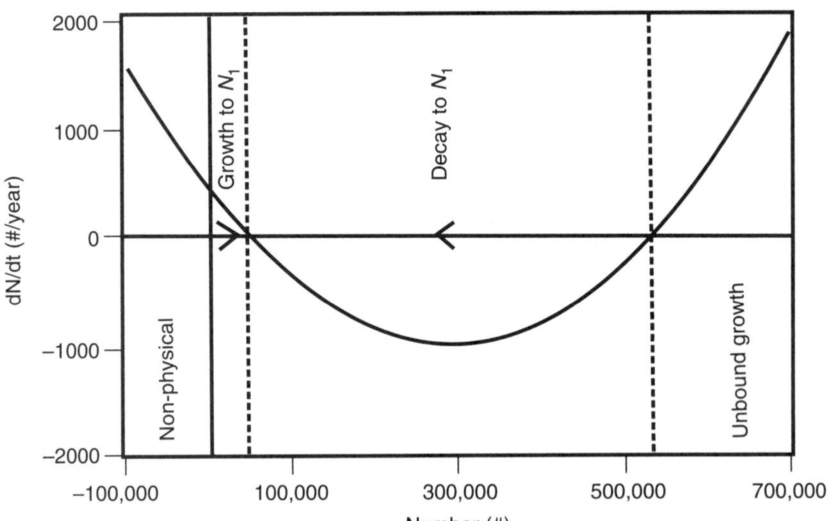

Fig. 4 Stability regimes for the particle in a box population model when $q > 0$.

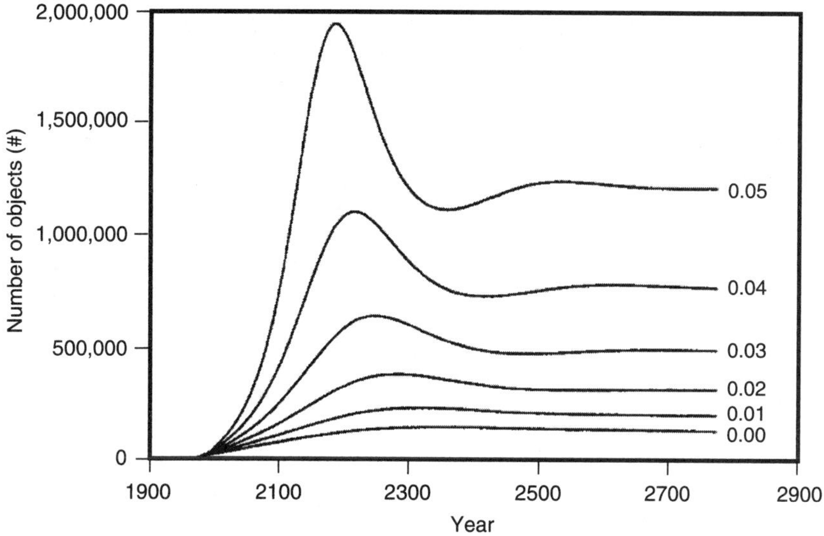

Fig. 5 Evolution of the debris population for different values of the coefficients.

the "*Kessler Syndrome.*" Finally for $q = 0$, the system exhibits a threshold behavior with a constant competition between sources and sinks.

The case of $q > 0$ will be examined in more details here. In this case there are two real equilibrium solutions $N_{1,2}$ in the system. The evolution of the population in this case is shown in Fig. 4. The end state depends on the initial debris population N_0. If the initial population is smaller than N_1 then it will increase until it reaches the value N_1, where it stabilizes. If N_0 is greater than N_1 but smaller than N_2, then it decreases until it reaches N_1 and stabilizes again. In the case $q > 0$ the equilibrium point N_1 acts as a stable focus. On the other hand, N_2 acts as an unstable focus and consequently if N_0 is greater than N_2 then the population undergoes an unbounded growth.

Although the environment in the case $q > 0$ is stable, there are a few considerations we should also take into account. Because of catastrophic collisions the value of N could exhibit abrupt increases. The model predicts that the environment would return to the equilibrium state N_1 if the value of N does not surpass the value N_2. To have a resilient to collisions environment, the separation between N_1 and N_2 should be large enough, so that under no circumstances the value of the population will exceed N_2. Another consideration is that, even though N_1 is a stable equilibrium that doesn't make the environment viable. For example, in the case that the value of N1 becomes large enough, then the mean time for a collision to happen is smaller than the typical operational lifetime; a scenario in which space activities will be impossible.

Let us consider now a specific instance of the LEO environment, with fixed A, B, and C [3]. In Fig. 5 six evolutionary cases are presented. Each one is characterized by a rate of growth of launches until the year 2020. After that a steady launch rate per year is assumed. In all cases the population of debris has the same qualitative behavior. It increases rapidly for the first hundreds of years, until the sink term kicks in and finally the population stabilizes at the steady state N_1.

In conclusion, the PIB is a simple but descriptive model for the space debris population. Provided a careful calculation of the coefficients A, B, and C, then the model can give us helpful predictions for the future debris environment. It should also be mentioned that the values of the coefficients are not only static, but also evolve in time, making the problem even more challenging.

B. LOTKA–VOLTERRA MODEL

In the same period when Talent was working on the PIB model, a team from the University of Pisa in Italy developed another model for the debris population [4], inspired by the predator–pray models used in ecology. The main difference in their approach is that instead of using just one variable for describing the whole LEO population, two different populations of orbiting bodies are considered. Then a system of two coupled linear differential equations is used to describe the evolution of the system in time.

In this approach, the population is divided into two distinct categories. The first category consists of the large satellites orbiting in LEO, which are denoted with N. The second category is the small debris fragments that are able to cause catastrophic collisions and are denoted with n. In analogy with the ecosystems, debris play the role of the predators and satellites play the role of the prey. The satellite population is characterized by object cross section of a few square meters and mass of hundreds of kilograms. The debris population is characterized by a typical size of a few centimeters and masses of a few grams. The time evolution of the populations can be described by the system:

$$\frac{\mathrm{d}N}{\mathrm{d}t} = A - xnN \tag{4}$$

$$\frac{\mathrm{d}n}{\mathrm{d}t} = \beta A + axnN \tag{5}$$

where in the Eq. (5) the parameters are

1. A: deposition coefficient
2. x: collision coefficient
3. β: primary fragments
4. a: fragments per collision

Coefficient A is describing the growth rate of the satellite population N. It is calculated from the number of launched objects per years, if we subtract the number of object that have reentered in the atmosphere. The coefficient x is defined such that xnN represents the number of collisions between projectiles and targets. The quantity xnN also represents the number of satellites that were destroyed by collisions. The value of x can be calculated by a *particles-in-a-box* approximation, similar to that described in the previous section. The parameter β represents the number of primary fragments generated by explosions or operational debris. Finally, a is the number of fragments produced in a typical collision.

Given a set of initial populations of satellites and debris, one can numerically integrate the equations of motion and predict the future debris environment. This is the typical approach, however the above set of equations admits also an analytic solution. First, the system of the two equations will be reduced to a single differential equation. Multiplying the first of Eq. (5) with a and adding them yields

$$a\frac{\mathrm{d}N}{\mathrm{d}t} + \frac{\mathrm{d}n}{\mathrm{d}t} = (a+\beta)A \Rightarrow \frac{\mathrm{d}(aN+n)}{\mathrm{d}t} = (a+\beta)A \tag{6}$$

Now the left-hand side of the equation is equal to a constant. And so, integrating with respect to time results to the relation

$$aN(t) + n(t) = (a+\beta)At + C \tag{7}$$

where C is an arbitrary constant of the integration and its value is determined from the initial conditions. For time t equal to zero it is

$$C = aN(0) + n(0) \tag{8}$$

Substituting the value of the constant C in the original equation, an expression for the time evolution of the sum of the satellites and debris is derived

$$\alpha N(t) + n(t) = (a+\beta)At + aN(0) + n(0) \tag{9}$$

From Eq. (9) the number of debris n can be expressed as a function of the number of satellites N and time. Now, plugging this expression in the original equation for the number of satellites, results in the equation

$$\frac{\mathrm{d}N}{\mathrm{d}t} = A + axN^2 - (a+\beta)xANt - (aN(0)+n(0))xN \tag{10}$$

It's enough to solve Eq. (10) for the number of satellites, then the number of debris is calculated from Eq. (9). The analytic solution can be expressed as

$$N(t) = -\frac{x\left(\sqrt{\pi}C2^{(a/a+\beta)}(a+\beta) - a\Gamma\left(-(a/2(a+\beta))\right)\right)\,{}_1F_1\left((\beta/2(a+\beta)); \frac{3}{2}; \Delta\right)\Sigma}{K\cdot(a+\beta)\cdot\Gamma\left(-(a/2(a+\beta))\right)} + \cdots$$
$$+\frac{\sqrt{\pi}\sqrt{A}C\sqrt{x}2^{(a/a+\beta)-(1/2)}\sqrt{a+\beta}\,{}_1F_1\left(-(a/2(a+\beta)); 1/2; \Delta\right)}{K\cdot\Gamma\left((\beta/2(a+\beta))\right)} \tag{11}$$

where K and Δ are

$$K = ax\left(CH_{-(\beta/a+\beta)}\left(\sqrt{\Delta}\right) + {}_1F_1\left(\frac{\beta}{2(a+\beta)}; \frac{1}{2}; \Delta\right)\right), \quad \Delta = \frac{x\cdot\Sigma^2}{2A(a+\beta)}$$

and Σ is equal to

$$\Sigma = (At(a+\beta) + a(aN(0) + n(0)))$$

Let us explain in a little more detail the symbols in Eq. (11). C is an arbitrary constant of the integration and $\Gamma(z)$ is the Euler Gamma function defined as

$$\Gamma(z) = \int_0^\infty t^{z-1}\,e^{-t}\,\mathrm{d}t \tag{12}$$

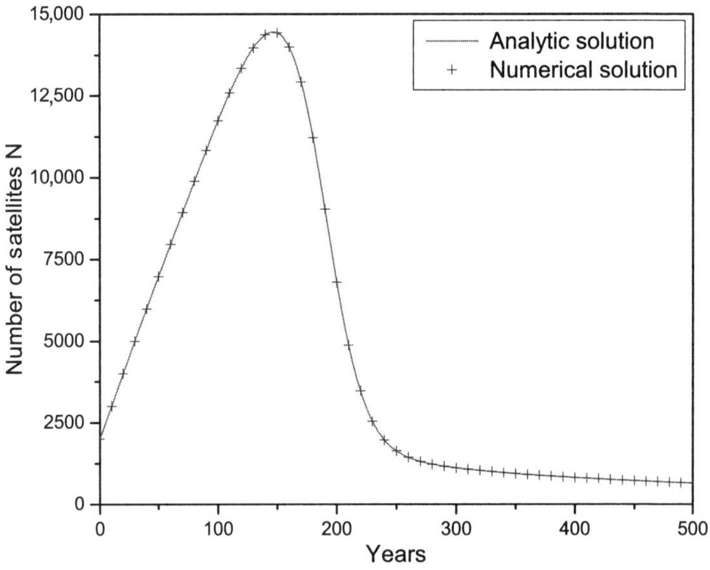

Fig. 6 Comparison between analytic and numerical solution of the Lotka–Voltera equations for given values of parameters A, a, x, β and a set of initial conditions $n(0), N(0)$.

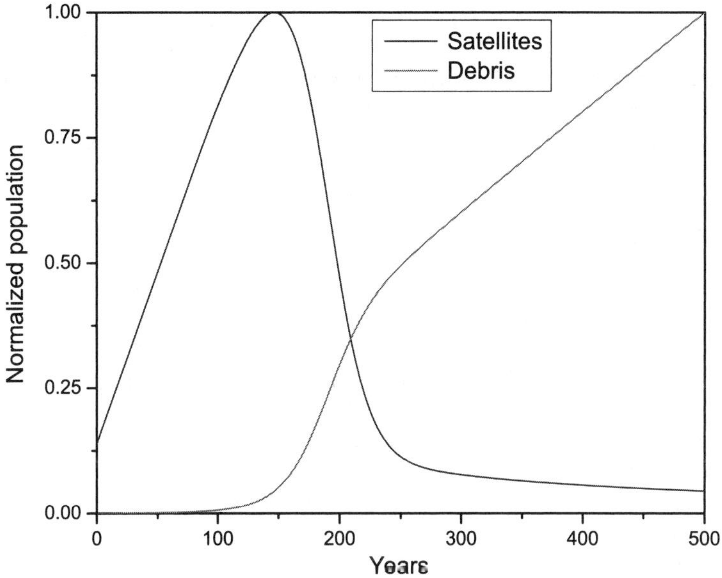

Fig. 7 Evolution of the normalized satellite and debris populations according to the Lotka–Voltera model.

The function $H_n(x)$ corresponds to the value of the nth-order Hermite polynomial calculated at x and ${}_1F_1(a; b; c)$ is the Kummer confluent hypergeometric function defined as follows:

$$ {}_1F_1(a; b; z) = \sum_{k=0}^{\infty} \frac{a^{(n)} z^n}{b^{(n)} n!} \tag{13} $$

where the notation $a^{(n)}$ is used for the rising factorial defined as

$$ a^{(0)} = 1, \quad a^{(n)} = a \cdot (a+1) \cdot (a+2) \cdots (a+n-1) \tag{14} $$

For a given set of the parameters A, a, β, x and for some initial conditions $n(0), N(0)$, Eq. (11) can be solved in order to calculate the value of the arbitrary constant C. In Fig. 6 both the analytic and the numerical solutions are plotted and as it is expected the two solutions are in complete agreement.

Finally, some reasonable values for the parameters $A = 100$, $a = 10^4$, $\beta = 70$, and $x = 3 \times 10^{-10}$ are chosen according to Ref. [4]. For initial populations $N(0) = 2 \times 10^3$ and $n(0) = 5 \times 10^4$, we follow the evolution of the normalized populations ($N/N_{\max}$, $n/n_{\max}$) of both satellites and debris according to the Lotka–Voltera model. In our case $N_{\max}$ is of order 10^4, whereas $n_{\max}$ is of order 10^8. In Fig. 7 the analytic solution is presented for 500 years of evolution. At the beginning the collisions are not important and both populations grow almost linearly in time. At about $t = 150$ years the abundance of satellites reaches its maximum, and afterward it rapidly falls, implying that the number of satellites having a catastrophic collision with debris is greater than those launched. Finally, at $t = 300$ years the population of satellites stabilizes in a quasi-steady state. In this state all the material launched into orbit is totally converted into fragments, which is totally unacceptable for the LEO environment.

III. MODELING WITH PDEs

In this section, a different model for the LEO population is discussed. It is based on an unusual tool in the field of astrodynamics, the PDEs. The model was developed in 2000 by McInnes [5] and was inspired by the PIB model of Talent. The main concept is to extend the PIB model by adding a spatial dimension to the problem.

This is possible if a continuum PDE model is used, similar to those used in fluid dynamics. In this approach, the population of orbital debris is modeled, not with a single variable such as in the PIB model, but with a number density $n(r, t)$ which is a function of both the orbit radius r and time t. Taking into account the inflow of debris due to air drag, as well as artificial sink and source terms, the PDE that governs the evolution of the debris population is derived assuming a continuity problem.

But, before we proceed with the derivation and the solution of the PDEs, a few physical considerations will be discussed. First of all, let us assume a 2-D planar debris population, which forms a near equatorial disk with inner radius r and outer radius $r + \Delta r$. This disk could be produced by a fragmentation event. The same discussion holds also for a spherically symmetric population, but the 2-D case will only be discussed for simplicity. The timescale required to obtain a uniform azimuthal distribution of debris could be calculated as follows. For a debris in a circular orbit of radius r, the angular velocity is given by

$$\omega = \sqrt{\frac{\mu}{r^3}} \tag{15}$$

Also the azimuthal rate of another debris in a circular orbit separated by Δr is

$$\Delta\omega = \frac{\partial\omega}{\partial r}\Delta r \tag{16}$$

The time needed to form a uniform distribution of debris could be estimated from the synodic period of the two debris, one at the inner edge of the disk at radius r, and the other one at the outer edge of the disk at radius $r + \Delta r$. This time is defined as $T_\theta|\Delta\omega| = 2\pi$ and could be calculated from the formula:

$$T_\theta \sim \frac{4\pi}{3\sqrt{\mu}}\frac{r^{5/2}}{\Delta r} \tag{17}$$

In Fig. 8 the timescale needed to get a uniform distribution for a range of disk widths Δr is plotted. Even for narrow disks, the timescale to get a uniform distribution is of order of tens of days. This timescale is significantly shorter than the

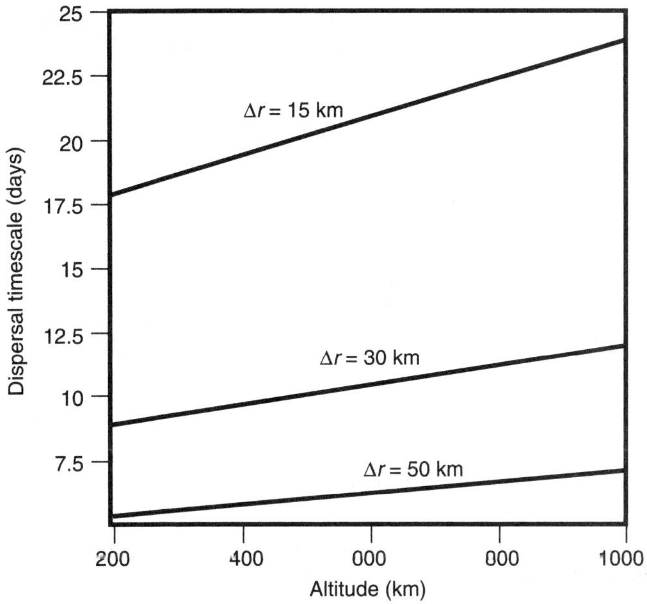

Fig. 8 Azimuthal dispersal timescales.

time needed for the shearing effect of air drag to act on the population. Consequently, after a fragmentation event, a uniform ring is formed in a relatively short time. Therefore, in order to study the long-term azimuthal evolution of the distribution, it is safe to assume that the mean number density of debris is only a function of time t and orbit radius r.

Until now, we have made no assumptions about the shape and size of space debris. And this will be the case in the rest of our analysis. Under this prism, the dynamical behavior of each particle in the debris cloud will be qualitatively the same. And because the motion of the debris could be considered quasi-circular, then a local Keplerian motion could be assumed.

In Fig. 9, a 2-D debris disk is considered. In this case, the vertical component of the particle velocity v_z is zero. Also the transverse velocity v_θ will be near Keplerian and the radial velocity v_r will be a function of orbit radius r and a ballistic coefficient B, simulating in this way the radial inflow due to the air drag.

$$v_r = -f(r, B) \tag{18}$$

$$v_\theta \sim \sqrt{\frac{\mu}{r^3}} \tag{19}$$

$$v_z \sim 0 \tag{20}$$

The functional form of the radial drift could be estimated under the assumption of an isothermal exponential atmosphere. In this case the drag acceleration will be

$$a = \frac{1}{2} B\rho(r)v^2 \tag{21}$$

where

$$\rho(r) = \rho_0 \exp\left(-\frac{r - R}{H}\right) \tag{22}$$

is the air density at altitude r, for a given scale height H and a base density ρ_0 at some reference radius R. Now the inflow speed could be determined, assuming a quasi-circular motion

$$v_r(r) = -\sqrt{\mu r} B\rho_0 \exp\left(-\frac{r - R}{H}\right) \tag{23}$$

The considerations for the population dynamics are summed up in Fig. 10. The ideas involved are more or less the same as those incorporated by the PIB model. Given an initial debris distribution function, the rules implied from the population dynamics are applied and after some transient time the population relaxes to an asymptotic population. For the population dynamics the standard sources and sinks model is used. A deposition term $\dot{n}^+$, as well as a sink term $\dot{n}^-$ are introduced. The air drag effect

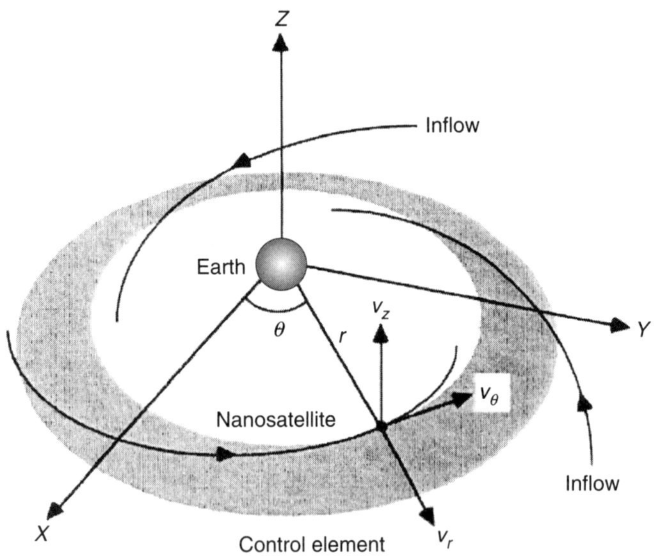

Fig. 9 Description of planar constellation dynamics.

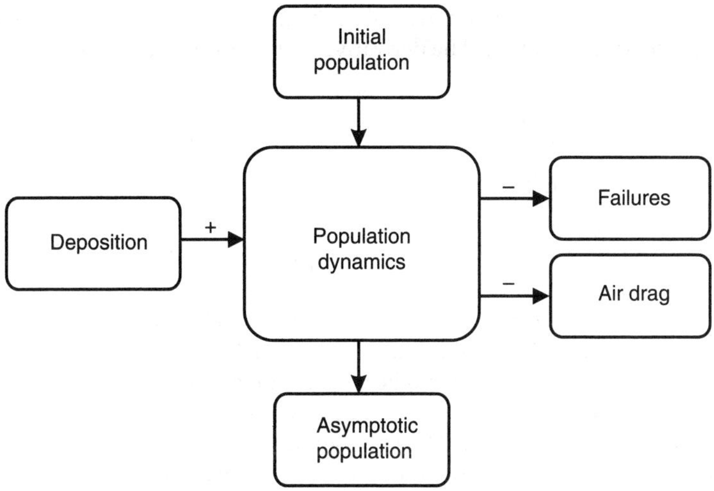

Fig. 10 Flow-chart of the population dynamics.

and the sink term, which models the effect of a possible debris sweeper or random on orbit failures, are the main reasons for the reduction in the space debris number.

Taking into account the above population dynamics, let us consider the inflow of debris through a control element at radius r and of width Δr. If we assume a uniform mean number density n, then in a control area of the disk $2\pi r \Delta r$, there is a total number of debris particles $2\pi r \Delta r n(r, t)$. The rate of change for the number of debris in the control area is obtained by considering the inflow and the outflow from neighboring disks, although allowing the source and sink term also to act. In this context, the continuity equation can be derived:

$$\frac{\partial}{\partial t}(2\pi r \Delta r n(r,t)) = \begin{array}{ll} v_r(r) 2\pi r n(r,t) & \text{flow in} \\ -v_r(r+\Delta r) 2\pi (r+\Delta r) n(r+\Delta r, t) & \text{flow out} \\ +2\pi r \Delta r(\dot{n}^+(r,t) - \dot{n}^-(r,t)). & \text{deposition-sweeper} \end{array} \tag{24}$$

Taking the limit of the control element width tending to zero $\Delta r \to 0$, we obtain a partial differential equation for the continuity problem

$$\frac{\partial n(r,t)}{\partial t} + \frac{1}{r}\frac{\partial}{\partial r}(r v_r(r) n(r,t)) = \dot{n}^+(r,t) - \dot{n}^-(r,t) \tag{25}$$

The solution of this equation describes the time evolution of the radial structure of the debris population. The derivative of the second term is computed as

$$\frac{1}{r}\frac{\partial}{\partial r}(r v_r(r) n(r,t)) = \frac{\partial n(r,t)}{\partial r} + n(r,t)\left(\frac{v_r'(r)}{v_r(r)} + \frac{1}{r}\right) v_r(r) \tag{26}$$

and the equation could be rewritten as

$$\frac{\partial n(r,t)}{\partial t} + \frac{\partial n(r,t)}{\partial r} + n(r,t)\left(\frac{v_r'(r)}{v_r(r)} + \frac{1}{r}\right) v_r(r) = \dot{n}^+(r,t) - \dot{n}^-(r,t) \tag{27}$$

We should mention that, in the case of PDE modeling an initial distribution is required. This is analog to the initial values used in the ODE modeling. The initial population distribution could be a Gaussian, a top-hat (i.e., a uniform distribution in a selected interval of altitudes) or even an experimental distribution that fits the observational data. The population of space debris is now described by a linear PDE, therefore, it can be solved using the method of characteristics. In this approach, a single PDE could be reduced into two ODEs. If the solution for the ODEs could be derived in closed form then a solution for the whole PDE can be found.

Applying the method of characteristics in our PDE gives us the following set of equations:

$$\frac{\mathrm{d}n}{-n(r,t)\left((v_r'(r)/v_r(r)) + (1/r)\right) v_r(r) + \dot{n}^+(r,t) - \dot{n}^-(r,t)} = \frac{\mathrm{d}r}{v_r(r)} = \mathrm{d}t \tag{28}$$

which provides two ODEs, namely

$$\frac{\mathrm{d}r}{\mathrm{d}t} = v_r(r) \tag{29}$$

$$\frac{\mathrm{d}n(r,t)}{\mathrm{d}r} + n(r,t)\left(\frac{v_r'(r)}{v_r(r)} + \frac{1}{r}\right) = \frac{q}{v_r(r)}(\dot{n}^+(r,t) - \dot{n}^-(r,t)) \tag{30}$$

The first thing to do is to try to solve Eq. (29). The integration of this equation will provide us a family of curves relating the orbit radius r and time t. These curves are called the characteristics of the PDE, and in this particular example they have a physical meaning: they represent the inflow trajectories of orbital debris due to air drag. The solution of the characteristic equation will be used later in order to solve the evolution equation Eq. (30). Taking into account the considerations for the inflow drag, Eq. (23), we have

$$\frac{\mathrm{d}r}{\mathrm{d}t} = v_r(r) = -\sqrt{\mu r} B \rho_0 \exp\left(-\frac{r-R}{H}\right) \Rightarrow \tag{31}$$

$$\frac{1}{a}\int \frac{\exp(r/H)}{\sqrt{r}}\mathrm{d}r + t = -C \tag{32}$$

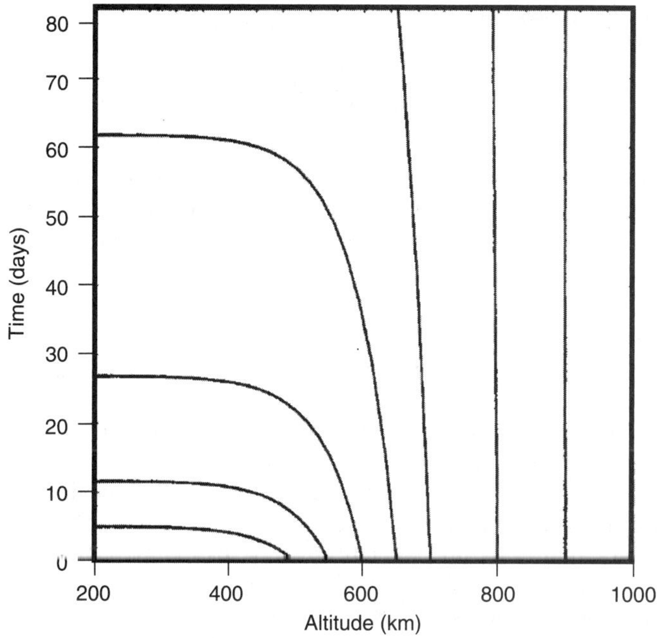

Fig. 11 Family of characteristics for space debris.

where $a = \sqrt{\mu} B \rho_0 \exp(r/H)$. Integrating the above equation results in

$$\exp\left(\frac{r}{H}\right) + \frac{a\sqrt{R}}{H} t = \tilde{C} \tag{33}$$

with $\tilde{C} = -(a\sqrt{R}/H)C$.

In Fig. 11 the family of characteristic is shown. It is clear that the differential air drag plays an important role in the evolution of the debris trajectories. Specifically, the shearing effect caused by it has a major influence on the long-term evolution of the space debris population.

Now that the characteristics of the PDE have been computed, the evolution equation can be solved. Firstly, it is assumed that there is neither deposition of new debris in the disk, nor any removal of debris due to sweepers or random failures. This is equivalent to setting the terms $\dot{n}^+$ and $\dot{n}^-$ equal to zero. Then, the following equation can be directly solved in order to obtain the free evolution of the debris cloud

$$\frac{\mathrm{d}n(r,t)}{\mathrm{d}r} + n(r,t)\left(\frac{v_r'(r)}{v_r(r)} + \frac{1}{r}\right) = 0 \tag{34}$$

The solution yields

$$ln(n(r,t)) = \frac{1}{ln(rv_r(r))} + \Psi(G(r,t)) \quad (35)$$

where $\Psi(G(r,t))$ is an arbitrary function of the characteristics. This arbitrary function is equivalent to the arbitrary constants that appear in the solutions of ODEs. The arbitrary function can be determined from the initial distribution of the debris population $n(r,0)$. For illustration, a Guassian initial distribution will be used of the form

$$n(r,0) = n_m \exp\left(-\lambda(r-r_m)^2\right) \quad (36)$$

where the parameters n_m and λ describe the shape of the initial distribution with a peak at orbit radius r_m. As already discussed, other types of distributions can also be used.

In Fig. 12 the solution is shown for two initial Gaussian distributions of debris, a narrow one and a broader one. The peak of both distributions is chosen at a height of 700 km above the Earth. The shearing effect of the air drag is observed in both distributions. This is expected, because the debris in the inner part of the disk have radial velocities larger than those the debris in the outer part of the disk have. Also, one would expect the peak of the population distribution to move toward lower altitudes. Although this is the case for the narrow Gaussian distribution, a different behavior is observed for the broader one. Namely, the peak of the debris distribution is moving outward. This could be explained by the fact that the debris at low altitude decay faster than the distribution can be replenished from previous one.

In the case that the deposition term is also present, a final steady-state is possible for the population. This is the result of the fact that the removal of

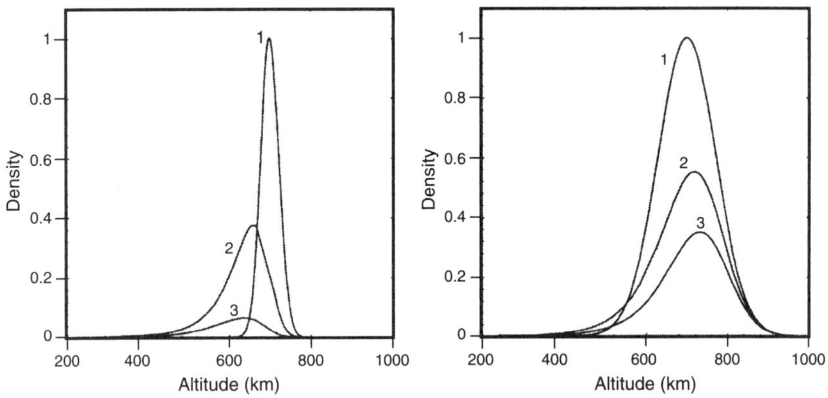

Fig. 12 Free evolution of a narrow (*left panel*) and a broad (*right panel*) Gaussian distribution of debris in a planar disk.

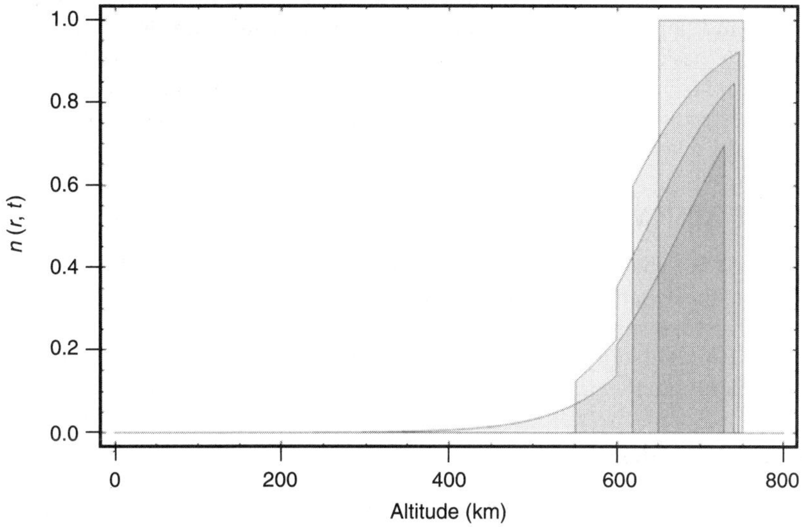

Fig. 13 Evolution of top-hat initial debris population under the effect of a debris sweeper.

debris by air drag can be balanced by the new debris deposited in the population. Then an asymptotic solution can be derived for the continuity equation given by the relation

$$n_\infty(r) = \frac{1}{r^2 v_r(r)} \int r^2 \dot{n}^+(r)\,\mathrm{d}r + D \tag{37}$$

where D is a constant of integration. The inverse problem can also be addressed: finding the mean deposition rate required to obtain a desired steady-state distribution of debris.

Finally, the effect of an active debris sweeper on a fixed circular orbit would be discussed. The debris sweeper can be modeled with a Dirac delta function at a single orbit radius. Assuming that the sweeper removes a fixed fraction λ of debris passing its orbit, the sweeper term is

$$\dot{n}^-(r,t) = -2\pi r v_r(r) n(r,t) \lambda \delta(r - \bar{r}) \tag{38}$$

In Fig. 13 the evolution of an initial top-hat debris population is presented, affected by a sweeper placed at 600-km altitude. Again, the shearing effect due to the differential radial velocity is present. Moreover, an effective reduction of the debris population is observed as they pass from the sweeper altitude.

In conclusion, in this section a simple analytical model was presented, suitable to investigate the long-term evolution of debris population under the effect of air drag, sweepers, and deposition of new debris. It is shown that due to air-drag, an initial narrow disk of debris caused by a fragmentation event will eventually broaden and populate lower altitudes. However, depending on the initial distribution, the peak of the number density of the population may move to higher altitudes. It is shown also that under the constant deposition of debris on the population, a steady-state population could be reached. Finally, the effect of a debris sweeper at a given altitude was discussed.

IV. CONCLUSION

The aim of this chapter was to present simple analytical tools, which allow us to investigate the dynamics of orbital debris populations. Three different approaches were discussed, two based on ODEs and one based on PDEs. All three models are based on simple assumptions and provide us quick estimations in order to have a better insight into the problem.

ACKNOWLEDGMENTS

I would like to thank A. Celletti for useful suggestions that improved this manuscript. Ioannis Gkolias was supported by the European Grant MC-ITN Stardust.

REFERENCES

[1] IADC, "Space Debris Mitigation Guidelines," Technical Report, Issue 1, Rev. 1, IADC, 2002.

[2] Cour-Palais, B. G., and Kessler, D. J., "Collision Frequency of Artificial Satellites: The Creation of a Debris Belt," *Journal of Geophysical Research*, Vol. 83, 1978, pp. 2637–2646.

[3] Talent, D. L., "Analytic Model for Orbital Debris Environmental Management," *Journal of Spacecraft and Rockets*, Vol. 29, 1992, pp. 508–513.

[4] Cordelli, A., and Farinella, P., "The Proliferation of Orbiting Fragments: A Simple Mathematical Model," *Science and Global Security*, Vol. 2, 1991, pp. 365–378.

[5] McInnnes, C. R., "Simple Analytic Model of the Long-Term Evolution of Nanosatellite Constellations," *Journal of Guidance, Control and Dynamics*, Vol. 23, 2000, pp. 332–338.

CHAPTER 3

Advanced Orbit Propagation Methods and Application to Space Debris Collision Avoidance

Davide Amato* **and Claudio Bombardelli***

Technical University of Madrid, Madrid, Spain

I. THE ORBIT PROPAGATION PROBLEM

The most straightforward way to propagate an orbit is to simply write down the equations of motion in Cartesian form, Eq. (1) and integrate them through a numerical technique.

$$\ddot{\boldsymbol{r}} = -\frac{\mu}{r^3}\boldsymbol{r} + \boldsymbol{a}_p \tag{1}$$

This is called *Cowell's formulation.*† Although it is very easy to understand, implement, and integrate in a software, it has two main drawbacks: it is *singular* for $r = 0$ and it exhibits a rapid propagation of the numerical error.

A. NUMERICAL ERROR AND ITS MITIGATION

Numerical error can be defined as the difference between the numerically computed and the exact orbit motion corresponding to a given mathematical model. This is due to the accumulation of

1. *Round-off error*: It arises when trying to represent a number with a finite number of digits in a computer, and can be reduced by moving to a higher-precision format.
2. *Truncation error*: It arises when discretizing a continuous function, and can be reduced by decreasing the step size or increasing the order of the numerical integrator.

At each instant t^* in the numerical propagation the total numerical error can be seen as an error on the initial conditions for the propagation for $t > t^*$, as

*Space Dynamics Group.

†Not to be confused with Cowell's *method*, which is an implicit multistep integration scheme for second-order ODEs.

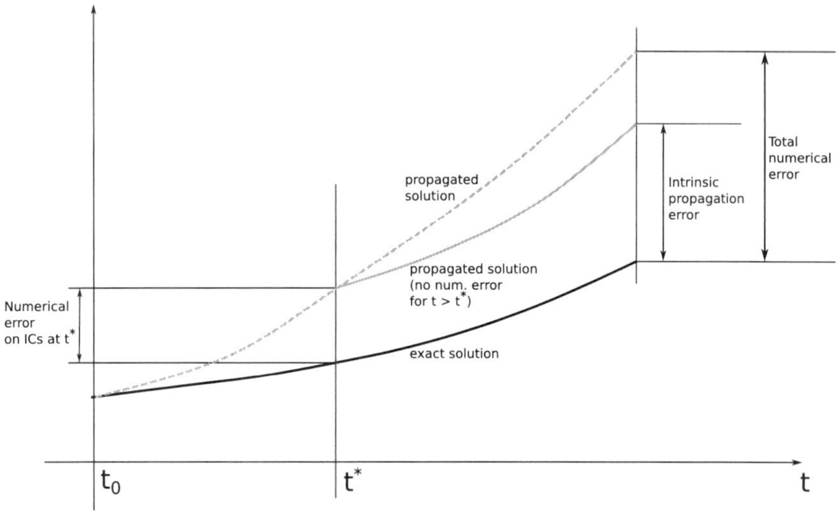

Fig. 1 Evolution of the numerical and intrinsic errors.

shown in Fig. 1. As a *Gedankenexperiment*, consider an ideal case in which the propagation is exact from t^* onwards. Even in this hypothesis, the error on the initial conditions will generally be amplified due to the dynamics of the problem, i.e. it is *intrinsic* to the propagation. Therefore, reducing the numerical error at each step is of utmost importance.

Supposing to represent Keplerian motion by Cowell's formulation, starting from initial conditions $\boldsymbol{r}_0$ and $\dot{\boldsymbol{r}}_0$,

$$\ddot{\boldsymbol{r}}(t) = -\frac{\mu}{r^3}\boldsymbol{r}(t); \quad \boldsymbol{r}(0) = \boldsymbol{r}_0, \quad \dot{\boldsymbol{r}}(0) = \dot{\boldsymbol{r}}_0 \tag{2}$$

Suppose now that the initial conditions are affected by an initial error $(\delta\boldsymbol{r}_0, \delta\dot{\boldsymbol{r}}_0)^\top$. Linearizing the equations around the initial conditions yields an equation which describes the evolution of the propagation error

$$\begin{bmatrix} \delta\dot{\boldsymbol{r}} \\ \delta\ddot{\boldsymbol{r}} \end{bmatrix} = \begin{bmatrix} \boldsymbol{0} & \boldsymbol{I} \\ \mathbf{G}(t) & \boldsymbol{0} \end{bmatrix} \begin{bmatrix} \delta\boldsymbol{r}_0 \\ \delta\dot{\boldsymbol{r}}_0 \end{bmatrix} = \boldsymbol{H}(t) \begin{bmatrix} \delta\boldsymbol{r}_0 \\ \delta\dot{\boldsymbol{r}}_0 \end{bmatrix} \tag{3}$$

where $\mathbf{G}(t)$ is the *gravity gradient matrix*

$$\mathbf{G}(t) = \frac{\mathrm{d}}{\mathrm{d}\boldsymbol{r}}\left(-\frac{\mu}{r^3}\boldsymbol{r}\right) = \frac{\mu}{r^5}(3\boldsymbol{r}\boldsymbol{r}^\top - r^2\boldsymbol{I}) \tag{4}$$

Integrating this system of first-order linear ordinary differential equations (ODEs) yields

$$\begin{bmatrix}\delta \boldsymbol{r}\\ \delta \dot{\boldsymbol{r}}\end{bmatrix} = \left[\exp\left(\int_{t_0}^{t} \boldsymbol{H}(t)\mathrm{d}t\right)\right]\begin{bmatrix}\delta \boldsymbol{r}_0\\ \delta \dot{\boldsymbol{r}}_0\end{bmatrix} = \boldsymbol{\Phi}(t_0, t)\begin{bmatrix}\delta \boldsymbol{r}_0\\ \delta \dot{\boldsymbol{r}}_0\end{bmatrix} \tag{5}$$

It can be shown that the eigenvalues of the matrix $\boldsymbol{H}(t)$ are

$$\lambda = \sqrt{\frac{\mu}{r^3}}\left[\pm\sqrt{2}, \pm i, \mp i\right]^{\top} \tag{6}$$

One of these eigenvalues is real positive, and proportional to $\sqrt{1/r^3}$. Therefore, even without considering any perturbation, Cowell's formulation exhibits fast error propagation. Moreover, it is faster the closer we get to the collision condition, that is, $r \to 0$.

For weakly perturbed problems, which are quite common in astrodynamics, more efficient formulations of dynamics can be employed, such as *element methods*. These formulations express the state of the body through quantities (*orbital elements*) which are linear functions of the independent variable in unperturbed motion.

B. REGULARIZATION AND LINEARIZATION

An often-used technique in element methods is *regularization*, that is, the elimination of singularities occurring *in the equations of motion* by selecting a proper change of variables. This is generally accomplished in two steps:

1. A change of the independent variable from time t to a fictitious time τ through a function of the state $\boldsymbol{x}$ and radius r (Sundman transformation):

$$\frac{\mathrm{d}t}{\mathrm{d}\tau} = c(\boldsymbol{x})r^{\alpha} \tag{7}$$

2. A further transformation using different methods such as projective decomposition, embedding, and so on.

As to improve the stability properties of the equations, it is also possible to *linearize* them. Linearization is often a result of the regularization process.

II. ELEMENT METHODS

In the history of celestial mechanics, several orbital element methods have been developed, often with different aims. *Classical* orbital elements were developed first, and allow to visualize the shape, size, and orientation of an orbit at a glance. However, Lagrange and Gauss planetary equations, which describe the variation of such elements consequent to perturbations, are often unwieldy and present singularities. Therefore, it is often convenient to use different methods, which usually allow to simplify the treatment of the equations of motion.

Hereby, we present some of the most-known element methods in astrodynamics. This assortment is by no means extensive, and the reader is invited to consult reviews such as the one by Hintz [1] to examine the subject in depth. However, a complete survey is probably impossible, and several formulations have to be left behind. One of these is the one first proposed by Milanković and modified by Rosengren and Scheeres [2], which allows a particularly compact expression of the equations of motion through the choice of the angular momentum and eccentricity vectors as elements.

A. CLASSICAL ORBITAL ELEMENTS

In this method, the classical orbital elements $(a, e, i, \Omega, \omega, M)$ are used as state variables and time is the independent variable. Gauss' equations, which define the rate of change of orbital elements, are nonlinear and suffer from singularities when $e = 0$ or $i = 0$. Especially for this last reason, they are not often employed for numerical propagation.

Using notation which is standard in celestial mechanics Gauss' equations are [3]

$$\frac{\mathrm{d}a}{\mathrm{d}t} = \frac{2}{n\sqrt{1-e^2}}\left\{e\sin(\nu)F_\mathrm{R} + \frac{p}{r}F_\mathrm{S}\right\} \tag{8a}$$

$$\frac{\mathrm{d}e}{\mathrm{d}t} = \frac{\sqrt{1-e^2}}{na}\left\{\sin(\nu)F_\mathrm{R} + \left(\cos(\nu) + \frac{e+\cos(\nu)}{1+e\cos(\nu)}\right)F_\mathrm{S}\right\} \tag{8b}$$

$$\frac{\mathrm{d}i}{\mathrm{d}t} = \frac{r\cos(u)}{na^2\sqrt{1-e^2}}F_\mathrm{W} \tag{8c}$$

$$\frac{\mathrm{d}\Omega}{\mathrm{d}t} = \frac{r\sin(u)}{na^2\sqrt{1-e^2}\sin(i)}F_\mathrm{W} \tag{8d}$$

$$\frac{\mathrm{d}\omega}{\mathrm{d}t} = \frac{\sqrt{1-e^2}}{nae}\left\{-\cos(\nu)F_\mathrm{R} + \sin(\nu)\left(1+\frac{r}{p}\right)F_\mathrm{S}\right\} - \frac{r\cot(i)\sin(u)}{h}F_\mathrm{W} \tag{8e}$$

$$\frac{\mathrm{d}M_0}{\mathrm{d}t} = \frac{1}{na^2e}\left\{\left(p\cos(\nu) - 2er\right)F_\mathrm{R} - (p+r)\sin(\nu)F_\mathrm{s}\right\} - \frac{\mathrm{d}n}{\mathrm{d}t}(t-t_0) \tag{8f}$$

B. EQUINOCTIAL ELEMENTS

As first shown by Lagrange and then popularized by Broucke and Cefola [4], classical orbital elements can be manipulated to remove the $e = 0$ and $i = 0$

singularities, obtaining a set of singularity-free equations with time as the independent variable. The state vector is then composed of the semimajor axis a and the elements

$$P_1 = e\sin(\Omega + \omega)$$
$$P_2 = e\cos(\Omega + \omega)$$
$$Q_1 = \tan\left(\frac{i}{2}\right)\sin\Omega$$
$$Q_2 = \tan\left(\frac{i}{2}\right)\cos\Omega$$
$$L = \Omega + \omega + \nu$$

We now suppose to have a perturbing acceleration acting on the body, which in radial, tangential, and normal components reads

$$\boldsymbol{a_d} = (a_r, a_t, a_n)^\top = (\varepsilon\cos\alpha\cos\beta, \varepsilon\sin\alpha\cos\beta, \varepsilon\sin\beta)^\top \tag{9}$$

Then, the rates of change of the equinoctial elements are obtained as [5, 6]

$$\frac{da}{dt} = \frac{2a^2}{h}\left[(P_2\sin L - P_1\cos L)\,\varepsilon\cos\beta\cos\alpha + \frac{p}{r}\varepsilon\cos\beta\sin\alpha\right] \tag{10a}$$

$$\frac{dP_1}{dt} = \frac{r}{h}\left\{-\frac{p}{r}\cos L\cdot\varepsilon\cos\beta\cos\alpha + \left[P_1 + \left(1 + \frac{p}{r}\right)\sin L\right]\varepsilon\cos\beta\sin\alpha \right.$$
$$\left. - P_2\,(Q_1\cos L - Q_2\sin L)\,\varepsilon\sin\beta\right\} \tag{10b}$$

$$\frac{dP_2}{dt} = \frac{r}{h}\left\{\frac{p}{r}\sin L\cdot\varepsilon\cos\beta\cos\alpha + \left[P_2 + \left(1 + \frac{p}{r}\right)\cos L\right]\varepsilon\cos\beta\sin\alpha \right.$$
$$\left. + P_1\,(Q_1\cos L - Q_2\sin L)\,\varepsilon\sin\beta\right\} \tag{10c}$$

$$\frac{dQ_1}{dt} = \frac{r}{2h}\left(1 + Q_1^2 + Q_2^2\right)\sin L\cdot\varepsilon\sin\beta \tag{10d}$$

$$\frac{dQ_2}{dt} = \frac{r}{2h}\left(1 + Q_1^2 + Q_2^2\right)\cos L\cdot\varepsilon\sin\beta \tag{10e}$$

$$\frac{dL}{dt} = \frac{h}{r^2} - \frac{r}{h}(Q_1\cos L - Q_2\sin L)\,\varepsilon\sin\beta \tag{10f}$$

C. STIEFEL–SCHEIFELE ELEMENTS

In this formulation, the equations are linearized and regularized. It derives from the Kustanheeimo-Stiefel (K–S) regularization scheme, and it is extensively described in Ref. [7]. It is based on a Sundman transformation of first order,[‡] that is, in Eq. (7) $\alpha = 1$ and $c = 1$.

The position of the spacecraft is represented through a 4-vector $\boldsymbol{u}$, which is related to the actual position of the spacecraft $\boldsymbol{x}$ by the matrix relation[§] $\boldsymbol{x} = \boldsymbol{L}(\boldsymbol{u})\boldsymbol{u}$.

$\boldsymbol{L}(\boldsymbol{u})$ is the *K–S-matrix*, a four-dimensional generalization of Levi–Civita's matrix:

$$\boldsymbol{L}(\boldsymbol{u}) = \begin{pmatrix} u_1 & -u_2 & -u_3 & u_4 \\ u_2 & u_1 & -u_4 & -u_3 \\ u_3 & u_4 & u_1 & u_2 \\ u_4 & -u_3 & u_2 & -u_1 \end{pmatrix} \tag{11}$$

The independent variable in this method is a generalized eccentric anomaly E, while the elements are the frequency ω, the time element τ, and two 4-vectors $\boldsymbol{\alpha}$, $\boldsymbol{\beta}$. In particular, ω, τ, and the physical time t are connected by the relation

$$t = \tau - \frac{1}{\omega}(\boldsymbol{u}, \boldsymbol{u}^*) \tag{12}$$

Vector $\boldsymbol{u}$ and its derivative with respect to E read

$$\boldsymbol{u} = (u_1, u_2, u_3, u_4)^\top = \boldsymbol{\alpha} \cos\frac{E}{2} + \boldsymbol{\beta} \sin\frac{E}{2} \tag{13}$$

$$\boldsymbol{u}^* = \frac{\mathrm{d}\boldsymbol{u}}{\mathrm{d}E} = -\frac{\boldsymbol{\alpha}}{2} \sin\frac{E}{2} + \frac{\boldsymbol{\beta}}{2} \cos\frac{E}{2} \tag{14}$$

Then, the equations for the rates of change of the elements are obtained as

$$\frac{\mathrm{d}\omega}{\mathrm{d}E} = -\frac{r}{8\omega^2}\frac{\partial V}{\partial t} - \frac{1}{2\omega}\left(\boldsymbol{u}^*, \boldsymbol{L}^\top \boldsymbol{P}\right) \tag{15a}$$

$$\frac{\mathrm{d}\tau}{\mathrm{d}E} = \frac{1}{8\omega^3}(K^2 - 2rV) - \frac{r}{16\omega^3}\left(\boldsymbol{u}, \frac{\partial V}{\partial \boldsymbol{u}} - 2\boldsymbol{L}^\top \boldsymbol{P}\right) - \frac{2}{\omega^2}\frac{\mathrm{d}\omega}{\mathrm{d}E}(\boldsymbol{u}, \boldsymbol{u}^*) \tag{15b}$$

$$\frac{\mathrm{d}\boldsymbol{\alpha}}{\mathrm{d}E} = \left\{\frac{1}{2\omega^2}\left[\frac{V}{2}\boldsymbol{u} + \frac{r}{4}\left(\frac{\partial V}{\partial \boldsymbol{u}} - 2\boldsymbol{L}^\top \boldsymbol{P}\right)\right] + \frac{2}{\omega}\frac{\mathrm{d}\omega}{\mathrm{d}E}\boldsymbol{u}^*\right\} \sin\frac{E}{2} \tag{15c}$$

$$\frac{\mathrm{d}\boldsymbol{\beta}}{\mathrm{d}E} = -\left\{\frac{1}{2\omega^2}\left[\frac{V}{2}\boldsymbol{u} + \frac{r}{4}\left(\frac{\partial V}{\partial \boldsymbol{u}} - 2\boldsymbol{L}^\top \boldsymbol{P}\right)\right] + \frac{2}{\omega}\frac{\mathrm{d}\omega}{\mathrm{d}E}\boldsymbol{u}^*\right\} \cos\frac{E}{2} \tag{15d}$$

where V is a perturbation potential and $\boldsymbol{P}$ is an additional, nonconservative perturbing force. The scalar product is denoted with $(\cdot, \cdot)$.

[‡]One should note that the same time--transformation had been already adopted by the Italian mathematician Tullio Levi-Civita, together with a change of the spatial variables for regularizing the restricted three-body problem.

[§]While a position vector in 3-D space has three components, the vector $\boldsymbol{x}$ has got actually four. This ambiguity is readily solved when considering that the fourth component of $\boldsymbol{x}$ is identically zero $\forall \boldsymbol{u}$.

Propagation using these elements is quite efficient, although requiring the solution of a system of 10 first-order ODEs.

D. SPERLING–BURDET ELEMENTS

Another regularized and linearized formulation which allows efficient propagation is the one by Sperling and Burdet, shown in Ref. [8]. It uses a fictitious time s as an independent variable, and Stumpff functions of jth order c_j to account for elliptic/hyperbolic transition. The state is described by two sets of elements: $(\alpha, \beta, \delta, \sigma, \alpha_J)$ describe the orbit evolution in space and (a, b, γ, τ) in time, for a total of 15 scalar quantities.

Full regularization is accomplished by *embedding* the eccentricity vector and Keplerian energy into the equations of motion.

The formulation consists of 15 first-order ODEs:

$$\boldsymbol{\alpha}' = -\mathbf{Q}sc_1 - \mu\boldsymbol{\varepsilon}'s^2c_2 - \alpha_J'\left[\boldsymbol{\alpha}s^2c_2 + 2\boldsymbol{\beta}s^3\tilde{c}_3 + \frac{1}{2}\boldsymbol{\delta}s^4c_2^2\right] \tag{16a}$$

$$\boldsymbol{\beta}' = \mathbf{Q}c_0 + \mu\boldsymbol{\varepsilon}'sc_1 + \alpha_J'\left[\boldsymbol{\alpha}sc_1 + \boldsymbol{\beta}s^2\tilde{c}_2 - \boldsymbol{\delta}s^3(2\tilde{c}_3 - c_1c_2)\right] \tag{16b}$$

$$\boldsymbol{\delta}' = \mathbf{Q}\alpha_J sc_1 - \mu\boldsymbol{\varepsilon}'c_0 + \alpha_J'\left[-\boldsymbol{\alpha}c_0 + 2\boldsymbol{\beta}\alpha_J s^3\tilde{c}_3 + \frac{1}{2}\boldsymbol{\delta}\alpha_J s^4c_2^2\right] \tag{16c}$$

$$a' = -\frac{1}{r}(\mathbf{Q}, \mathbf{r})sc_1 - \alpha_J'\left[as^2c_2 + 2bs^3\tilde{c}_3 + \frac{1}{2}\gamma s^4c_2^2\right] \tag{16d}$$

$$b' = \frac{1}{r}(\mathbf{Q}, \mathbf{r})c_0 + \alpha_J'\left[asc_1 + bs^2\tilde{c}_2 - \gamma s^3(2\tilde{c}_3 - c_1c_2)\right] \tag{16e}$$

$$\gamma' = \frac{1}{r}(\mathbf{Q}, \mathbf{r})\alpha_J sc_1 + \alpha_J'\left[-ac_0 + 2b\alpha_J s^3\tilde{c}_3 + \frac{1}{2}\gamma\alpha_J s^4c_2^2\right] \tag{16f}$$

$$\tau' = \frac{1}{r}(\mathbf{Q}, \mathbf{r})s^2c_2 + \alpha_J'\left[as^3c_3 + \frac{1}{2}bs^4c_2^2 - 2\gamma s^5(c_5 - 4\tilde{c}_5)\right] \tag{16g}$$

$$\alpha_J' = (2(-\mathbf{r}' + r\boldsymbol{\omega} \times \mathbf{r}), \mathbf{P}) \tag{16h}$$

$$\sigma' = r\boldsymbol{\omega} \cdot \mathbf{r} \times F \tag{16i}$$

where

$$\mu\boldsymbol{\varepsilon}' = 2(\mathbf{r}', \mathbf{F})\mathbf{r} - (\mathbf{r}, \mathbf{F})\mathbf{r}' - (\mathbf{r}', \mathbf{r})\mathbf{F} \tag{17a}$$

$$\mathbf{Q} = r^2\mathbf{F} + 2(\sigma - V)\mathbf{r} \tag{17b}$$

The Stumpff functions take as argument $z = \alpha_J s^2$, in particular $c_n = c_n(z)$, $\tilde{c}_n = \tilde{c}_n(4z)$. The perturbation $\mathbf{F}$ is split into potential and non-potential terms, $\mathbf{F} = \mathbf{P} - \partial V(\mathbf{r}, t)/\partial\mathbf{r}$. System (16) can be reduced to 14 ODEs through the relation $\gamma + a\alpha_J = \mu = \text{constant}$.

E. PELÁEZ ELEMENTS (DROMO)

Finally, we present the *Dromo* formulation, developed by Peláez et al. [9] on the basis of previous work by Deprit [10], Ferràndiz [11], and Sharaf et al. [12].

This formulation uses a fictitious time σ as the independent variable, using the second-order Sundman transformation:

$$\frac{\mathrm{d}t}{\mathrm{d}\sigma} = \frac{r^2}{h} \tag{18}$$

This variable differs from the true anomaly by a perturbation-induced angular drift, $\Delta\sigma = \Delta\nu + \gamma$. The drift can be related to the classical orbital elements evolution by

$$\gamma = \Delta\omega + \int_{\Omega_0}^{\Omega} \cos i \, \mathrm{d}\Omega \tag{19}$$

Then, a *projective decomposition* of the equations of motion along the orbit local vertical is employed to achieve almost-full regularization and linearization; a singularity is still present for rectilinear motion.

The state is then expressed through seven elements q_i. A first set of three elements describes the orbit evolution in its instantaneous plane and is related to the classical orbital elements as

$$q_1 = \frac{e}{h}\cos\gamma, \quad q_2 = \frac{e}{h}\sin\gamma, \quad q_3 = \frac{1}{h} \tag{20}$$

The remaining four elements are the Euler parameters describing, together with $\Delta\sigma$, the rotation between inertial space and local vertical local horizontal (LVLH) orbital axes. Thus, the rotation matrix $Q_{\mathcal{RI}}$ is obtained as

$$Q_{\mathcal{RI}} = Q_0 M_\sigma \tag{21}$$

where

$$M_\sigma = \begin{bmatrix} \cos\Delta\sigma & -\sin\Delta\sigma & 0 \\ \sin\Delta\sigma & \cos\Delta\sigma & 0 \\ 0 & 0 & 1 \end{bmatrix} \tag{22}$$

$$Q_0 = \begin{bmatrix} 1 - 2(q_5^2 + q_6^2) & 2(q_4 q_5 - q_6 q_7) & 2(q_4 q_6 + q_5 q_7) \\ 2(q_4 q_5 + q_6 q_7) & 1 - 2(q_4^2 + q_6^2) & 2(q_5 q_6 - q_4 q_7) \\ 2(q_4 q_6 - q_5 q_7) & 2(q_5 q_6 + q_4 q_7) & 1 - 2(q_4^2 + q_5^2) \end{bmatrix} \tag{23}$$

The evolution of the elements $q_1, \ldots, q_7$ and of time t is described by the following set of eight ODEs:

$$\frac{\mathrm{d}t}{\mathrm{d}\sigma} = \frac{r^2}{h} \tag{24a}$$

$$\frac{\mathrm{d}}{\mathrm{d}\sigma}\begin{pmatrix} q_1 \\ q_2 \\ q_3 \end{pmatrix} = \frac{1}{q_3 s^3}\begin{pmatrix} s\sin\sigma & (s+q_3)\cos\sigma \\ -s\cos\sigma & (s+q_3)\sin\sigma \\ 0 & -q_3 \end{pmatrix}\begin{pmatrix} a_r \\ a_\theta \end{pmatrix} \tag{24b}$$

$$\frac{\mathrm{d}}{\mathrm{d}\sigma}\begin{pmatrix} q_4 \\ q_5 \\ q_6 \\ q_7 \end{pmatrix} = \frac{a_z}{2q_3 s^3}\begin{pmatrix} 0 & 0 & -\sin\Delta\sigma & \cos\Delta\sigma \\ 0 & 0 & \cos\Delta\sigma & \sin\Delta\sigma \\ \sin\Delta\sigma & -\cos\Delta\sigma & 0 & 0 \\ -\cos\Delta\sigma & -\sin\Delta\sigma & 0 & 0 \end{pmatrix}\begin{pmatrix} q_4 \\ q_5 \\ q_6 \\ q_7 \end{pmatrix} \tag{24c}$$

where $s = q_1\cos\sigma + q_2\sin\sigma + q_3$ is the nondimensional transversal velocity and the angular momentum is $h = 1/q_3$.

III. ANALYTICAL ORBIT PROPAGATION

In this section, we will show two orbit propagation problems for which an approximate analytical solution can be found by using the previously shown Dromo formulation.

A. CONSTANT RADIAL ACCELERATION

In this problem, we will seek an asymptotic solution for the problem of a spacecraft in an elliptical orbit perturbed by a radial acceleration of constant magnitude [13].

1. PERTURBATION METHOD

Before starting the exposition of the problem, it is necessary to revise a few concepts of perturbation methods applied to ODEs. Suppose we have a first-order, regular (i.e., nonsingular) ODE which can be written in a nondimensional form as

$$\frac{\mathrm{d}\boldsymbol{q}}{\mathrm{d}\sigma} = f(\boldsymbol{q}, \varepsilon) \tag{25}$$

where ε is a small parameter.

Then, we can hope that for a large-enough interval $\Delta\sigma$, the solution of the equation can be approximated by its *asymptotic expansion*

$$\boldsymbol{q}(\sigma, \varepsilon) = \boldsymbol{q}_0(\sigma) + \varepsilon\boldsymbol{q}_1(\sigma) + \varepsilon^2\boldsymbol{q}_2(\sigma) + \cdots \tag{26}$$

where q_i are the solutions, in cascade, of the differential equations

$$\begin{aligned}
\frac{\mathrm{d}\boldsymbol{q}}{\mathrm{d}\sigma} &= f(\boldsymbol{q})\big|_{\varepsilon=0} \\
\frac{\mathrm{d}\boldsymbol{q}_1}{\mathrm{d}\sigma} &= \left.\frac{\mathrm{d}f(\boldsymbol{q})}{\mathrm{d}\varepsilon}\right|_{\varepsilon=0} \\
\frac{\mathrm{d}\boldsymbol{q}_2}{\mathrm{d}\sigma} &= \frac{1}{2}\frac{d^2 f(\boldsymbol{q})}{\mathrm{d}\varepsilon^2}\Big|_{\varepsilon=0} \\
&\vdots
\end{aligned} \tag{27}$$

Typically, the asymptotic solution breaks down (i.e., it loses validity) for $\Delta\sigma \approx (1/\varepsilon)$.

2. SOLUTION THROUGH ASYMPTOTIC EXPANSION

Setting ε to be the ratio between the perturbing radial and local gravitational accelerations and particularizing Eq. (24b) to the problem, we get

$$\frac{\mathrm{d}}{\mathrm{d}\theta}\begin{pmatrix} q_1 \\ q_2 \\ q_3 \end{pmatrix} = \frac{\varepsilon}{q_3 s^2}\begin{pmatrix} \sin\theta \\ -\cos\theta \\ 0 \end{pmatrix} \tag{28}$$

where θ is the angular position of the spacecraft from the initial eccentricity vector. As expected, element $q_3 = 1/h$ stays constant, due to the perturbation being a central force.

We seek a solution to Eq. (28) of the form shown in Eqs. (26) and (27). At zeroth order we obtain

$$\frac{\mathrm{d}}{\mathrm{d}\theta}\begin{pmatrix} q_{10} \\ q_{20} \end{pmatrix} = 0 \Rightarrow \begin{pmatrix} q_{10}(\theta) \\ q_{20}(\theta) \end{pmatrix} = \begin{pmatrix} q_{1i} \\ 0 \end{pmatrix} \tag{29}$$

The elements remain constant and equal to their initial values. This result could be anticipated, since the unperturbed case corresponds to Keplerian motion. Using this we have, at first order

$$\frac{\mathrm{d}}{\mathrm{d}\theta}\begin{pmatrix} q_{11} \\ q_{21} \end{pmatrix} = \frac{1}{q_{3i}(q_{3i} + q_{10}\cos\theta + q_{20}\sin\theta)^2}\begin{pmatrix} \sin\theta \\ -\cos\theta \end{pmatrix} \tag{30}$$

Equation (30) can be solved by simple quadrature, obtaining the solution at first order in ε:

$$q_{11} = \frac{1}{q_{1i}q_{3i}(q_{3i} + q_{1i}\cos\theta)} - \frac{1}{q_{1i}q_{3i}(q_{3i} + q_{1i})}$$

$$q_{21} = \frac{-\sin\theta}{(q_{3i} + q_{1i}\cos\theta)(q_{3i}^2 - q_{1i}^2)} + \frac{2q_{1i}}{q_{3i}(q_{3i}^2 - q_{1i}^2)^{3/2}}\left(\frac{\theta}{2} + \arctan\mathcal{K}(\theta)\right) \tag{31}$$

$$\mathcal{K}(\theta) = -\frac{\sin\theta\left(-q_{3i} + q_{1i} + \sqrt{q_{3i}^2 - q_{1i}^2}\right)}{(1 + \cos\theta)\sqrt{q_{3i}^2 - q_{1i}^2} + (1 - \cos\theta)(q_{3i} - q_{1i})}$$

A plot of $q_1(\theta) = q_{10} + \varepsilon q_{11}$ is shown in Fig. 2, along with a solution obtained through the method of multiple time scales. It can be seen that this asymptotic solution starts breaking down after 3–4 orbits, when $\theta \sim 1/\varepsilon$.

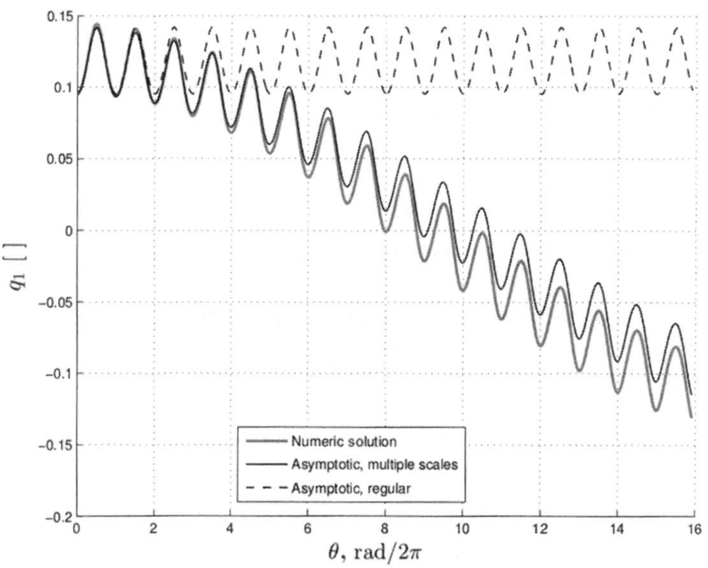

Fig. 2 Comparison of solutions for the dimensionless Dromo element q_1 in the radially perturbed two-body problem, for $e_0 = 0.2$, $\varepsilon = 5 \times 10^{-3}$. The dashed line is the solution obtained through a regular asymptotic expansion as in Sec. 2, which is compared with a numerical solution and the one obtained through the method of multiple scales. (Taken from Ref. [13].)

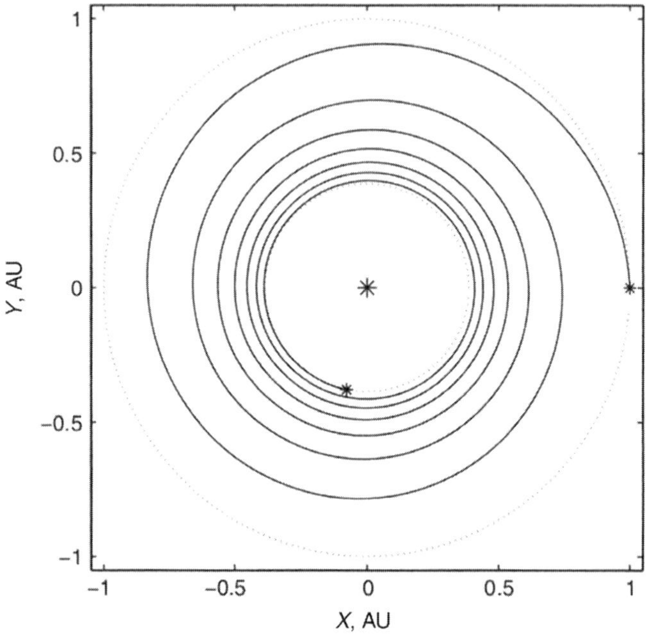

Fig. 3 Schematic of a low-thrust transfer from Earth to Mercury. Planets' orbits are assumed to be circular and coplanar. (Taken from Ref. [14].)

B. CONSTANT TANGENTIAL ACCELERATION

An asymptotic solution for the two-body problem perturbed by a constant tangential acceleration was obtained in Ref. [14].

The approach used is analogous to the one of the previous section, in which perturbation methods are used to find approximate, closed-form solutions to the Dromo Eqs. (24b) and (24c). These solutions are *rectified* (i.e., they are iterated) once every few orbits, before they start to break down.

Plots of the solution to the problem particularized to an Earth–Mercury low-thrust transfer (Fig. 3) are presented in Fig. 4.

IV. COLLISION AVOIDANCE PROBLEM

One of the applications of these methods and techniques for orbit propagation is the *collision avoidance* problem, which can be stated as follows.

Suppose that a collision is predicted between two satellites S_1 and S_2 within a timespan Δt, and that S_1 can perform an impulsive avoidance maneuver $\Delta \boldsymbol{v}$. One is asked to determine the maneuver's *direction* and *location* (along the orbit of S_1)

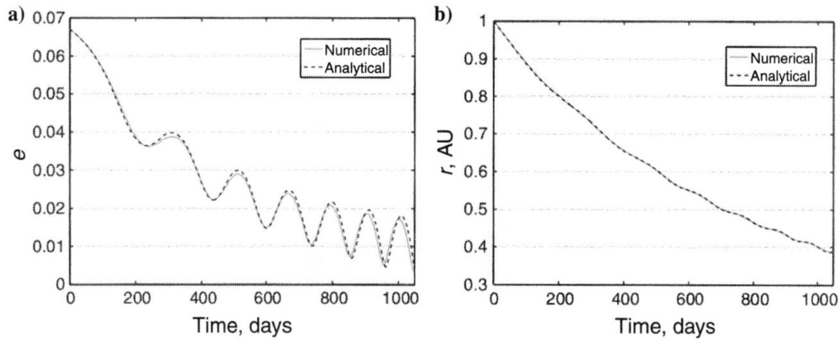

Fig. 4 Comparison between an analytical and numerical solution for an Earth-to-Mercury transfer. Here, the analytical formulas are rectified three times per orbital revolution. (Taken from Ref. [14].)

which either minimize the collision probability or, almost equivalently, maximize the collision miss distance.

The following development is taken from Ref. [15].

A. GEOMETRY OF THE PROBLEM

The orbit of the maneuverable satellite S_1 is taken as a reference. Therefore, we locate both the collision and the maneuver positions on S_1's orbit through the position angles θ_c and θ_m respectively (Fig. 5). Moreover, we define a_0, e_0 to be the initial semimajor axis and eccentricity of S_1.

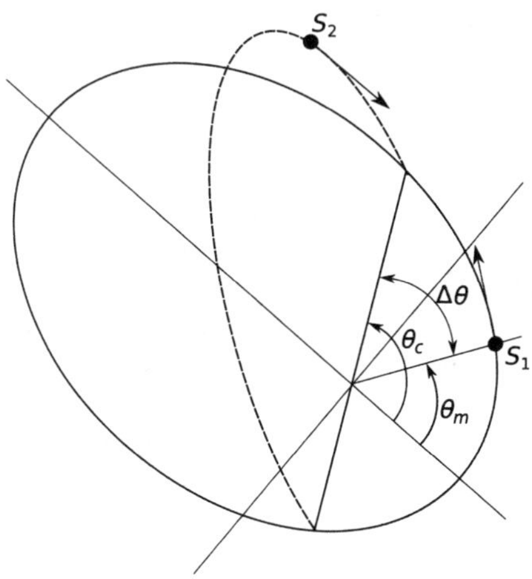

Fig. 5 Collision and maneuver geometry.

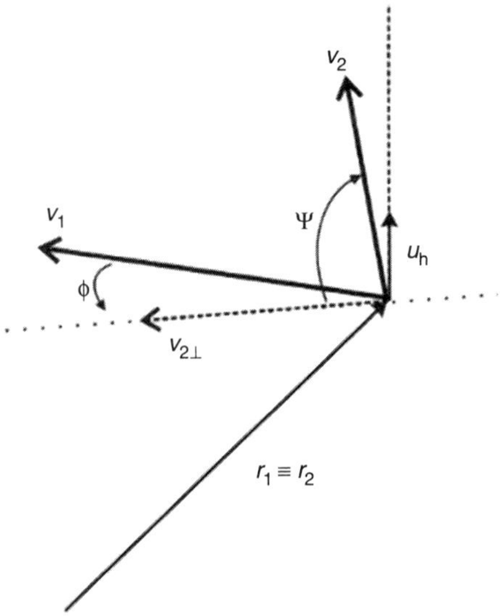

Fig. 6 Relation between the satellite velocity vectors v_1, v_2.

At collision, the velocity vector of S_2 is obtained from the one of S_1 through an in-plane rotation ϕ followed by an out-of-plane rotation ψ and a magnitude scaling χ (Fig. 6).

With these assumptions, the collision is completely described by the six scalar quantities $(a_0, e_0, \theta_c, \phi, \psi, \chi)$, while the maneuver is completely described by the four scalars $(\theta_m, \Delta \boldsymbol{v})$, where $\theta_m = \theta_c - \Delta\theta$, $\Delta \boldsymbol{v} = (\Delta v_r, \Delta v_\theta, \Delta v_h)$.

B. COLLISION KINEMATICS

A key assumption when examining the kinematics of the problem is that *the displacement of S_1 from the impact point as a result of the maneuver is small relative to its radial orbital distance.* This assumption is usually satisfied in all practical cases.

Moreover, within a small interval of time Δt around the impact event, the relative motion of S_1, S_2 can be considered rectilinear.

Given these assumptions, we can use the velocity vectors $\boldsymbol{v}_1$, $\boldsymbol{v}_2$ at collision to construct a local reference system $\langle x, y, z \rangle$, centred at S_2 at the impact event, as displayed in Fig. 7.

$$\boldsymbol{u}_x = \frac{\boldsymbol{v}_1}{\|\boldsymbol{v}_1\|}, \quad \boldsymbol{u}_z = \frac{\boldsymbol{v}_1 \times \boldsymbol{v}_2}{\|\boldsymbol{v}_1 \times \boldsymbol{v}_2\|}, \quad \boldsymbol{u}_y = \boldsymbol{u}_z \times \boldsymbol{u}_x \tag{32}$$

With reference to Fig. 7, the displacement of S_1 *on the b-plane of* S_2 [16] due to the maneuver will be given by

$$\begin{cases} \xi = -\delta z \\ \zeta = -\delta x \sin\beta - \delta y \cos\beta \end{cases} \tag{33}$$

where β is the angle between the inertial velocity of S_1 and the velocity of S_1 relative to S_2, which is obtained as

$$\cos\beta = \frac{(\boldsymbol{v}_1 - \boldsymbol{v}_2)\cdot \boldsymbol{v}_1}{\|\boldsymbol{v}_1\|\|\boldsymbol{v}_1 - \boldsymbol{v}_2\|} \tag{34}$$

The total collision miss distance results in

$$\rho = \sqrt{\xi^2 + \zeta^2} \tag{35}$$

Now, one needs to relate the displacement $(\delta x, \delta y, \delta z)^{\top}$ to the characteristics of the perturbed orbital motion of S_1. It can be shown that the most general outcome of a maneuver performed at $\theta = \theta_m$ is to produce, at $\theta = \theta_c$

1. A radial displacement δr
2. A time delay δt along the trajectory
3. An out-of-plane displacement δw

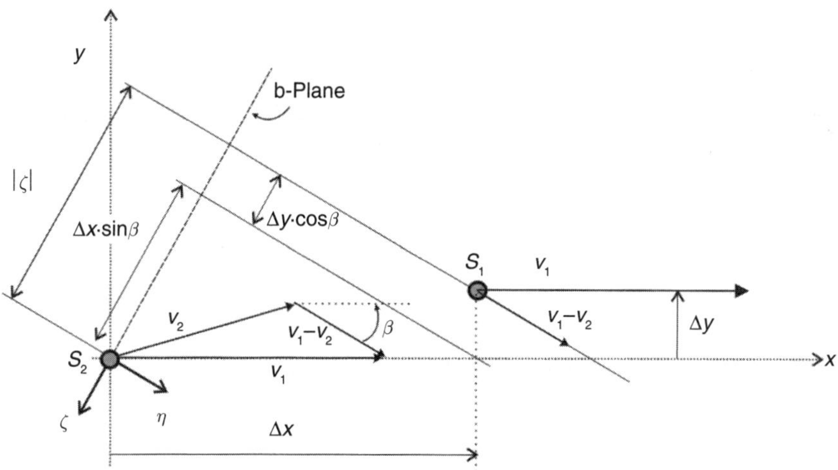

Fig. 7 Snapshot of the $S_1 - S_2$ encounter geometry in the *x*, *y* plane after the maneuver. (Taken from Ref. [15].)

with respect to the unperturbed orbit. Once these are known, the linear collision approximation leads to the linear relationship

$$\begin{pmatrix} \delta x \\ \delta y \\ \delta z \end{pmatrix} = \begin{pmatrix} -v_1 & \sin\alpha \sin\theta_c & 0 \\ 0 & -\dfrac{\cos\alpha \sin\phi \cos\psi}{\sqrt{1-\cos^2\psi \cos^2\phi}} & \dfrac{\sin\psi}{\sqrt{1-\cos^2\psi \cos^2\phi}} \\ 0 & \dfrac{\cos\alpha \sin\psi}{\sqrt{1-\cos^2\psi \cos^2\phi}} & \dfrac{\sin\phi \cos\psi}{\sqrt{1-\cos^2\psi \cos^2\phi}} \end{pmatrix} \begin{pmatrix} \delta t \\ \delta r \\ \delta w \end{pmatrix} \tag{36}$$

where α is the flight path angle of S_1 at collision, which obeys

$$\sin\alpha = \frac{q_{10}\sin\theta_c}{v_1}, \qquad \cos\alpha = \frac{q_{30} + q_{10}\cos\theta_c}{v_1} \tag{37}$$

In this expression, the first and second Peláez elements [9] evaluated at collision appear. It can be shown that

$$q_{10} = \frac{e_0}{\sqrt{1 + e_0\cos\theta_c}}, \qquad q_{30} = \frac{1}{\sqrt{1 + e_0\cos\theta_c}} \tag{38}$$

C. COLLISION DYNAMICS

The study of the dynamics of S_1 after the impulsive maneuver $\Delta\boldsymbol{v}$ is made easier by the use of Peláez elements, which allow to derive approximated relations for the radial shift, time delay, and out-of-plane displacement.

1. PLANAR DYNAMICS

The relation between the applied impulsive $\Delta\boldsymbol{v}$ and the consequent $(\delta r, \delta t)$ can also be achieved by using Pelàez elements. For the radial shift we have that

$$\delta r \approx C_{r\theta}\Delta v_\theta + C_{rr}\Delta v_r \tag{39}$$

where the constants $C_{r\theta}$, C_{rr} are fixed from the geometry of the collision avoidance event

$$\begin{aligned} C_{r\theta} &= \frac{2q_{30}[1-\cos(\theta_c - \theta_m)] - q_{10}\sin(\theta_m)\sin(\theta_c - \theta_m)}{q_{30}(q_{30} + q_{10}\cos\theta_m)(q_{30} + q_{10}\cos\theta_c)^2} \\ C_{rr} &= \frac{\sin(\theta_c - \theta_m)}{q_{30}(q_{30} + q_{10}\cos\theta_c)^2} \end{aligned} \tag{40}$$

Analogously, it may be demonstrated that the time delay δt due to the maneuver is

$$\delta t \approx C_{t\theta}\Delta v_\theta + C_{tr}\Delta v_r \tag{41}$$

where

$$
\begin{aligned}
C_{t\theta} = &\frac{1}{q_{30}(q_{30}^2 - q_{10}^2)^{5/2}(q_{30} - q_{10}\cos E_m)}[K_{\theta 1}(E_c - E_m) + K_{\theta 2}(\sin E_c - \sin E_m) \\
&+ K_{\theta 3}(\sin 2E_c - \sin 2E_m) + K_{\theta 4}(\cos E_c - \cos E_m) \\
&+ K_{\theta 5}(\cos 2E_c - \cos 2E_m] \qquad (42)
\end{aligned}
$$

$$
\begin{aligned}
C_{tr} = &\frac{1}{q_{30}(q_{30}^2 - q_{10}^2)^{2}(q_{30} - q_{10}\cos E_m)}[K_{r1}(E_c - E_m) + K_{r2}(\sin E_c - \sin E_m) \\
&+ K_{r3}(\sin 2E_c - \sin 2E_m) + K_{r4}(\cos E_c - \cos E_m) \\
&+ K_{r5}(\cos 2E_c - \cos 2E_m] \qquad (43)
\end{aligned}
$$

The constants $K_{\theta i}$, K_{ri} also depend on the geometry of the problem through the elements (q_{10}, q_{30}) and are listed in Ref. [15].

2. OUT-OF-PLANE DYNAMICS

Out-of-plane motion is decoupled from the planar one and can be described with sufficient accuracy by linearizing gravity. After rescaling and nondimensionalizing, it is possible to obtain the out-of-plane displacement due to an out-of-plane Δv as

$$
\delta w = \frac{\sqrt{q_{30}^2 + q_{10}q_{30}\cos\theta_c}}{q_{30} + q_{10}\cos\theta_m}\sin(\theta_c - \theta_m)\Delta v_h \qquad (44)
$$

V. MANEUVER OPTIMIZATION

The analytical expression for the miss distance follows directly from the previous equations. This closed-form solution allows for full optimization of the maneuver by solving an eigenvalue problem [17], which will be performed for two examples: the 2009 Iridium–Cosmos collision one in which the maneuverable satellite is in a highly elliptical orbit.

The maneuver orientation is parameterized with respect to the tangential direction as

$$
\begin{aligned}
\Delta v_r &= \Delta v \cos\gamma \sin(\sigma + \alpha) \\
\Delta v_\theta &= \Delta v \cos\gamma \cos(\sigma + \alpha) \qquad (45) \\
\Delta v_h &= \Delta v \sin\gamma
\end{aligned}
$$

where α is the flight path angle, and the optimization variables are σ, the in-plane angle and γ, the out-of-plane angle.

TABLE 1 PARAMETERS FOR THE 2009 IRIDIUM–COSMOS COLLISION

a_0, km	e_0	ϕ, deg	ψ, deg	θ_c, deg	χ
7155.8	2×10^{-4}	180.0	77.5	−16.85	1.0

A. IRIDIUM–COSMOS COLLISION

The 2009 Iridium–Cosmos collision took place on 10th February 2009 and saw the active Iridium 33 spacecraft impacting the disabled Cosmos 2251 satellite at roughly 788.6 km. The collision is fully characterized by the parameters in Table 1. The magnitude of the maneuver Δv is 1 m/s.

The results from the optimization are interesting as they point out that there exists a non-obvious, locally optimal location along the orbit of S_1 where the collision avoidance maneuver should be performed (Fig. 8a). Moreover, Fig. 8b shows that it is in general necessary to thrust out of the orbital plane to achieve the maximum miss distance, especially for last-minute maneuvers with $\Delta\theta < 2\pi$.

B. HIGH-ECCENTRICITY COLLISION

In this case, we suppose the satellite S_1 to be in a highly elliptical orbit, with the collision characterized by the parameters in Table 2. The magnitude of the maneuver Δv is 1 cm/s.

Figure 9a highlights the importance of performing the maneuver near periapsis, as it is well known. Also, and similarly to the Iridium–Cosmos example, the

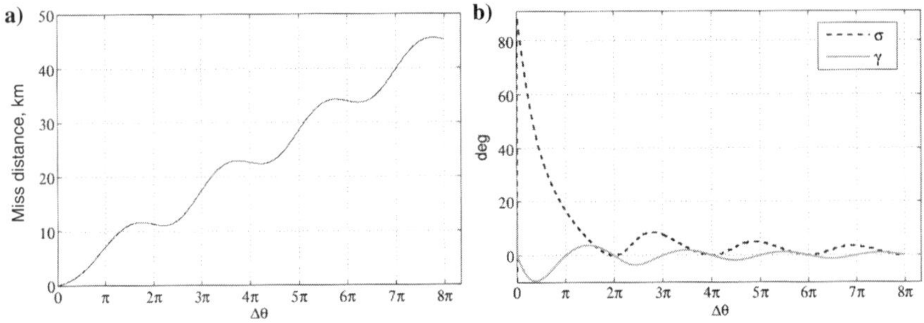

Fig. 8 Optimal miss distance and maneuver direction for the Iridium–Cosmos collision. a) Maximum achievable miss distance km as a function of the maneuver separation arc for the Iridium–Cosmos collision. A Δv of 1 m/s is assumed. b) Optimal maneuver direction angles for the Iridium–Cosmos collision. (Taken from Ref. [15].)

TABLE 2 PARAMETERS FOR THE HIGH-ECCENTRICITY COLLISION

a_0, km	e_0	ϕ, deg	ψ, deg	θ_c, deg	χ
133560.0	0.95	180.0	77.5	0	1.0

optimal maneuver orientation is nearly tangential when applied more than one orbit before the impact and far from tangential when applied during the last orbit (Fig. 9b).

C. B-PLANE SHIFT AFTER MANEUVER

Finally, we make two final observations.

First, by examining the components of the deflection in the b-plane (Fig. 7), we note that the ζ component of the deflection dominates for a long lead time.

Second, generally, the achievement of the maximum miss distance also yields the minimum impact probability. However, for last-minute maneuvers, the difference between these two optimization approaches is not negligible [18].

VI. CONCLUSIONS

In this chapter, several advanced orbit propagation formulations have been described. Their usage can be extremely advantageous for long-term accurate and efficient propagation, since they enable to decrease or eliminate the intrinsic propagation error in several cases. Thus, efficiency and speed of numerical implementations can be greatly increased.

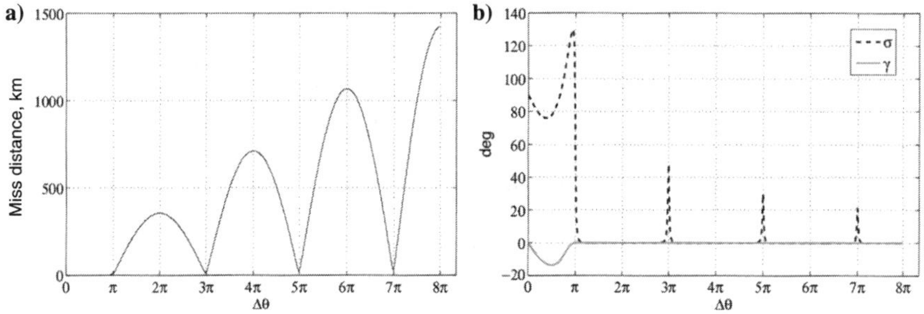

Fig. 9 Optimal miss distance and maneuver direction for the highly eccentric collision. a) Maximum achievable miss distance, km as a function of the maneuver separation arc for the highly eccentric collision. A Δv of 1 cm/s is assumed. b) Optimal maneuver direction angles for the highly eccentric collision. (Taken from Ref. [15].)

Additionally, particular formulations like Dromo allow to find fully analytical asymptotic solutions for weakly perturbed orbits, for instance, the ones related to low-thrust spacecraft.

Finally, the Dromo formulation was applied to the problem of planning and optimizing of a collision avoidance maneuver.

REFERENCES

[1] Hintz, G. R., "Survey of Orbit Element Sets," *Journal of Guidance, Control and Dynamics*, Vol. 31, May–June 2008, pp. 785–790.

[2] Rosengren, A. J., and Scheeres, D. J., "On the Milankovitch Orbital Elements for Perturbed Keplerian Motion," *Celestial Mechanics and Dynamical Astronomy*, Vol. 118, Jan. 2014, pp. 197–220.

[3] Vallado, D. A., *Fundamentals of Astrodynamics and Applications*, 4th ed., Microcosm Press, Hawthorne, CA, 2013.

[4] Broucke, R. A., and Cefola, P., "On the Equinoctial Orbit Elements," *Celestial Mechanics*, Vol. 5, 1972, pp. 303–310.

[5] Battin, R. H., *An Introduction to the Mathematics and Methods of Astrodynamics*, Revised edition, AIAA, Reston, VA, 1999.

[6] Walker, M. J. H., Owens, J., and Ireland, B., "A Set of Modified Equinoctial Elements," *Celestial Mechanics*, Vol. 36, Aug. 1985, pp. 409–419.

[7] Stiefel, E. L., and Scheifele, G., *Linear and Regular Celestial Mechanics*, Springer-Verlag, New York, 1971.

[8] Bond, V. R., and Allman, M. C., *Modern Astrodynamics: Fundamentals and Perturbation Methods*, Princeton University Press, Princeton, NJ, USA, 1996.

[9] Peláez, J., Hedo, J. M., and de Andrés, P. R., "A Special Perturbation Method in Orbital Dynamics," *Celestial Mechanics and Dynamical Astronomy*, Vol. 97, Feb. 2007, pp. 131–150.

[10] Deprit, A., "Ideal Elements for Perturbed Keplerian Motions," *Journal of Research of the National Bureau of Standards, Section B: Mathematical Sciences*, Vol. 79B, Jan. 1975, pp. 1–15.

[11] Ferrándiz, J. M., "A General Canonical Transformation Increasing the Number of Variables with Application in the Two-Body Problem," *Celestial Mechanics*, Vol. 41, 1987/88, pp. 343–357.

[12] Sharaf, M. A., Awad, M. El-S., and Najmuldeen, S. Al-S. A., "Motion of Artificial Satellites in the Set of Eulerian Redundant Parameters," *Earth, Moon, and Planets*, Vol. 56, Feb. 1992, pp. 141–164.

[13] Gonzalo, J. L., and Bombardelli, C., "Asymptotic Solution for the Two-Body Problem with Radial Perturbing Acceleration," *Advances in the Astronautical Sciences*, Vol. 152, 2014, pp. 359–377.

[14] Bombardelli, C., Baù, G., and Peláez, J., "Asymptotic Solution for the Two-Body Problem with Constant Tangential Thrust Acceleration," *Celestial Mechanics and Dynamical Astronomy*, Vol. 110, July 2011, pp. 239–256.

[15] Bombardelli, C., "Analytical Formulation of Impulsive Collision Avoidance Dynamics," *Celestial Mechanics and Dynamical Astronomy*, Vol. 118, No. 2, 2014, pp. 99–114.

[16] Valsecchi, G., Milani, A., Gronchi, G., and Chesley, S., "Resonant Returns to Close Approaches: Analytical Rheory," *Astronomy and Astrophysics*, Vol. 408, No. 3, Sep. 2003, pp. 1179–1196.

[17] Bombardelli, C., and Hernando-Ayuso, J., "Optimal Impulsive Collision Avoidance in Low Earth Orbit," *Journal of Guidance, Control, and Dynamics*, Vol. 38, No. 2, 2015, pp. 217–225.

[18] Bombardelli, C., Hernando Ayuso, J., and García Pelayo, R., "Collision Avoidance Maneuver Optimization," *Advances in the Astronautical Sciences*, Vol. 152, 2014, pp. 1857–1869.

CHAPTER 4

The Accessibility of the Near-Earth Asteroids

Ettore Perozzi*
Deimos Space, Madrid, Spain

Stefano Marò†
University of Pisa, Pisa, Italy

I. INTRODUCTION

In 1925, the German engineer Walter Hohmann published a book *The Attainability of Celestial Bodies* in which he lays the foundations of modern interplanetary traveling. He was the first to consider the problem of reaching a planet, and specifically Venus and Mars because at that time space travel was conceivable only with humans on board a spaceship. The basic problem dealt with was the choice of the trajectory that the spacecraft had to follow in order to reach the target planet. Hohmann noticed that Keplerian motion could be used to this end, thus finding "a new use for an old object: the ellipse" [1]. The so-called "Hohmann transfer strategy" consists of connecting the orbit of the Earth to the orbit of the target planet with an elliptic trajectory tangent to both orbits. To actually carry out the orbital transfer a spacecraft has to perform two maneuvers, one to leave Earth's orbit and the other upon arrival at the planet's orbit. Because a maneuver produces a change of velocity, Hohmann introduced a quantity, the total ΔV, given by the sum of the velocity variations resulting from the two orbital maneuvers, which is a measure of the accessibility of a celestial body: the higher the ΔV, the less accessible the body.

Hohmann focused on the accessibility of the planets and therefore he assumed that the target orbits were coplanar to that of the Earth and that the planets moved on circular orbits. In what follows the possibility of applying the Hohmann transfer strategy to the study of the accessibility of the near-Earth asteroids (NEAs) is discussed. Note that in this case the Hohmann original assumptions are not met because NEAs can have large inclinations and eccentricities. The interest on the accessibility of NEAs has grown in the last decades due to the operation of all-sky surveys, which increased dramatically the NEA discovery rate allowing to

*ESA NEO Coordination Centre (ESRIN, Frascati, Italy).
†Department of Mathematics.

detect also small objects with physical properties interesting for both scientific and exploration missions.

After giving an overview of the classical Hohmann transfer and of its possible extension, we will see how it can be used in NEA space mission design. More details and results can be found in [4–8].

II. HOHMANN TRANSFER

The whole accessibility problem deals with velocities. If one wants to reach a celestial body not only for flying past it but for entering in orbit around it or attempting to land on its surface, the first thing to worry about is how to make a spacecraft reach the same velocity of the target body. This is called a "rendez-vous" mission, as opposed to a short duration "flyby mission." Table 1 compares some well-known reference speeds (e.g., car, airplane, sound) to those typical of space trajectories.

Note the difference between the velocity needed to get into space on a suborbital flight and the much larger value for orbiting our planet. When translating velocity changes into fuel masses and considering that Hohmann assumed manned spacecrafts, it is clear why he tried to find transfer trajectories, which minimized the ΔV.

And so let us start describing the Hohmann transfer referring to Fig. 1.

Consider a spacecraft orbiting on a circular heliocentric orbit with radius r_1. The Hohmann problem consists of sending the spacecraft to meet another circular heliocentric orbit with radius r_2 that is supposed to be coplanar with the first one.

TABLE 1 REFERENCE SPEEDS

Craft	km/s	Reference
Car	0.10	Formula 1
Airplane	0.25	Jet airline
Sound	0.34	Dry air at 20 C
Supersonic airplane	0.65	Concorde
Suborbital flight	0.98	SpaceShipONE
Rocket plane	2.30	X-15
First cosmic velocity	7.82	LEO
Second cosmic velocity	11.20	Earth escape
Third cosmic velocity	16.60	Solar system escape
Earth	29.74	Orbital velocity
Meteor	>70.00	At atmospheric entry
Light	299.792.458.00	Through void

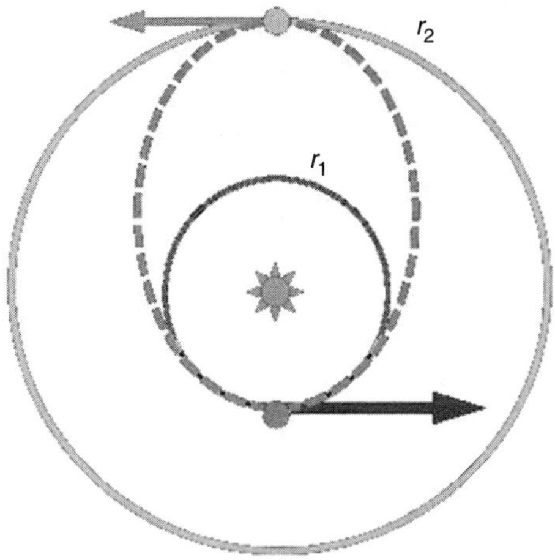

Fig. 1 Hohmann transfer scheme.

The idea is to connect them with a heliocentric ellipse whose aphelion and perihelion are tangency points with respect to the larger and smaller orbit, respectively (assuming for simplicity that the target orbit is farther away than the departure orbit). The geometry is very simple and straightforward, two-body computations, gives that the semi-major axis of the ellipse has to be

$$a = \frac{r_1 + r_2}{2}$$

Being a Keplerian motion we also have the orbital period from which one can compute the transfer time and the orbital velocity at perihelion and aphelion. One gets

$$V = \sqrt{m}\left(\frac{2}{r} - \frac{1}{a}\right)^{1/2}$$

where $r = r_1$ corresponds to the velocity at the tangency point with the starting orbit (perihelion) and $r = r_2$ corresponds to the velocity at the tangency point with the larger orbit (aphelion). Remembering that the circular velocity is given by

$$V_C = \sqrt{m}\left(\frac{1}{r}\right)^{1/2}$$

we obtain the change of velocity needed to insert the spacecraft into the elliptic transfer orbit from the departure circular orbit:

$$\Delta V_1 = \sqrt{m}\left[\left(\frac{2}{r_1} - \frac{1}{a}\right)^{1/2} - \left(\frac{1}{r_1}\right)^{1/2}\right]$$

Analogously the change of velocity needed to get into the second circular orbit from the transfer orbit:

$$\Delta V_2 = \sqrt{m}\left[\left(\frac{1}{r_2}\right)^{1/2} - \left(\frac{2}{r_2} - \frac{1}{a}\right)^{1/2}\right]$$

TABLE 2 TABLE OF ΔV

Celestial body	ΔV, km/s	Min distance from Earth, AU	ΔV_1, km/s	ΔV_2, km/s	Transfer time
Venus	5.2	0.27	2.5	2.7	5 months
Mars	5.6	0.52	2.9	2.7	8.5 months
Jupiter	14.4	4.20	8.8	5.6	2.7 years
Saturn	15.7	8.54	10.3	5.4	6 years
Neptune	15.7	29.06	11.7	4.1	30.5 years
Uranus	15.9	18.18	11.3	4.7	16 years
Mercury	17.2	0.61	7.5	9.6	3.5 months
Sun	29.8	1.00	29.8	0	7 days

So that the total change of velocity is given by

$$\Delta V = \Delta V_1 + \Delta V_2$$

which is a measure of the accessibility of a celestial body; if the departure orbit is that of the Earth one can then compare the ΔV value found with the one that can be achieved at the present technological level, or resulting from mission constraints on cost and performances. Applying the Hohmann transfer strategy throughout the solar system, it is possible to have an idea of the ΔV corresponding to the planets, as reported in Table 2.

Note that the accessibility does not increase monotonically with the distance—as evident for the cases of Neptune and Mercury. This can be better understood by using an H-plot representation, as shown in the Fig. 2.

The black curve represents the ΔV_1, that is, the change in velocity needed to escape from the initial Earth-like orbit and being inserted into a transfer elliptic orbit of increasing/decreasing aphelion/perihelion distance. This curve corresponds to the ΔV needed for a flyby mission, because the spacecraft is flying past the target body and then moves back on the return branch of the transfer ellipse. In this case the ΔV increases with the distance and tends to the solar system escape velocity. On the other hand, the light grey line corresponds to the case of the total ΔV, that is, we are summing to the previous ΔV_1 the quantity ΔV_2 needed to parse the target body circular orbit velocity, thus leaving the elliptic transfer trajectory. This ΔV gives an estimate of the accessibility of a body for a rendezvous mission. We point out that this curve has a maximum, after which it decreases toward the solar system escape velocity. Note also that the difference between the two curves is given by ΔV_2 and therefore it gives the relative velocity at flyby. If one sets the radius of the orbit of the Earth $r_1 = 1$ whereas $r_2 = r$ is the radius of the target

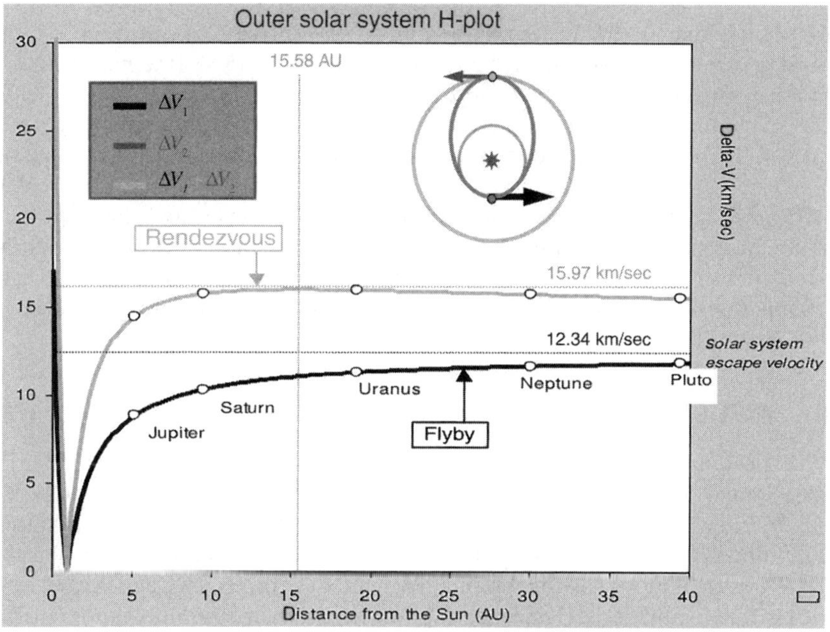

Fig. 2 Solar system H-plot.

orbit, the total ΔV is given by

$$\Delta V = \sqrt{\mu}\left[\sqrt{2}\left(1 - \frac{1}{1+r}\right)^{1/2} - 1 + \left(\frac{1}{r}\right)^{1/2} - \sqrt{2}\left(\frac{1}{r} - \frac{1}{1+r}\right)^{1/2}\right]$$

and

$$\lim_{r\to\infty} \Delta V = \sqrt{\mu}(\sqrt{2} - 1) = 12.34\,\text{km/s}$$

corresponding to the solar system escape velocity. Moreover, differentiating with respect to r and equating to zero

$$\frac{\partial}{\partial r}\Delta V = 0$$

one obtains the critical value $r = 15.58$ corresponding to the maximum $\Delta V_{\max} = 15.97$ of the upper light grey curve. In practice this means that, independently from the distance there is an upper limit to the ΔV needed to rendezvous a solar system body orbiting in a circular orbit coplanar to that of the Earth. Moreover, this value has also a meaning for optimization theory because it has been demonstrated that the Hohmann transfer represents an optimal strategy (in terms of ΔV) only if the ratio between the radius of the target and the radius of the departure orbit is less than 11.94. Exceeding this value, the choice of a suitable bielliptic transfer is more convenient. If the ratio is greater that 15.58 (the maximum of ΔV) than any bielliptical orbit is more convenient.

Concluding, the Hohmann transfer strategy gives a good approximation of the accessibility of solar system bodies even in a simplified mission scenario. First of all, the spacecraft is supposed to be already in interplanetary space on a circular 1 AU orbit, far from the gravity field of the Earth. Moreover, no "phasing," that is, synchronization of the initial positions of the planet and of the spacecraft, is taken into account. As shown in Table 2 the duration of Hohmann transfers can in some cases become unrealistic, thus needing the use of faster transfers, even if at the price of larger ΔVs.

III. APPLICATIONS TO NEAs

When dealing with missions to the NEAs the classical Hohmann transfer scenario is no longer valid. NEAs are fragments generated by collisions among asteroids in the Main Belt, subsequently injected through mean motion resonances with Jupiter into planet crossing orbits allowing close approaches with the Earth. Therefore, their motion is chaotic and in general their orbits have non zero eccentricities and inclinations. Hechler et al. [2] try to generalize the Hohmann transfer strategy to evaluate the case of the eccentric and inclined orbits of NEAs. Let us consider the case of an eccentric target orbit with zero inclination. Such an orbit can be found in three different geometries with respect to the circular orbit of the Earth. This gives the distinction of NEAs into the three well-known categories (see Fig. 3).

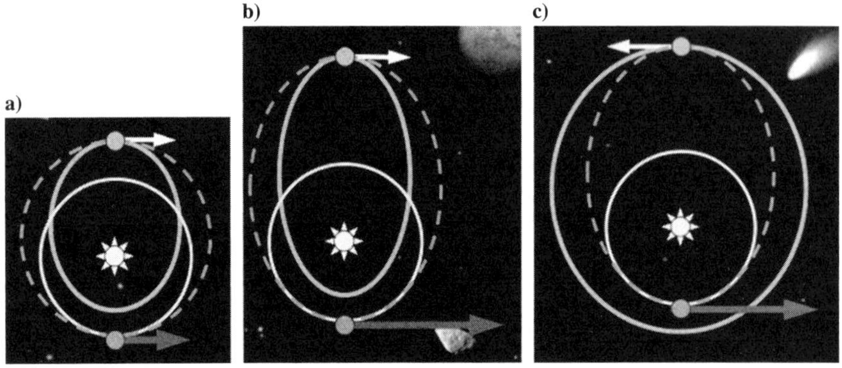

Fig. 3 Classification of NEAs: a) ATENs: objects on orbits crossing the orbit of the Earth and having a semimajor axis less than 1 AU, b) APOLLOs: objects on orbits crossing the orbit of the Earth and having a semimajor axis larger than 1 AU, and c) AMORs: objects on orbits approaching (within 1.3 AU) but not crossing the orbit of the Earth.

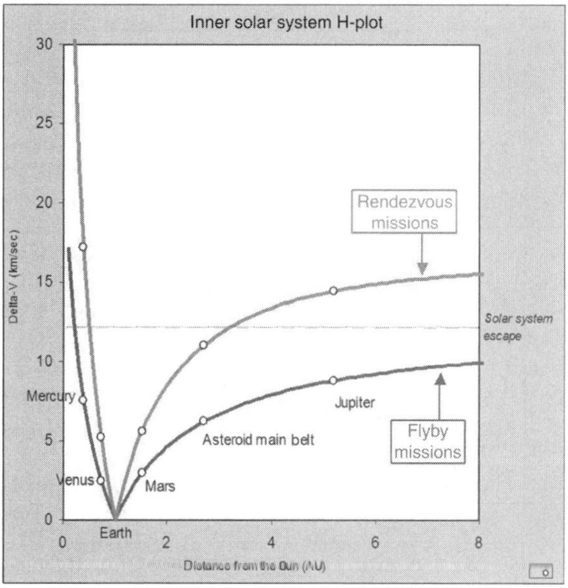

Fig. 4 Inner solar system H-plot.

With some additional assumptions on the orientation of the eccentric orbit it is possible to apply a Hohmann-like strategy for finding an elliptic transfer orbit to the eccentric orbit of an NEA and compute the corresponding ΔV. Also, in this cases one has to perform two maneuvers (see Fig. 3). Note that if the orbit of an NEA is tangent at perihelion or aphelion to the orbit of the Earth, then the transfer ellipse corresponds to the target orbit and only the first maneuver is required. But in general the results reflect the dynamical variety of the NEO population. In Fig. 4 an H-plot representation of the inner solar system is reported.

Following Fig. 5 it is possible to superimpose to this plot the NEA population provided that the aphelion distance and the ΔV are reported on the axis.

In this plot full dots correspond to Amors asteroids, empty dots to Apollos, and triangles to ATENs. Figure 6 represents the same plot when inclination is also considered.

In this last case objects are much more dispersed in ΔV because changing the inclination is extremely demanding. A general picture summarizing the accessibility of a sample NEA population is shown in Fig. 7 where also the location of several mission targets is remarked.

Using this representation it is possible to clearly see that some of the most interesting objects for science (e.g., dormant comets) are challenging because of the large ΔV required. Of course gravity assist missions or multiple maneuvers could be considered but this would imply more complex and lengthy mission

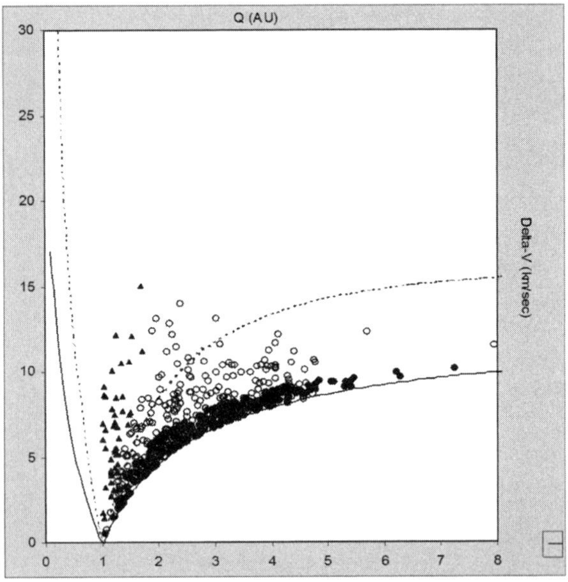

Fig. 5 Sample NEA population disregarding the contribution of the inclinations.

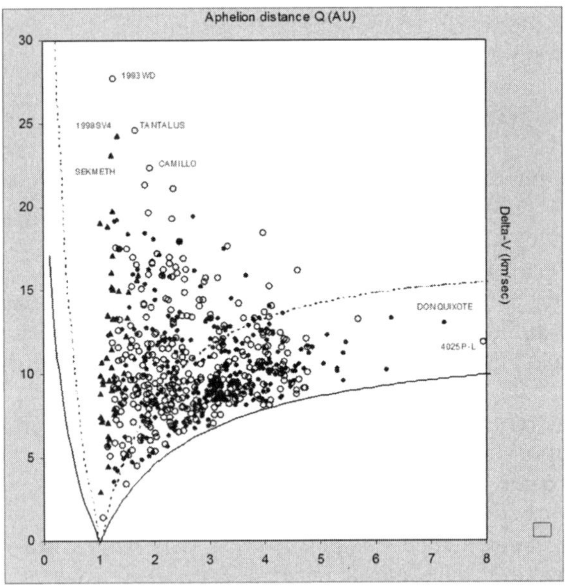

Fig. 6 Sample NEA population with inclinations.

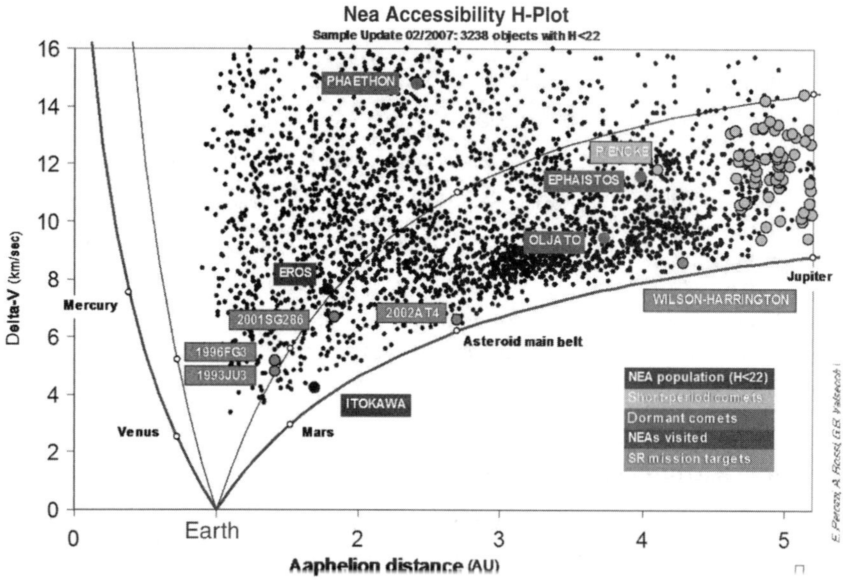

Fig. 7 Sample NEA population H-plot.

profiles. As an example, asteroid Eros was actually reached by the U.S. NEAR mission back in 2000, notwithstanding the relatively high Hohmann ΔV, which was largely due to an orbital inclination of about 10 deg. An Earth gravity assist was then needed to change the inclination of the spacecraft orbit by the required amount. On the other hand, one can see that asteroid Itokawa, which was targeted by the extremely challenging Japanese Hayabusa sample return mission, is one of the most accessible NEAs.

In passing, the plot in Fig. 7 also explains the difficulties involved in exploring the short period comets population. These comets are under the control of Jupiter

TABLE 3 AMUN

	Element value	1-Sigma variation
a, AU	0.973586	8.389-0.8
Eccentricity	0.280375	8.394e-07
Inclination, deg	23.364	3.019e-05
Ascending node	358.688	3.017e-05
Arg perihelion	359.355	0.0001298
M, deg	249.576	0.0008051

a)

Departure	14/08/2007
First flyby	10/10/2009
Travel time	788.1 days
Relative velocity	16.3 km/sec
Final mass	99.2 kg
ΔV	4.60 km/sec

b)

Departure	05/08/2007
First flyby	19/04/2009
Travel time	622.7 days
Relative velocity	9.1 km/sec
Final mass	98.5 kg
ΔV	4.87 km/sec

Fig. 8 Flybys of SIMONE mission: a) Ascending node first and b) descending node first (see Ref. [3]).

and have in general a large aphelia and significant inclinations. The European Rosetta mission needed to undergo several gravity assists by the Earth and Venus and it took more than seven years to reach its distant target, comet P/Churyumov Gerasimenko.

IV. APPLICATIONS TO MISSIONS

The H-plot representation has been fruitfully exploited for the selection of NEA targets within the framework of several mission proposals. A peculiar case is that

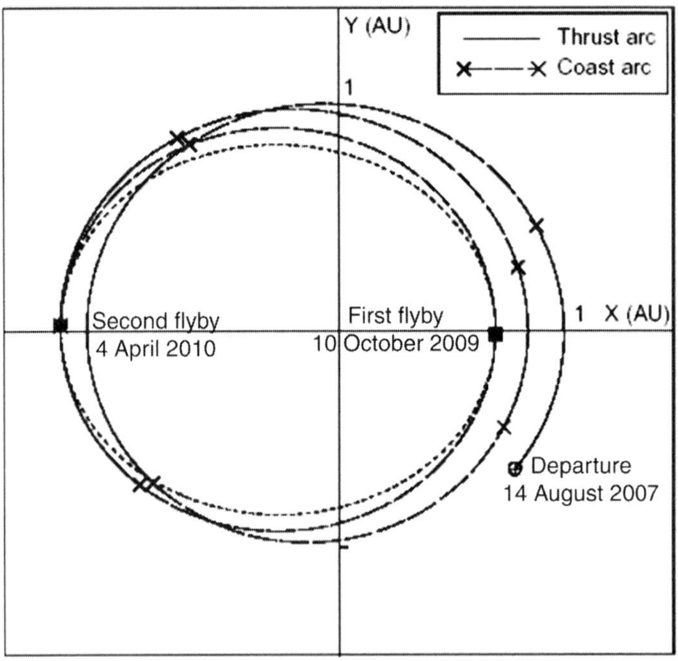

Fig. 9 SIMONE mission to Amun.

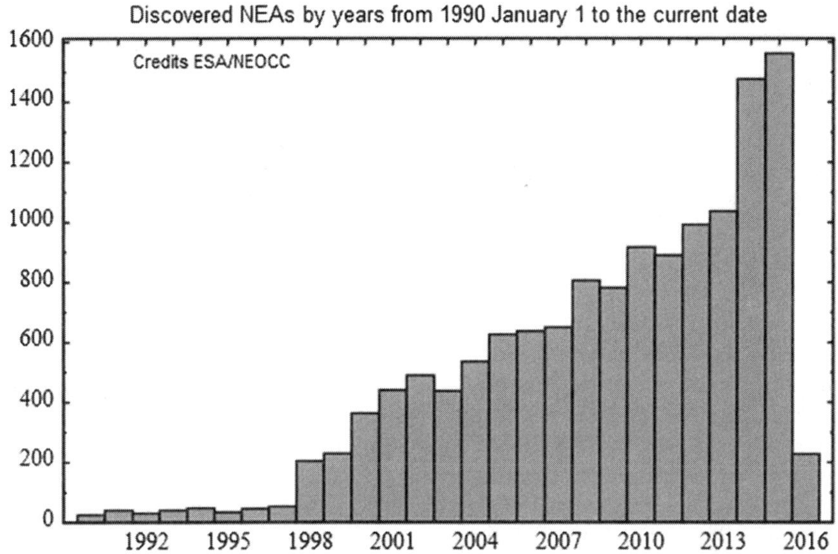

Fig. 10 NEA discovery statistics.

of the Simone mission [3], which has been studied within the framework of the ESA NEA Small Mission Initiative.

The proposed mission profile aims to visit objects with very high inclination, that is, those occupying the upper region of the H-plot and for which a rendezvous mission does not represent a feasible option. The possibility of fly-bying these objects when they cross the ecliptic (i.e., at the nodes of their orbits) has been therefore studied (Fig. 8). In this way a zero inclination transfer orbit could be selected, which is technically far less demanding than a rendezvous mission. For the Simone mission suitable targets were chosen among those NEAs having a nodal distance close to 1 AU and periods of revolution not much larger than one year, in order to allow a sufficiently short time between subsequent flybys. The best choice turned out to be asteroid Amun, because of its peculiar orbital elements, summarized in Table 3, allowed the spacecraft to encounter the asteroid at both nodes at six-month intervals.

A detailed outline of a possible mission to Amun is shown in Fig. 9 and Table 3.

A. THE ESA NEO COORDINATION CENTER

Determining the accessibility of the NEA population has recently gained momentum because of several reasons. First of all the increasing performances of the U.S.-based NEA surveys, which make use of high-sensitivity wide-field telescopes, have dramatically raised the discovery rate, as shown in Fig. 10. Among the newly discovered NEAs an increasing fraction is represented by small and accessible objects. This proved to be timely fitting the NASA plans for aiming to a near-Earth asteroid, the first manned mission beyond the Moon, as well as to capture and redirect a small NEA to the vicinity of the Earth (ARM—Asteroid Retrieval Mission). Finally, the inauguration of the European NEO Coordination Center, in May 2013, has allowed ESA to bring a significant contribution to the worldwide efforts in NEA hazard monitoring and mitigation.

The NEO Coordination Centre (NEOCC) has been realized within the framework of the ESA Space Situational Awareness (SSA) program, which aims to gain a European independence in the understanding of the risks characterizing the space environment (i.e., Space Weather, NEOs, and Space Debris). NEOCC is located at ESRIN, the ESA center close to Frascati (Italy). The operations of NEOCC (presently in the precursor services phase) foresee two major activities: 1) making available a system able to provide public data on the near-Earth objects population and on the related hazard, and 2) contributing to NEO detection by prioritizing follow-up observations and coordinating existing assets.

These activities are carried out under ESA contract by an industrial team led by Deimos Space. The NEO Coordination Centre services are available through a web portal accessible at http://neo.ssa.esa.int/

REFERENCES

[1] McLaughlin, W. I., "Walter Hohmann's Roads in Space," *Journal of Space Mission Architecture*, Vol. 2, No. 2, 2000, pp. 1–14.

[2] Hechler, M., Cano, J. L., and Yanez, A., "Smart-1 Mission Analysis: Asteroid Option," ESOC Mission Analysis Section Working Paper, 403, 1998.

[3] Perozzi, E., Casalino, L., Colasurdo, G., Rossi, A., and Valsecchi, G. B., "Resonant Fly-by Missions to Near Earth Asteroids," *Celestial Mechanics and Dynamical Astronomy* , Vol. 83, No. 1, 2002, pp. 49–62.

[4] Perozzi, E., Rossi, A., and Valsecchi, G. B., "Basic Targeting Strategies for Rendezvous and Flyby Missions to the Near Earth Asteroids," *Planetary and Space Science* , Vol. 49, No. 1, 2001, pp. 3–22.

[5] Binzel, R. P., Perozzi, E., Rivkin, A. S., Rossi, A., Harris, A. W., Bus, S. J., and Valsecchi, G. B., "Dynamical and Compositional Assessment of Near Earth Object Mission Targets," *Meteoritics and Planetary Science*, Vol. 39, No. 3, 2004, pp. 351–366.

[6] Carusi, A., Perozzi, E., and Scholl, H., "Neo Mitigation Strategy," *Compets Rendus Physique*, Vol. 6, No. 3, 2005, pp. 367–374.

[7] Perozzi, E., Binzel, R.P., Rossi, A., and Valsecchi, G. B., "Asteroids More Accessible Than the Moon," *European Planetary Science Congress Abstract*, 2010, p. 750.

[8] Perozzi, E., "The Near Earth Object Hazard and Mitigation," *Mathematical Methods for Planet Earth*, edited by A. Celletti, U. Locatelli, and E. Strickland, Springer, 2014, Volume 6, Springer INdAM Series, pp. 87–97.

CHAPTER 5

Physical Properties of Near-Earth Asteroids

Georgios Tsirvoulis*
Astronomical Observatory, Belgrade, Serbia

Patrick Michel†
University of Nice Sophia-Antipolis, Nice, France

I. INTRODUCTION

As we have discussed in the previous chapter, our solar system consists of far more than what meets the eye. What does indeed meet the eye of the keen observer, are the largest or closest members of the solar system. From the magnificent Moon that seduces all of us, to the bright planets Venus, Mars and Jupiter, and even beyond, with a small technological aid from a pair of binoculars, to Saturn and the Galilean moons of Jupiter. So, without any doubt, there are many beautiful things out there. To the trained astronomers and researchers though, even seemingly insignificant objects, such as rocks, bear a special kind of beauty. Truth be told, there is a huge number of "just rocks" in the solar system, leading to an ever-growing interest on their study: how did they form, how did they evolve, what are they made of, and what does their future look like. With the advancement of technology, more and more of these questions are finally able to be approached and partly answered.

The questions that will bother us most throughout this chapter have to do with the physical properties of those "rocks" that live somewhat close to Earth. But since scientists started devoting more and more time to this study, they felt a bit uncomfortable having to work on what an everyday person would call "just rocks." So they came up with a special name: *Asteroids*, derived from the Greek meaning of "alike stars." Back then, they were just faint moving spots on the night sky, looking like the brighter spots, stars, so the name seemed well fitted. So nowadays, many scientists are studying asteroids, in many aspects, from orbital or rotational dynamics, to composition and physical properties, but it is worth bearing in mind that they are still larger or smaller rocks orbiting the Sun.

Asteroids are the remnants of the material that was used to build the planets ~4.6 billion years ago. The protoplanetary disk, a thin disk of gas and dust

*Researcher.
†Lagrange Laboratory and Cote d'Azur Observatory.

particles that was orbiting the Sun soon after its formation, started clumping from dust grains to kilometer-sized planetesimals, as the building blocks of planets are called. Then with a bit more clumping and accretion the planets were formed. The gaseous giant planets were also able to accrete gas from the disk before the radiation of the Sun managed to photoevaporate it, hence the current situation of the planets. But as it happens with every construction site, not all building materials make it into the final product. So, it happened with the formation of the solar system, where a lot of planetesimals did not manage to fulfill their destiny to form planets. Instead they remained as small rocks, orbiting the Sun. But still they do have their special kind of "life." The environment they live in is a rather hectic one, with big planets orbiting around, and other asteroids in their vicinity. The consequences are a literal chaos. Planets induce resonant phenomena to their orbits and other asteroids tend to lead to collisions which force any specific asteroid to a far-from-peaceful life.

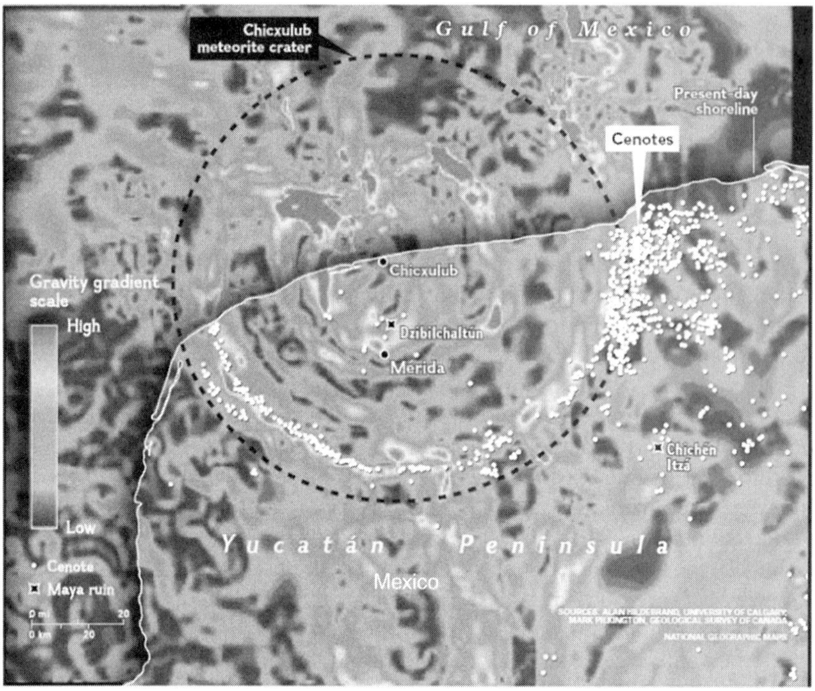

Fig. 1 Gravity gradient map of the Chicxulub crater, Mexico, which is believed to have caused global environmental effects and the extinction of dinosaurs. (Credit: http://www.uqac.ca/miac/chicxulub.htm.)

As we explained in the previous chapter on Asteroids, the main population of them is the Main Asteroid Belt, residing between the orbits of Mars and Jupiter. The main belt is the most stable region for asteroids to live in, although it has its unstable parts as well. The asteroids we will have our focus on throughout this chapter are indeed those that originated from those unstable areas of the main belt. Because these asteroids, when perturbed from their initial orbits, tend to pay a visit to the inner solar system, becoming crossers, or even establishing new orbits close to the Earth, bearing the special designation Near-Earth Asteroids (NEAs). Now because "near Earth" makes people nervous, bearing in mind what would happen, and has happened in the past (Fig. 1), if "near" becomes "nearer" and then "kaboom," a lot of effort is being paid to their study. From extensive surveys to acquire detailed orbits of the population of NEAs to the research for mitigation methods in case a future collision is detected, the problem is approached from all sides. We will describe the actual physical characteristics of NEAs in the following sections. It is very useful to try and broaden our understanding on how do these asteroids look like, what do they consist of, what is their internal structure, and how do they spin.

We will begin this chapter by presenting some of the current knowledge on the basic physical properties of NEAs that are being acquired from ground observations as well as numerical modeling. Next, we will outline some space missions that revealed great details about the visited asteroids. We will then argue on the importance of this knowledge of their physical properties as far as mitigation of possible hazardous asteroids is concerned, and present some concluding remarks.

II. PROPERTIES OF NEAs DERIVED FROM OBSERVATIONS

A. SIZE DISTRIBUTION OF NEAs

Ground-based observations of NEAs are able to reveal quite some different properties of the subjects. First of all, the size of the asteroids is being determined with quite good accuracy using optical as well as infrared observations. Knowing the sizes of the different NEAs is crucial for all the consequent research regarding their other properties as well as possible mitigation parameters. The potentially hazardous asteroids range in size from the smallest ones that are able to survive the entry to the atmosphere, to a few tens of kilometers, where a global catastrophe would occur. A very powerful statistical tool concerning the sizes of NEAs that is being used is the size distribution, essentially a representation of how many NEAs of each size exist. Such a distribution is obtained by observation as well as different models. Indeed, observations are restricted to the bigger members of the NEA population, down to hundreds of meters in diameter. For smaller asteroids, different models are developed, to get a grasp on the actual distribution. The overall distribution seems to follow a power-law function, with slight differences between different models. Such a size distribution can be seen in Fig. 2, where results from observations and models are combined.

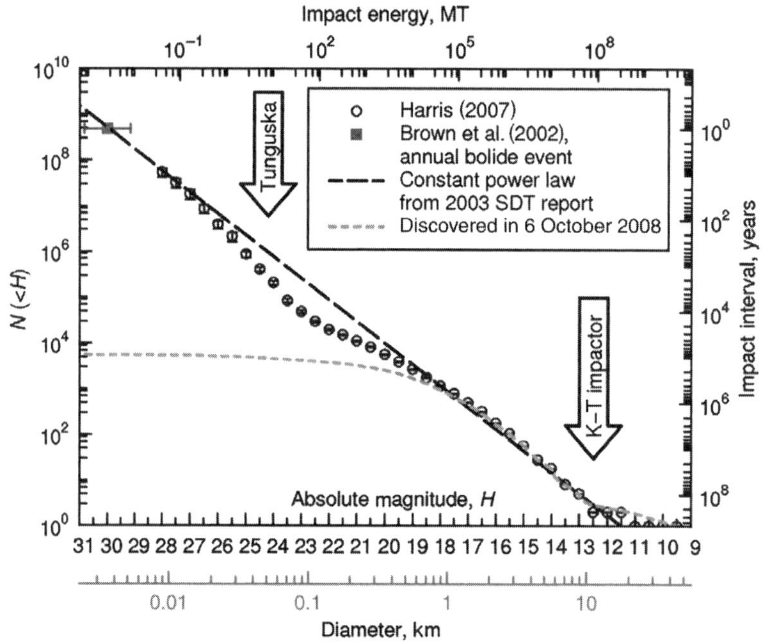

Fig. 2 Open black circles represent the cumulative number of NEAs brighter than a given absolute magnitude H, defined as the visual magnitude V that an asteroid would have in the sky if observed at 1 AU distance from both the Earth and the Sun, at zero-phase angle. A power-law function (dashed black line) is shown for comparison. Ancillary scales give impact interval (right), impact energy in megatons TNT for the mean impact velocity of 20 km/s (top), and the estimated diameter corresponding to the absolute magnitude H. (Credit: P. Michel 2012 after A. W. Harris 2007.)

B. SPIN RATES OF NEAs

Apart from the sizes of NEAs, observations can reveal details concerning the shape of each asteroid, and the spin rate. To determine the spin rate of an asteroid, the lightcurve is obtained, where the reflected light is plotted against time. If the asteroid has an irregular shape, which is true in most of the cases, the density of the reflected light is not constant, because as the asteroid spins a different side is facing the observer. The irregularity of the shape means that different sides of the asteroid have a different area and/or different surface structure, thus leading to the differences in the reflected light. If such a lightcurve has a periodicity in its shape, the only reasonable conclusion is that this periodicity is caused by the spin of the asteroid, a spin with the same period as the lightcurve reveals. To statistically study the spins of NEAs, the distribution of spin versus size of each body is produced. Such a distribution of approximately 1500 asteroids is shown in Fig. 3.

The spin rates of asteroids vary a lot, from the order of minutes up to several days. The observed distribution of spin rates reveals some details about the internal structure of the asteroids. For example, we note that on the right-hand side of Fig. 3 where the largest asteroids in diameter are, the upper limit of their spin rate is defined by the horizontal grey line denoted as "gravity limit." This means that for larger objects, it is gravity that holds them together as one body, and this limit is the maximum spin where gravity is strong enough to hold the material of the asteroid together. On the contrary, smaller asteroids at the left-hand side of the plot, go above the gravity limit, because it is the tensile/cohesive strength of the material itself that holds them together, which allows for faster spin rates before breakup would occur. We have to clarify a certain aspect of this claim though.

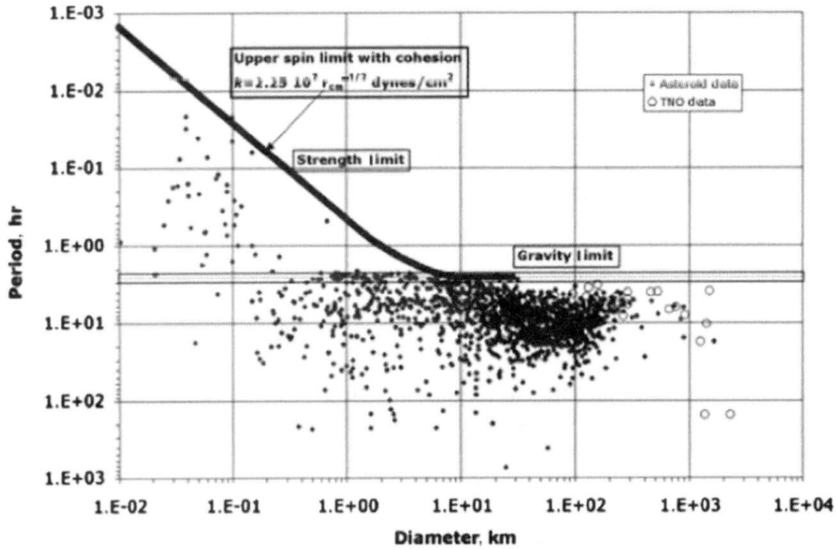

Fig. 3 Spin limits and data for small solar system bodies. The dark-sloped line assumes a size-dependent strength; it joins the horizontal grey band for materials without cohesion. On the left, the spin limit is determined by the cohesive/tensile strength of the bodies and defines a strength regime. The horizontal asymptote on the right characterizes a gravity regime where tensile/cohesive strength is of no consequence. These gravity regime values actually depend on the shape and friction angle of the material composing those bodies, so average values have been assumed. The data in the upper- left triangular region are for the fast-spinning NEAs. The triangular points for the large-diameter bodies on the right are trans-neptunian objects. (Credit: P. Michel 2012.)

The fact that larger asteroids are held together by gravity is not a proof by itself that those asteroids are *rubble pile*. With rubble pile, we mean asteroids that are agglomerates of numerous pieces of rock held together by gravity, as opposed to *monolithic* where one body is a single rock. This is because the tensile strength of a rock decreases as the rocks' sizes increase, and eventually it becomes so small that it is negligible with respect to gravity, even if the body was monolithic. So, the body may be a rubble pile, but it does need to be one, as even a monolithic one would have a very small strength because of this decrease in strength with increasing size. Similarly, the cohesive force that supports smaller asteroids is not proof of them being strictly monolithic.

C. SURFACE PROPERTIES

As far as surface properties are concerned, ground-based observations can only contribute up to a certain level of detail. For example, we might be able to determine whether a given asteroid's surface is bare rock or has smaller-level textures caused by stones and so on, called *regolith*, but the actual properties of this regolith, such as the size distribution of the grains it consists of, or its depth, porosity, and so on, are not available. In addition, measurements of the thermal flux density of a given NEA can give insight about its thermal properties such as the surface temperature. This kind of information can be used to constrain the magnitude of the Yarkovsky effect, thus helping to better study the orbital evolution of the asteroid.

D. COMPANION OBJECTS

Another interesting feature that ground observations using radar and/or lightcurves can reveal is the possible existence of a small companion orbiting around an asteroid. An example of this is the asteroid 1999 KW_4. This asteroid was found to have such a small companion by radar observations as shown in Fig. 4.

The peculiar shape of the asteroid itself and the existence of the companion were studied by Walsh et al. [1], who were able to reveal the mechanism that led to the observed picture. According to them, the object was first rather spherical but its spin rate increased due to the *YORP* effect. The YORP effect refers to the change of spin rate caused by the anisotropical emission of thermal photons from the rough surface of the asteroid, in analogy to how Yarkovsky changes the orbital energy. This spin up first caused the material located at the pole to migrate down to the equator forming the bulge that is visible in Fig. 4. But YORP continued increasing the spin rate until the particles located at the equator (where the centrifugal force is maximal) could escape and reaccumulate under certain conditions to form the small companion that we see.

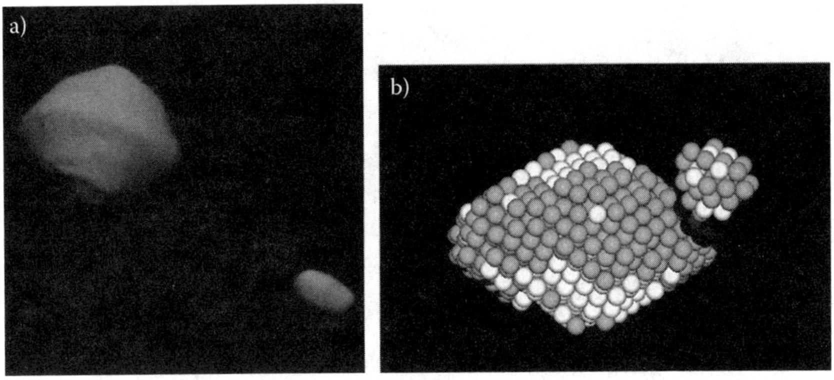

Fig. 4 Asteroid 1999 KW_4 and companion. a) Radar-derived model by the observations and b) simulation model by Walsh et al. [1]. (Credits: (a) S. Ostro et al. 2005 and (b) K. Walsh et al. 2008 [1].)

III. SPACE MISSIONS TO NEAs

As our everyday experience suggests, to tackle a problem in great detail, one has to closely observe it. As NEAs are small, faint objects in the night sky, it is really challenging to extract any details from ground observations. The only alternative, and one that improves observations by far, are space missions that either fly by, or land, or even return a sample from an NEA. Such missions have been launched already and more are being planned for the future, as the only tools to really verify theoretical models about asteroids and determine the role of small bodies in the grand scheme of things that is the history of the solar system.

A. GALILEO SPACE MISSION (1989–2003)

Galileo was a spacecraft designed by National Aeronautics and Space Administration (NASA), with a primary goal to visit and study Jupiter and its moons. Along its journey to Jupiter, it performed two fly-bys at asteroids 951 Gaspra and 243 Ida, and both were S-type main belt asteroids. The first fly-by of 951 Gaspra, was the first-ever visit of a small body and took place in 1991. Galileo sent images of the asteroid that reveal its shape as well as its surface characteristics in great detail. Following that, in 1993, Galileo passed within 3000 km of 243 Ida. The first striking discovery of the photographs taken was the existence of a small moon, later named Dactyl, orbiting Ida. This was the first discovery of an asteroid having a moon, enabling the determination of the asteroid's mass and density. Further analysis of the data retrieved from Galileo's instruments revealed a difference in their spectra, leading to the belief that Dactyl does not originate from Ida itself but was rather captured by it, for instance during the catastrophic disruption that possibly produced both of them.

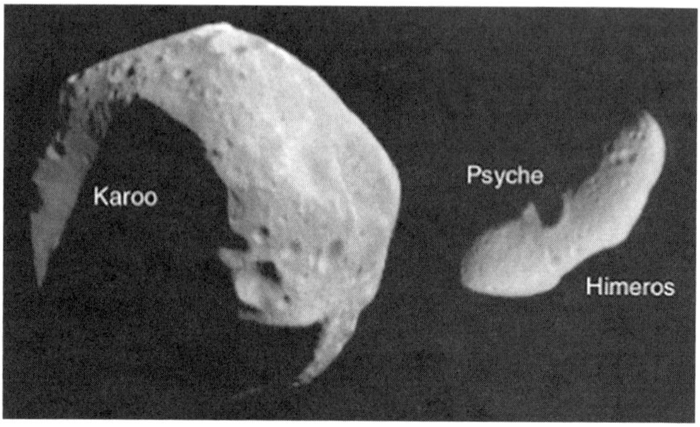

Fig. 5 Asteroids visited by NEAR: Mathilde (left) and Eros (right). The images are in the approximate relative scale, but the C asteroid Mathilde is much darker than shown relative to Eros. On Mathilde, the giant crater Karoo, diameter 33 km, is marked. On Eros, the 5-km crater Psyche is at the terminator and is labeled. At the bright limb of Eros, the saddle-shaped depression Himeros is labeled. (Credit: Cheng [4].)

B. NEAR-SHOEMAKER MISSION (1996–2001)

The NEA Rendezvous space mission was the first one designed with the primary objective of studying a small body of the solar system, S-type NEA 433 Eros. Eros is the second-largest NEA, with its largest diameter being 34 km.

On its way to orbit the insertion around Eros, NEAR spacecraft also performed a fly-by of the main belt asteroid 253 Mathilde, producing over 500 photographs Fig. 5, covering almost 60% of its surface, the first close ones of a dark, C-type asteroid. This fly-by also provided data, from gravitational perturbation suffered by the spacecraft, concerning the mass of Mathilde, which was the first directly determined. The measured mass of 1.03×10^{20} g and the estimated volume of 78,000 km^3 imply a mean density of $1.3 \pm 0.2\,g\,cm^{-3}$ [2]. Carbonaceous chondrite meteorites, which are spectrally similar to C-type asteroids, have more than double the density of Mathilde. This leads to the conclusion that Mathilde has a very high porosity, 50% or even more [3]. The images obtained reveal enormous craters on the surface of Mathilde, at least five of which have diameters comparable to the mean radius of the asteroid itself. The high porosity is what allowed so many strong impacts without disruption of the asteroid, because it causes an efficient shock damping [4].

The main target of the NEAR mission, Eros, was reached after a rather eventful journey. The first scheduled maneuver was problematic, leading to almost the loss of the spacecraft. Hopefully control was regained, and a new plan for orbital insertion was devised and followed through, until 14 February 2000, when orbital insertion was achieved, a year later than originally planned. The data produced by the spacecraft led to a very accurate determination of the mass of Eros at $(6.687 \pm 0.003) \times 10^{15}$ kg, and its bulk density at (2.67 ± 0.03) g cm^{-3}. This density is also lower than that of ordinary chondrite meteorites, indicating that Eros is also porous, but not as much as Mathilde is. Measurements of the gravity field suggest a rather uniform interior density of Eros, with a margin of a few percent of the one expected from a uniform body of the same shape [4].

Images of Eros Fig. 5 led to the interpretation that the asteroid is a consolidated cohesive body, not an agglomeration of smaller-component bodies bound by gravity [4]. This is based on the structural strength necessary for the various features that are observed, like a variety of ridges, grooves, and chains of pits or craters, that display coherent alignments over several kilometers across Eros [4]. The conclusion drawn by the linear structural features, the tectonic features, the joined craters, and the internal coherence is that Eros is a collisional fragment from a larger parent body with a regolith cover of about 100 m [4]. However, other interpretations are possible such as the possibility that the majority of these faults occur just in the upper tens of meters, where cohesion exceeds gravitational stress even for loose piles of lunar-like regolith, implying that Eros is a rubble pile [5].

The surface of Eros was found to be generally old and close to equilibrium saturation by craters >200 m, with only some regions being extensively resurfaced, thus relatively younger. A striking result is the absence of small craters (<100 m). The surface is very rough, with a fractal structure that spans from 1 m to hundreds of meters in scale. Steep slopes in crater walls have led to downslope motion of boulders, which in turn resulted in bright albedo features at those areas. The darkened, space-weathered material that slides down exposes the underlying layers which are less weathered, and thus brighter. On the surface, seismic shaking from impacts has led to the formation of ponds, extremely low level areas of very fine material, at gravitational lows.

The composition of Eros was found to be overall consistent with that of chondritic meteorites. There was no spatial compositional variation detected. Furthermore, the silicate mineralogy inferred from visible and infrared spectra suggests that melting or differentiation may have occurred only partially, if at all [4]. Finally, after the main phase of its operations, NEAR descended on Eros, despite having not been designed for it. During the descent, measurements for possible magnetization were made, with none found, in contrast with the meteorites which are generally magnetized.

C. HAYABUSA MISSION (2003–2010)

In May 2003, the Japan Aerospace Exploration Agency (JAXA) launched *Hayabusa*, the first ever sample return mission to an NEA. In contrast with NEAR, which visited the second-largest NEA, Hayabusa aimed for a much-smaller one. The target was asteroid 1998 SF_{36}, named *Itokawa* in August 2003, after the japanese rocket science pioneer Hideo Itokawa. As the first of its kind, Hayabusa faced a multitude of complications during its operational time, involving losing contact with Earth and some severe system malfunctions. Nevertheless, we will not expand our discussion on the eventful journey in this chapter. It is sufficient to say that it successfully completed all of its objectives, including the return of sample grains from Itokawa back to Earth for examination.

As it always happens, the first, most anticipated, and most-appealing data returns of space missions, were close-up images. In Fig. 6, an image of Itokawa is shown, revealing its rubble-pile structure and surface characteristics.

Even at first look, Itokawa's shape and surface characteristics give hints of its structure. Indeed, it seems like it is formed up by two big blocks merged together, with a lot of overlying material. Previous knowledge based on observations and modeling suggests that most of the asteroids in the size range from a few hundred meters to several tens of kilometers are not monolithic, but rather rubble-pile bodies. As such, Itokawa was found to have a low mean density of only $\sim 2\,\mathrm{g\,cm^{-3}}$. The scenario of Itokawa's formation was studied by Michel and Richardson [6], as a gravitationally reaccumulated body from the remains of a parent body suffering a collision. They performed a numerical simulation featuring the disruption of a 25-km asteroid and the reaccumulation of the resulting fragments under the force of gravity. In Fig. 7, the sequence of events leading from the parent body to the Itokawa-like final asteroid is shown.

The parent body is shown in the first snapshot at the end of the fragmentation phase when all fragments (grey dots) are about 200 m in diameter, followed

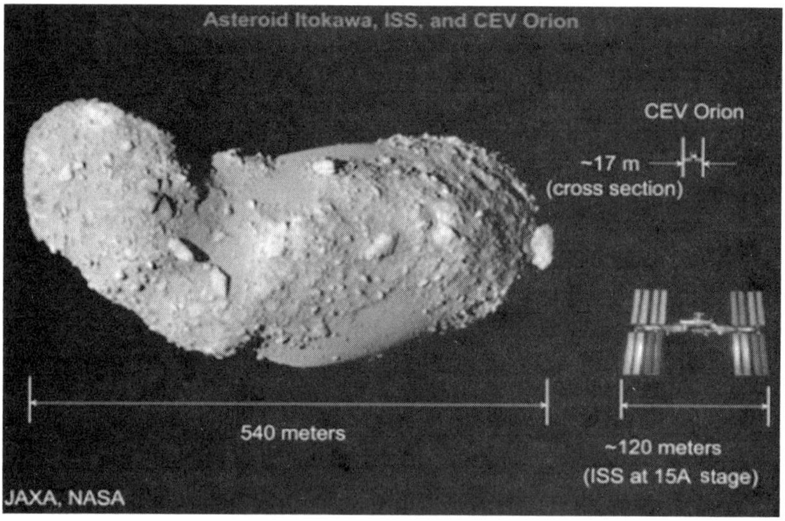

Fig. 6 Picture of NEA Itokawa as captured by Hayabusa. Sketches of the International Space Station and CEV Orion are depicted for size reference. (Credit: JAXA [11], NASA.)

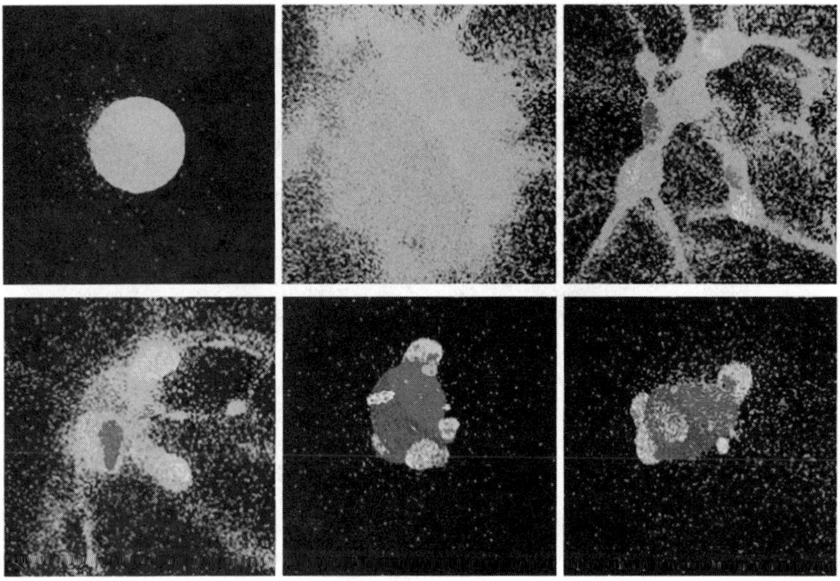

Fig. 7 Snapshots of the reaccumulation process following the disruption of a 25-km-diam asteroid. (Credit: Michel, P. and Richardson, D.C. 2013 [6].)

by the ejection of those fragments a few seconds later. Then we witness the first reaccumulations that occur because of the slow relative speed between some fragments, showing the formation of a few aggregates represented by different colors. Going on, we see the formation of the largest fragment of this disruption by reaccumulation of several aggregates into a single body, and finally the resulting agglomeration is shown at two different instants: the boulders on its surface and its overall shape are reminiscent of Itokawa [6].

The surface of Itokawa has features across a wide range of dimensions. The most evident are the big boulders visible in Fig. 6 which are of the order of tens of meters in diameter, and loosely lie on the surface of the main body. Also visible are areas looking extremely smooth, revealing ponding in gravitational lows. Indeed, a simple comparison shows that the smooth areas in Fig. 6 correspond to the purple gravitational lows of Fig. 8, the gravity map of Itokawa.

An interesting result of the Itokawa images is that even in the smooth areas, there is no fine-grain material found. Indeed, the lowest-scale components are gravel-sized rocks, and there is no sand or dust-level material whatsoever. The absence of fine material can be explained by segregation of the regolith caused by seismic shaking or even avalanche-like effects during mass flow on the surface. Furthermore, the layout of the rocky material on the surface of Itokawa seems to be well set. There are no gravels laying on top of bigger rocks, or rocks standing up. Everything seems to be in a lay-down position. The cause of this surface condition

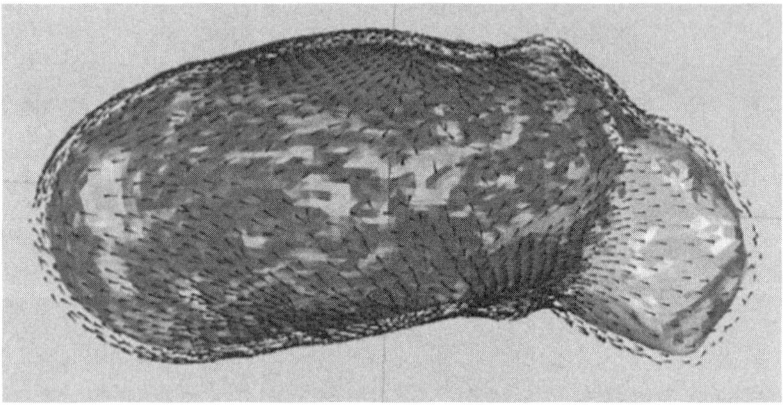

Fig. 8 Gravitational field of Itokawa as measured by Hayabusa. Black areas denote gravitational lows whereas gray areas denote gravitational highs. (Credit: University of Tokyo and JAXA.)

seems to be the reallocation of the gravels after deposition due to global-scale vibrations of the asteroid. Such vibrations are easily caused on small asteroids by small impacts.

The samples of Itokawa that returned to Earth for examination, were the first-ever material coming directly from an asteroid without suffering the reentry stress that meteorites go through. Although the operation to collect the samples on Itokawa went wrong both times that was attempted, some dust managed to sneak into the capsule. After the analysis of the samples, over 1500 dust grains were confirmed as originating from Itokawa. Figure 9 shows an electron microscope photograph of the material found inside Hayabusa's sample return capsule. Gray arrows point to dust particles from Itokawa. The study of those particles showed that the mineral chemistry of Itokawa, a silica-rich S-type asteroid, is almost identical to a particular class of meteorites, called low-total-iron, low metal ordinary chondrites (LL chondrites).

D. ROSETTA (2004–2015)

Rosetta is a spacecraft launched by the European Space Agency (ESA), to be the first mission to study a comet. Indeed, Rosetta being a cornerstone mission of ESA will be orbiting comet 67P/Churyumov–Gerasimenko for 17 months during the comets next perihelion passage from May 2014 until sometime in 2015. It is expected to make a radical contribution to our knowledge on comets, perhaps the most primitive members of our solar system.

On its way to Churyumov–Gerasimenko, Rosetta has performed two flybys of main belt asteroids. In the September of 2008, it flew by asteroid (2867)

Steins, a rare E-type asteroid with a diameter of around 5 km. By looking at the images of Steins captured by Rosetta's OSIRIS camera Fig. 10, we see that it has a diamond-like shape. The science team that studied the images [7] conclude that the cause of this shape is exactly the same mechanism as we discussed previously for 1999 KW_4. The YORP effect caused a spin up of the asteroid which in turn resulted in some material moving toward the equator. Keller et al. [7] also attribute the surface reshaping and the absence of smaller craters to the YORP effect. On the contrary, the biggest crater on Steins and chains of pits linked to it, are an evidence of a big impact that caused fracturing of the asteroid body, resulting in its rubble-pile structure.

The next asteroid visited by Rosetta in July 2010, was the much-larger (21) Lutetia, with a diameter of 120 km. The OSIRIS camera onboard Rosetta revealed an irregular shape with many craters. By studying the images, more than 350 craters with diameters from several hundreds of meters up to several tens of kilometers were identified across the side facing Rosetta. Differences in the crater density revealed that regions with the largest number of craters were clearly older than those with fewer craters. The study of the craters suggests that the surface of Lutetia is 3.6 billion years old, according to Sierks et al. [8], suggesting that it is a remaining planetesimal from the formation of the solar system, which despite having suffered a lot of impacts that are responsible for its irregular shape, was

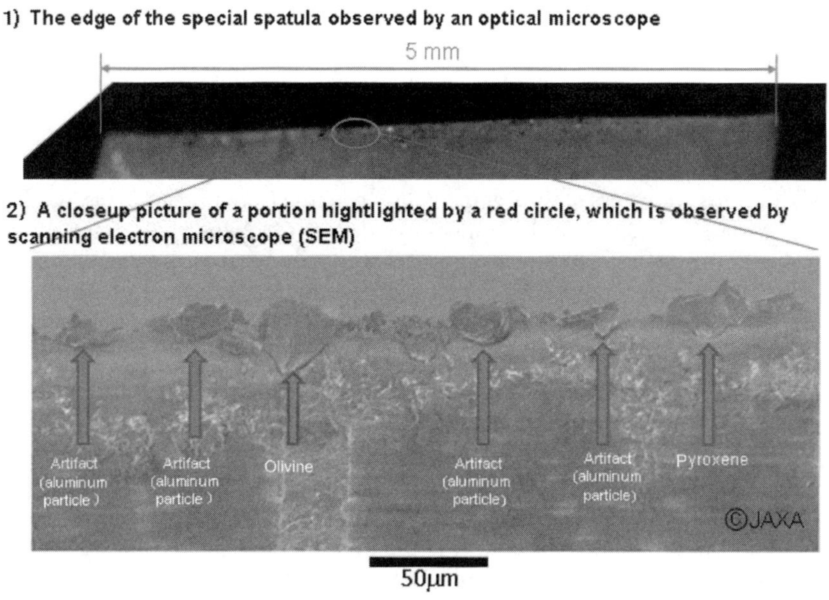

Fig. 9 The special spatula that collected dust samples from the Hayabusa return capsule, observed by an optical (top) and scanning electron microscope (bottom). (Credit: JAXA [11].)

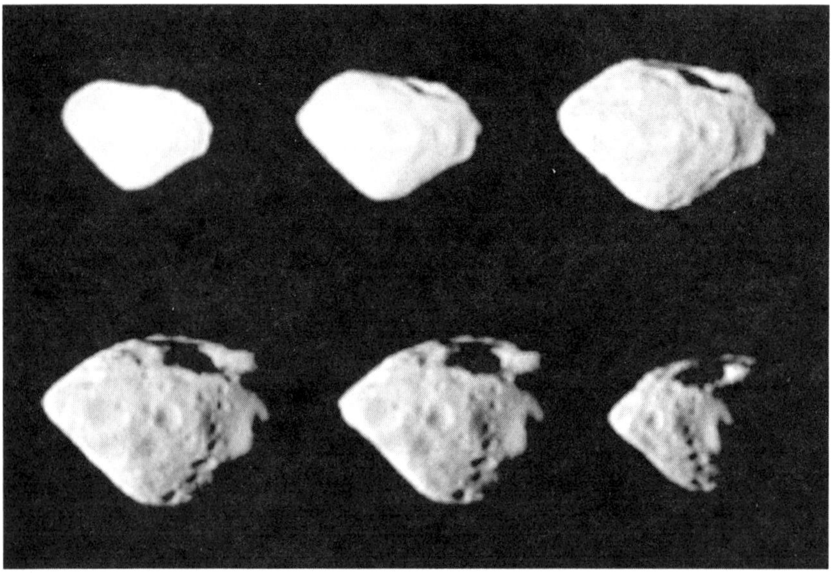

Fig. 10 Images of asteroid (2867) Steins taken by the OSIRIS Wide Angle Camera on Rosetta during the fly-by of 5 September 2008. (Credit: ESA.)

able to survive intact ever since. On the surface of Lutetia, many different features have been identified. Landslides which are visible at some regions, have caused albedo variations, by exposing fresh, unweathered material, and have also caused some craters to be partially or even completely buried. A network of grooves, fractures and rifts, and scars of the fragmentation suffered from impacts is also visible on the surface of Lutetia. The high-measured density of Lutetia ($\sim$3.4 g cm^3) combined with the modest porosity mean that it is enriched in metallic elements such as iron, and might be indications of partial differentiation of its interior. The spectral analysis of Lutetia suggests a composition similar to iron-poor carbonaceous or enstatite chondrites [9].

E. DAWN MISSION (2007)

DAWN is a satellite built and launched by NASA in September 2007, with a scientific goal to study the two most massive objects of the main asteroid belt, asteroid Vesta and the dwarf planet Ceres. It is the first-ever mission to study such big asteroids. Before it, Lutetia was the largest one visited by Rosetta. According to DAWN's science team "The three principal scientific drivers for the mission are first that it captures the earliest moments in the origin of the solar system enabling us to understand the conditions under which these objects formed. Second, Dawn determines the nature of the building blocks from which the terrestrial planets formed, improving our understanding of this formation. Finally, it contrasts the formation and evolution of two small planets that followed very different evolutionary paths so that we understand what controls that evolution."

DAWN entered Vesta orbit in July 2011, and remained there for more than 400 days, carrying out a series of operations. Mapping of the surface of Vesta revealed many planetary characteristics, such as mountains and valleys, large graben-like tectonic systems, and craters [10], with an absolute relief of ~40 km, comparable to that of the much-larger Mars and the Moon. Near the south pole, two massive ancient basins were found, with multiple troughs Fig. 11. Vesta's iron core is believed to be the reason why these huge impacts did not disrupt Vesta completely [10]. These impacts have erased previous smaller craters at the region of the basins, as opposed to the rest of the surface which is nearly saturated with craters.

Surface analysis further strengthened the belief that howardite–eucrite–diogenite (HED) achondritic meteorites found on Earth originated from Vesta. Geochemical analysis of HED meteorites therefore leads to the conclusion that Vesta melted and formed an iron core, with a dense olivine-rich mantle and a crust

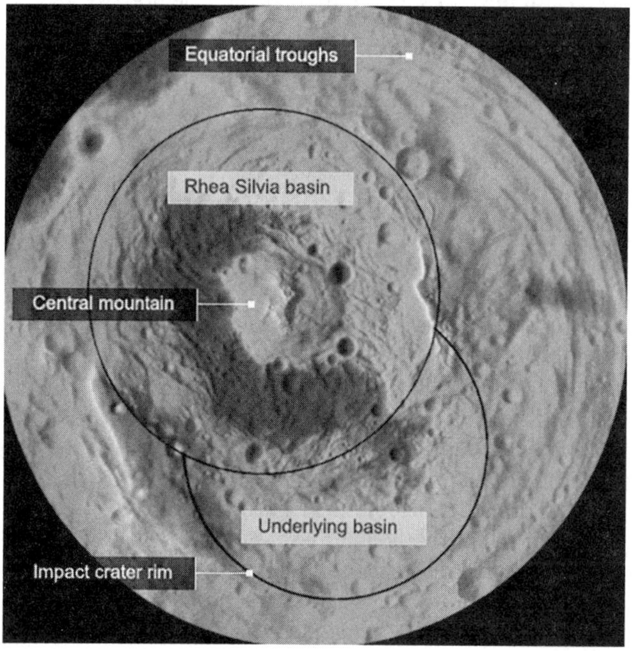

Fig. 11 Relief map of the south pole of Vesta, showing the two huge basins, Rheasilvia and the underlying Veneneia. (Credit: JPL/NASA.)

of a lighter eucritic material [10]. The infall of carbonaceous chondrites on Vesta has delivered a dark, hydrous material which was detected in dark-rayed craters. Some dark deposits however may be endogenic, recently exposed by impact melt. The observations of signatures of volatiles indicate that Vesta together with the whole inner solar system have suffered a significant amount of mass delivery from the outer main belt, where C-type asteroids, associated with carbonaceous chondrites, are the dominant species.

DAWN is continuing its journey toward the dwarf planet Ceres, where it will carry out a series of observations giving further insight into our understanding of the early solar system and the formation of the terrestrial planets.

F. FUTURE PLANS FOR MISSIONS

1. HAYABUSA 2

Following the great success of its first asteroid exploration mission, Hayabusa, JAXA decided to build on the acquired knowledge and technological expertise, and launch Hayabusa-2, its successor. The first-proposed launch window for Hayabusa-2 is in July 2014, and its target asteroid is 1999 JU_3, a C-type asteroid, which is a more primitive body than Itokawa, its predecessor's target. The spacecraft will have almost the same configuration as before with an updated technology and novel technologies. It will also feature a device designed to create a crater on the surface of the target asteroid artificially, and study the process as well as gather samples for the return part of the mission [11].

2. OSIRIS-REX

Asteroidal exploration has gained more and more ground in the scientific community, greatly aided by the success of Hayabusa and the other missions presented. As a consequence, NASA could not stay out of it, and is launching the Origins Spectral Interpretation Resource Identification Security—Regolith Explorer (OSIRIS-REx) spacecraft in 2016, to study near-Earth B-type asteroid "Bennu." It will also be a sample return mission targeting to return more than 60 g of material. It will also carry out observations to determine the morphology, texture geochemistry, and spectral properties of the target asteroid, so that the geological context of the samples will be known in great detail, something that is lacking to meteorites. It will also attempt to measure the Yarkovsky effect that acts upon the asteroid directly [12].

IV. PROPERTIES RELEVANT TO MITIGATION OF POTENTIAL TARGETS

As we discussed, a lot of effort is being paid to study all the different properties of asteroids and NEAs in particular. Although much of it is basic research, trying to improve our understanding of the solar system and the universe in general, a possible discovery of a future impact with Earth will be far from irrelevant to our everyday lives. To be prepared for such an event, numerous concepts of mitigation have been proposed and extensively studied, and are demonstrated in other chapters of this book. Our goal is to present how the physical properties of asteroids affect the design of a given mitigation method.

Naturally the most-deciding factor when it comes to mitigation is the mass of the target body. All non-Bruce-Willis mitigation methods revolve around artificially altering the orbit of the incoming asteroid so that it misses Earth, by applying force through some mechanism. The net amount of force required is directly dependent on the mass, as Newton informed us. If we were to be impacted by Ceres, for example, the sheer size of it would mean there is nothing to do other than enjoy the rest of our lives. But for smaller targets, which are the only realistic possible impactors, the exact determination of the mass is key to define the details of the mitigation method. Additionally, the shape and rotational properties are key to all mitigation concepts. A spherical, almost-still asteroid, is much easier to handle than an irregular fast spinner. Kinetic impactor and nuclear-based concepts depend on the ejection of material from the surface of the asteroid, resulting in an opposite momentum transfer on the body of the asteroid. As such, the properties of the surface and the subsurface are important. For small targets where global disruption is the goal, the internal structure is also key. Methods that require surface interaction, such as anchoring, also require knowledge of the mechanical properties of the surface material. Finally the thermal parameters such as conductivity, heat capacity, and composition, are determining factors for methods involving laser ablation of material or Yarkovsky-driven deflection of the orbit.

V. CONCLUDING REMARKS

Although ground-based observations are far easier and far less expensive to carry out, the deliverable results are very limited when compared to space-mission data. Ground telescopes may inform us on the absolute magnitude, the albedo, a rough-shape estimation, the spin rate, and global spectral properties, but if a mitigation mission is necessary, this is just too few information. As presented previously in the section about previous missions, the accurate shape, mass, density, surface topography and properties, as well as composition and internal structure are all achievable results. Of course, we simply cannot visit all NEAs, but the more data we collect, the better our understanding of their properties as a population; therefore, we can easily put constraints on their properties based on the ample ground-based observations.

REFERENCES

[1] Walsh, K. J., Richardson, D. C., and Michel, P., "Rotational Breakup as the Origin of Small Binary Asteroids," *LPI Contributions*, Vol. 1405, 2008, p. 8138.

[2] Yeomans, D. K., Barriot, J.-P., Dunham, D. W., Farquhar, R. W., Giorgini, J. D., Helfrich, C. E., Konopliv, A. S., McAdams, J. V., Miller, J. K., Owen, W. M. Jr., Scheeres, D. J., Synnott, S. P., and Williams, B. G., "Estimating the Mass of Asteroid 253 Mathilde from Tracking Data During the NEAR Flyby," *Science*, Vol. 278, Dec. 1997, p. 2106.

[3] Veverka, J., Thomas, P., Harch, A., Clark, B., Bell, J. F., Carcich, B., Joseph, J., Murchie, S., Izenberg, N., Chapman, C., Merline, W., Malin, M., McFadden, L., and Robinson, M., "NEAR Encounter with Asteroid 253 Mathilde: Overview," *Icarus*, Vol. 140, July 1999, pp. 3–16.

[4] Cheng, A. F., "Near Earth Asteroid Rendezvous: Mission Summary," *Asteroids III*, 2002, pp. 351–366.

[5] Asphaug, E., "Eros Is a Rubble Pile," *AAS/Division for Planetary Sciences Meeting Abstracts #40, Bulletin of the American Astronomical Society*, University of Arizona Press, Vol. 40, Sept. 2008, p. 431.

[6] Michel, P., and Richardson, D. C., "Collision and Gravitational Reaccumulation: Possible Formation Mechanism of the Asteroid Itokawa," *Astronomy and Astrophysics*, Vol. 554, June 2013, p. L1.

[7] Keller, H. U., Barbieri, C., Koschny, D., Lamy, P., Rickman, H., Rodrigo, R., Sierks, H., A'Hearn, M. F., Angrilli, F., Barucci, M. A., Bertaux, J.-L., Cremonese, G., Da Deppo, V., Davidsson, B., De Cecco, M., Debei, S., Fornasier, S., Fulle, M., Groussin, O., Gutierrez, P. J., Hviid, S. F., Ip, W.-H., Jorda, L., Knollenberg, J., Kramm, J. R., Kührt, E., Küppers, M., Lara, L.-M., Lazzarin, M., Moreno, J. L., Marzari, F., Michalik, H., Naletto, G., Sabau, L., Thomas, N., Wenzel, K.-P., Bertini, I., Besse, S., Ferri, F., Kaasalainen, M., Lowry, S., Marchi, S., Mottola, S., Sabolo, W., Schröder, S. E., Spjuth, S., and Vernazza, P., "E-Type Asteroid (2867) Steins as Imaged by OSIRIS on Board Rosetta," *Science*, Vol. 327, Jan. 2010, pp. 190–193.

[8] Sierks, H., Lamy, P., Barbieri, C., Koschny, D., Rickman, H., Rodrigo, R., A'Hearn, M. F., Angrilli, F., Barucci, M. A., Bertaux, J.-L., Bertini, I., Besse, S., Carry, B., Cremonese, G., Da Deppo, V., Davidsson, B., Debei, S., De Cecco, M., De Leon, J., Ferri, F., Fornasier, S., Fulle, M., Hviid, S. F., Gaskell, R. W., Groussin, O., Gutierrez, P., Ip, W., Jorda, L., Kaasalainen, M., Keller, H. U., Knollenberg, J., Kramm, R., Kührt, E., Küppers, M., Lara, L., Lazzarin, M., Leyrat, C., Moreno, J. J. L., Magrin, S., Marchi, S., Marzari, F., Massironi, M., Michalik, H., Moissl, R., Naletto, G., Preusker, F., Sabau, L., Sabolo, W., Scholten, F., Snodgrass, C., Thomas, N., Tubiana, C., Vernazza, P., Vincent, J.-B., Wenzel, K.-P., Andert, T., Pätzold, M., and Weiss, B. P., "Images of Asteroid 21 Lutetia: A Remnant Planetesimal from the Early Solar System," *Science*, Vol. 334, Oct. 2011, pp. 487–490.

[9] Coradini, A., Capaccioni, F., Erard, S., Arnold, G., De Sanctis, M. C., Filacchione, G., Tosi, F., Barucci, M. A., Capria, M. T., Ammannito, E., Grassi, D., Piccioni, G., Giuppi, S., Bellucci, G., Benkhoff, J., Bibring, J. P., Blanco, A., Blecka, M., Bockelee-Morvan, D., Carraro, F., Carlson, R., Carsenty, U., Cerroni, P., Colangeli, L., Combes, M., Combi, M., Crovisier, J., Drossart, P., Encrenaz, E. T., Federico, C., Fink, U., Fonti, S., Giacomini, L., Ip, W. H., Jaumann, R., Kuehrt, E., Langevin, Y., Magni, G., McCord, T., Mennella, V., Mottola, S., Neukum, G., Orofino, V., Palumbo, P., Schade, U., Schmitt, B., Taylor, F., Tiphene, D., and Tozzi, G., "The Surface Composition and Temperature of

Asteroid 21 Lutetia as Observed by Rosetta/VIRTIS," *Science*, Vol. 334, Oct. 2011, pp. 492–494.

[10] Russell, C. T., Raymond, C. A., Jaumann, R., McSween, H. Y., De Sanctis, M. C., Nathues, A., Prettyman, T. H., Ammannito, E., Reddy, V., Preusker, F., O'Brien, D. P., Marchi, S., Denevi, B. W., Buczkowski, D. L., Pieters, C. M., McCord, T. B., Li, J.-Y., Mittlefehldt, D. W., Combe, J.-P., Williams, D. A., Hiesinger, H., Yingst, R. A., Polanskey, C. A., and Joy, S. P., "Dawn Completes Its Mission at 4 Vesta," *Meteoritics and Planetary Science*, Vol. 48, Nov. 2013, pp. 2076–2089.

[11] JAXA, b612.jspec.jaxa.jp/hayabusa2/e/index_e.html. *Hayabusa 2 official webpage.*

[12] University of Arizona, http://osiris-rex.lpl.arizona.edu/. *OSIRIS-REx official webpage.*

CHAPTER 6

Linking Near-Earth Asteroids to Their Main-Belt Source Regions

Georgios Tsirvoulis*
Astronomical Observatory, Belgrade, Serbia

Bojan Novakovic†
University of Belgrade, Belgrade, Serbia

I. INTRODUCTION

Everyone is familiar with the greater image of our solar system, the planets orbiting around the Sun in a peaceful way. Some are also excited when noticing Venus bright as it is during dusk or dawn, or Jupiter in the night sky. Some fewer might have gone a step further, to try to observe with a pair of binoculars, discovering that there is more to it. The Galilean moons of Jupiter are indeed visible, yet so far away. It is then that one realizes the solar system is much more populous than our everyday experience reveals. The entire region within Saturn's orbit is full of small rocky bodies orbiting around the Sun, the minor planets or Asteroids. Moreover, there are also the various families of comets going outward, until we reach the Oort cloud, a massive reservoir of comets at the outskirts of our solar system. In total, there are a huge number of small bodies in our solar system compared to the "big" planets. But their total mass is not even worth mentioning. All the asteroids, for example, account for as much as 5% of the mass of the Moon. To put things more into perspective, Jupiter is twice as massive as everything else in our solar system combined (excluding the Sun). Still it is common knowledge that astronomers have a tendency to try to explore every little detail. Doing so over many years, they realized that these details are actually meaningful, aiding the quest for answers on some really difficult astronomical questions: How exactly was the solar system formed? Why is it the way we observe it now?

We focus our study of small bodies in the inner part of the solar system (interior to Jupiter's orbit), namely the asteroids. More specifically we are going to study the dynamical evolution of asteroids, the change of their orbits around the Sun. Because "within Saturn's orbit" as stated previously, there is also our planet, the Earth, we naturally expect some of this big number of asteroids to live close to us. Those asteroids are our main focus during the rest of the chapter, trying

*Researcher.
†Faculty of Mathematics.

to understand when, how, and why are they here, “close” to us. The process to address these questions is motivated basically by two reasons: first is the primal instinct of survival. The Earth, as well as the other terrestrial planets, have visible scars on their surfaces of what an impact with an asteroid can result to. Chicxulub crater in Yukatan, Mexico, was formed during the impact of a 10 km large object that hit the Earth some 65 million years ago. This event formed the 180 km crater visible today and is believed to have caused the extinction of dinosaurs along many other species, as well as global scale environmental changes. And this is just the most famous example of such an event, not even the biggest one. Even much smaller events can have a significant impact, such as the recent ~17 m Chelyabinsk meteor, which although exploded at around 25–30 km from the surface, caused injuries and damage due to the shock wave it generated. Having this in mind the need to study, predict, and utterly try to prevent such events from occurring again becomes pretty obvious. The second reason for which astronomers study those small objects that seem insignificant in the great picture of the universe is linked to another instinct. They do it just for the heck of and that’s all about it.

The goal of this chapter is to present the methodology and the prerequisites needed, to establish links between specific near-Earth asteroids (NEAs) and the regions they originated from. Our discussion begin from the very beginning, the formation of the solar system and the asteroids. Then we jump forward in time more than 4 billion years, to the present, and examine their current state. We divide them into groups corresponding to the different regions they reside, and after giving some basic insight on the dynamics, we address the main topic of this chapter, finding ways to establish links between the so-called NEAs and their source regions in the main asteroid belt, providing also with some example results of recent studies.

II. FORMATION OF THE SOLAR SYSTEM

The story of our solar system begins about 4.6 billion years ago. This was when the solar nebula, a rotating cloud of gas and dust, starts collapsing under its own gravity. A nearby supernova explosion may have triggered this collapse [1], a fact that also explains the enrichment of the protostellar disk with short lived radioactive isotopes (^{26}Al), daughter products of which have been detected in meteoritic samples. As the rotating molecular cloud collapses, it forms a disk, which will give birth to the planetary system.

A. FORMATION OF THE PLANETS

At the center of the disk, where the density and temperature are highest, the so-called protosun is formed. The protosun is heating the disk from inside leading to a gradient of the protoplanetary disk’s temperature, which is hotter toward its center and colder toward the outskirts as can be seen in Fig. 2. At the innermost

Fig. 1 Artist's impression of the solar system formation sequence. (Credit http://oz.plymouth.edu/~sci_ed/Turski/Courses/Earth_Science/Intro.html.)

part of the disk, where hydrogen compounds stay vaporized, dust particles are rocky and metallic, whereas beyond the frost line the limit where water can form ice, hydrogen compounds also condense into dust particles. Dust particles collide with each other and start accreting into larger clumps, utterly forming planetesimals, planetary embryos, and finally planets (see Fig. 1). Although the process of going from dust to planetesimals and then planets is not so trivial, it is out of the scope of this chapter to discuss those details. The reader is referred to Refs. [2, 3] for more details about these two steps in planet formation.

Depending on their distance from the protosun, planets are either rocky (terrestrial planets) or icy/gaseous (giant planets), exactly due to the composition of the planetesimals that existed in each specific region. Before nuclear activity set off at the core of the Sun, the outer planets also had enough time to accrete gas from the disk. This gas was later blown away by the solar radiation, leaving the solar system as it is today, at least in terms of its members.

B. FORMATION OF ASTEROIDS AND COMETS

With the process discussed previously, the dust of the protoplanetary disk formed planetesimals, which in turn formed the planets. But not all the planetesimals were able to accrete onto the planetary embryos. Indeed, a significant fraction of the initial dust, stopped accumulating at smaller dimensioned objects. The remnants of these "unused" planetesimals are what we see today as asteroids and comets [4]. Exactly as described previously, the rocky, metallic planetesimals interior to the frost line correspond to the asteroids, and the icy planetesimals beyond it correspond to the comets. Although there is no strict distinction between the two, that is, asteroids may have some ice, and comets do have rocky cores, we focus on the first group, which resides at the inner solar system.

As our everyday experience suggests, there is a huge variety of rocky/metallic materials on the Earth. Exactly the same is the situation in the total population of asteroids. The gradient in the temperature of the protoplanetary disk as shown in Fig. 2, also lead to differences of what kind of rocky dust grains condensed at the various distances [5]. Figure 3 shows the distributions of different compositions of asteroids across the solar system.

Although this figure shows the general trend of primitive asteroid compositions across the solar system, it fails to catch the details. Throughout the history of the solar system, dramatic dynamical events took place that had an impact on asteroids, that is, caused by some mixing of the different types, resulting in a much more complex distribution [7].

The main tools used to study the different asteroids by composition are their albedos and spectra. Albedo or reflection coefficient is the fraction of sunlight that is reflected from a body's surface. Using albedos we can distinguish the different types of asteroids, for example, carbonaceous or *C*-type are dark, with albedoes between 4 and 8% due to their composition being mostly of carbon-compounds, whereas *E*-type are the whitest due to the presence of enstatite ($MgSiO_3$) [5].

To obtain a spectrum, a multitude of reflectivity measurements is carried out at different narrow bands of the electromagnetic spectrum, giving the surface's reflectivity as a function of wavelength. Spectra give much more detailed information about an asteroid's surface

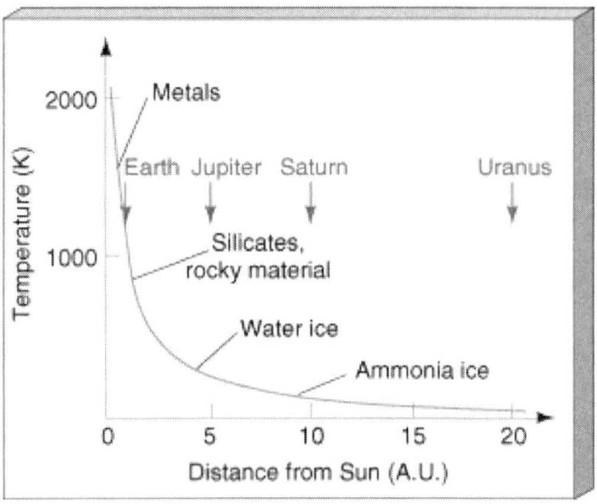

Fig. 2 Temperature gradient of the protoplanetary disk. (Credit www.astro.virginia.edu.)

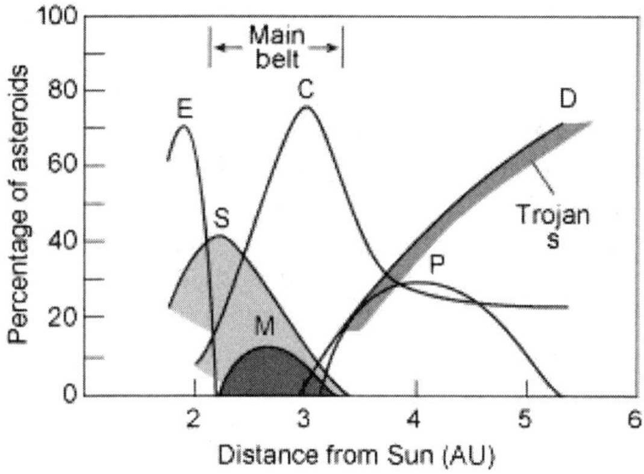

Fig. 3 Composition of asteroids varying with heliocentric distance. The different letters correspond to: C-carbonaceous, S-silicate, E-enstatitic, D&P-red asteroids with low albedos, M-rare metallic. (Credit Gradie and Tedesco [6].)

composition but are far more difficult to obtain than albedos. Both spectra and albedos of asteroids are compared with those of meteoritic samples that have been gathered to determine the actual compositions of asteroids and their properties [5]. Figure 4 shows the spectra of different types of asteroids in comparison to such meteoritic samples, where the matches for main types are evident.

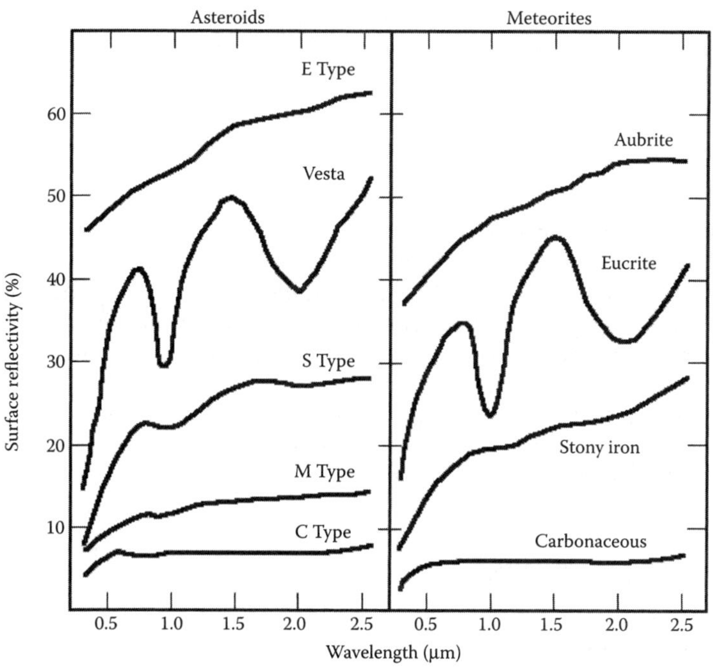

Fig. 4 Comparison of spectra between asteroids and meteoritic samples, used to derive asteroids' surface compositions. (Credit C. Chapman.)

III. DYNAMICS OF ASTEROIDS

A. ORBITAL ELEMENTS

To study the dynamics of asteroids, we first have to define a set of coordinates that are used to specify the motion of an asteroid around the Sun. Because, the Sun is by far the most massive body in the solar system, the motion of asteroids at first approximation is well simulated by the two-body problem. This means that asteroids will have elliptic orbits around the Sun, although the influences from the other planets, being so small compared to the Sun, are measured in as perturbations of the simple two-body solution. Although this is the general rule, there are some exceptions, when the asteroid is very close to a planet, in reality Jupiter only, where its orbit is defined by the three-body problem. Those cases are mentioned shortly.

In 3-D space where asteroids move, one needs a set of three coordinates and three velocities to have their motion defined. Cartesian coordinates and velocities, spherical ones, or even cylindrical do work, but they are not the most convenient. Because the symmetry of the problem results in an ellipse as a solution of the motion, it is more convenient to make use of this fact and use the Keplerian elliptic orbital elements, see Fig. 5. Indeed, because the solution of the two-body problem is the motion on a fixed ellipse, the orbital elements define this ellipse in 3-D space, plus the position of the orbiting body on the ellipse as follows [8]: the shape of the ellipse is defined by its semimajor axis (α) and its eccentricity (e). Then we use three angles to define the orientation of the ellipse in space. The inclination (i) is the angle of the orbital plane with respect to the reference plane, usually taken as the ecliptic plane. This way the orbit intersects the reference plane at two points called the *nodes*. We distinguish the *ascending node* (Ω), where the orbiting body crosses the reference plane toward the positive z direction and the *descending node* toward the negative z direction. Then we define the *longitude of the ascending node* (Ω) as the angle formed on the reference plane between a reference direction, usually the equinox point γ, and the ascending node. The final angle used to define the orientation of the ellipse is the *argument of perihelion* (ω), the angle measured on the orbital plane between the ascending node and the pericenter of the ellipse. Finally, we define the position of the orbiting body along its elliptic orbit by the angle formed between its current position and the pericenter with respect to the main focus of the ellipse, called *true anomaly* (ν).

B. GRAVITATIONAL DYNAMICS

It is obvious that the orbital motion of asteroids, and all celestial bodies, is almost completely governed by gravity. The simple keplerian solution of the two-body problem presented previously, is not the exact one in the solar system, exactly because there is no two-body problem. In reality, due to the existence of all the planets, and even neighboring asteroids, the general elliptic solution is all but

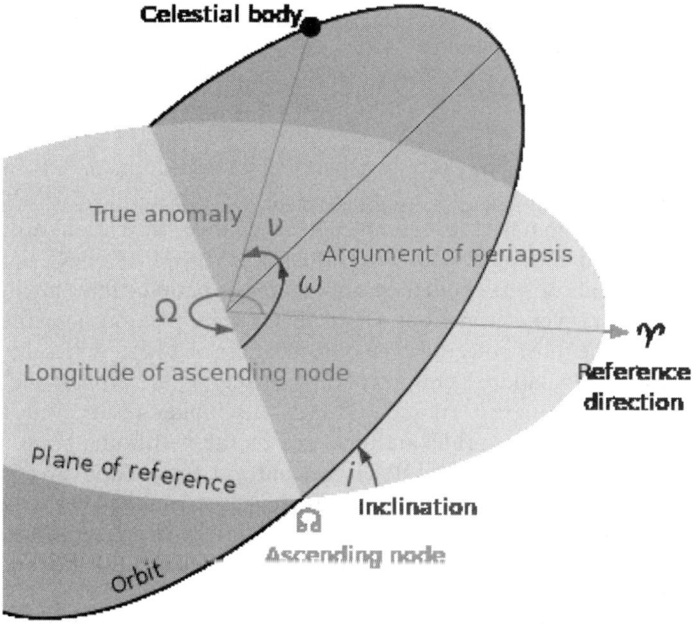

Fig. 5 Keplerian orbital elements. (Credit L. Snyder.)

stable. We will present the main manifestations of gravitational perturbations on small bodies.

1. RESONANCES

Although asteroids are in general sufficiently far from planets, and so the force exerted onto them is small compared to that of the Sun, there are certain circumstances when these small effects are amplified. Such phenomena are the various resonances that occur between the asteroid and one or even more planets [8]. First and foremost, there are the various *mean motion resonances* (hereafter MMRs). These occur when the orbital periods of the asteroid and the planet in case are in a ratio of small integers as shown in Eq. (1).

$$\frac{n_{\text{ast}}}{n_{\text{pl}}} = \frac{p+q}{p} \tag{1}$$

where n_{ast} and n_{pl} are the mean motions of the asteroid and the planet, respectively, and p, q are small integers. q is referred to as the *order* of the resonance, and the smaller it is, the more powerful the resonance. We must note here that as it is known by Kepler's law

$$T^2 = \frac{4\pi^2}{GM}\alpha^3 \tag{2}$$

the period of an orbiting body is only depending on its semimajor axis and none of its other orbital elements. For example, when we mention the 3:1 MMR with Jupiter, we refer to an object that completes exactly three revolutions around the Sun in one period of Jupiter. This is an order 2 MMR. MMRs have significant effects on the orbits of small bodies because there is a periodic force applied onto them by the planet every few periods, at exactly the same position along their orbit [8]. MMRs are the most powerful type of resonances and have noticeable effects on the asteroids population. Examples of these effects are the *Trojans* and *Hildas*, groups of asteroids that are in 1:1 and 3:2 MMR with Jupiter, respectively. In these cases the resonant bodies' orbits are regulated by the perturbing body, Jupiter, effectively being *shepherded* by it [9]. On the contrary, the regions corresponding to the 2:1 and 3:1 MMRs with Jupiter in the main asteroid belt are depleted of asteroids exactly due to the strong influence of Jupiter, which quickly removes them. MMRs alter the shape of the orbits of asteroids, and because they are caused by the orbital motion they are acting fast.

The orientation of an asteroid's orbit changes in time, in two manners corresponding to the precession of the perihelion and of the nodal line. In the stable case the argument of perihelion and the longitude of the ascending node precess at a constant rate over thousands of years. But, because the same precessions occur on the planets also, there can be resonances of those precession rates. Such resonances are called *secular* due to the fact that they are much slower than MMRs [8]. In this case the resonances are described by a more complicated formulation, as they can involve linear combinations of both frequencies of more than one planet. For example, the following combination:

$$g - g_5 + s - s_6 \tag{3}$$

where $g = \dot{\omega}$, $s = \dot{\Omega}$, and subscripts refer to the corresponding planet (5-Jupiter, 6-Saturn), refers to the secular resonance where the sum of the precession frequencies of the asteroid's perihelion and ascending node, is equal to Jupiter's perihelion precession rate plus Saturn's Nodal precession rate. We refer to those resonances later when we investigate dynamical paths that lead asteroids to the near-Earth region. Furthermore, we will just mention two more kinds of resonances for completeness reasons. The first is the *Kozai* resonance, an effect of the three-body problem, where a periodic exchange between the orbiting body's eccentricity and inclination occurs. A prerequisite for *Kozai* resonance to be possible to happen is the third, small body, to have substantial inclination (usually greater than 40 deg) [10]. The second, there is the *spin–orbit resonance* where a body's spin and orbital periods are in a ratio of small integers. Known examples of this include the Moon (1:1 spin–orbit resonance) and Mars (3:2 spin–orbit resonance)

2. CLOSE ENCOUNTERS

As can be seen in Fig. 8, asteroids are numerous and well spread across the inner solar system. Keeping this in mind it is rational to assume that some asteroids will not orbit peacefully during their lifetime, affected only by the effects described previously. There will also be times that the circumstances are exactly right for an asteroid to pass really close to another body: a neighboring asteroid or even a planet. Such occurrences are called close encounters and completely alter the picture of the simple two-body solution plus some perturbation [11]. When an asteroid comes really close to another body, its dynamics are heavily dependent on their mutual gravitational interaction. Of course close encounters with planets are far more effective due to their mass, than with their other counterpart asteroids.

The general scheme adopted for handling close encounters is considering for the small duration when the encounter takes place, the two bodies as isolated, and treating it as a hyperbolic deflection as shown in Fig. 6 [11]. Then the resulting change in instantaneous velocity is incorporated to the orbital motion at that point. This method works well, provided that the relative velocity of the bodies is sufficiently high compared to their orbital velocities [11].

C. NONGRAVITATIONAL PHENOMENA

As far as the nongravitational forces are concerned, the only one that has significant impact on the orbits of asteroids is the *Yarkovsky effect.* This phenomenon is related to the anisotropic re-emission of the absorbed solar radiation by the asteroid's surface [13]. The yarkovsky effect has two components: the *Diurnal* and the *Seasonal* one, the difference being that the diurnal is caused by the asteroid's rotation about its axis, whereas the seasonal is caused by the orbital motion of the asteroid around the sun. Both components are governed by the same principle, that is, the place on the body's surface with the highest temperature at any given

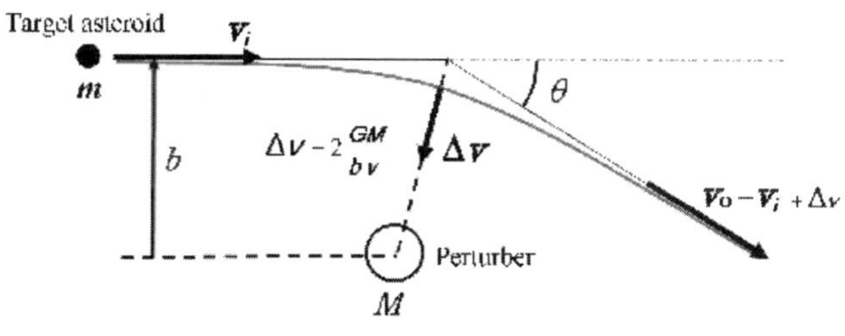

Fig. 6 Representation of a hyperbolic deflection. (Credit Mouret et al. [12].)

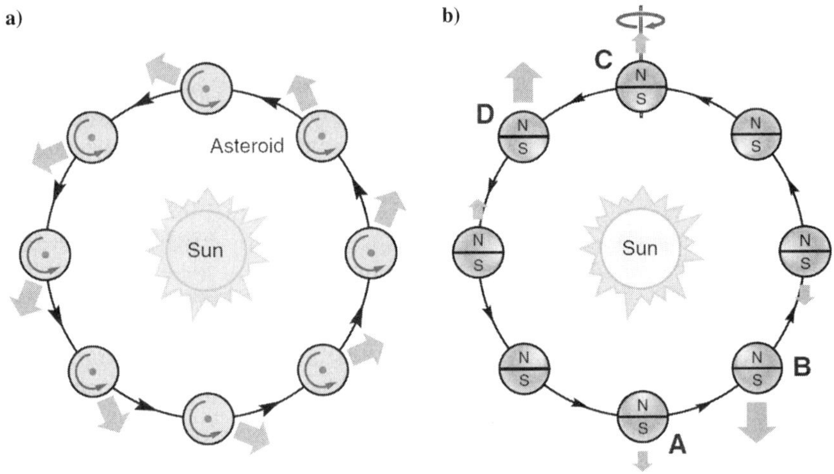

Fig. 7 Schematic representation of the Yarkovsky effect: a) The diurnal effect for prograde rotation and b) the seasonal effect. Arrows indicate the net force exerted on the asteroid at any given time. (Credit Bottke et al. [13].)

time is shifted by the rotation of the asteroid compared to the place directly facing the Sun. Thus, more infrared photons are emitted by this area of the surface. For the diurnal effect the rotation of the body is the regulator and the maximum effect is achieved when the rotation axis is normal to the orbital plane. An asteroid with a *prograde* rotation (the rotational and orbital axes are parallel) will emit more photons "behind" it, which carry momentum, leading to a net force pushing it along its orbit faster. A *retrograde* asteroid will emit more thermal photons ahead of it, leading to a net force trying to slow it down. Those forces will increase, or decrease the semimajor axis of the asteroid, respectively. The seasonal effect, which is caused by the orbital motion of the asteroid, is much more subtle, f thus takes its maximum strength when the rotation axis of the asteroid is parallel to the orbital plane, because in this case there is no diurnal effect. The seasonal effect always tries to slow the asteroid down, due to the geometry of the problem, resulting in a decrease of the semimajor axis. A schematic representation of the two components can be seen in Fig. 7.

Because the force exerted by the Yarkovsky effect on the asteroid is actually a reaction to the emission of thermal photons, as described previously there is a simple relation between the size of the asteroid and the strength of the effect [13]. To have a greater force, more photons need to be emitted, thus a larger emitting area is required, provided that the solar radiation and surface characteristics. are the same. On the contrary, for the same amount of momentum transfer to exert a greater force, the mass receiving the force has to be small. Assuming that the density of asteroids is almost constant, the strength of the Yarkovsky effect is inversely proportional to the asteroid's size, because for any close surface the ratio of the surface area to the enclosed volume is increasing with decreasing linear dimension of the body. In reality only asteroids with diameter less than ~30 km have a significant effect, which is still pretty small but not negligible during millions of years. To give a grasp on the magnitude of the effect, an asteroid with a diameter of 1 km experiences a drift of $1\text{–}5 \times 10^{-4}$ AU/Myr on its semimajor axis. As we demonstrate later, the Yarkovsky effect combined with resonances is a key mechanism of transporting asteroids across the different regions of the solar system [13].

D. ASTEROID POPULATIONS

Having described the most important dynamical phenomena governing the orbits of asteroids, we can now see what consequences these phenomena have on the total population of the asteroids. Indeed, because at the various regions of the solar system different phenomena are more dominant than others, asteroids are divided in distinct populations, dictated by them. We have mentioned previously two such populations, the *Trojans* and *Hildas*, asteroids locked in MMRs with Jupiter. These two groups show in a clear way, how impactful an MMR can be

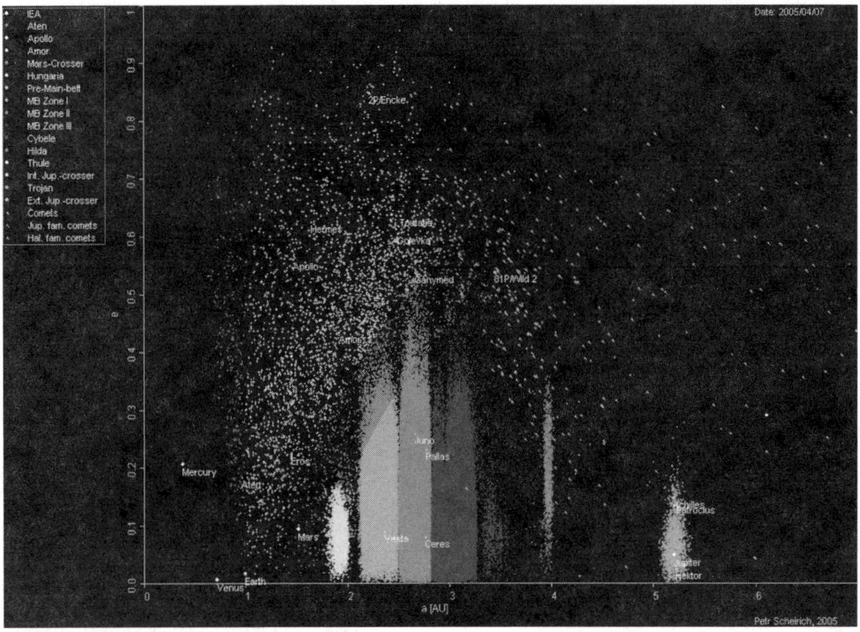

Fig. 8 Semimajor axis vs eccentricity plot of the different asteroid groups. (Credit P. Scheirich.)

on an asteroid's orbit, completely defining its dynamics. The vast majority of the asteroids resides in the region between the orbits of Mars and Jupiter called the *main asteroid belt*. We explain more on the main belt in a separate subsection. Furthermore, there are asteroids that reach the region of the terrestrial planets. Mars crossers are asteroids with perihelia close to the orbit of Mars. Interior to them are the *Near Earth Asteroids* or *NEAs*, the target group of our study. The different asteroid groups are shown in Fig. 8.

IV. NEAR-EARTH ASTEROIDS

A. NEA GROUPS

Near-Earth asteroids, as their name obviously suggests, are those asteroids that at some point of their orbits come relatively close to the orbit of the Earth. But in this wide definition, many different kinds of orbits are plausible, so to better distinguish them we divide them in smaller groups with somewhat more strict criteria [14]. Asteroids with perihelia (q) in the range $1.02\,\mathrm{AU} < q \leqslant 1.3\,\mathrm{AU}$ are called *Amors*. Their perihelia are close to the aphelion of Earth, so these asteroids come close to Earth but they do not cross its orbit. Asteroids with smaller perihelia than Amors, $q \leqslant 1.02\,\mathrm{AU}$, but still greater semimajor axis than Earth are called *Apollos* and are Earth crossers. Those with semimajor axis less than 1 AU, but with aphelia $Q \leqslant 1.02\,\mathrm{AU}$ are also earth crossers called *Atens*. Finally there are the *Atira* or *Apohele* asteroids whose orbits lie entirely interior to Earth's orbit. The definitions of those groups are summarized in Fig. 9 and their positions in the solar system are visible in Fig. 8.

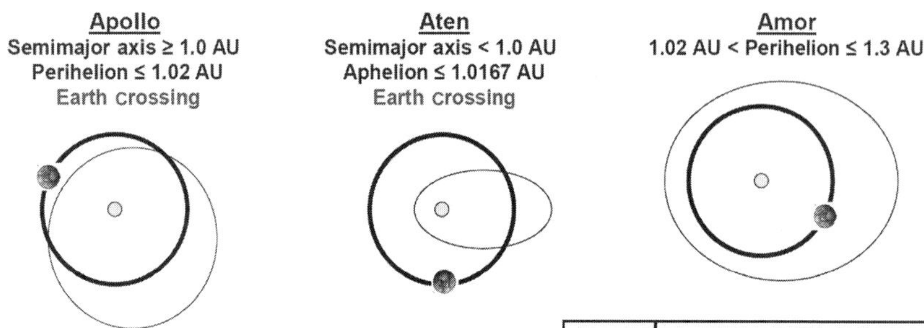

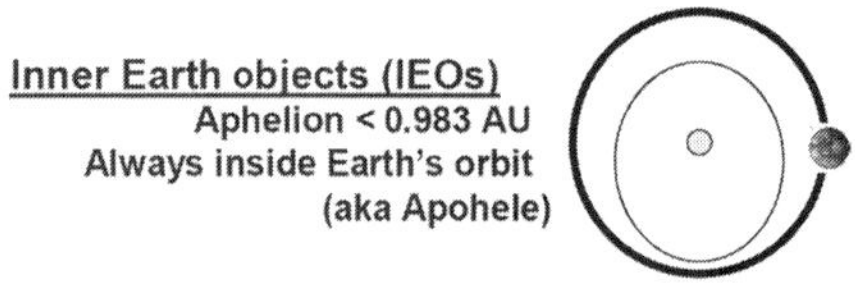

Type	Near-Earth population
Apollo	62% of known asteroids
Aten	6% of known asteroids
Amor	32% of known asteroids
IEO	6 known asteroids

Fig. 9 The different NEA groups. (Credit neo.jpl.nasa.gov.)

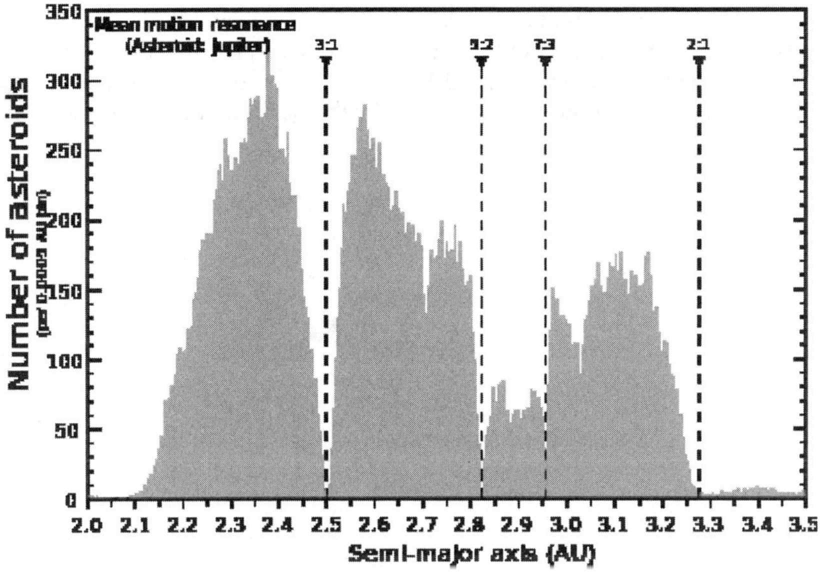

Fig. 10 The Kirkwood gaps of the main asteroid belt. (Credit Alan Chamberlain.)

B. THE MAIN ASTEROID BELT AS A SOURCE FOR NEAs

Near-Earth asteroids orbit in a very unpleasant environment, dynamically speaking. They cohabitate the same region with the Earth and Mars and this makes the perturbations caused by the close encounters as well as the resonances very powerful, enough to destabilize their orbits substantially, forcing them onto the Sun or ejecting them in hyperbolic orbits. Numerical integrations of NEAs reveal that their typical dynamical lifetime is of the order of 10 Myrs. Although this may sound like a lot compared to our everyday experience, it is not but a very brief duration compared to the age of our solar system, ~4.6 Gyrs. So it becomes evident that those objects cannot have been around for long. And because they cannot live for long, but are currently present, the only possibility is whether they are steady-state population, balanced between a supply–removal interplay. Then immediately rises the question: Where do those objects come from? This one has a rather obvious answer, as there is only one contender for the supplier. The *Main Asteroid Belt* has many asteroids, orbiting in a far more peaceful environment where they can stay for much longer. In fact, the present main belt is there since its formation, but has lost a lot of members. The most obvious scars are the *Kirkwood gaps*, regions significantly depleted of asteroids due to the most powerful MMRs with Jupiter [15], see Fig. 10. Those regions, together with the large gap caused by the ν_6 secular resonance with Saturn are believed to be the main sources for the NEAs [16, 17]. Indeed, those powerful resonances are able to excite the eccentricities of asteroids enough to force them into planet crossing orbits, where after close encounters with the terrestrial planets they are captured as NEAs. In Fig. 11 we see the results of a simulation of this mechanism. The asteroids are initially excited in eccentricity by the ν_6, 3:1 and 5:2 resonances into

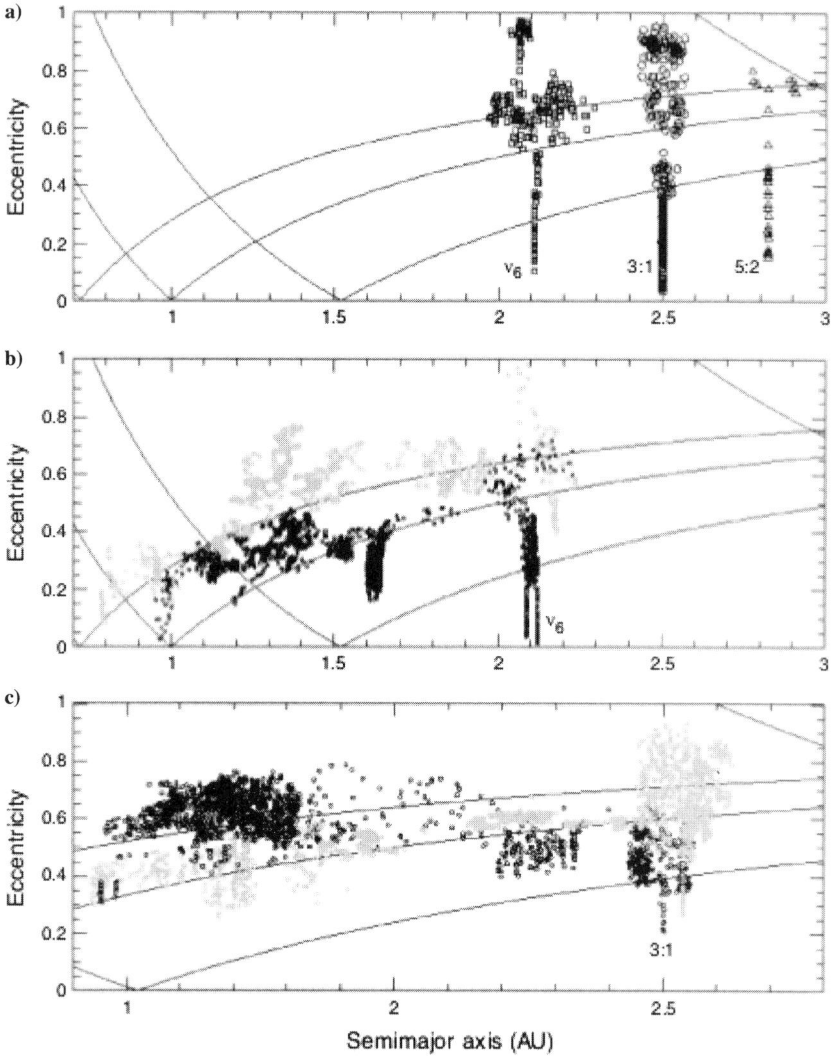

Fig. 11 Transfer mechanism of asteroids to the near-Earth region due to the interplay between resonances and close encounters. (Credit Morbidelli et al. [16].)

the planet crossing regions, and eventually their semimajor axes are shortened by the close encounters, ending up as NEAs [16].

C. ASTEROID FAMILIES AS SOURCES OF NEAs

Yet those source regions are already witnessed to be quite depleted, there also has to be some mechanism feeding asteroids into them, so the process can continue going. A way to overcome this problem is by considering that some random phenomena take place in the main belt, drastic enough to inject asteroids into the resonances. Such phenomena are collisions between asteroids and formation of asteroid families.

Asteroid families are groups of asteroids in the main belt that have very similar orbital characteristics [18], namely their *proper elements* [19]: proper semimajor axis (a_p), eccentricity (e_p), and inclination (i_p), as well as physical characteristics, at least in general. The definition of proper elements is out of the scope of this chapter but it is sufficient to say that they are quasi-integrals of the motion of asteroids, remaining almost constant overtime. Indeed, asteroids with seemingly random distributions of instantaneous orbital elements, appear as clusters in the proper elements space [19]. This is due to their common origin, as ejecta produced in the past by either a catastrophic or a cratering collision between the parent asteroid and an impactor. These collisions throw away numerous smaller asteroids with some ejection velocity. If such an event happens close to one of the aforementioned resonances this velocity is sufficient to inject the body in case, into the resonance, where it begins a journey to the near-Earth region [16,17]. But, even if the collision does not take place close enough to a resonance, to inject bodies directly into it, those small fragments are still being subjected to the Yarkovsky effect as described previously. So, now this mechanism gives them the additional push, a gentle long push to be precise, toward the resonances. If a collision is really significant, and close to a resonance, a large family is expected to be followed by a shower of incoming NEAs, leading to intense craterization of the terrestrial planets.

D. LINKING NEAs TO THEIR ORIGINS

Previously we described the general scenario of transporting asteroids from the main belt, which acts as a tank, to the near-Earth region. But there is much more to investigate, much more detail than just the big picture. The transport routes are themselves not so simplistic or efficient as we may think. Different possibilities exist for different source regions, with different inclinations or size of the NEAs. And so the study of individual NEAs is also important, trying to combine their exact current orbit and physical characteristics with a source region with pinpoint accuracy. To do this, extensive observational campaigns are underway, in order to extract as much data as possible, such as obtained albedoes, spectra, SDSS colors, and photometric phase curves of near-Earth and main-belt asteroids. Combining

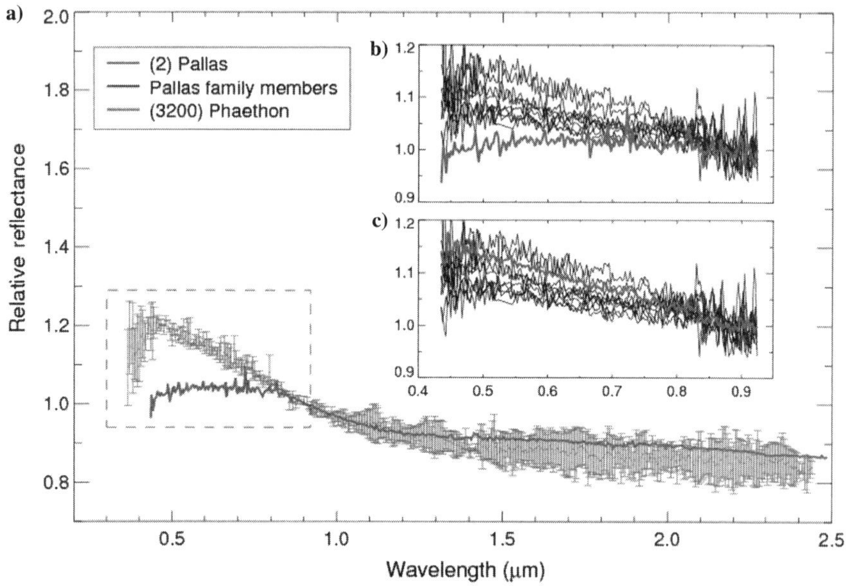

Fig. 12 Spectral comparison between (3200) Phaethon, (2) Pallas, and Pallas family members: a) Phaethon (thick gray line with error bars) compared to (2) Pallas (thick gray line), b) (2) Pallas (thick gray line) compared to Pallas family members (black lines), and c) (3200) Phaethon (thick gray line) compared to Pallas family members (black lines). (Credit de Leon et al. [20].)

all this information with dynamical simulations, we can verify a specific region as a source for a given asteroid by comparing their inferred physical characteristics. Moreover, one can also study the possible future of such an object, through this chaotic near-Earth region. Section 1 presents some examples of such studies, establishing links of specific NEAs to their origins.

1. (3200) PHAETHON AND THE PALLAS FAMILY

The (3200) Phaethon is a very interesting near-Earth asteroid due to its high orbital inclination ($i = 22.18$ deg) and low perihelion distance ($q = 1.4$ AU). It was associated with the Geminid stream, a meteor shower. It is classified as a B-type object. The (2) Pallas on the other hand is one of the largest asteroids of the main belt with diameter of ~550 km, also following a highly inclined $i \sim 34$ deg, and highly eccentric $e = 0.23$ orbit. Also, (2) Pallas is the parent body of the homonymous high inclination asteroid family. The link between the two was established by de León et al. [20] spectral comparison of (3200) Phaethon, Pallas family members, and asteroid (2) Pallas itself, as well as by dynamical modeling of the time evolution of Pallas fragments.

The spectral comparison presented in Fig. 12 reveal a difference of Phaethon with (2) Pallas as expected between a parent body and a fragment, and a perfect match of Phaethon with other Pallas family members, strengthening the claim of this origin from a spectroscopic point of view. Furthermore, numerical integrations of 1000 fictitious Pallas family members were carried out, with positive

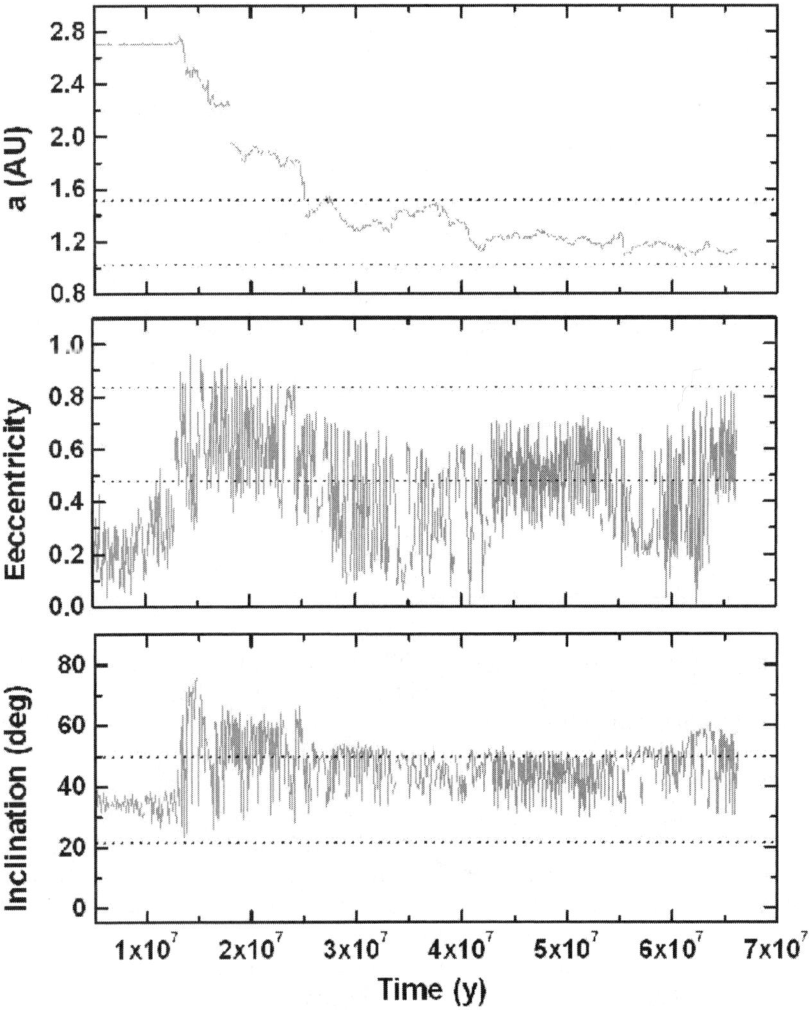

Fig. 13 Dynamical evolution of semimajor axis, eccentricity and inclination of a fictitious Pallas family member toward a Phaethon-like orbit. Dashed lines denote a ~1σ deviation from the corresponding mean values of (3200) Phaethon. (Credit de Leon et al. [20].)

results. As we see in Fig. 13 the time evolution of the orbital elements of an example particle. The object is injected into the 8:3 MMR with Jupiter, resulting in its eccentricity getting excited After repeated close encounters with Earth it was extracted from the resonance and placed close to the terrestrial planets, reaching the near-Earth region after ~18 Myrs, and spending ~40 Myrs in a Phaethon like orbit before falling onto the Sun. In total 21 out of the initial 1000 fictitious particles showed similar evolution. To prove the uniqueness of this link, any other possible parent candidates were also evaluated both spectroscopically and dynamically. The spectral study yielded 17 main-belt asteroids and one Mars-crosser as matches. Six of those belong to the Themis asteroid family, which is very close to the 2:1 MMR with Jupiter and thus very unlikely to transport bodies to the near-Earth region. The rest main-belt asteroids were also excluded as possible parent bodies due to dynamical reasons. The remaining mars-crosser (2629) Rudra, was also rejected, but was surprisingly found to possibly also originate from the Pallas family. It is most likely following a similar route to Phaethon but at an earlier stage of its journey. The final verdict that points out (2) Pallas as the parent body of Phaethon is therefore very robust. This study is a concrete example of the use of different methods, observations and simulations to establish a link of an NEA to its source region in the main belt.

2. ASTEROID 101955 (1999*RQ*$_{36}$) AND THE POLANA FAMILY

As a second example we present a study on asteroid 101955 (1999RQ_{36}). RQ_{36} is particularly interesting, being the target asteroid of NASA's *OSIRIS–REx* sample return mission, also being a potentially hazardous asteroid. This asteroid was studied both dynamically and spectroscopically as with the previous example of Phaethon, by H. Campins et al. [21]. Previous studies by Bottke et al. [22], of all the dynamical source regions of NEAs, suggest the ν_6 resonance as the source for 1999RQ_{36} with a 95% probability. Thus, the inner-belt (2.15 AU $< a <$ 2.5 AU), low inclination ($i <$ 10 deg) is the most likely source region. This region is dominated by the *Nysa* and *Polana* families. The spectroscopic examination showed that both (142) Polana and 1999RQ_{36} are B-type asteroids, whereas Nysa is totally inconsistent, thus ruled out. The numerical simulation of fictitious objects revealed that the transport route consists of Polana family members drifting into the ν_6 secular resonance by the Yarkovsky effect, due to their retrograde rotation, matching that of 1999RQ_{36}. The orbital eccentricities are then increased, becoming Mars- or even Earth-crossers, stored to their current orbits after a set of close encounters. ν_6 does not act on the inclinations of the transported bodies (both (142) Polana and 1999RQ_{36} have low inclinations), giving another confirmation as being the most plausible transport route compared to the other ones, which are MMRs and do alter the inclination. This case is also a neat example of how both dynamical and physical properties of NEAs are used to narrow down in their own way the source candidates, giving solid evidence of a specific link.

V. CONCLUSIONS

As with all scientific domains, the future is promising for far more detailed studies and expanding the knowledge on NEAs. Technological advance will enable better dynamical simulations of the complex main belt to near-Earth region transfer routes. Moreover, the growing database of observational data will aid the quest to link specific NEAs to potential source regions in the main belt as well as to fine tune key models such as the Yarkovsky effect. Although a substantial amount of work has been done on the field with remarkable results, some of which we presented previously, there are many more cases worth studying closely. The benefits from establishing ever-more such links are valuable not only because of the practical part, of the protection of Earth from potential hazards, but the insight gained on these dynamical phenomena broaden our understanding of how our solar system works.

REFERENCES

[1] Gritschneder, M., Lin, D. N. C., Murray, S. D., Yin, Q. Z., and Gong, M. N., "The Supernova Triggered Formation and Enrichment of Our Solar System," *Astrophysical Journal*, Vol. 745, Jan. 2012, p. 22.

[2] Garaud, P., Meru, F., Galvagni, M., and Olczak, C., "From Dust to Planetesimals: An Improved Model for Collisional Growth in Protoplanetary Disks," *Astrophysical Journal*, Vol. 764, Feb. 2013, p. 146.

[3] Levison, H., "The Formation of the Cores of the Giant Planets," Sept. 2013. http://www.boulder.swri.edu/~hal/talks/planet-form/thesis.pdf.

[4] Carr, M. H., "Asteroids, Comets and Planet Formation," Technical Report, June 1984.

[5] Nelson, M. L., Britt, D. T., and Lebofsky, L. A., "Review of Asteroid Compositions," *Resources of Near-Earth Space*, edited by J. S. Lewis, and M. S. Matthews and M. L. Guerrieri, Provided by the SAO/NASA Astrophysics Data System, 1993, pp. 493–522. Available: http://adsabs.harvard.edu/abs/1993rnes.book..493N.

[6] Gradie, J., and Tedesco, E., "Compositional Structure of the Asteroid Belt," *Science*, Vol. 216, June 1982, pp. 1405–1407.

[7] DeMeo, F. E., and Carry, B., "Solar System Evolution from Compositional Mapping of the Asteroid Belt," *Nature*, Vol. 505, Jan. 2014, pp. 629–634.

[8] Morbidelli, A., *Modern Celestial Mechanics: Aspects of Solar System Dynamics*, Taylor & Francis, London, 2002.

[9] Robutel, P., and Souchay, J., "An Introduction to the Dynamics of Trojan Asteroids," *Lecture Notes in Physics*, edited by J. Souchay and R. Dvorak, Vol. 790, Springer-Verlag, Berlin, March 2010, pp. 195–227.

[10] Kokubo, E., "Kozai Mechanism ('Resonance')," *Over the Resonance, 35th Symposium on Celestial Mechanics*, edited by E. Kokubo, H. Arakida, and T. Yamamoto, Izu-Nagaoka-Onsen, Shizuoka, Japan, 2003, pp. 52–59.

[11] Rickman, H., "Cometary Dynamics," *Lecture Notes in Physics* edited by J. Souchay and R. Dvorak, Vol. 790, Springer-Verlag, Berlin, March 2010, pp. 341–399.

[12] Mouret, S., Hestroffer, D., and Mignard, F, "Asteroid Masses and Improvement with Gaia," *Astronomy and Astrophysics*, Vol. 472, Sept. 2007, pp. 1017–1027.

[13] Bottke, W. F. Jr., Vokrouhlický, D., Rubincam, D. P., and Nesvorný, D., "The Yarkovsky and Yorp Effects: Implications for Asteroid Dynamics," *Annual Review of Earth and Planetary Sciences*, Vol. 34, May 2006, pp. 157–191.

[14] NASA Near-Earth Object Program, Near-Earth object groups, http://neo.jpl.nasa.gov/neo/groups.html [last accessed 1 March 2014].

[15] Kirkwood, D., "The Asteroids, or Minor Planets Between Mars and Jupiter," *Bulletin Astronomique, Serie I*, Vol. 5, 1888, pp. 214–214.

[16] Morbidelli, A., Bottke, W. F. Jr., Froeschlé, C., and Michel, P., "Origin and Evolution of Near-Earth Objects," *Asteroids III*, 2002, pp. 409–422.

[17] Zappalà, V., Cellino, A., Gladman, B. J., Manley, S., and Migliorini, F., "NOTE: Asteroid Showers on Earth after Family Breakup Events," *Icarus*, Vol. 134, July 1998, pp. 176–179.

[18] Zappala, V., Cellino, A., Farinella, P., and Knezevic, Z., "Asteroid Families. I—Identification by Hierarchical Clustering and Reliability Assessment," *Astronomical Journal*, Vol. 100, Dec. 1990, pp. 2030–2046.

[19] Milani, A., and Knezevic, Z., "Secular Perturbation Theory and Computation of Asteroid Proper Elements," *Celestial Mechanics and Dynamical Astronomy*, Vol. 49, Dec. 1990, pp. 347–411.

[20] de León, J., Campins, H., Tsiganis, K., Morbidelli, A., and Licandro, J., "Origin of the Near-Earth Asteroid Phaethon and the Geminids Meteor Shower," *Astronomy and Astrophysics*, Vol. 513, April 2010, p. A26.

[21] Campins, H., Morbidelli, A., Tsiganis, K., de León, J., Licandro, J., and Lauretta, D., "The Origin of Asteroid 101955 (1999 RQ_{36})," *Astrophysical Journal Letters*, Vol. 721, Sept. 2010, pp. L53–L57.

[22] Bottke, W. F., Morbidelli, A., Jedicke, R., Petit, J.-M., Levison, H. F., Michel, P., and Metcalfe, T. S., "Debiased Orbital and Absolute Magnitude Distribution of the Near-Earth Objects," *Icarus*, Vol. 156, April 2002, pp. 399–433.

CHAPTER 7

Regular and Chaotic Motions in Dynamical Systems with Applications to Asteroid and Debris Dynamics

Alessandra Celletti* and Fabien Gachet*

University of Rome "Tor Vergata," Rome, Italy

I. INTRODUCTION

It is nowadays well established that there is a strong interaction between dynamical systems theory and astrodynamics. Concepts like periodic orbits, quasiperiodic motions, and chaos trajectories have become very common while designing space missions, most notably around the Lagrangian equilibrium points of the restricted three-body problem. This novel approach opened a new way to deal with the dynamics of natural and artificial bodies: small natural bodies are strongly influenced by different forces whose interaction might provoke an irregular behavior, whereas chaotic trajectories might be used to design low-cost interplanetary missions.

Among the bodies of a small size in the solar system, we will be mainly interested in near-Earth objects (NEOs) and space debris, which are current threats for human asset. In particular, we are interested in investigating the orbital and rotational dynamics of such objects. This task requires the knowledge of the possible dynamical behaviors, most notably the periodic motions (when the solution of the equations of motion comes back to the initial conditions after a time-called period), quasiperiodic motions (when the solution is not periodic, but it comes arbitrarily close to the initial conditions after regular intervals of time), and chaotic trajectories (which are characterized by an extreme sensitivity to slight changes of the initial conditions). We will also need to distinguish between conservative systems (for which the phase space volume is preserved) and dissipative systems (when the volume either contracts or expands).

To explain these concepts, we will use a paradigmatic model known as the *standard map;* we will present a conservative and a dissipative version of such mapping, which helps to introduce the main notions of dynamical systems. This model is a *discrete* system, which will turn out to be strictly related to a model describing the rotational dynamics, known as the *spin–orbit problem*. Indeed,

*Department of Mathematics.

astrodynamical models suitable to investigate the motion of NEOs are presented in Sec. III. There, we start by introducing the previously mentioned spin–orbit problem, which describes the rotational motion of a triaxially shaped body, rotating about a spin–axis and moving on a Keplerian orbit around a primary body. Then, we introduce the restricted three-body problem, which provides the motion of a small body under the gravitational influence of two larger primaries. In both cases, we also mention possible dissipative forces that might act on the systems, like the tidal torque in rotational dynamics, and Stokes drag or the Yarkowski effect in orbital dynamics.

The analysis of the dynamics of space debris requires a different approach, since it is necessary to have an accurate modeling of the geopotential, due to the fact that these objects orbit very close to the Earth. The effects of Sun and Moon are also very important, as well as the solar radiation pressure and the atmospheric drag, the latter one for debris orbiting at an altitude of less than 1500 km above the Earth's surface.

To investigate the dynamical character of such models, we review numerical and analytical tools. As for the numerical techniques, we shortly outline how to compute a Poincaré map (we provide a graphical description of the phase space associated to some model problems); we also give the formulae for the computation of the Lyapunov exponents and the fast Lyapunov indicators (FLIs), in order to discern between the regular and chaotic character of the dynamics. Finally, we introduce perturbation theory, which is a very effective analytical tool, allowing to obtain a good approximation of the solution in nearly integrable systems, like the spin–orbit model or the three-body problem. The numerical and analytical techniques presented in this chapter provide the basic tools to investigate the orbital and rotational dynamics of NEOs and space debris.

This chapter is organized as follows. In Sec. II, we introduce the main concepts of dynamical systems like regular and chaotic motions, using the conservative and dissipative standard maps to make these notions concrete. Some models for the description of rotational and orbital dynamics are introduced in Sec. III, which also includes the presentation of models for studying space debris. Numerical methods to detect regular and chaotic motions are provided in Sec. IV, whereas perturbation theory is reviewed in Sec. V.

II. PERIODIC, QUASIPERIODIC, AND CHAOTIC DYNAMICS

Conservative and dissipative systems are characterized by different behaviors, which can typically be classified as periodic, quasiperiodic, and chaotic. In this section, we provide the definition of regular (i.e., periodic and quasiperiodic) trajectories as well as chaotic motions. We will exploit both discrete and continuous systems. However, we will make the previous concepts concrete by using a paradigmatic discrete model, known as the *standard map*, either in the conservative or dissipative setting.

A. DYNAMICAL BEHAVIORS

To define the different dynamical behaviors, we introduce discrete and continuous systems as follows.

An n-dimensional *discrete* system is governed by equations of motion of the form

$$\mathbf{Z}' = \mathbf{f}(\mathbf{Z}), \quad \mathbf{Z} \in \mathbb{R}^n \tag{1}$$

where $\mathbf{f} : \mathbb{R}^n \to \mathbb{R}^n$ is a regular vector function and the prime denotes the iterate under the mapping $\mathbf{f}$. An n-dimensional *continuous* system is governed by equations of motion of the form

$$\dot{\mathbf{Z}} = \mathbf{F}(\mathbf{Z}), \quad \mathbf{Z} \in \mathbb{R}^n$$

where $\mathbf{F} : \mathbb{R}^n \to \mathbb{R}^n$ denotes the *vector field* and the dot represents the derivative with respect to time. Below, we give qualitative definitions of the different motions.

Definition 1: We refer to a *periodic motion* as a solution of the equations of motion, which retraces its own steps after a given interval of time, called *period*.

Definition 2: We refer to a *quasiperiodic motion* as a solution of the equations of motion, which comes indefinitely close to its initial conditions at regular intervals of time, though never exactly retracing itself.

Definition 3: We refer to periodic or quasiperiodic orbits as *regular* motions.

Definition 4: We refer to *chaotic dynamics* as an irregular motion showing an *extreme sensitivity to the choice of the initial conditions.*

We address the reader to Refs. [1,2] for the formal definitions.

Remark 1: It is important to stress that chaotic motions are unpredictable; in particular, the divergence of the orbits will be measured by chaos indicators, like the *Lyapunov exponents* or the *FLIs*. We will provide the definition of such indicators in Sec. IV.

B. STANDARD MAP

To make the previous concepts concrete, we start by analyzing a paradigmatic model known as the *standard map*, which enters the class of discrete systems described by equations of type (1). Despite its simplicity, the standard map can exhibit all kinds of dynamical behaviors, from regular to chaotic motions. We will present a conservative version of the mapping (see Sec. 1) as well as a dissipative version (see Sec. 2).

1. CONSERVATIVE STANDARD MAP

We consider the discrete map defined by the following equations:

$$
\begin{aligned}
y' &= y + \varepsilon\, g(x), \quad y \in \mathbb{R},\ x \in \mathbb{T} \\
x' &= x + y'
\end{aligned}
\tag{2}
$$

with $\varepsilon > 0$ the *perturbing parameter* and with $g = g(x)$ an analytic function. The most common choice is to take $g(x) = \sin(x)$; this mapping was introduced in Ref. [3] and we will see that it has a strong relation with astrodynamics, being the discrete analog of a model of rotational dynamics known as the *spin–orbit problem* (see Sec. A). This map is *conservative* as it is shown next, which is why we refer to it as the *conservative standard map.*

Using an equivalent notation for discrete systems, we can write Eq. (2) as

$$
\begin{aligned}
y_{j+1} &= y_j + \varepsilon g(x_j) \\
x_{j+1} &= x_j + y_{j+1} = x_j + y_j + \varepsilon g(x_j) \quad \text{for } j \geqslant 0
\end{aligned}
\tag{3}
$$

We now list some important properties of the conservative standard map.

1. The map (3) is *nonintegrable* for $\varepsilon \neq 0$, whereas it is *integrable* for $\varepsilon = 0$. In the latter case we have

$$
\begin{aligned}
y_{j+1} &= y_j = y_0 \\
x_{j+1} &= x_j + y_{j+1} = x_j + y_j = x_0 + j y_0 \quad \text{for } j \geqslant 0
\end{aligned}
$$

   namely y_j is constant and x_j increases at each iteration by a constant quantity y_0.

2. In the integrable case, if $y_0 = 2\pi(p/q)$ with p, q integers and with $q \neq 0$, then on the line $y = y_0$ we obtain that the iterates of the angle are given by

$$
x_1 = x_0 + 2\pi \frac{p}{q}, \quad x_2 = x_0 + 4\pi \frac{p}{q}, \ldots, x_q = x_0 + 2\pi p
$$

   The last iterate shows that $x_q = x_0$ and therefore the orbit is *periodic* with period $2\pi q$; in particular, we see that the interval $[0, 2\pi)$ is spanned p times.

3. In the integrable case, if y_0 is equal to 2π times an irrational number, then on the line $y = y_0$ the iterates of x_0 densely fill the line $y = y_0$, which indicates a quasiperiodic motion: the iterates never come back to the initial condition, but as close as one wishes after a sufficient number of iterations.

4. The *frequency* of the mapping is defined as the quantity

$$
\omega = \lim_{j \to \infty} \frac{x_j - x_0}{j}
\tag{4}
$$

In the integrable case, one easily finds that $\omega = y_0$; thus, we can rephrase the previous remarks by saying that the motion is periodic if $\omega/2\pi$ is a rational number, whereas it is quasiperiodic if $\omega/2\pi$ is irrational.

5. The mapping (3) is *conservative*, since the determinant of the corresponding Jacobian is equal to one; in fact, setting $g_x(x_j) \equiv (\partial g(x_j)/\partial x)$, one has

$$\det \begin{pmatrix} 1 & \varepsilon g_x(x_j) \\ 1 & 1 + \varepsilon g_x(x_j) \end{pmatrix} = 1$$

6. The *fixed points* of Eq. (3) are obtained by solving the equations

$$y_{j+1} = y_j$$
$$x_{j+1} = x_j$$

We can write the first equation as $y_{j+1} = y_j + \varepsilon g(x_j)$, from which we get that it must be $g(x_j) = 0$; from the second equation, we obtain $x_j = x_j + y_{j+1}$, which gives $y_{j+1} = 0 = y_0$. If we specify that the function g is set as $g(x) = \sin x$, then the fixed points are $(y_0, x_0) = (0, 0)$ and $(y_0, x_0) = (0, \pi)$.

7. We can investigate the *linear stability* of the fixed points that we found at the previous item by computing the first variation $(\delta y_{j+1}, \delta x_{j+1})^T$, which is provided by the equations

$$\begin{pmatrix} \delta y_{j+1} \\ \delta x_{j+1} \end{pmatrix} = \begin{pmatrix} 1 & \varepsilon g_x(x_j) \\ 1 & 1 + \varepsilon g_x(x_j) \end{pmatrix} \begin{pmatrix} \delta y_j \\ \delta x_j \end{pmatrix}$$

For $g(x) = \sin(x)$ the eigenvalues of the linearized system are determined by solving the characteristic equation

$$\lambda^2 - (2 \pm \varepsilon)\lambda + 1 = 0$$

where the positive sign holds for $(0, 0)$, whereas the negative sign holds for $(0, \pi)$. Therefore, we obtain that the eigenvalue associated to $(0, 0)$ is greater than one, so that we conclude that the fixed point is unstable. For $\varepsilon < 4$ the eigenvalues associated to $(0, \pi)$ are complex conjugated with a real part less than one; therefore, $(0, \pi)$ is a stable fixed point (compare with Refs. [4, 5]).

In Fig. 1, we provide the graph of the conservative standard map with $g(x) = \sin x$ for different values of ε. We take 20 different initial conditions and we let them evolve, drawing the corresponding plot in the (x, y) plane. We start from $\varepsilon = 0$, namely the integrable case, where only periodic and quasiperiodic rotational motions (extending around the torus) are possible. When we switch on the

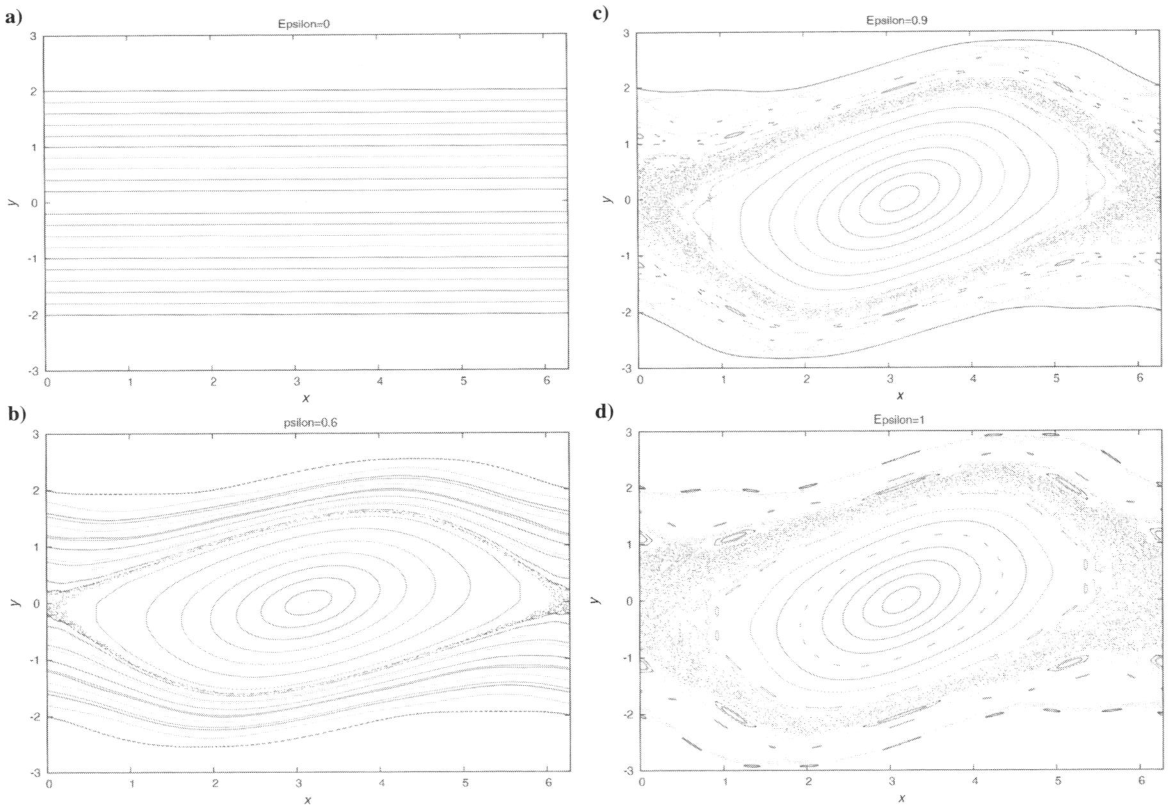

Fig. 1 Graphs of the conservative standard map with $g(x) = \sin x$ for 20 different initial conditions and for different values of ε. a) $\varepsilon = 0$: the system is integrable; there is a stable equilibrium point at $(0, \pi)$ and an unstable point at $(0, 0)$, b) $\varepsilon = 0.6$: the perturbation is switched on and chaotic regions appear, c) $\varepsilon = 0.9$: for a large perturbation, more chaos appears, only a few quasiperiodic curves remain, and larger islands are found around higher-order periodic orbits, and d) $\varepsilon = 1$: for very large perturbations, no more quasiperiodic curves are present and a large chaotic region appears.

perturbation, librational islands appear around the stable point at $(0, \pi)$, whereas the quasiperiodic curves are deformed and displaced. The amplitude of such librational islands increases as we take ε larger; at the same time, the number of quasiperiodic curves decreases, since they break down as the perturbing parameter increases. Around the separatrix chaotic motions appear (see Ref. [2]). For larger values of ε there is a birth of higher-order periodic orbits, surrounded by librational curves as well as by the corresponding chaotic separatrix. Above a given threshold, no more quasiperiodic rotational curves exist. Such a threshold has been numerically evaluated in Ref. [6] and it amounts to $\varepsilon = 0.971635$.

Remark 2: The existence of quasiperiodic tori is the content of the celebrated Kolmogorov–Arnold–Moser theory, known as KAM theory (Refs. [7–9]). Under suitable nondegeneracy conditions, provided that the frequency of motion satisfies a Diophantine inequality (to guarantee that it is sufficiently far from rationals), for ε, sufficiently small KAM theory provides the persistence of an invariant curve that is run by a quasiperiodic motion. We refer to Refs. [7–9] for further details.

2. DISSIPATIVE STANDARD MAP

We now consider a variation of Eq. (2) to encompass the dissipative context. Precisely, we multiply the first term of the first equation by a factor $0 < \lambda < 1$, to which we refer as the *dissipative constant*, and in the first equation, we add a constant μ, to which we refer as the *drift term*. The new mapping is known as the *dissipative standard map* and it is described by the following equations:

$$\begin{aligned} y' &= \lambda y + \mu + \varepsilon g(x) \quad y \in \mathbb{R},\ x \in \mathbb{T} \\ x' &= x + y' \quad \lambda, \mu, \varepsilon \in \mathbb{R},\ \varepsilon \geqslant 0 \end{aligned} \tag{5}$$

Also this map has a connection with astrodynamics, since it is related to the discretization of the spin–orbit problem with tidal friction. Again, we can use an equivalent notation and write Eq. (5) as

$$\begin{aligned} y_{j+1} &= \lambda y_j + \mu + \varepsilon g(x_j) \\ x_{j+1} &= x_j + y_{j+1} \quad \text{for } j \geqslant 0 \end{aligned} \tag{6}$$

The main properties of the dissipative standard map are the following:

1. When $\lambda = 1$ and $\mu = 0$, the mapping (6) reduces to the conservative standard map.
2. When $\lambda \neq 1$, the map is dissipative, since the determinant of the Jacobian is equal to λ.
3. The drift μ plays a crucial role in the definition of the dynamics. In fact, consider $\varepsilon = 0$ and look for an invariant solution, such that $y' = y$; this condition

implies that $\lambda y + \mu = y$, which yields

$$y = \frac{\mu}{1 - \lambda}$$

Therefore, there is a close relation between the frequency of motion [which is equal to y in the standard map thanks to Eq. (4)] and the drift μ; in particular, if $\mu = 0$, then we have the solution $y = 0$.

The dynamics of dissipative systems is very different from that of conservative systems. In fact, in dissipative systems, we have a set of attractors, which might be periodic or quasiperiodic, whereas chaotic motions are replaced by the so-called *strange attractors*. In Fig. 2, we show some graphs concerning the dissipative standard map with $g(x) = \sin x$. As in Fig. 1, we take different values of ε from 0 to 1, and we select 20 different initial conditions. All such conditions end up on the attractor(s) after a sufficient number of iterations (see Ref. [10]). The graphs in Fig. 2 are obtained plotting the evolution after such transient number of iterations. The dynamics shows a quasiperiodic attractor for low values of the perturbing parameter; increasing ε we get the coexistence of such a quasiperiodic attractor with a periodic orbit. As far as ε gets sufficiently large, the quasiperiodic orbit breaks down and it leaves place to another periodic orbit.

Remark 3: The persistence of quasiperiodic attractors has been established in Ref. [11] using a KAM approach; numerical methods to evaluate the break-down threshold of quasiperiodic attractors have been presented in Ref. [12].

C. CONTINUOUS SYSTEMS AND HAMILTONIAN FORMALISM

Many conservative models of celestial mechanics are described by *nearly integrable* Hamiltonian systems. The typical example is the three-body problem, which can be thought of as a two-body problem, which is integrable, perturbed by the gravitational action of the third body.

In Sec. III, we will introduce the rotational and orbital dynamics of celestial bodies in continuous systems modeling. A convenient approach in the conservative case is to use the Hamiltonian formalism. In particular, a nearly integrable system can be suitably described by action-angle variables (see Ref. [13]) and we can introduce a Hamiltonian of the form

$$\mathcal{H}(y, x) = h(y) + \varepsilon f(y, x) \tag{7}$$

where we denote the number of degrees of freedom by n, $y \in \mathbb{R}^n$ are the actions, $x \in \mathbb{T}^n$ are the angles, and $\varepsilon > 0$ is a small positive parameter. The function $h = h(y)$ represents the integrable part (e.g., it describes the two-body problem modeling the motion of an asteroid around the Sun), the function $f = f(y, x)$ is

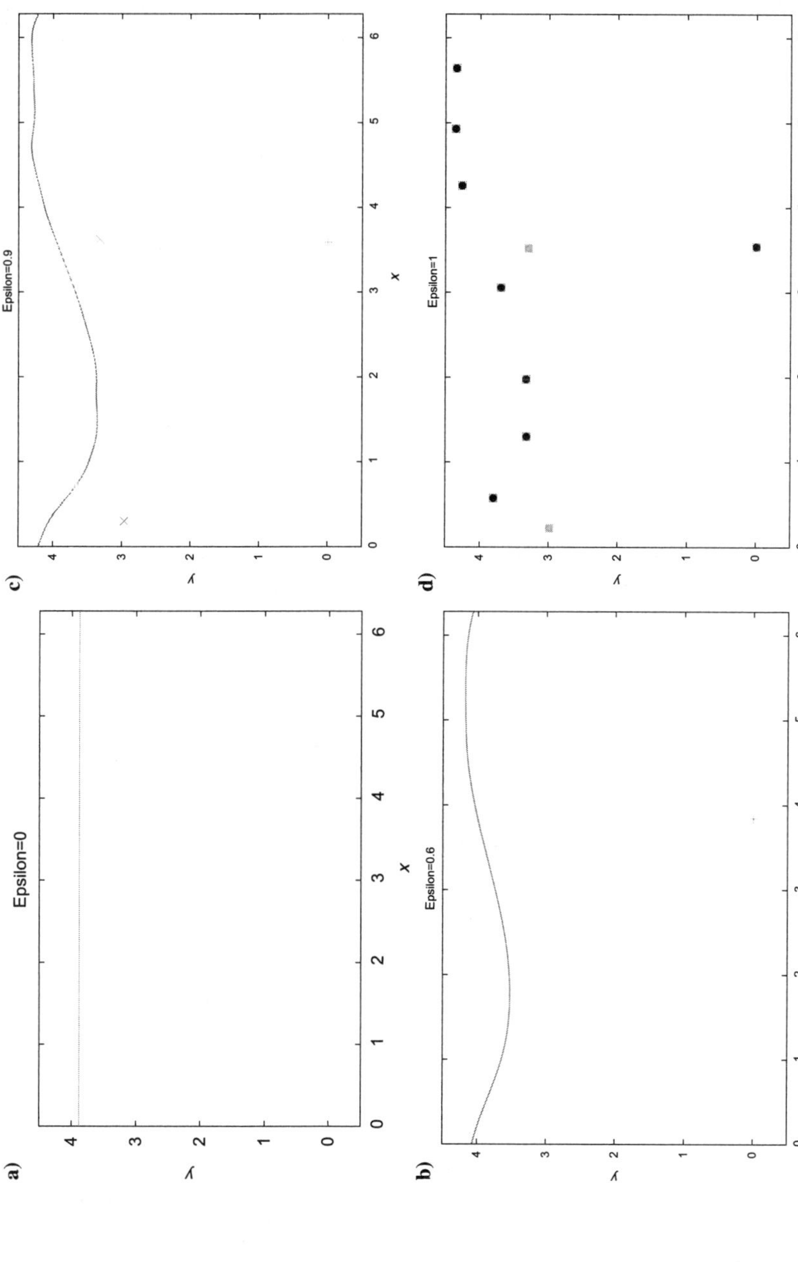

Fig. 2 **Graphs of the dissipative standard map with $g(x) = \sin x$ for 20 different initial conditions and for different values of ε. a) $\varepsilon = 0$: there exists just a quasiperiodic attractor, b) $\varepsilon = 0.6$: there exists a quasiperiodic attractor and a periodic orbit, c) $\varepsilon = 0.9$: there are two periodic orbits coexisting with a quasiperiodic attractor, and d) $\varepsilon = 1$: the quasiperiodic attractor breaks down and leaves place to a periodic orbit.**

called the *perturbing function* (e.g., it represents the gravitational attraction of Jupiter on the asteroid), and the perturbing parameter ε provides the strength of the perturbation (in this example it is equal to the Jupiter–Sun mass ratio).

In the integrable approximation $\varepsilon = 0$, Hamilton's equations associated to Eq. (7) are solved as

$$\dot{y} = -\frac{\partial h(y)}{\partial x} = 0 \Rightarrow y(t) = y(0) = \text{const.}$$

$$\dot{x} = \frac{\partial h(y)}{\partial y} \equiv \omega(y) \Rightarrow x(t) = \omega\left(y(0)\right) t + x(0)$$

where $(y(0), x(0))$ denote the initial conditions. If $\varepsilon \neq 0$, then the system is nonintegrable and Hamilton's equations read as

$$\dot{y} = -\varepsilon \frac{\partial f(y,x)}{\partial x}$$

$$\dot{x} = \omega(y) + \varepsilon \frac{\partial f(y,x)}{\partial y}$$

Many problems of celestial mechanics and astrodynamics are affected by dissipations. For example, to describe the orbital motion of an asteroid or space debris, one should consider dissipative contributions like the Yarkowski effect or the atmospheric drag; in rotational dynamics, it is essential to evaluate the importance of tidal torques or YORP effect. Many of such problems are described by *nearly integrable dissipative* systems; an example is given by the following equations of motion:

$$\dot{y} = -\varepsilon \frac{\partial f(y,x)}{\partial x} - \lambda(y - \mu)$$

$$\dot{x} = \omega(y) + \varepsilon \frac{\partial f(y,x)}{\partial y}$$

where we added the contribution of a linear dissipation with dissipative constant $\lambda > 0$ and drift term μ.

III. ROTATIONAL AND ORBITAL DYNAMICS

The dynamics of a celestial body is determined by the orbital motion around a primary and by the rotational motion of the body about itself. In this section, we describe the rotational and orbital motions, providing some simple models that allow us to get a good description of the dynamics. Precisely, we present the spin–orbit problem describing the motion of an oblate body (e.g., an asteroid) around an internal spin–axis. We start by introducing a conservative model and then we consider a dissipative model, where the tidal torque—due to the nonrigidity of the body—is taken into account. We will show that the spin–orbit problem, both in its conservative and dissipative formulation, has a close link with the conservative and dissipative standard maps introduced in Secs. 1 and 2.

To describe the orbital motion, we will introduce a model providing the dynamics of a body (e.g., an asteroid) under the gravitational attraction of two primaries (e.g., Sun and Jupiter). This model goes under the name of *a restricted three-body problem*. We will also consider some dissipative forces, which might enter the equations of motion.

We also provide a model to describe the dynamics of space debris; in this case, we need to consider the attraction of the Earth, its shape, the lunar attraction, the gravitational influence of the Earth, and solar radiation pressure. For debris close to the Earth, we also need to introduce the atmospheric drag, which represents a dissipative effect.

A. ROTATIONAL DYNAMICS

We consider the rotational motion of a celestial body around an internal spin–axis; we will introduce a simplified model, the spin–orbit problem, where we assume that the center of mass of the body moves on a Keplerian orbit around a primary. The rotational motion can be affected by important dissipative contributions, for example

1. A nonrigid body can be subject to a *tidal torque* due to its internal nonrigidity.
2. The joint action of solar lighting and rotation of the body causes the reemission of the absorbed solar radiation that occurs along a direction different from that of the Sun, thus provoking a variation of the angular momentum and therefore of the orbit. The corresponding effects on the orbital and rotational motions are called the *Yarkowski* and *YORP* effects.
3. Artificial satellites are affected by several dissipative effects like an internal sloshing, mass consumption, the flexibility of the structure, and even crew's motion.

B. SPIN–ORBIT PROBLEM

The spin–orbit model describes the rotational dynamics of a triaxial body S (e.g., an asteroid) with principal axes of inertia $A < B < C$, moving on a *Keplerian* orbit around a central body P (e.g., the Sun). We assume that the spin–axis is *perpendicular* to the orbit plane and that it coincides with the *shortest physical axis*.

At first, we neglect the dissipative forces; denoting the rotation angle by x formed by the longest axis of the ellipsoid and the periapsis line (see Fig. 3), the

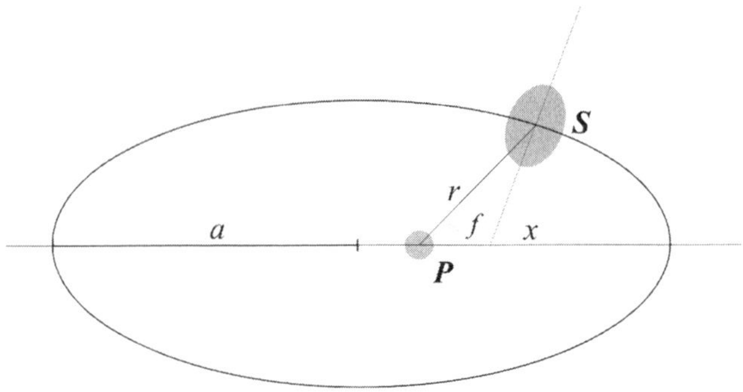

Fig. 3 Geometry of the spin–orbit problem describing the motion of a body *S* around a primary *P*: *a* is the semimajor axis, *r* is the orbital radius, *f* is the true anomaly, and *x* is the rotation angle.

equations of motion are the following:

$$\ddot{x} + \varepsilon \left(\frac{a}{r}\right)^3 \sin(2x - 2f) = 0, \qquad \varepsilon = \frac{3}{2}\frac{B - A}{C} \tag{8}$$

where a is the semimajor axis, whereas the position of S on its orbit is determined by the orbital radius r and the true anomaly f. Equation (8) is associated to the one-dimensional, time-dependent Hamiltonian

$$\mathcal{H}(y, x, t) = \frac{y^2}{2} - \frac{\varepsilon}{2}\left(\frac{a}{r(t)}\right)^3 \cos(2x - 2f(t))$$

We now introduce the definition of *spin–orbit resonance* of order $p:q$ with p, q nonzero integers, whenever the ratio of the period of revolution $T_{\rm rev}$ and the period of rotation $T_{\rm rot}$ is equal to p/q

$$\frac{T_{\rm rev}}{T_{\rm rot}} = \frac{p}{q}$$

Denoting the corresponding frequencies by $\omega_{\rm rev}$, $\omega_{\rm rot}$ [i.e., $T_{\rm rev} = (2\pi/\omega_{\rm rev})$, $T_{\rm rot} = (2\pi/\omega_{\rm rot})$], we have

$$\frac{\omega_{\rm rot}}{\omega_{\rm rev}} = \frac{p}{q}$$

In terms of the variation of the rotation angle x and of the mean anomaly[†] ℓ, we have

$$\frac{\dot{x}}{\dot{\ell}} = \frac{p}{q}$$

[†]The mean anomaly ℓ is the angle that S would span if the ellipse was run with uniform velocity, namely $\dot{\ell} = \omega_{\rm rev}$; see, for example, Ref. [4].

Remark 4:

1. The Moon and all evolved satellites of the solar system always point the same face to the host planet; this means that they are in a 1:1 (or *synchronous*) resonance, which corresponds to the fact that they experience one rotation per one revolution around their primary.
2. The only nonsynchronous spin–orbit resonance is provided by Mercury, which is found in a 3:2 spin–orbit resonance: it makes three rotations around its spin–axis per two revolutions around the Sun.
3. Asteroids and NEOs are not necessarily in a spin–orbit resonance.

As noted before, some dissipative effects might affect the rotational motion; among the dissipative contributions concerning natural bodies like asteroids, the most important one is the tidal torque, provoked by the nonrigidity of the body. When including the tidal torque, the equations of motion describing the spin–orbit problem must be modified as follows (see Ref. [4]):

$$\ddot{x} + \varepsilon \left(\frac{a}{r}\right)^3 \sin(2x - 2f) = -K\,\lambda(e)(\dot{x} - \mu(e)) \tag{9}$$

where K is the *dissipative constant*, depending on the physical properties of the body and λ, μ are functions of the eccentricity of the orbit

$$\lambda(e) \equiv \frac{1}{(1-e^2)^{9/2}}\left(1 + 3e^2 + \frac{3}{8}e^4\right)$$
$$\mu(e) \equiv \frac{1}{(1-e^2)^6\lambda(e)}\left(1 + \frac{15}{2}e^2 + \frac{45}{8}e^4 + \frac{5}{16}e^6\right)$$

1. RELATION BETWEEN THE STANDARD MAP AND THE SPIN–ORBIT MODEL

The link between the conservative standard map and the spin–orbit model that was mentioned in Sec. B is explained here.

Let us write the conservative spin–orbit Eq. (8) as

$$\ddot{x} = \varepsilon g(x,t), \quad g(x,t) = -\left(\frac{a}{r(t)}\right)^3 \sin(2x - 2f(t)) \tag{10}$$

We can write Eq. (10) as the first-order differential system:

$$\dot{y} = \varepsilon g(x,t)$$
$$\dot{x} = y$$

We proceed to integrate the previous equations with a symplectic first-order Euler's method with step-size h; we obtain the discrete system

$$\begin{aligned} y_{j+1} &= y_j + \varepsilon g(x_j, t_j)h \\ x_{j+1} &= x_j + y_{j+1}h \end{aligned} \tag{11}$$

where the time evolves as $t_{j+1} = t_j + h$. It is readily recognized that Eq. (11) is equivalent to the conservative standard map.

The link between the dissipative standard map and spin–orbit model is obtained as follows. Let us write Eq. (9) in the form

$$\ddot{x} = \varepsilon g(x,t) - K\lambda(e)(\dot{x} - \mu(e)), \quad g(x,t) = -\left(\frac{a}{r(t)}\right)^3 \sin(2x - 2f(t))$$

which can be written as the first-order differential system:

$$\begin{aligned} \dot{y} &= \varepsilon g(x,t) - K\lambda(e)(y - \mu(e)) \\ \dot{x} &= y \end{aligned} \tag{12}$$

Integrating Eq. (12) with a symplectic first-order Euler's method with step size h, we have

$$\begin{aligned} y_{j+1} &= y_j + \varepsilon g(x_j,t)h - K\lambda(y_j - \mu) = (1 - K\lambda)y_j + K\lambda\mu + \varepsilon g(x_j,t)h \\ x_{j+1} &= x_j + y_{j+1}h \end{aligned}$$

again with $t_{j+1} = t_j + h$. The previous system is equivalent to the dissipative standard map.

C. RESTRICTED THREE-BODY PROBLEM

To describe the orbital motion of celestial objects of a small size when compared to the host body, like the asteroids or artificial satellites, we introduce the *restricted three-body problem*. Consider the motion of a small body S with negligible mass, subject to the gravitational influence of two primaries, say P_1, P_2 with masses m_1, m_2; due to the fact that the mass of the small body is negligible with respect to that of the primaries, we assume that the larger bodies move on Keplerian orbits about their common barycenter; in this case we speak of a *restricted* problem. Furthermore, we assume that the orbits of the primaries are circular and that all bodies move on the same plane; in this case we speak of the *planar, circular,* and *restricted* three-body problem. For the moment, we do not consider the action of dissipative forces.

Adopting suitable normalized units and action-angle Delaunay variables $(L, G) \in \mathbb{R}^2$, $(\ell, g) \in \mathbb{T}^2$ (see Ref. [4]), we obtain the following 2 degrees of freedom Hamiltonian function:

$$\mathcal{H}(L, G, \ell, g) = -\frac{1}{2L^2} - G + \varepsilon R(L, G, \ell, g)$$

with the following meaning of the quantities involved:

1. ε represents the primaries' mass ratio m_2/m_1
2. For $\varepsilon = 0$ the Hamiltonian describes the Keplerian motion of the small body around the bigger primary, say P_1
3. (L, G) are the action variables, which are related to the elliptic elements by $L = \sqrt{a}$, $G = L\sqrt{1-e^2}$ with a the semimajor axis and e the eccentricity
4. (ℓ, g) are the angle variables, where ℓ is the mean anomaly and $g = \tilde{\omega} - t$ with $\tilde{\omega}$ the argument of perihelion
5. The function $R = R(L, G, \ell, g)$ represents the gravitational influence of P_2 on S

An equivalent formulation can be given using Cartesian variables. Precisely, let (O, ξ, η) be an inertial reference frame in the plane of motion. Let (ξ, η) be the coordinates of S and let (ξ_1, η_1), (ξ_2, η_2) be the coordinates of P_1, P_2. Assume that the units of measure are such that $m_1 + m_2 = 1$ and let us introduce the quantity m such that $m_1 = 1 - m$ and $m_2 = m$. Then, the equations of motion can be written as

$$\ddot{\xi} = m\frac{\xi_1 - \xi}{r_1^3} + (1-m)\frac{\xi_2 - \xi}{r_2^3}$$
$$\ddot{\eta} = m\frac{\eta_1 - \eta}{r_1^3} + (1-m)\frac{\eta_2 - \eta}{r_2^3}$$

where $r_1 = \sqrt{(\xi_1 - \xi)^2 + (\eta_1 - \eta)^2}$ and $r_2 = \sqrt{(\xi_2 - \xi)^2 + (\eta_2 - \eta)^2}$ are the distances of S from the primaries.

Let us now introduce a *synodic* reference frame (O, x, y), rotating with the angular velocity of the primaries. Let (x, y) be the coordinates of S, which are related to those of the inertial frame by the expressions

$$\xi = x\cos t - y\sin t$$
$$\eta = x\sin t + y\cos t$$

yielding the new distances from the primaries

$$r_1 = \sqrt{(x+m)^2 + y^2}, \quad r_2 = \sqrt{(x-1+m)^2 + y^2}$$

In this frame, the primaries are located, respectively, at $(-m, 0)$ for P_1 and at $(1-m, 0)$ for P_2. The equations of motion of the small body in the synodic frame are given by

$$\begin{aligned} \ddot{x} &= 2\dot{y} + x - (1-m)\,\frac{x+m}{r_1^3} - m\,\frac{x-1+m}{r_2^3} \\ \ddot{y} &= -2\dot{x} + y - (1-m)\,\frac{y}{r_1^3} - m\,\frac{y}{r_2^3} \end{aligned} \tag{13}$$

1. DISSIPATIVE FORCES

In this section, we modify Eqs. (13) in order to introduce a dissipative contribution. Precisely, we consider the following dissipative forces:

1. The *Stokes drag*, which models the collision of particles with the molecules of the gas nebula during the formation of the planetary system.
2. The *Poynting–Robertson effect*, which is due to the absorption and reemission of the solar radiation on small particles, thus provoking a decrease of the velocity.

The equations of motion in the synodic frame, taking into account the dissipative effects, are the following:

$$\ddot{x} = 2\dot{y} + x - (1-m)\frac{x+m}{r_1^3} - m\frac{x-1+m}{r_2^3} + F_x$$

$$\ddot{y} = -2\dot{x} + y - (1-m)\frac{y}{r_1^3} - m\frac{y}{r_2^3} + F_y$$

where (F_x, F_y) denote the components of the dissipative force. For Stokes drag and Poynting–Robertson effect we have the following expressions:

$$(F_x, F_y) = -K(\dot{x} - y + \alpha\Omega\, y, \dot{y} + x - \alpha\Omega\, x) \quad \text{(Stokes)}$$

$$(F_x, F_y) = -\frac{K}{r_1{}^2}(\dot{x} - y, \dot{y} + x) \quad \text{(Poynting–Robertson)}$$

where K denotes the dissipative constant, $\Omega = \Omega(r) \equiv r^{-3/2}$ is the Keplerian angular velocity at distance $r = \sqrt{x^2 + y^2}$, and $\alpha \in [0, 1)$ is the ratio between the gas velocity and the Keplerian velocity.

D. SPACE DEBRIS DYNAMICS

A description of the dynamics of space debris must take into account several factors, whose importance varies according to the altitude of the object from the Earth's surface. At all altitudes, space debris experience the influence of the Earth's gravity field as well as Earth's oblateness, which adds zonal and spherical harmonics to the gravitational potential, whose effect decreases as the altitude grows; moreover, one has to take into account the solar radiation pressure, the gravity of the Moon and the Sun, which takes more importance as the altitude grows, and the atmospheric drag, which is important only for low Earth orbits (say next to 1500 km from Earth's surface).

Remark 5: Global navigation satellite systems such as the GPS, GLONASS, Galileo, or Beidou, orbit at around 26,560 km from Earth's center with a period of 12^h, locked in a 2:1 resonance. The geosynchronous orbit is at

42,164 km from Earth's center; satellites take 24^h to do one revolution and they are locked in a 1:1 resonance with the rotation of the Earth.

1. EQUATIONS OF MOTION FOR SPACE DEBRIS

In a quasi-inertial reference frame with origin at the center of the Earth, the equations of motion are provided by the sum of the contributions of the Earth's gravitational influence, including the oblateness effect, the lunar and solar attraction, as well as the sum of the contributions of the non-gravitational forces (namely, solar radiation pressure and atmospheric drag) to which we refer as $\mathbf{a}_{\rm ng}$; denoting the position vector of the space debris by $\mathbf{r}$, the equations of motion are given by

$$\ddot{\mathbf{r}} = -G \int_{V_E} \rho(\mathbf{r}_p) \frac{\mathbf{r}-\mathbf{r}_p}{|\mathbf{r}-\mathbf{r}_p|^3} \, dV_E - G m_S \left(\frac{\mathbf{r}-\mathbf{r}_S}{|\mathbf{r}-\mathbf{r}_S|^3} + \frac{\mathbf{r}_S}{|\mathbf{r}_S|^3} \right) - G m_M \left(\frac{\mathbf{r}-\mathbf{r}_M}{|\mathbf{r}-\mathbf{r}_M|^3} + \frac{\mathbf{r}_M}{|\mathbf{r}_M|^3} \right) + \mathbf{a}_{\rm ng}$$

where G is the gravitational constant, $\rho(\mathbf{r}_p)$ is the density at some point $\mathbf{r}_p$ inside the volume V_E of the Earth, m_S, m_M are the masses of the Sun and the Moon, respectively, and $\mathbf{r}_S$, $\mathbf{r}_M$ are the distance vectors of the Sun and the Moon with respect to the Earth's center. In Cartesian coordinates the components are

$$\ddot{x} = V_{{\rm geo},x}(x,y,z) - G m_S \left(\frac{x-x_S}{|\mathbf{r}-\mathbf{r}_S|^3} + \frac{x_S}{r_S^3} \right) - G m_M \left(\frac{x-x_M}{|\mathbf{r}-\mathbf{r}_M|^3} + \frac{x_M}{r_M^3} \right) + a_{1ng}$$

$$\ddot{y} = V_{{\rm geo},y}(x,y,z) - G m_S \left(\frac{y-y_S}{|\mathbf{r}-\mathbf{r}_S|^3} + \frac{y_S}{r_S^3} \right) - G m_M \left(\frac{y-y_M}{|\mathbf{r}-\mathbf{r}_M|^3} + \frac{y_M}{r_M^3} \right) + a_{2ng}$$

$$\ddot{z} = V_{{\rm geo},z}(x,y,z) - G m_S \left(\frac{z-z_S}{|\mathbf{r}-\mathbf{r}_S|^3} + \frac{z_S}{r_S^3} \right) - G m_M \left(\frac{z-z_M}{|\mathbf{r}-\mathbf{r}_M|^3} + \frac{z_M}{r_M^3} \right) + a_{3ng}$$

where M_E is the mass of the Earth and $V_{\rm geo}$ is the Earth's potential, which is written in terms of spherical coordinates (r, ϕ, λ) as

$$V_{\rm geo}(r,\phi,\lambda) = \frac{GM_E}{r} \sum_{i=0}^{\infty} \left(\frac{R_E}{r} \right)^i \sum_{k=0}^{i} P_i^k(\sin\phi) \, (C_{ik}\cos k\lambda + S_{ik}\sin k\lambda)$$

In the previous expression, R_E denotes the radius of the Earth, and the quantities P_i^k are defined in terms of the Legendre polynomials

$$P_n(x) \equiv \frac{1}{2^n n!} \frac{d^n}{dx^n}\{(x^2-1)^n\}$$

$$P_n^m(x) \equiv (1-x^2)^{m/2} \frac{d^m}{dx^m}\{P_n(x)\}$$

whereas C_{ik} and S_{ik} are defined as

$$C_{nm} \equiv \frac{2-\delta_{0m}}{M_E} \frac{(n-m)!}{(n+m)!} \int_{V_E} (r_p R_E)^n \, P_n^m(\sin\phi_p) \cos(m\lambda_p) \rho(\mathbf{r}_p) \, dV_E$$

$$S_{nm} \equiv \frac{2-\delta_{0m}}{M_E} \frac{(n-m)!}{(n+m)!} \int_{V_E} (r_p R_E)^n \, P_n^m(\cos\phi_p) \cos(m\lambda_p) \rho(\mathbf{r}_p) \, dV_E$$

with (r_p, λ_p, ϕ_p) being the spherical coordinates of a point P inside the Earth and δ_{0m} is the Kronecker symbol.

We define $J_2 = -C_{20}$ and $J_{22} = \sqrt{C_{22}^2 + S_{22}^2}$. These terms play the biggest role in the potential, and for the first approach they can be used to approximate Earth's oblateness.

Concerning the nongravitational effects included in $\mathbf{a}_{\text{ng}}$, we can express them as follows. The solar radiation pressure is given by

$$\mathbf{F}_{\text{srp}} = C_r \, P_r \, a_S^2 \left(\frac{A}{m}\right) \frac{\mathbf{r} - \mathbf{r}_S}{|\mathbf{r} - \mathbf{r}_S|^3}$$

where C_r is the reflectivity coefficient, depending on the optical properties of the space debris surface, P_r is the radiation pressure for an object located at $a_S = 1$ AU, A/m is the area-to-mass ratio with A being the cross section of the space debris, and $\mathbf{r}_S$ is (as previously) the geocentric position of the Sun.

The acceleration of the satellite due to the atmospheric drag can be modeled as

$$\mathbf{a}_d = -\frac{C_D}{2} \rho(\mathbf{r}) \, \frac{A}{m} \, V^2 \, \mathbf{e}_v$$

where V is the velocity of the debris relative to the atmosphere, $\mathbf{e}_v$ is the unit vector of the debris velocity relative to the atmosphere, and C_D is the drag coefficient that can be assumed within $2 \leqslant C_D \leqslant 2.5$, where $C_D = 2$ holds for spherical satellites. The density ρ varies with the altitude above the Earth's surface; setting $h = |\mathbf{r}|$, the density is given by the barometric formula

$$\rho(h) = \rho_0 \, e^{-(h-h_0)/H_0}$$

where ρ_0 represents the density at a reference altitude h_0, whereas H_0 is the scaling height at h_0 (see Ref. [14]).

IV. DYNAMICAL NUMERICAL METHODS

Even simple dynamical systems can exhibit a very complex behavior as seen in the previous section (compare with Sec. B); therefore, the implementation of numerical methods to study the dynamics can prove to be very useful. In this context, we show some tools that provide an evidence of periodic and quasiperiodic orbits, as well as chaos trajectories. The most common numerical methods consist of the computation of the so-called *Poincaré maps* (used to provide a graphical description of the dynamics) and of the *Lyapunov exponents* (which give a measure of the divergence of the orbits). A variant of the latter method consists of the computation of the *FLIs*, which yield an efficient information on the chaoticity of a given system.

A. POINCARÉ MAPS

The Poincaré map allows to reduce the study of a continuous system to that of a discrete mapping. This method consists of computing the trajectory in the phase space of a dynamical system and of drawing its intersections with a specific surface of section. The most interesting properties of the trajectory (including periodicity, stability, and ergodicity) are then available.

Let us consider an n-dimensional continuous system (see Sec. A) described by the equations

$$\dot{\underline{z}} = \underline{f}(\underline{z}), \quad \underline{z} \in \mathbb{R}^n$$

where $\underline{f} = \underline{f}(\underline{z})$ is a generic regular vector field. Let $\underline{\Phi}(t; \underline{z}_0)$ be the flow at time t with the initial condition $\underline{z}_0$. Let Σ be an $(n-1)$-dimensional hypersurface, the *Poincaré section*, transverse to the flow, which means that if $\underline{\nu}(\underline{z})$ denotes the unit normal to Σ at $\underline{z}$, then $\underline{f}(\underline{z}) \cdot \underline{\nu}(\underline{z}) \neq 0$ for any $\underline{z}$ in Σ.

For a periodic orbit, let $\underline{z}_p$ be the intersection of the periodic orbit with Σ; let U be a neighborhood of $\underline{z}_p$ on Σ. Then, for any $\underline{z} \in U$ we define the *Poincaré map* as $\underline{\Phi}' = \underline{\Phi}(T; \underline{z})$, where T is the first return time of the flow on Σ.

In practical terms, if we consider a system described with Cartesian coordinates (x, y), as it is shown in the following example, the trajectories are in a four-dimensional phase space $(x, y, \dot{x}, \dot{y})$. In the two problems considered hereafter, there is a first integral (typically the energy), and either $\dot{x}$ or $\dot{y}$ can be expressed in terms of the other variables. Thus, if we fix the value of this first integral, the trajectory lies in a three-dimensional space, for example x, y, and $\dot{x}$. Let us now consider the intersections of this trajectory with the plane $(x, \dot{x})$ with a *positive* direction, which means to consider the points satisfying

$$y = 0, \quad \dot{y} > 0$$

In the configuration space (x, y), these points correspond to the intersection with the x-axis. If a periodic orbit crosses the x-axis $2n$ times, it will be represented as a set of n points. As said previously, the set of points in the $(x, \dot{x})$ plane exhibits interesting properties easier to study than the orbit in the (x, y) plane. We provide an evidence of this statement through the illustration of specific examples: the Hénon–Heiles problem (Sec. 1), the restricted three-body problem (Sec. 2), and the spin–orbit problem (Sec. 3).

1. THE HÉNON–HEILES PROBLEM

As a first example, we study the Hénon–Heiles problem, which is a 2-degrees-of-freedom, non-integrable system describing the motion of stars around the galactic center. Beside retaining a physical interest, the model is quite easy to analyze, since the vector field is polynomial in the coordinates. The equations of motion describing the Hénon–Heiles model are the following:

$$\begin{aligned} \ddot{x} &= -Ax - 2xy \\ \ddot{y} &= -By - x^2 + \epsilon y^2 \end{aligned} \tag{14}$$

where $A, B > 0$, $\epsilon \in \mathbb{R}$ are constants; Eqs. (14) are associated to the following Hamiltonian function:

$$H(\dot{x}, \dot{y}, x, y) = \frac{1}{2}(\dot{x}^2 + \dot{y}^2 + Ax^2 + By^2) + x^2 y - \frac{1}{3}\epsilon y^3$$

Figures 4a and 4b provide the Poincaré sections associated to Eqs. (14). To obtain these figures, we select the surface of section $x = 0$, we fix the initial conditions y_0, $\dot{y}_0$, and we compute $\dot{x}_0$ from a fixed value of E_0, the energy of the system. Then, we take the crossings of the solution with the surface of section for $\dot{x}_0 > 0$.

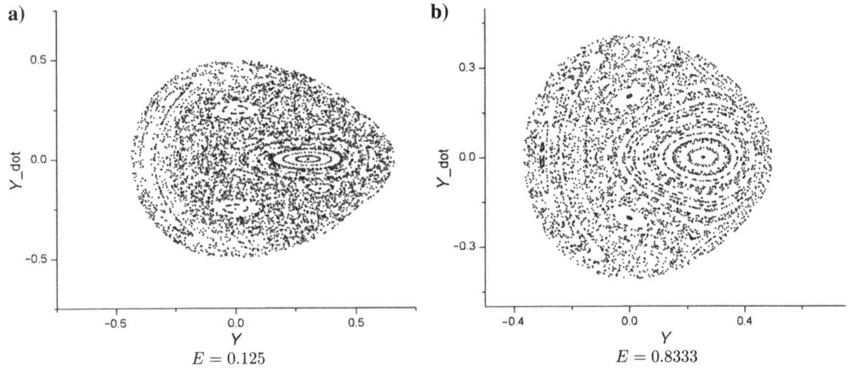

Fig. 4 Poincaré surfaces of section of the Hénon–Heiles problem for different values of the energy and $\epsilon = 1$, $A = 1$, and $B = 1$. a) $E = 0.125$ and b) $E = 0.8333$.

By discretizing the $(y, \dot{y})$-plane and by taking various values of $(y_0, \dot{y}_0)$, we obtain the final figures. As an integration method, we used a fourth-order Runge–Kutta method with a fixed integration step equal to 0.1.

2. THE RESTRICTED THREE-BODY PROBLEM

Using Eqs. (13), and acknowledging that these equations admit the first integral, called *Jacobi constant*, whose expression is given by

$$C = x^2 + y^2 + \frac{2(1-m)}{r_1} + \frac{2m}{r_2} - \dot{x}^2 - \dot{y}^2$$

we compute the Poincaré sections associated to the equations of motion. We choose $m = 0.5$, which corresponds to say that the masses of the primaries are equal. The result is shown in Fig. 5. To get these plots, we fix the surface of section $y = 0$, we select initial conditions x_0, $\dot{x}_0$, and we compute $\dot{y}_0$ from a fixed value of C, the Jacobi constant. Then, we take the crossings of the trajectories whenever the condition $\dot{y}_0 > 0$ is satisfied. By discretizing the $(x, \dot{x})$ plane and by taking different values of $(x_0, \dot{x}_0)$, we obtain the panels shown in Fig. 5. The integration method is again a fourth- order Runge–Kutta algorithm with a fixed integration step equal to 10^{-6}. A filtering is also applied to take into account close encounters with the primaries and to check that the Jacobi constant is preserved.

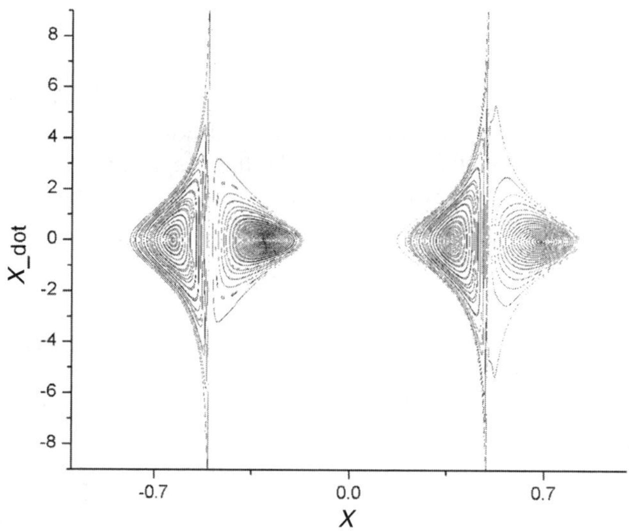

Fig. 5 Poincaré surfaces of section of the restricted three-body problem for a value of the Jacobi constant equal to 4.5.

3. THE SPIN–ORBIT PROBLEM

With reference to Sec. B, we recall the equations of motion describing the spin–orbit problem, which we write in the form

$$\dot{x} = p_x$$
$$\dot{p}_x = -\varepsilon \left(\frac{a}{r}\right)^3 \sin(2x - 2f)$$

complemented with the following equations:

$$r = a(1 - e\cos u)$$
$$\tan\frac{f}{2} = \sqrt{\frac{1+e}{1-e}}\tan\frac{u}{2}$$
$$\ell = u - e\sin u$$
$$\ell = nt + \ell_0$$

where ℓ is the mean anomaly, with an initial value ℓ_0, n is the mean motion, and u denotes the eccentric anomaly. Here, we fix $t = 0$ modulus 2π as the surface of section; for given initial conditions (x_0, p_{x0}), we compute the Poincaré mapping as the intersection of the solutions with the surface of section under the condition $p_x > 0$. For different values of ε we obtain the Poincaré sections provided in Figs. 6a–6c.

B. LYAPUNOV EXPONENTS

A widely used method to determine if a system is chaotic consists of studying the trajectories of two nearby orbits and of looking at how their distance diverges. A measure of such divergence is provided by the *Lyapunov exponents*. Quantitatively, let us consider two nearby trajectories at an initial distance $\delta\underline{z}(0)$; such orbits diverge at a rate given by

$$|\delta\underline{z}(t)| \approx e^{\lambda t}\,|\delta\underline{z}(0)|$$

where λ is the Lyapunov exponent. The rate of separation can assume different values in different directions, depending on the dimension of the phase space; as a consequence, there exists a whole spectrum of Lyapunov exponents equal in number to such a dimension. The largest Lyapunov exponent is called *maximal Lyapunov exponent* (hereafter MLE); a positive value of the MLE gives an indication of the chaotic character of the dynamical system. Let λ_M denote the MLE; it can be computed as

$$\lambda_M = \lim_{t\to\infty}\ \lim_{\delta\underline{z}(0)\to 0} \frac{1}{t}\ln\frac{|\delta\underline{z}(t)|}{|\delta\underline{z}(0)|}$$

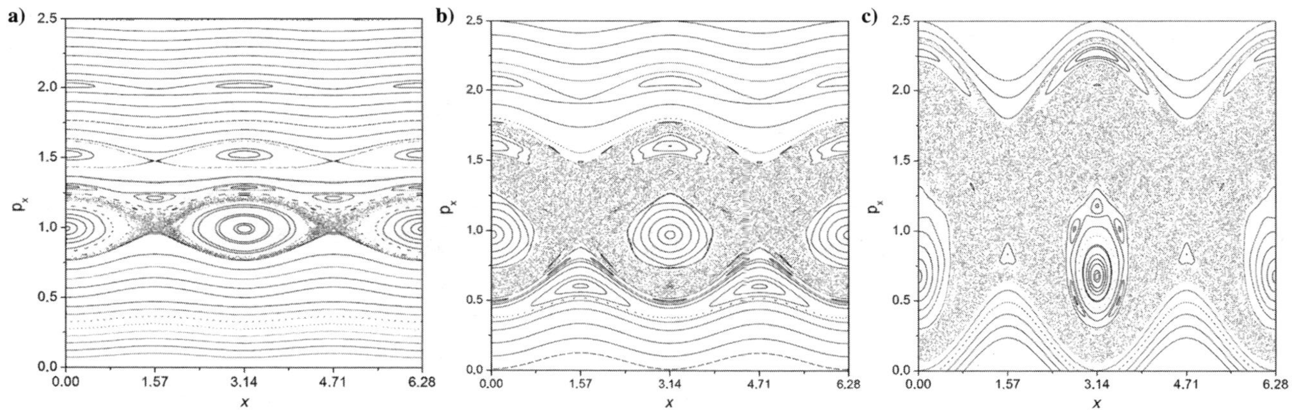

Fig. 6 Poincaré surfaces of section of the spin–orbit problem for different values of ε. a) $\varepsilon = 0.024$, b) $\varepsilon = 0.1$, and c) $\varepsilon = 0.4$.

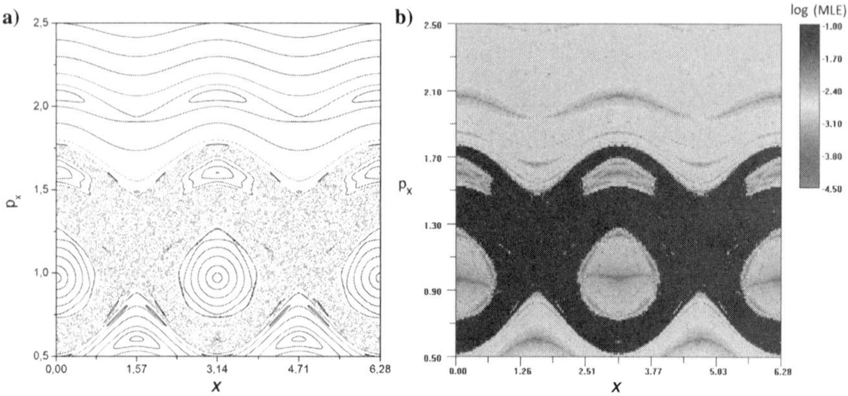

Fig. 7 Comparison of a Poincaré surface of section and the MLE for the spin–orbit problem; here $\varepsilon = 0.1$. a) Poincaré surface of section and b) MLE.

Figures 7a and 7b provide a comparison between the Poincaré surface of section and the MLE, calculated for the spin–orbit problem for an identical given value of ε; we remark that the overall behavior of the dynamics is the same. The Poincaré section provides a clear distinction between regular (periodic or quasiperiodic) and chaotic trajectories; on the other hand, the MLE yields a quantitative information on the divergence of the trajectories, which is provided by the color bar on a logarithmic scale.

C. FAST LYAPUNOV INDICATOR

The FLI is obtained as the value of the MLE at a fixed time, say T. For a vector field $\dot{\underline{z}} = \underline{f}(\underline{z})$, $\underline{z} \in \mathbb{R}^n$, we write the variational equations as

$$\dot{\underline{v}} = \left(\frac{\partial \underline{f}(\underline{z})}{\partial \underline{z}} \right) \underline{v}$$

The FLI is defined as follows: given the initial conditions $\underline{z}(0) \in \mathbb{R}^n$, $\underline{v}(0) \in \mathbb{R}^n$, the FLI at time $T \geqslant 0$ is provided by the expression

$$FLI(\underline{z}(0), \underline{v}(0), T) \equiv \sup_{0<t\leqslant T} \log ||\underline{v}(t)||$$

A comparison of the FLIs as the initial conditions are varied allows one to discern among all kinds of motion, namely regular, resonant, or chaotic (see Ref. [15]). The FLI has been used in Ref. [16] for the study of the spin–orbit problem with tidal torque and in Ref. [17] for a three-body problem with Poynting–Robertson effect or Stokes drag.

V. BASICS OF PERTURBATION THEORY

Most of the models describing the dynamics of asteroids, NEOs, and space debris are provided by a nearly integrable system. However, the integrable part provides only a rough approximation and one may want to build better approximations. This can be done by implementing the so-called *perturbation theory* that we are going to sketch next, referring to the wide existing literature for more details (see, e.g., Refs. [4,18]). We remark that the procedure outlined next gives a constructive algorithm to determine the transformation of coordinates that allows us to get a better approximation.

For an n-dimensional nearly integrable Hamiltonian function

$$\mathcal{H}(\underline{I}, \underline{\varphi}) = h(\underline{I}) + \varepsilon f(\underline{I}, \underline{\varphi}) \tag{15}$$

let the *frequency vector* be defined as

$$\underline{\omega}(\underline{I}) \equiv \frac{\partial h(\underline{I})}{\partial \underline{I}}$$

We aim to construct a canonical transformation $\mathcal{C} : (\underline{I}, \underline{\varphi}) \to (\underline{I}', \underline{\varphi}')$, such that the transformed Hamiltonian takes the form

$$\mathcal{H}'(\underline{I}', \underline{\varphi}') = \mathcal{H} \circ \mathcal{C}(\underline{I}, \underline{\varphi}) \equiv h'(\underline{I}') + \varepsilon^2 f'(\underline{I}', \underline{\varphi}')$$

where h' and f' denote, respectively, the new unperturbed Hamiltonian and the new perturbing function.

Such transformation is obtained through the following steps, detailed hereafter:

1. Define a suitable canonical transformation close to the identity
2. Perform a Taylor series expansion in the perturbing parameter
3. Require that the change of variables removes the dependence on the angles up to second-order terms
4. Expand in Fourier series to construct the explicit form of the Fourier coefficients of the canonical transformation

For the first step, one has to define a close-to-identity change of variables with the generating function $\underline{I}' \cdot \underline{\varphi} + \varepsilon \Phi(\underline{I}', \underline{\varphi})$ providing the change of coordinates

$$\begin{aligned} \underline{I} &= \underline{I}' + \varepsilon \frac{\partial \Phi(\underline{I}', \underline{\varphi})}{\partial \underline{\varphi}} \\ \underline{\varphi}' &= \underline{\varphi} + \varepsilon \frac{\partial \Phi(\underline{I}', \underline{\varphi})}{\partial \underline{I}'} \end{aligned} \tag{16}$$

where $\Phi = \Phi(\underline{I}', \underline{\varphi})$ is the unknown transformation.

Then, for the second step, we split f as $f(\underline{I},\underline{\varphi}) = \bar{f}(\underline{I}) + \tilde{f}(\underline{I},\underline{\varphi})$, where $\bar{f}(\underline{I})$ denotes the average of the function f over the angle variables and $\tilde{f}(\underline{I},\underline{\varphi})$ is the remainder function (i.e., $\tilde{f} = f - \bar{f}$). Inserting transformation (16) into Eq. (15) and expanding in Taylor series around $\varepsilon = 0$ up to the second order, one gets

$$h\left(\underline{I}' + \varepsilon \frac{\partial \Phi(\underline{I}',\underline{\varphi})}{\partial \underline{\varphi}}\right) + \varepsilon f\left(\underline{I}' + \varepsilon \frac{\partial \Phi(\underline{I}',\underline{\varphi})}{\partial \underline{\varphi}}, \underline{\varphi}\right) = h(\underline{I}') + \varepsilon\, \underline{\omega}(\underline{I}') \cdot \frac{\partial \Phi(\underline{I}',\underline{\varphi})}{\partial \underline{\varphi}}$$
$$+ \varepsilon \bar{f}(\underline{I}') + \varepsilon \tilde{f}(\underline{I}',\underline{\varphi}) + O(\varepsilon^2)$$

Concerning the third step, we replace the previous expansion in Eq. (15) to get

$$\mathcal{H}(\underline{I}',\underline{\varphi}) = h(\underline{I}') + \varepsilon \bar{f}(\underline{I}') + \varepsilon \left(\underline{\omega}(\underline{I}') \cdot \frac{\partial \Phi(\underline{I}',\underline{\varphi})}{\partial \underline{\varphi}} + \varepsilon \tilde{f}(\underline{I}',\underline{\varphi})\right) + O(\varepsilon^2)$$

From this expression, we see that the transformed Hamiltonian is integrable up to the second order in ε, provided that the function Φ satisfies

$$\underline{\omega}(\underline{I}') \cdot \frac{\partial \Phi(\underline{I}',\underline{\varphi})}{\partial \underline{\varphi}} + \tilde{f}(\underline{I}',\underline{\varphi}) = \underline{0} \tag{17}$$

In this case, the new unperturbed Hamiltonian becomes

$$h'(\underline{I}') = h(\underline{I}') + \varepsilon \bar{f}(\underline{I}')$$

For the fourth step, to obtain an explicit expression of Φ, we perform a Fourier series expansion of Φ and $\tilde{f}$ as

$$\Phi(\underline{I}',\underline{\varphi}) = \sum_{\underline{m} \in \mathbb{Z}^n \setminus \{\underline{0}\}} \hat{\Phi}_{\underline{m}}(\underline{I}')\, e^{i\underline{m}\cdot\underline{\varphi}}$$
$$\tilde{f}(\underline{I}',\underline{\varphi}) = \sum_{\underline{m} \in \mathcal{I}} \hat{f}_{\underline{m}}(\underline{I}')\, e^{i\underline{m}\cdot\underline{\varphi}}$$

where $\mathcal{I}$ is a suitable set of integer vectors associated to the Fourier indexes of $\tilde{f}$. Inserting the expansion in Eq. (17), we obtain

$$i \sum_{\underline{m} \in \mathbb{Z}^n \setminus \{\underline{0}\}} \underline{\omega}(\underline{I}') \cdot \underline{m}\, \hat{\Phi}_{\underline{m}}(\underline{I}')\, e^{i\underline{m}\cdot\underline{\varphi}} = -\sum_{\underline{m} \in \mathcal{I}} \hat{f}_{\underline{m}}(\underline{I}')\, e^{i\underline{m}\cdot\underline{\varphi}}$$

which provides the Fourier coefficients of Φ:

$$\hat{\Phi}_{\underline{m}}(\underline{I}') = -\frac{\hat{f}_{\underline{m}}(\underline{I}')}{i\, \underline{\omega}(\underline{I}') \cdot \underline{m}}$$

In conclusion, the generating function is given by

$$\Phi(\underline{I}', \underline{\varphi}) = i \sum_{\underline{m} \in \mathcal{I}} \frac{\hat{f}_{\underline{m}}(\underline{I}')}{\underline{\omega}(\underline{I}') \cdot \underline{m}} e^{i\underline{m} \cdot \underline{\varphi}}$$

We remark that the function Φ is well defined unless there exists an integer vector $\underline{m} \in \mathcal{I}$, such that

$$\underline{\omega}(\underline{I}') \cdot \underline{m} = 0$$

This is the classical problem of the *small divisors* and in order to properly define the generating function, we must assume that the following nonresonance condition is satisfied:

$$\underline{\omega}(\underline{I}') \cdot \underline{m} \neq 0 \quad \text{for all } \underline{m} \in \mathcal{I}$$

This concludes a short sketch of perturbation theory, for which we refer to Refs. [1, 2, 4, 13, 18].

VI. CONCLUSION

In this chapter, we introduced different dynamical notions, namely periodic, quasiperiodic, and chaotic behaviors. We gave examples of discrete systems exhibiting such different kinds of behaviors; in particular, we presented the paradigmatic model known as the standard map. We also introduced Hamiltonian formalism to describe continuous systems.

We showed analytical models that can be used to describe the rotational and orbital motions of asteroids and NEOs. In particular, we introduced the spin–orbit problem and the three-body problem. We also considered space debris dynamics, whose description requires a modeling of the shape of the Earth.

To study the dynamics of such models, we introduced some numerical methods that are used to discern between regular and chaotic motions. To this end, we described the computation of Poincaré maps, Lyapunov exponents, and FLIs.

In the final section, we introduced the reader to perturbation theory, which is a powerful analytical tool that can be implemented in nearly integrable systems to get a more accurate description of the dynamics.

ACKNOWLEDGMENTS

We deeply thank I. Gkolias and G. Pucacco for interesting suggestions and discussions. We are grateful to I. Gkolias for providing us with Figs. 6 and 7. A.C. was partially supported by GNFM-INdAM and by the European Grant MC-ITN Stardust. F.G. was supported by the European Grant MC-ITN Stardust.

REFERENCES

[1] Guckenheimer, J., and Holmes, P., *Nonlinear Oscillations, Dynamical Systems, and Bifurcations of Vector Fields*, Springer, Berlin, Heidelberg, 1983.

[2] Lichtenberg, A. J., and Lieberman, M. A., *Regular and Chaotic Dynamics*, 2nd ed., Springer-Verlag, Berlin, Heidelberg, New York, 1992.

[3] Chirikov, B. V., "A Universal Instability of Many-Dimensional Oscillator Systems," *Physics Reports*, Vol. 52, No. 5, 1979, pp. 264–379.

[4] Celletti, A., *Stability and Chaos in Celestial Mechanics*, Springer-Verlag, Berlin, 2010.

[5] Dvorak, R., and Lhotka, C., *Celestial Dynamics*, Wiley-VCH, Berlin, 2013.

[6] Greene, J. M., "A Method for Determining a Stochastic Transition," *Journal of Mathematics Physics*, Vol. 20, 1979, pp. 1183–1201.

[7] Kolmogorov, A. N., "On the Conservation of Conditionally Periodic Motions for a Small Change in Hamilton's Function," *Dokl. Akad. Nauk SSSR (N.S.)*, Vol. 98, 1954, pp. 527–530. English translation in "Stochastic Behavior in Classical and Quantum Hamiltonian Systems" (*Volta Memorial Conference*, Como, 1977), Lecture Notes in Physics, 93, pp. 51–56. Springer, Berlin, 1979.

[8] Arnol'd, V. I., "Proof of a Theorem of A. N. Kolmogorov on the Invariance of Quasi-Periodic Motions Under Small Perturbations," *Russian Mathematics Surveys*, Vol. 18, No. 5, 1963, pp. 9–36.

[9] Moser, J., "On Invariant Curves of Area-Preserving Mappings of an Annulus," *Nachr. Akad. Wiss. Göttingen Math.-Phys. Kl. II*, Vol. 1962, 1962, pp. 1–20.

[10] Celletti, A., and Lhotka, C., "Transient Times, Resonances and Drifts of Attractors in Dissipative Rotational Dynamics," *Communications Nonlinear Science Numerical Simulation*, Vol. 19, 2014, pp. 3399–3411, http://dx.doi.org/10.1016/j.cnsns.2014.01.013, 2014.

[11] Calleja, R., Celletti, A., and de la Llave, R., "A KAM Theory for Conformally Symplectic Systems: Efficient Algorithms and Their Validation," *Journal of Differential Equations*, Vol. 255, No. 5, 2013, pp. 978–1049.

[12] Calleja, R., and Celletti, A., "Breakdown of Invariant Attractors for the Dissipative Standard Map," *CHAOS*, Vol. 20, No. 1, 2010, p. 013121.

[13] Arnol'd, V. I., *Mathematical Methods of Classical Mechanics*, 2nd ed., Springer-Verlag, New York, 1989.

[14] Wertz, J. R., and Larson, W. J., "Space Mission Analysis and Design," *Space Mission Analysis and Design*, edited by J. R. Wertz, W. J. Larson, Wiley J. El Segundo, Calif.: Microcosm Press, Dordrecht, Boston: Kluwer, Space technology library, Vol. 8, 1999, p. 1.

[15] Froeschlé, C., Lega, E., and Gonczi, R., "Fast Lyapunov Indicators. Application to Asteroidal Motion," *Celestial Mechanics and Dynamical Astronomy*, Vol. 67, No. 1, 1997, pp. 41–62.

[16] Celletti, A., Lega, E., and Froeschlé, C., "Dynamics of the Conservative and Dissipative Spin–Orbit Problem," *Planetary and Space Science*, Vol. 55, 2007, pp. 889–899.

[17] Celletti, A., Stefanelli, L., Lega, E., and Froeschlé, C., "Some Results on the Global Dynamics of the Regularized Restricted Three-Body Problem with Dissipation," *Celestial Mechanics and Dynamical Astronomy*, Vol. 109, No. 3, 2011, pp. 265–284.

[18] Ferraz-Mello, S., *Canonical Perturbation Theories*, Springer, New York, 2007.

CHAPTER 8

Orbital Dynamics About Small Bodies

Juan Luis Cano*

Elecnor Deimos, Madrid, Spain

Davide Amato†

Technical University of Madrid, Madrid, Spain

I. EXPLORATION OF SMALL BODIES AND NEAR-EARTH ASTEROIDS

Small bodies encompass comets and most of the asteroids in the Solar System. Near-Earth asteroids (NEAs) are those having a perihelion of less than 1.3 AU and their size is usually smaller than a few kilometers (Fig. 1). These asteroids migrated from the Main Belt into the inner Solar System and, as the other asteroids, are remnants from the origin of the Solar System; thus they provide information on its history and dynamical evolution.

In recent years, knowledge of NEAs has boomed thanks to substantial increases of detection and observation programs and an improved understanding of the dynamics and the evolution of the Solar System. This has allowed to discover most of the NEAs bigger than 1 km.

A number of deep space missions aimed at NEAs have been performed at various degrees of complexity, such as NEAR, Hayabusa, and Rosetta (Table 1). The main objective of such missions has been of executing precise measurements of asteroid characteristics, some of which are difficult or impossible to obtain through ground observations.

Another motivator, which is driving interest into space missions to asteroids in the last decade, is the mitigation of impact threats. Most, if not all, asteroid deflection techniques rely on actual asteroid rendezvous and close in-orbit operations, which have to be tested through demonstration missions.

Finally, a very recent "trending topic" is the one of exploration and exploitation of asteroids. There is considerable activity in developing missions aimed at bringing an NEA within the Earth–Moon system, with the ultimate objective of performing manned exploration and mining of resources.

---

*Flight Systems Business Unit.

†Space Dynamics Group.

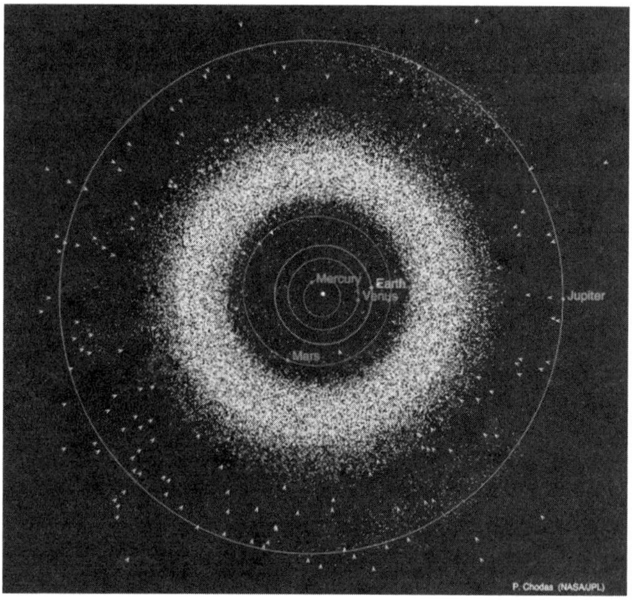

Fig. 1 Known population of small bodies in the inner Solar System. Asteroids are displayed as yellow dots and comets as Sun-pointing wedges. (Courtesy NASA/JPL-Caltech.)

TABLE 1 PAST AND CURRENT MISSIONS TO SMALL BODIES

Agency	Mission	Year	
ESA	Giotto	1985	Flew by comet Halley in 1986.
NASA	Galileo	1989	Flew by 951 Gaspra in 1991, 243 Ida in 1993.
NASA	NEAR	1996	Entered asteroid Eros' orbit in 2000, landed on Eros in 2001.
NASA	Deep Space 1	1998	Flew by 9969 Braille in 1999, comet Borrelly in 2001.
JAXA	Hayabusa	2003	Landed on asteroid Itokawa in 2005, returned samples to Earth in 2010
ESA	Rosetta	2004	Flew by asteroids 2867 Steins (2008) and 21 Lutetia (2010). Will enter orbit around comet 67P/Churyumov-Gerasimenko in 2014 and release a lander.
NASA	Deep Impact	2005	Impacted comet Tempel 1 in 2005.
NASA	Dawn	2007	Entered Vesta's orbit in 2011, will enter Ceres' orbit in 2015.

Source: Data taken from Refs. [1–7].

II. DYNAMICAL ENVIRONMENT ABOUT SMALL BODIES

Understanding the features of the dynamical environment in the vicinity of small bodies is of key importance in mission design.

Clearly, spacecraft dynamics about a small body are influenced by its gravity field. Nevertheless, one has to consider the effect of additional phenomena such as solar radiation pressure (SRP) and solar tide.

A. GRAVITY FIELD

Most small bodies exhibit irregular shapes and mass distributions, and their rotational state can also be complicated. Thus, their gravity field is usually not easily described. In particular, the morphology and rotational state have a prominent role in spacecraft dynamics either when the asteroid is large or when operating very close to the surface in small asteroids.

B. SOLAR RADIATION PRESSURE

Impacting photons from the Sun transfer momentum to the spacecraft, therefore giving rise to a net acceleration. This mainly depends on the spacecraft surfaces exposed to the Sun and on their optical properties; simple models assume a constant exposed surface and a uniform reflectivity parameter.

This perturbation can be especially relevant for spacecraft using electric propulsion, which have a large solar panel area that increases their area-to-mass ratio. This is even more relevant when considering that electric propulsion is a common choice for missions to NEAs.

Because the resulting acceleration is proportional to the amount of solar radiation reaching the spacecraft, its magnitude depends on the inverse square of the distance from the Sun. This means that when orbiting around NEAs in elliptic heliocentric orbits the acceleration will pulsate with the same period as the orbit's, passing from a minimum at aphelion to a maximum at perihelion (Fig. 2).

For small NEAs, the intensity of the SRP perturbation will be especially strong with respect to the gravity field, even considering spacecraft operating at close distances from the asteroid.

C. THIRD-BODY EFFECTS

1. SOLAR GRAVITY

The tidal effect due to solar gravity can be usually considered as a minor perturbation, except in cases where the spacecraft orbits at large distances from the asteroid. In these cases, this perturbation can compete with the solar radiation

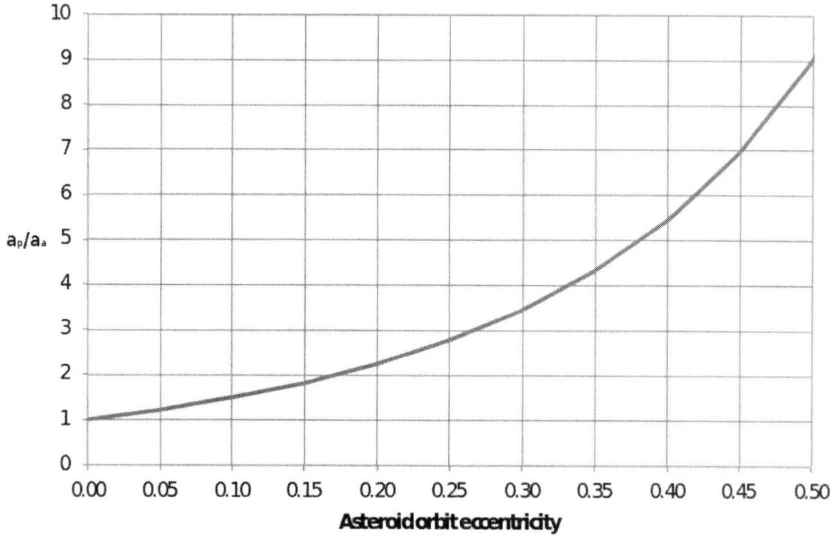

Fig. 2 Variation of the solar radiation pressure magnitude as a function of the asteroid's orbit eccentricity. Ratio between SRP at perihelion and aphelion is displayed on the ordinates.

pressure. Typically, as the required operational distances to the asteroids are small, this interaction is neglected.

2. MULTIPLE ASTEROID SYSTEMS

Gravitational perturbations can heavily influence dynamics when the spacecraft operates in a binary or multiple asteroid system. In this case, trajectory design requires modeling of the dynamics of the asteroid system in addition to the motion of the spacecraft about the system [8].

III. ORBITAL STABILITY ABOUT NEAs

Because of the complexity of the dynamical environment, the motion around an asteroid is strongly perturbed. Orbits can be heavily distorted, chaotic, and unstable, but in some particular cases they can be stable enough for a spacecraft to operate close to the asteroid.

The typical approach to the assessment of orbital stability is to study gravitational and solar radiation perturbations separately, and to sum them to the central gravity field of the asteroid.

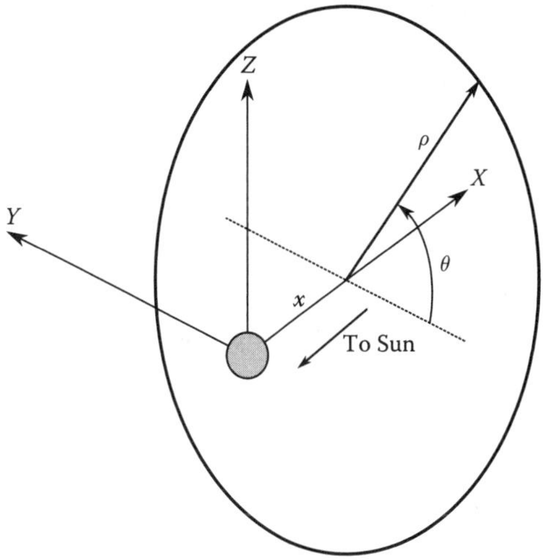

Fig. 3 Geometry of the orbit relative to the asteroid.

However, in some orbits only one of these perturbations may be dominating, which in the case of small NEAs is solar radiation pressure. We will review SRP-dominated orbits around single-asteroid NEAs.

The analyses in this section are taken from Ref. [8].

A. POINT MASSES AND CONSTANT ACCELERATION

The simplest model to analyze is that of a nonrotating, constant perturbing acceleration acting on a spacecraft orbiting about a point mass. These hypotheses imply a scenario in which the spacecraft orbits an asteroid in a quasi-circular orbit around the Sun, and it does so with an orbital period much smaller than the asteroid's.

Even if this model is quite approximate, it gives useful insight to analyze more complicated dynamical scenarios.

First of all, let the SRP acceleration be expressed as:

$$\mathbf{g} = \frac{\beta}{d^3}\mathbf{d}, \quad \beta = (1+\rho)\,\frac{G_1}{B} \tag{1}$$

where $\mathbf{d}$ is the Sun-to-asteroid vector (assumed stationary), ρ is the spacecraft's reflectivity, B is the ratio between its mass and the sunlit area and G_1 is a solar constant, $G_1 \approx 10^8\,\mathrm{kg\,km^3/(s^2\,m^2)}$.

It is convenient to use a cylindrical frame with its axis x directed along $\hat{d}$ and the radius ρ[‡] and polar angle θ measured perpendicular to this direction (Fig. 3). The distance of the spacecraft from the asteroid is then $r^2 = \rho^2 + x^2$.

[‡]From now on ρ will denote such radius and not the reflectivity of the spacecraft, as stated previously.

In such a reference frame the equations of motion are expressed as

$$\ddot{x} = -\frac{\mu x}{r^3} + g \tag{2}$$

$$\ddot{\rho} - \rho\dot{\theta}^2 = -\frac{\mu\rho}{r^3} \tag{3}$$

$$\rho\ddot{\theta} + 2\dot{\rho}\dot{\theta} = 0 \tag{4}$$

Note that integrating Eq. (4) gives that the total angular momentum in the direction of the Sun, $H_d = \rho^2\dot{\theta}$, is conserved. Moreover, introducing the potential U_d, it is possible to rewrite the equations in the form

$$\ddot{x} = \frac{\partial U_d}{\partial x} \tag{5}$$

$$\ddot{\rho} = \frac{\partial U_d}{\partial \rho} \tag{6}$$

$$U_d = \frac{\mu}{r} - \frac{1}{2}\frac{H_d^2}{\rho^2} + gx \tag{7}$$

This reduced set of equations still has a Jacobi integral:

$$J = \frac{1}{2}(\dot{x}^2 + \dot{\rho}^2) - U_d \tag{8}$$

Equations (5–7) exhibit an equilibrium solution that corresponds to a circular orbit, offset from the center of the point mass along the direction $\hat{\mathbf{x}}$ and perpendicular to this same direction. Solving for $\partial U_d/\partial x = 0$ and $\partial U_d/\partial \rho = 0$ yields such equilibrium point:

$$x_{\mathrm{eq}} = \frac{g}{\mu} r_{\mathrm{eq}}^3 \tag{9}$$

$$\rho_{\mathrm{eq}}^4 = \frac{H_d^2}{\mu} r_{\mathrm{eq}}^3 \tag{10}$$

where these two equations are coupled through the relation $r^2 = x^2 + \rho^2$. Here the crucial parameters are the orbit radius r and the ratio g/μ. Once these are specified, the values of x and ρ are fixed. By definition $x \leqslant r$, and thus leads to a fundamental limit on the orbit radius, $r \leqslant \sqrt{\mu/g}$. Conversely, as $\rho \leqslant r$ as well, a lower limit can be found as $H_d^2/\mu \leqslant r$, which implies a limit $H_d \leqslant \sqrt[4]{(\mu^3/g)}$.

To study the stability of this equilibrium, the equations of motion are linearized and the characteristic equation is computed as

$$\lambda^4 - \lambda^2(U_{\rho\rho} + U_{xx}) + U_{\rho\rho}U_{xx} - U_{\rho x}^2 = 0 \tag{11}$$

For stability the real part of each root has to be nonpositive. It is possible to show that this corresponds to the condition

$$x_{\mathrm{eq}} < \frac{1}{3\sqrt{3}}\sqrt{\frac{\mu}{g}} \tag{12}$$

In turn, this condition induces a limit on the osculating semimajor axis of an initial orbit evaluated at $x = 0$, which is

$$a \leqslant \frac{\sqrt{3}}{4}\sqrt{\frac{\mu}{g}} \tag{13}$$

This serves as a useful design parameter in constraining the maximum orbit size for mission design purposes.

We can mention that, accounting for both a rotating solar radiation pressure acceleration and the solar tide, it is possible to derive a sufficient condition on a for a spacecraft's trajectory to be bound to a small body, as demonstrated in Ref. [8]. In these hypotheses, it is also possible to find two equilibrium points located on the Sun line. The Sun-ward point can be exploited for monitoring of the body (e.g. during a perihelion pass), while the anti-Sun point provides a sufficient condition for escape.

B. GENERAL SRP PROBLEM WITH AVERAGING

Given that the acceleration from solar radiation pressure is usually small, it is possible to consider it as a perturbation whose magnitude varies with the asteroid's distance from the Sun. Therefore, we consider the corresponding perturbation potential, average it over one spacecraft orbit, and substitute it in the Lagrange planetary equations of the orbiter. It is then obtained that the averaged semimajor axis is constant; moreover, the resulting equations can be integrated in close form and it is possible to arrive at a linear differential equation for the angular momentum and eccentricity vectors:

$$\begin{bmatrix} \dot{\mathbf{e}} \\ \dot{\mathbf{h}} \end{bmatrix} = \frac{3g}{2}\sqrt{\frac{a}{\mu}} \begin{bmatrix} -\dfrac{2\dot{\nu}}{3g}\sqrt{\dfrac{\mu}{a}}\hat{\mathbf{z}} & \tilde{\hat{\mathbf{d}}} \\ \tilde{\hat{\mathbf{d}}} & -\dfrac{2\dot{\nu}}{3g}\sqrt{\dfrac{\mu}{a}}\hat{\mathbf{z}} \end{bmatrix} \begin{bmatrix} \mathbf{e} \\ \mathbf{h} \end{bmatrix} \tag{14}$$

where $\hat{\mathbf{z}}$ is the unit vector of the considered reference frame, which is rotating at the true anomaly rate $\dot{\nu}$, and $\tilde{\hat{\mathbf{d}}}$ is the averaged direction of the SRP acceleration. Because $\dot{\nu}/g$ is time-invariant, we can define

$$\frac{2\dot{\nu}}{3g}\sqrt{\frac{\mu}{a}} = \frac{2\sqrt{\mu_{\mathrm{Sun}}A(1-E^2)}}{3\beta}\sqrt{\frac{\mu}{a}} = \frac{1}{\tan\Lambda} = \text{constant} \tag{15}$$

and we recognize that $\Lambda \rightarrow \pi/2$ for a strong SRP perturbation, whereas $\Lambda \rightarrow 0$ if it is weak. A and E are the asteroid's semimajor axis and eccentricity, respectively.

By changing the independent variable from time to the asteroid's true anomaly about the Sun, a time-invariant formulation can be derived:

$$\begin{bmatrix} \mathbf{e}' \\ \mathbf{h}' \end{bmatrix} = \begin{bmatrix} -\tilde{\hat{\mathbf{z}}} & \tan\Lambda \tilde{\hat{\mathbf{d}}} \\ \tan\Lambda \tilde{\hat{\mathbf{d}}} & -\tilde{\hat{\mathbf{z}}} \end{bmatrix} \begin{bmatrix} \mathbf{e} \\ \mathbf{h} \end{bmatrix} \tag{16}$$

where the apex denotes differentiation with respect to ν.

The solution to Eq. (16) can be obtained in closed form and is periodic in $\psi = \nu/\cos\Lambda$, hence its period is inversely proportional to the relative strength of the SRP perturbation.

1. FROZEN ORBITS

Two kinds of frozen orbits (for which $\dot{\mathbf{e}} = \dot{\mathbf{h}} = 0$) exist in this problem.

Ecliptic plane frozen orbits are obtained with the conditions:

$$-(\hat{\mathbf{d}} \cdot \hat{\mathbf{e}})(\hat{\mathbf{z}} \cdot \hat{\mathbf{h}}) = 1 \tag{17}$$

$$e = \sin\Lambda \tag{18}$$

Such orbits lie in the same plane as the asteroid's heliocentric orbit. For strong perturbations these orbits are highly elliptic and not desirable; moreover, they suffer eclipses.

Terminator plane frozen orbits are obtained with the conditions:

$$-(\hat{\mathbf{z}} \cdot \hat{\mathbf{e}})(\hat{\mathbf{d}} \cdot \hat{\mathbf{h}}) = 1 \tag{19}$$

$$e = \cos\Lambda \tag{20}$$

These orbits lie in the terminator plane, which is orthogonal to the Sun line. For strong perturbations these orbits are more circular, which tends to stabilize them. They do not suffer eclipses, however, visibility conditions are not optimal due to the high solar aspect angle.

As for the stability of terminator plane orbits, one has to consider that the SRP magnitude varies significantly along the path of eccentric asteroid orbits, therefore, an initially stable orbit could be destabilized for the asteroid approaching perihelion. Asteroid oblateness and ellipticity combined may also destabilize the orbit through resonance effects. A criterion to avoid this destabilizing phenomenon is

$$a > \frac{3}{2}\left(\frac{T^2\mu}{4\pi^2}\right)^{1/3} \tag{21}$$

where T is the rotational period of the asteroid.

IV. APPLICATIONS TO SPACE MISSIONS

When designing or examining a mission it is crucial to consider the relative strength of the accelerations imposed on the spacecraft. Depending on this, the dynamics around the small body could be dominated by gravity, solar radiation pressure, or both.

Figure 4 shows the impact of these considerations on some past and future missions. NEAR, having operated at close distance from Eros and having landed on it, had dynamics that were dominated by gravity. This is not the case for Hayabusa and Osiris-REx, in which SRP dominates the mission design in many mission phases. The case of Rosetta falls in between both extremes.

As shown in the previous section, an actual stable terminator orbit for such missions would be at ~43% of the reported distances.

A. HAYABUSA

Being a low-thrust mission to Itokawa, Hayabusa's solar panels were comparatively large, thus the solar radiation perturbation acting on the spacecraft was not negligible.

For Hayabusa, $a_{\max} = (\sqrt{3}/4)\sqrt{(\mu/g)} = 1.6\,\text{km}$ and $\Lambda \approx 87\,\text{deg}$, evidencing the strong impact of solar radiation pressure in the dynamics. In fact, due to the uncertainty in the knowledge of the asteroid mass, it was decided to take a safe

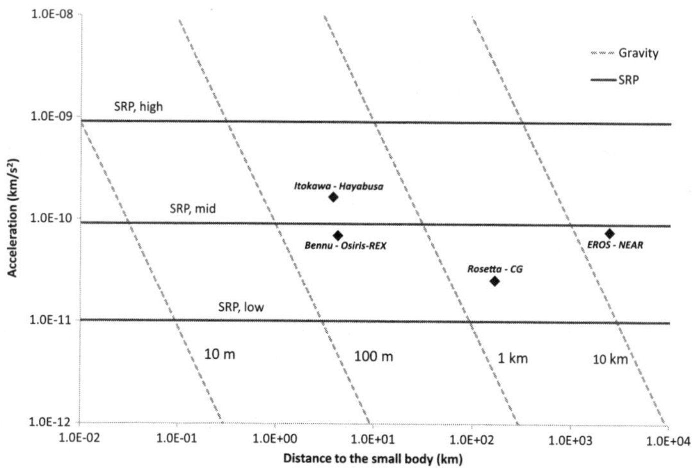

Fig. 4 Magnitudes of solar radiation perturbation and central gravity field for different asteroid diameters and levels of solar radiation pressure. Points of equal SRP and gravitational acceleration are displayed for some past and future missions.

TABLE 2 NOMINAL STABILITY BOUNDARIES FOR TERMINATOR ORBITS ABOUT 1989 ML AND 2002 AT_4

		Altitude Above		Altitude Above	
Asteroid Mass	**Asteroid Shape**	**Min**	**Max**	**Min**	**Max**
Minimum	Sphere	0.3	3.6	0.2	2.4
	Prolate spheroid	1.9	3.6	0.2	2.4
	Oblate spheroid	1.5	2.9	N/A	N/A
Maximum	Sphere	1.7	7.1	0.3	3.6
	Prolate spheroid	3.7	7.2	1.4	3.6
	Oblate spheroid	2.7	6.0	N/A	N/A

approach and design a hovering strategy, in which the spacecraft used thrust to null out perturbation accelerations.

An a posteriori analysis was done with the available information and it was determined that a terminator orbit would have been feasible with an SMA between 1.0 and 1.5 km [8].

B. DON QUIJOTE

Don Quijote was a mission concept designed to demonstrate and validate the kinetic impactor asteroid deflection technology. It consisted of sending an orbiter, followed by an impactor, to a small NEA. A number of stability assessments were done for asteroids 1989 ML and 2002 AT4.

Terminator orbits were required in order for the orbiter to perform a radio tracking experiment, and stable solutions were found for both asteroids [9]. Minimum and maximum boundaries for the semimajor axes of such orbits for both asteroids are reported in Table 2.

C. PROBA-IP

Proba-IP was an ESA phase 0 study for an interplanetary mission aimed at close-up reconnaissance of an NEA, chosen between 1989 UQ, 2001 CC21, and Apophis.

Terminator orbits were again required for the spacecraft, and stable solutions were found for the first two asteroids. However, it was not possible to find any long-term stable terminator orbit about Apophis because of its small size and its large rotation period, which was commensurate with the orbital period of the terminator orbit. This caused resonant perturbations, which led to orbit instabilities [10].

V. CONCLUSIONS

The dynamical environment about a small body presents a set of perturbations that make the motion of an object in its vicinity quite complex. In many cases the resulting trajectories will be unstable due particularly to the combination of a large SRP with other perturbations. Nevertheless, in some cases stable solutions can be found analytically, whose stability needs to be double-checked with full perturbation simulations. One of those examples are the so-called *terminator orbits*, which allow circling in an off-set orbit behind the asteroid. Stability of these orbits is mainly affected by the eccentricity of the asteroid orbit, the asteroid's gravity field, and its rotational state.

Moreover, lack of a priori knowledge of the asteroid properties is a major source of mission complexity and cost.

REFERENCES

[1] ESA, *Giotto Fact Sheet*, http://sci.esa.int/giotto/47355-fact-sheet/. Last updated 15 March 2013.

[2] NASA, *Galileo Legacy Site*, http://solarsystem.nasa.gov/galileo/. Last updated 28 June 2010.

[3] NASA, *Deep Space 1*, http://www.jpl.nasa.gov/missions/ details.php?id=5870. Retrieved 14th Feb. 2014.

[4] The Johns Hopkins Applied Physics Laboratory. *Near Earth Asteroid Rendezvous Mission*, http://near.jhuapl.edu/. Last updated 12 Dec. 2002.

[5] NASA, *Deep Impact Mission Timeline*, http://solarsystem.nasa.gov/deepimpact/ mission/timeline.cfm. Last updated 28 June 2010.

[6] JAXA, *Asteroid Explorer Hayabusa*, http://www.jaxa.jp/pr/brochure/pdf/04/ sat14.pdf. Retrieved 14 Feb. 2014.

[7] NASA, *Dawn Mission Timeline*, http://dawn.jpl.nasa.gov/mission/ timeline.asp. Retrieved 14 Feb. 2014.

[8] Scheeres, D. J., *Orbital Motion in Strongly Perturbed Environments*, Springer, Berlin, Germany, 2012.

[9] Cano, J. L., Sánchez, N., Sánchez, M., and Cornara, S., "Mission Analysis for the Don Quijote Phase a Study," *58th International Astronautical Congress*, Hyderabad, India, 2007. International Astronautical Federation, IAC-07-A3.5.08.

[10] Bellei, G., Cano, J. L., and Sánchez, M., "Operational Orbiting Strategies about Minor Bodies," *21st International Symposium on Spaceflight Dynamics*, Toulouse, France, 2009, http://issfd.org/ISSFD_2009/InterMissionDesignI/Cano.pdf.

CHAPTER 9

Brief Review of Numerical Methods for Entry Flow Simulations

Clemens Rumpf*
University of Southampton, Southampton, United Kingdom
Edmondo Minisci†
University of Strathclyde, Glasgow, United Kingdom

I. INTRODUCTION

When an object enters the atmosphere of a planet, it experiences a series of different phenomena, and consequent actions, whose complex nature is related to the involved very high speeds and changes of atmosphere density/composition. Since the high kinetic energy of entering objects leads to a strong heating phenomena around the object, the involved actions are called aerothermodynamic actions, and their description and modeling involves disciplines such as aerodynamics, thermodynamics, chemistry, as well as materials and structures.

This chapter provides a very brief overview of the aerothermodynamic phenomena encountered during the entry phase and points out the main challenges to numerically model and simulate them.

II. ENTRY FLOWS

An object that enters the atmosphere with orbital velocity experiences different flow regimes, depending on the flight velocity and atmosphere density. On the basis of the flight speed, the flight regimes, from fast to slow, are hypersonic, supersonic, and subsonic. This section describes the three entry flow regimes and expands on the hypersonic regime by subdividing it into continuum and rarefied flows.

A. SUBSONIC AND SUPERSONIC FLOWS

Sub- and supersonic flows are encountered toward the end of an atmospheric entry trajectory. Both regimes are introduced in the following two sections.

*Astronautics Research Group.
†Department of Mechanical and Aerospace Engineering.

1. SUBSONIC

The subsonic flight regime covers speeds from standstill to the local speed of sound, that is, for 0 < Mach number (Ma) < 1, with the Ma defined as

$$Ma = \frac{V}{a} \tag{1}$$

where V is the object speed, and a is the medium's local speed of sound. Flow information such as pressure propagates with the speed of sound through a medium. Consequently, even the flow upstream of a subsonic vehicle reacts to the presence of the vehicle. As the vehicle approaches the sonic speed, this behavior changes because the vehicle itself travels as fast as the information. Moreover, flight-dynamic and aerodynamic characteristics of the object experience a dramatic change as this critical speed is reached. For example, a sudden relocation of the aerodynamic center‡ along the chord line of an airfoil when it passes through the speed of sound occurs [1, 2], with heavy effects on aircraft stability and controllability. Furthermore, drag varies significantly in the transonic region around 1 Ma. A typical dependency of the drag coefficient C_D over Ma is shown in Fig. 1.

As the flow approaches Ma = 1, C_D increases sharply, reaching a maximum at the speed of sound, and then lowers levels; similar to faster flows, the drag coefficient returns to subsonic levels.

2. SUPERSONIC

In the supersonic flight regime, the vehicle travels at speeds for which $Ma > 1$. A dominant characteristic of supersonic flight is the occurrence of shock waves as shown in Fig. 2. Shock waves are regions where the flow is forced to

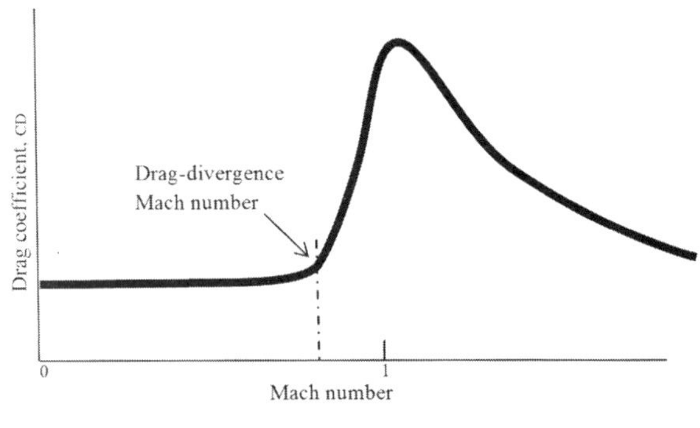

Fig. 1 Typical dependency of drag coefficient by Mach number.

‡The aerodynamic center is the location along an airfoil's chord line where the moment remains constant for varying angles of attack. This point is also frequently used as the location on which the aerodynamic forces lift, drag, and attack.

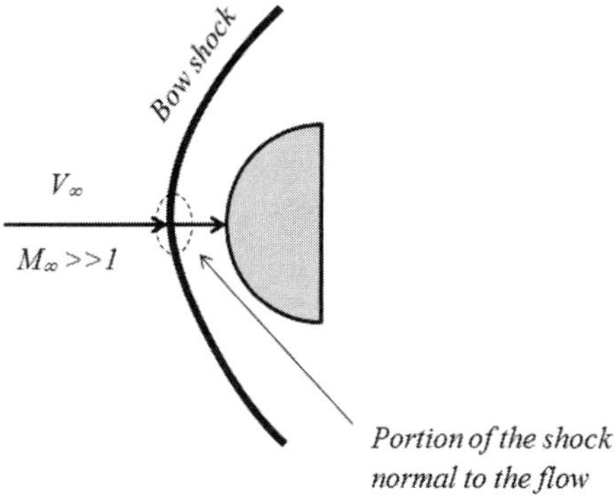

Fig. 2 Example of a normal shock wave in front of a blunt body [1].

decelerate to subsonic speeds, and pressure, temperature, and density experience a rapid increase across them. The ratio of flow properties across a normal shock is shown in Fig. 3 as a function of Mach number in front of the shock for a perfect gas $\gamma = 1.4$.

Subscript 1 denotes the physical condition before the shock and subscript 2 represents the conditions after the shock. The curve denoted with M shows the Ma of the flow after the normal shock. Flow speed after the shock is lower for higher initial velocities. The total pressure p_0 is made up of static and dynamic pressure and dynamic pressure is proportional to flow velocity. Total pressure decreases across the shock because of decreasing flow velocity (and thus a decreasing dynamic pressure component). On the other hand, static pressure p, temperature T, and density ρ increase across the shock.

B. HYPERSONIC FLOW

In contrast to the transition from subsonic to supersonic flow regimes, where the speed of sound represents a distinct boundary, the lower boundary of flight speed characterizing the start of the hypersonic flow regime is not defined by a specific velocity (or Ma), but rather by the occurrence of certain physical phenomena. Depending on the environment, these effects can start to be significant at speeds

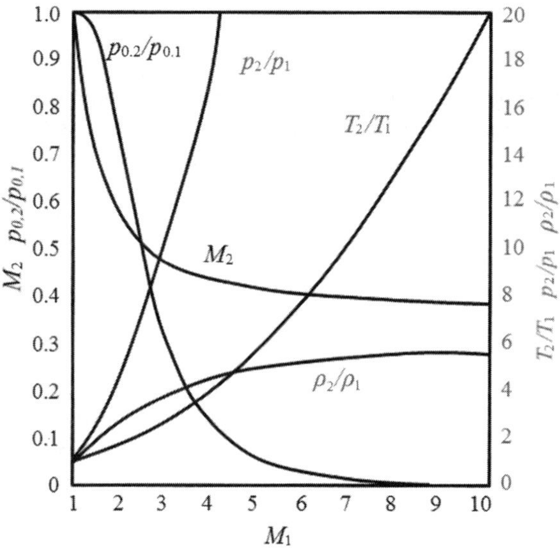

Fig. 3 Variation of flow characteristics across a normal shock in dependency of the *Ma* number before the shock. A perfect gas is assumed for this plot [1].

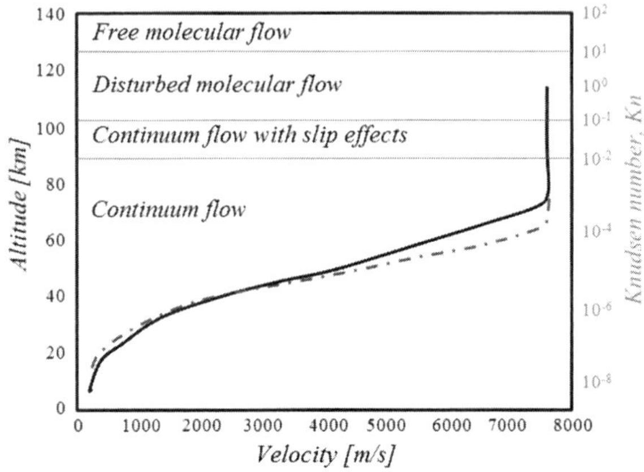

Fig. 4 Flow regimes over an altitude [3].

corresponding to *Ma* as low as 3, but as a rule of thumb, it can be said that the hypersonic regime begins at around $Ma = 5$ and is fully developed for $Ma \geqslant 7$.

During the hypersonic phase of the (re-) entry, the considered objects also experience considerable environmental changes. At the start of the entry, the density of the atmospheric medium is so low that interactions between atmospheric molecules and the object are singular distinct events. This flow regime is then called a rarefied flow regime. As the object penetrates deeper into the atmosphere, the density increases and the object's interaction with the atmospheric molecules becomes so frequent that single events are no longer distinguishable, and the medium can be considered as a continuum. It is a major challenge in today's modeling efforts to adequately represent the entire entry trajectory because of these profound shifts in environmental properties. Figure 4 visualizes an atmospheric entry trajectory and relates to the altitude, velocity, and Knudsen number.

1. RAREFIED FLOWS

In rarefied flows, the molecules that make up the flow (medium) are separated by a sufficiently long distance such that the medium can be regarded as a collection of individual particles. For objects entering the Earth atmosphere, this situation holds true for objects in a low planetary orbit down to an altitude of about 100 km. The parameter that is used to represent the degree of rarefaction is the Knudsen number, *Kn*, which is defined as

$$Kn = \frac{\lambda}{L} \tag{2}$$

where λ, mean molecular-free path is described, and *L* is the characteristic length (or dimension) of the analyzed object [4]. The mean-free path is the distance a molecule travels on average before interacting with another molecule. For rarefied flows, the mean-free path between molecules is on the same order or larger than the characteristic dimension of the object traveling through the medium. In this regime, the density of molecules is so low that software tools dedicated to this

TABLE 1 RAREFIED REGIMES BASED ON KNUDSEN NUMBER

Knudsen Number	Regime
$Kn < 0.01$	Continuum
$0.01 < Kn < 0.1$	Slip flow (Sec. H)
$0.1 < Kn < 10$	Transitional
$10 < Kn$	Free molecular

environment actually model each individual particle. Typically, the Boltzmann equations are used to describe this environment.

Based on Knudsen number, flow regimes can be classified as in Table 1 [5]. This classification is similar to the graphical representation in Fig. 4, with the extremes being the free molecular regime and the continuum regime.

At near-vacuum conditions, the molecular mean-free path becomes sufficiently long for the probability of gas molecule–molecule interaction to diminish. The molecules are assumed to move in straight lines until they hit an obstacle. Flow characteristic information such as velocity or pressure are impractical for the purpose of describing the flow because these characteristics are not shared between the molecules. In this free-molecular regime, it is sensible to describe each individual molecule rather than describing a volume of molecules as an entity.

2. CONTINUUM FLOW

As an object plunges through the atmosphere and encounters higher-density gases, the interaction between the object and the atmospheric molecules becomes so frequent that individual interactions are no longer distinguishable. In this environment, the continuum assumption holds true. The flow field around the object does not resemble a collection of individual particles but rather a volume of a homogeneous medium. In contrast to rarefied flows, in continuum flow, the molecule–molecule interactions within the gas have a significant effect on the behavior of the flow because they occur much more frequently compared to rarefied flows. For continuum flows, the mean-free path is smaller than the dimensions of the object traveling through the medium and the Knudsen number is (much) lower. The medium is efficiently modeled as a homogeneous medium rather than a collection of individual particles. This is sensible because due to the molecule–molecule interactions, flow information is shared among neighboring molecules and thus, the molecules within a sufficiently small subvolume of the medium can be assumed to have the same physical properties. Furthermore, the computational effort to model each molecule individually would be unacceptably high because the density of the medium is higher compared to rarefied flows and thus many more particles are present. In this regime, the Navier–Stokes equations apply.

III. HYPERSONIC FLOW PHENOMENA

The flow regime of hypersonics is characterized by the occurrence and domination of flow phenomena that are negligible at lower velocities. This section introduces these phenomena and provides more information about concepts employed in the aerothermodynamic modeling.

A. VISCOUS INTERACTION

As the high-velocity flow enters the boundary layer, it slows down by viscous effects (mainly shear stress due to molecule–molecule interaction). High kinetic energy of the flow is partially transformed into internal energy. Because pressure is constant across a boundary layer, in the direction normal to the surface, the increase in temperature T results in a decrease in density. To pass the required mass flow through the boundary layer at reduced density, the boundary-layer thickness must be larger. This phenomenon lets hypersonic boundary layers grow more rapidly compared to what is expected at lower speeds. The thicker boundary layer induces a virtual change of shape to the (thin) hypersonic object. The external flow reacts according to the new virtual geometry and the flow field is different relative to the same object with a thinner boundary layer. Hypersonic modeling needs to account for thicker boundary layers.

B. THIN SHOCK LAYER

Shock waves are a characteristic of supersonic flows. Together with the body surface, the shock waves enclose the shock layer. For hypersonic flows, this shock layer can be very thin (Fig. 5). One physical phenomenon that occurs at high Mach numbers is the interaction of the shock wave itself with a thick, viscous boundary layer that grows from the body surface. Future research needs to address the accurate modeling of this interaction.

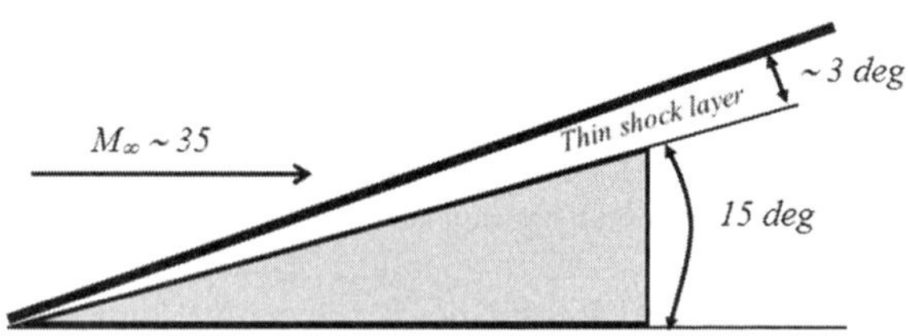

Fig. 5 Thin shock layer example for a hypersonic flow over a sharp body [4].

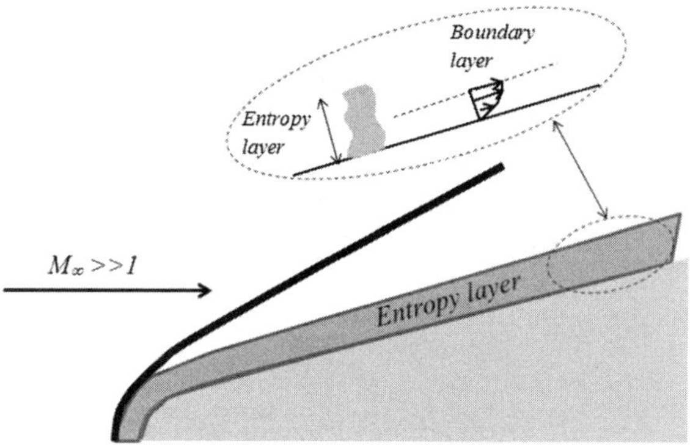

Fig. 6 Entropy layer around a blunt body [4].

C. ENTROPY LAYER

Across a sonic shock, the entropy of the flow increases radically. For a blunt body, the entropy gradient is the highest at the stagnation point[§] in front of the body where the flow passes in the normal direction through the shock. The gradient decreases along the downstream direction as the flow has more time and space to adjust to the conditions close to the body. The entropy layer flows downstream and essentially wets the body for large distances from the nose. As shown in Fig. 6, the boundary layer is located within the entropy layer and grows within it. Both layers also interact and are affected by each other. This phenomenon has to be considered in numerical efforts to model the flow.

D. HIGH-TEMPERATURE FLOWS

1. HEAT CAPACITY

With increasing temperature, the molecules in the flow start to vibrate with higher amplitudes. This causes the specific heat capacities c_p (gas kept at constant pressure) and c_v (gas kept at constant volume) to become functions of temperature. Suppose a gas is being heated. Heat capacity is a measure of how much heat energy is required to increase the gas temperature by 1 deg. Temperature, in a physical sense, is the vibrational energy of gas molecules where we can think of gas molecules as individual particles. As gas temperature increases further with more heat energy being delivered to it, the atoms, that collectively form the molecule, start to vibrate themselves against the attracting forces of their bonds. This type of energy absorption does not contribute to gas temperature increase and, therefore, more energy is needed to heat the gas. The result is higher heat capacities at higher temperatures. Consequently, the ratio of specific heats, $\gamma = c_p/c_v$, also becomes temperature dependent. Thermodynamic effects like these need to be taken into account when modeling hypersonic flows. In Fig. 7, a comparison between the temperatures of the shock layer over a spectrum of entry speeds is given for a medium that is modeled as a perfect gas and one that is modeled to be

[§]The stagnation point is the point on the leading edge of the object where the flow decelerated to zero velocity.

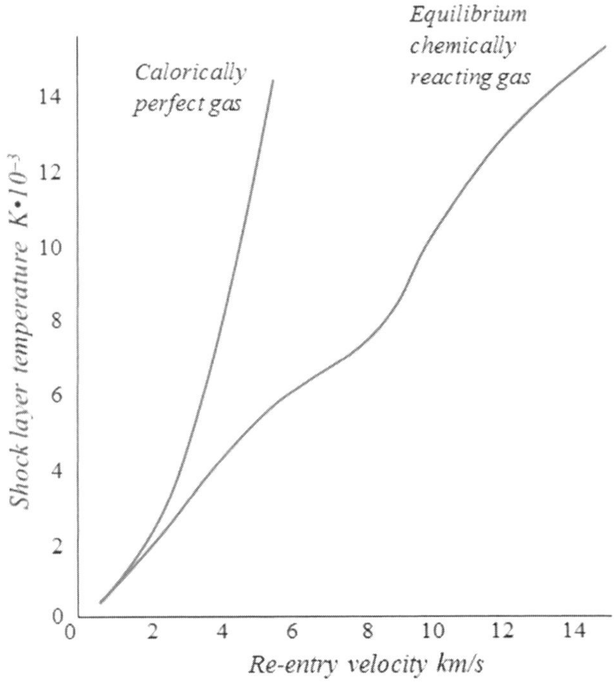

Fig. 7 Comparison of gas temperature in the shock layer when modeled as an ideal gas and gas in chemical equilibrium [4].

in chemical equilibrium. The perfect gas assumption yields conservative results where the temperatures are too high. Taking into account the energy absorption due to chemical reactions taking place in the flow, yields lower and hence more realistic temperatures [1].

2. AERODYNAMIC CONSIDERATIONS

The gas temperature behind the strong shock wave can be enormous at hypersonic speeds. The high-temperature hypersonic flow also has an effect on aerodynamic characteristics such as lift, drag, and moments because the pressure distribution over the object changes at hypersonic speeds [4]. In a practical example, the Space Shuttle's body-flap deflection had to be increased to twice the amount initially predicted. This situation occurred because the chemical reactions taking place in a high-temperature flow were originally neglected and the resulting pressure distribution was inaccurately modeled.

With increasing speeds, the chemical compounds of a gas have insufficient time to react. Consequently, the gas is in chemical and thermal nonequilibrium and the ideal gas assumption no longer applies. Consider the following definitions:

Definition—Thermal nonequilibrium: A gas is in thermal nonequilibrium if its internal energy cannot be characterized by a single temperature.

Definition—Chemical nonequilibrium: A gas is in chemical nonequilibrium if its chemical state does not satisfy chemical equilibrium conditions (state in which both reactants and products are present at concentrations which have no further tendency to change with time).

E. DAMKOHLER NUMBER

The Damkohler number gives an understanding for the potential of chemical nonequilibrium. It represents the ratio between the fluid motion time scale τ_f and the chemical reaction time scale τ_c and is defined as

$$Da = \frac{\tau_f}{\tau_c} \tag{3}$$

When $Da \to \infty$, the flow takes a longer time to cover a specific distance, while the chemical reaction time is relatively short. The flow reacts immediately to changes in chemical composition.

When $Da \to 0$, the flow is very fast in relation to the chemical reaction time. In this case, parts of the flow are in chemical nonequilibrium as the reactants are carried downstream before they can react. The gas is called frozen in this state.

F. FINITE-RATE WALL CATALYSIS

Connected with the flow speed and its specifics, chemical reaction time is the capability of the flow to chemically react with the vehicle surface. The potential to chemically react with the vehicle surface also depends on whether the wall is catalytic. The wall is catalytic when it facilitates a chemical reaction among gas molecules themselves or among gas molecules and the wall. When the recombination rate parameter is high (as used in Fig. 8—chemical reactions occur quickly—the flow molecules have the opportunity to react with the vehicle wall molecules before they move on). This also implies a catalytic wall. Heat transfer occurs where a portion of the molecule's internal energy is emitted to the body wall.

Figure 8 shows a plot of heat-transfer rate parameter $Nu/\sqrt{Re}$ (where Nu is the Nusselt number and Re denotes the Reynolds number) as a function of recombination rate parameter. Graph number 1 in Fig. 8 shows the total heat transfer for a catalytic wall. It becomes clear that heat-transfer rate is not a function of recombination rate parameter. However, as can be seen by graph number 3, the ratio of conductive and diffusive heat-transfer rate for a catalytic wall is a function of the recombination rate parameter. All heat transfer is due to conduction when the gas in the boundary layer is in chemical equilibrium (high recombination rate parameter) and can react with the wall. On the other hand, when the boundary layer is chemically frozen, a large portion of the total heat transfer is due to diffusion.

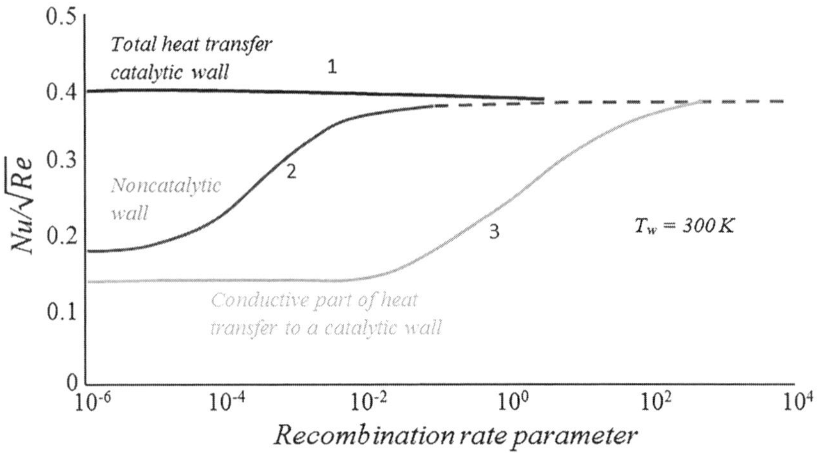

Fig. 8 Heat transfer rate ($Nu/\sqrt{Re}$) as a function of recombination rate parameter [4].

The boundary layer is a region around the object where viscous effects (shear stress) of the medium dominate the flow behavior. In continuum flow, the flow layer directly adjacent to the wall experiences shear stress due to the wall. This effectively slows down these flow molecules to zero velocity. Moving away in the normal direction from the wall, the flow speed gradually increases until it reaches a full flow speed. The region from the wall to this point is called boundary layer and is visualized in Fig. 6.

When the flow is frozen (small recombination rate parameter) and the wall is noncatalytic (curve number 2 in Fig. 8), heat transfer is reduced by up to 50%. This is due to the fact that, because the flow and the wall cannot react chemically, a big portion of the molecules' internal energy cannot be transferred to the wall as heat. It should also be noted that recombination rate scales with the square of the density ($\propto \rho^2$). Therefore, recombination is slow in the upper parts of the atmosphere during reentry. The chemical reaction, on the other hand, is less affected by density but strongly dependent on the flow molecule's diffusion capability through the wall. A challenge to the correct modeling and design of chemical interaction between the flow and the body wall is the uncertainty of the material properties. Generally, reaction behavior of the wall material is only known with a selection of reaction partners. Additionally, this knowledge is tied to specific temperatures. However, the reaction behavior changes with the temperature regime.

G. NONEQUILIBRIUM THERMAL RADIATION

Thermal radiation is also affected by the nonequilibrium state of the flow. However, this process is very hard to model because of complex rate-dependent processes that determine the thermal energy produced by the shock wave. The temperature behind the shock wave can significantly exceed the equilibrium-state temperature. Thus, a large population of overexcited molecules exist that will emit energy in the form of photons (radiative heating) to return to their equilibrium energy state. The chemical composition can also be different from the equilibrium state which means that the energy levels of the molecules present are different from the ones in the flow before the shock wave. Generally speaking, radiative heating is important for high-energy flows because a significant increase in radiative heating due to nonequilibrium is expected to occur. The modeling of flow thermal radiation remains a major challenge.

H. SLIP EFFECT

In continuum flow, the flow velocity continuously decreases along the thickness of the boundary layer toward the vehicle wall. At the vehicle wall, the flow velocity is zero. This no-slip condition does not hold true for flows approaching the rarefied environment. In rarefied flows, the molecule vehicle interaction becomes sufficiently scarce for nonzero flow velocities to be observable at the vehicle wall [6].

I. VIBRATION-DISSOCIATION COUPLING

When a gas is heated, its population of molecules becomes vibrationally excited. This leads to elevated dissociation rates. The mechanisms of this process are not fully understood. Simple models that can be employed in complex numerical simulations largely lack validation. Their results are therefore subject to uncertainties. This is crucial because even small changes in dissociation rate affect the flow field significantly. Therefore, local and integrated flow characteristics are only known with low confidence.

1. MOLECULAR DISSOCIATION

Eventually, the internal molecular vibrations become violent enough to break their bonds. Dissociation of the molecule ensues. For example, O_2 starts to dissociate to 2 O at about 2000 K and is completely dissociated at 4000 K. At this temperature N_2 dissociates to 2 N with complete dissociation occurring at 9000 K.

2. MOLECULAR IONIZATION

Ionization of atoms is the next step along increasing temperature. Over 9000 K, nitrogen ionizes into $N^+ + e^-$ and oxygen ionizes into $O^+ + e^-$. At this state, the gas is a partially ionized plasma.

3. HIGH HEAT-TRANSFER RATES

The extreme heat gradient from the hot-boundary layer to the vehicle's surface causes high heat-transfer rates. Heat is transferred via two mechanisms; convective heating q_c occurs where the flow directly touches the surface. In this case, the flow and vehicle molecules interact directly. This is the first mechanism. Convective heat transfer is linearly proportional to the temperature gradient, $q_c \propto T_1 - T_2$. The second mechanism is thermal radiation. In contrast to convective heating, radiative heat transfer is proportional to the difference of the two temperatures raised to their fourth power respectively, $q_d \propto T_1^4 - T_2^4$. This type of heat transfer becomes predominant at higher-temperature gradients. In fact, radiative heat transfer made up 30% of Apollo's total heating during reentry. In the case of a space probe entering Jupiter's atmosphere, radiative heating amounts to 95% of the total heat rate.

IV. FLOW EQUATIONS AND MODELING

Numerical methods have proven to accurately predict flow characteristics of moving objects, provided that the essential physics is considered. The selection of adequate models/equations depends on the flow environment (speed, density, . . .); for example, an overview of the applicable equations depending on the Knudsen number is provided in Fig. 9.

Continuum models may be employed at lower speeds and higher-density regions relative to discrete particle models. The latter are more adequate for the rarefied regimes. In theory, discrete particle models could be employed over the entire flow speed spectrum. However, in high-density flows, the computational requirements forbid this.

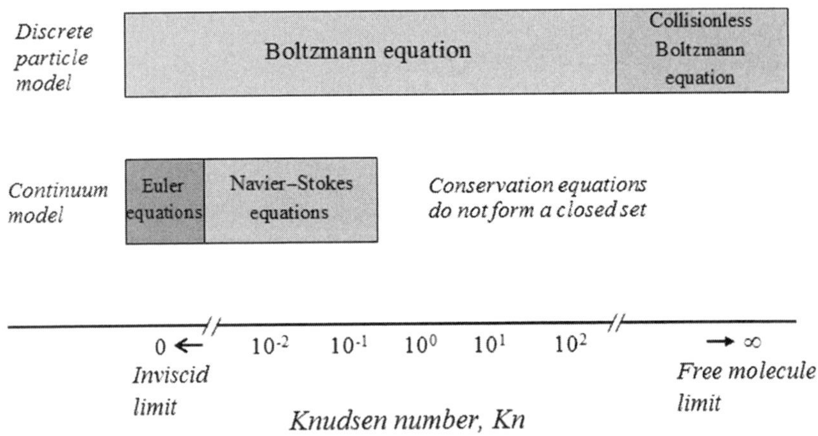

Fig. 9 Applicable equations for Knudsen number regimes [4].

A. NUMERICAL APPROACH FOR HIGH-ENERGY FLOWS

Computational modeling of flows faces unique challenges when high-temperature and hypersonic effects must be considered.

The standard equations to solve for continuum flow are the Navier–Stokes equations. For accurate high-energy flow modeling considering chemical nonequilibrium, the set of equations needs to be extended. They need to incorporate mass and energy conservation. In addition, a turbulence model as well as chemical species continuity equations need to be employed. To account for thermal nonequilibrium, more modeling on the molecular and nuclear level is required. Equations describing the energy conservation of vibrationally excited molecules and the energy equations for electron modes are to be implemented.

In the rarefied regime, the continuum assumptions are no longer valid, and nonequilibrium effects dominate. The Boltzmann equation provides an effective means of modeling this regime with high fidelity

$$\frac{\partial(nf)}{\partial t} + c\frac{\partial(nf)}{\partial r} + F\frac{\partial(nf)}{\partial c} = J(f, f^*) \tag{4}$$

$$J(f, f^*) = \int_{-\infty}^{\infty}\int_{0}^{4\pi} n^2(f^* f_1^* - f f_1)c_r \xi \, \mathrm{d}\Omega \, \mathrm{d}c_1 \tag{5}$$

Solving the equation requires advanced methods because of the nonlinearity of the collision integral. Numerical methods are employed to find solutions to the Boltzmann equation. These methods are also called direct simulation methods because they try to model the physics of the dynamics directly rather than transforming the problem into analytical models. Consequently, a common approach is to directly represent real molecules as particles in simulations. The molecules' positions, velocities, and energies are stored and updated at each time step. A vastly successful numerical method for solving rarefied gas flows in the transitional regimes has been the direct simulation Monte Carlo method (DSMC) [7].

A peculiarity of hypersonic flow research is that the analyzed problems not only entail hypersonic flows but also exhibit super- and subsonic regions locally. It is due to this flow phenomena diversity that an efficiency increase in modeling may be achieved by applying a variation of equations across the problem space. For example, equations taking into account the slip effect only need to be applied in regions where the slip effect occurs. Similarly, a set of Boltzmann equations might be more useful for the regions in front of the shock waves whereas Navier–Stokes equations are more adequate for the slower, higher-density flows postshock. This approach increases the complexity of setting up the model but it saves computational resources.

V. VALIDATION

Validation is essential in order to gain any practical value from simulations. Simulations need to be tested against and adjusted to real-life test results. This is the only way one can hope to get accurate predictions out of computer models. Two validation methods shall be mentioned here.

A. WIND TUNNELS

Wind tunnels are facilities in which particular flow conditions are artificially created. Their configuration differs depending on the targeted flow velocities (Reynolds number). All test installations have in common that they cannot reproduce the entire flight envelope but rather only cover specific parts of it. Additionally, test facilities always idealize the real environment. Testing only improves our confidence in believing that a vehicle will behave as intended. The only way to gain certainty is by flying the actual flight hardware.

1. LOW-TO-MEDIUM-SPEED WIND TUNNEL

Test facilities for lower Reynolds numbers or subsonic velocities are often closed circular tube installations with a turbine accelerating the gas medium [8].

These facilities are able to produce laminar flow fields for practically indefinite durations of time. Medium-speed wind tunnels achieve velocities of up to 5 Ma. The faster the flow, the harder it is to build a wind tunnel that can continuously maintain the high-velocity flow field. This is why in the supersonic range of medium-speed wind tunnels, only a small fraction is capable of supporting a constant high-velocity flow field. Wind tunnels that provide a supersonic test environment for up to a few seconds are more common.

2. HIGH-ENERGY WIND TUNNEL

High-velocity wind tunnels, on the other hand, are usually "open-loop" installations in which the flow field can only be maintained for up to a few milliseconds. They can cover flow regimes from 3 to 30 Ma.

a. SCIROCCO Plasma Wind Tunnel

A notable example of a hypersonic wind tunnel is the European SCIROCCO [9]. It was built in 2001 in collaboration with the European Space Agency (ESA) as a special wind tunnel capable of simulating the thermal conditions of space vehicles reentering from low Earth or interplanetary orbits (from the Moon or Mars), thereby improving flight safety in this very sensitive phase. At the time of its conception, the Hermes spaceplane program was under way and the need for this type of testing facility was identified [10]. The facility typically focused on the development and qualification of thermal protection systems for aerospace use. Electric arc heaters are used to generate high air temperatures of up to 10,000°C. The dimensions of the test chamber, the 70-MW electric arc, the size of the jet, and the level of automation make SCIROCCO the largest and most-advanced plasma tunnel in the world. The system is capable of producing test conditions for up to 30 min that correspond to the mission profile of reentry vehicles [9].

b. Shock Tunnels

Shock tunnels can recreate flight conditions for about 50 ms and aim at providing very-high-speed flows. This is achieved by using a driver gas to compress a gas volume called test gas. Upon compression, the test gas is expanded over the test object recreating in flight conditions. There are three common shock tunnel designs: 1) the reflected shock tunnel, 2) the gun tunnel, and 3) the longshot tunnel. In the reflected shock tunnel, the driver gas is put under high pressure (HP reservoir) until a diaphragm ruptures. The sudden expansion generates a shock wave in the test gas that travels down the tube. The reflection of the wave from the downstream end of the shock tube produces a quasi steady test gas reservoir. Gun tunnels and longshot tunnels operate on similar principles. However, in these designs, the driver gas is separated from the test gas by means of a piston. The existence of a piston introduces oscillations in the flow due to the mass inertia of the piston. In the longshot design, these oscillations are thought to be minimized by a heavier piston and nonreturn valves in front of the nozzle.

B. IN-FLIGHT TESTS

Out of all validation efforts, flight tests come closest to real hardware operation conditions. However, only a limited portion of the entire flight envelope may usually be covered during the test. Still, this kind of testing is always desirable because of its representative potential. However, in-flight tests are scarcely employed because they are expensive and require long lead times. Another issue is the general nonrepeatability of test conditions because environmental influences change from test to test.

VI. PREVIOUS 25-YEARS' MODELING PROGRESSION

In the last 25 years, computational fluid dynamics (CFD) modeling has become a research and industry standard for flow analysis. In this time frame, the discipline experienced the introduction of new technologies that changed the way modeling is done. However, some aspects remained constant throughout the period [11].

A. CHEMISTRY MODELS

The way underlying chemical processes, such as reaction rates, vibration-dissociation coupling, and other phenomena, are modeled have largely remained the same from the early 1990s to the 2010s [12]

B. STAGNATION POINT CHARACTERISTICS

One of the major challenges that remain unresolved to a satisfactory degree is the prediction of stagnation point heating. This shortcoming is more pronounced for very blunt, capsule-like geometries.

C. POTENT COMPUTER HARDWARE

The increase in computational capabilities continues to follow a rapid growth. Processing power doubles approximately every two years.[¶] As a result, one off-the-shelf server has roughly 100 times the computing power compared to a world-class supercomputer from 20 years ago. A similar development has been observable for memory and hard-disk capacity. In practical terms, this process necessitated massive rewriting of the computational methods. To take full advantage of the new capabilities, the code needed to become parallel-computing compatible. Today, both improvements in numerical methods and advances in computing hardware that affect simulation fidelity potential increase equally.

D. GRID GENERATION

With the technological means of conducting simulations steadily improving, the problem of grid generation emerges as a dominant problem. Commercially available grid generators approach their operational limit at grid sizes of 100 million elements. It is generally difficult to manage the grid generation process for massive grids. Subtle problems in grid generation degrade simulation quality significantly. For example, the spatial distribution and density of grid nodes is usually adjusted to the local needs of different regions of the modeled problem. In regions where flow characteristics (temperature, pressure, density, and chemical nonequilibrium) change fast (a high gradient), the analysis resolution has to be finer. Typically, the region around shocks requires very high node densities to accurately capture the flow effects. Skillful node placement is required and is usually the result of an iterative process. Just one poorly located element has the potential to ruin an entire simulation.

E. DATA MANAGEMENT AND VISUALIZATION

Another area that emerges as one of the primary concerns for CFD specialists is data management and visualisation of results. The latter process can be very time consuming.

[¶] Moore's law.

F. UNSTEADY CFD PROBLEMS

Today, CFD models primarily address steady-state problems. This means that the flow conditions are stable and do not change over time. However, an object entering the atmosphere from orbital velocities experiences a wide range of flow conditions. Thus, the problem of atmospheric entry is inherently unsteady. One of the next steps in hypersonic modeling is the routine solution of unsteady problems. In this type of problem, the flow does not approach a steady equilibrium state over time but rather adapts to changing conditions. Unsteady CFD problems increase simulation time excessively and cause severe solution analysis problems.

G. MODEL VALIDATION

In summary, it can be noted that major advances on the technological side of CFD modeling allow for increasingly detailed simulations. However, the achieved progress of validation procedures and mechanisms for these ever-more complex simulations have lagged behind the computational advances. This is an area that needs to be addressed in the future to increase confidence in simulation accuracy.

H. FUTURE PROSPECTS

The development of CFD algorithms is advancing. New schemes such as an upwind for improved shock capturing are researched. Already, advancements in multigrid algorithms have significantly accelerated the convergence to a steady state compared to single-grid algorithms. There is still ample room for improvement for the chemical source terms in these algorithms. Prospects of future development exist to address this shortcoming. Problems associated with convergence and truncation errors in very-low-velocity flows, are sought to be improved by measures such as local preconditioning. This method will prove to be valuable in regions such as the stagnation point's vicinity of a blunt body.

REFERENCES

[1] Anderson, J. D., *Fundamentals of Aerodynamics*, 5th ed., McGraw-Hill Higher Education, New York, 2011.

[2] Cook, M., *Flight Dynamic Principles*, Butterworth-Heinemann, Oxford, UK, 2007.

[3] Weiland, C., *Computational Space Flight Mechanics*, Springer, New York, USA, 2010.

[4] Anderson, J. D., *Hypersonic and High Temperature Gas Dynamics*, 2nd ed., AIAA Education, AIAA, Reston, VA, 2006.

[5] Josyula, E., and Burt, J., "Review of Rarefied Gas Effects in Hypersonic Applications," Technical Report, NATO, 2012.

[6] Arkilic, E. B., Schmidt, M. A., and Breuer, K. S., "Slip Flow in Micro Channels," *Proceedings of Rarefied Gas Dynamics Symposium*, Massachusetts Institute of Technology, MA, Oxford, UK, 1994, p. 7.

[7] Bird, G. A., *The DSMC Method*, CreateSpace Independent Publishing Platform, 2013.

[8] Pereira, J. D., *Wind Tunnels: Aerodynamics, Models and Experiments. Engineering Tools, Techniques and Tables*, Nova Science Publishers, Inc., New York, 2011.

[9] Borrelli, R., and Martucci, A., *SCIROCCO Plasma Wind Tunnel: Synergy Between Numerical and Experimental Activities for Tests on Aerospace Structures. Wind Tunnels and Experimental Fluid Dynamics Research*, InTech, 2011.

[10] Lu, F. K., and Marren, D. E., *Advanced Hypersonic Test Facilities*. Progress in Astronautics and Aeronautics, AIAA, Reston, VA, 2002.

[11] Candler, G. V., Mavriplis, D. J., and Trevio, L., "Current Status and Future Prospects for the Numerical Simulation of Hypersonic Flows," *Proceedings of 47th AIAA Aerospace Sciences Meeting Including the New Horizons Forum and Aerospace Exposition*, AIAA, Reston, VA, 2009, pp. 1755–1765.

[12] Riedel, U., Maas, U., and Warnatz, J., "Detailed Numerical Modeling of Chemical and Thermal Nonequilibrium in Hypersonic Flows," *Impact of Computing in Science and Engineering*, Vol. 5, 1993, pp. 20–52.

CHAPTER 10

Classical Methods of Orbit Determination

Giovanni F. Gronchi* **and Stefano Marò***
University of Pisa, Pisa, Italy

I. INTRODUCTION

One important date in the history of celestial mechanics is 1 January 1801, when G. Piazzi detected Ceres, the first asteroid. He could follow it up in the sky for about one month, collecting 21 observations that formed an arc of about 3 deg. After that he wasn't able to observe it any more, thus the problem arose of predicting in which part of the sky one should look to recover it. This problem is nowadays known as the orbit determination problem and consists of computing the orbital elements of a celestial body given its observations. Ceres was recovered in 1802 by H. W. Olbers and F. Von Zach following the computations of C. F. Gauss who determined an orbit with Piazzi's observations. Given at least three observations of a solar system body, his method consists of two steps: first compute a preliminary orbit, then apply the least squares method (differential corrections) using the preliminary orbit as a starting guess. A preliminary orbit determination was already studied by Laplace in 1780 [1], whereas Gauss published his method in 1809 [2].

The scheme proposed by Gauss is still used today and we are going to describe it in details. More precisely, we present both Laplace's and Gauss' methods to find a preliminary orbit. Then we present the least squares method. We stress the fact that Gauss' and Laplace's methods have been studied and modified during the years: one can find some recent reviews in Refs. [3, 4]. Here we are going to present a modern version of both methods.

Laplace's method makes the approximation of having geocentric observations. It leads to an 8-deg polynomial equation to be solved. Gauss' method naturally deals with topocentric observations. Similar to Laplace's case, it leads to an 8-deg polynomial equation. This equation has the same algebraic structure as Laplace's and contains it as a particular case. Both equations may have multiple solutions, corresponding to different preliminary orbits. This problem is of great importance because it can lead to severe misunderstandings. For example, one can confuse

*Department of Mathematics.

a near-Earth object (NEO) with a main belt asteroid (MBA). The occurrence of multiple solutions was first studied by Charlier [5,6] in the simpler case of Laplace. We present how he was able to prove that the number of solutions depend in a simple way on the position where the object is observed. Recently, a generalization of Gauss' method was given in Ref. [7] and we summarize it.

Once a preliminary orbit is computed, one can improve it by adding other observations. The idea of Gauss was to use a method to minimize the difference between the observations and the computed orbit. This is known as the least squares method, and the solution is found by an iterative algorithm called differential corrections.

This chapter is organized as follows. In Sec. II, we are concerned with the preliminary orbit problem. We give a formal introduction to the problem and describe the methods by Laplace and Gauss. In Sec. III, we explain the least squares method and the related differential corrections. Finally in Sec. IV, we deal with the existence of multiple solutions: we present the classical theory by Charlier and its generalization.

II. PRELIMINARY ORBITS

We describe the topocentric position $\boldsymbol{\rho}$ of a celestial body in spherical coordinates ρ, α, δ representing, respectively, topocentric radial distance, right ascension, and declination.

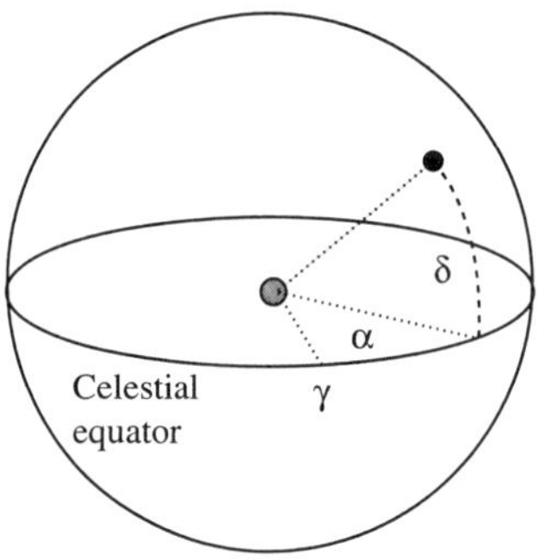

Notice that the observations are taken from the Earth, that is not the only influencing body because of the presence of the Sun. Let us introduce the following notation. We indicate by $\boldsymbol{q}_{\oplus}$ the heliocentric distance of the center of the Earth and by $\boldsymbol{p}_{\text{obs}}$ the position of the observer. Thus, the heliocentric distance of the observer is $\boldsymbol{q} = \boldsymbol{q}_{\oplus} + \boldsymbol{p}_{\text{obs}}$. The heliocentric distance $\boldsymbol{r}$ of the body is given by $\boldsymbol{r} = \boldsymbol{q} + \boldsymbol{\rho}$. The angle ϵ between $\boldsymbol{q}$ and $\boldsymbol{\rho}$ is called coelongation and satisfies the geometric equation

$$r^2 = q^2 + \rho^2 + 2q\rho\cos\epsilon \tag{1}$$

where $r = ||\boldsymbol{r}||$ and $q = ||\boldsymbol{q}||$.

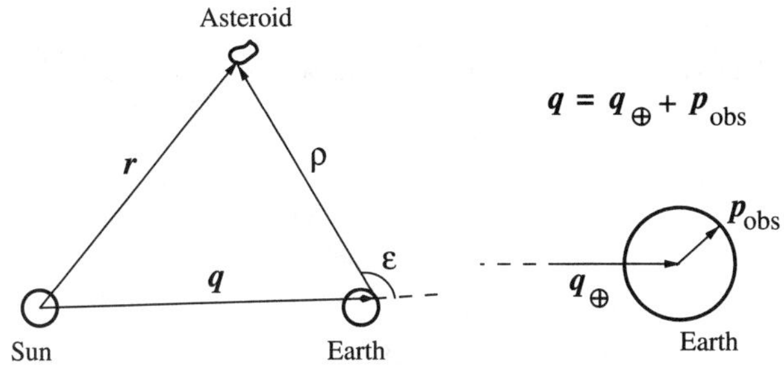

From optical telescopes we can obtain information on the angular coordinates α and δ and their corresponding velocities $\dot{\alpha}$ and $\dot{\delta}$. Generally, the radial coordinate ρ and the related velocity $\dot{\rho}$ cannot be determined. Preliminary orbit determination methods consist of finding a value for the unknown ρ and $\dot{\rho}$. These values are used to get the set $(\boldsymbol{r}_i, \dot{\boldsymbol{r}}_i)$ that allows to compute orbital elements.

A. LAPLACE'S METHOD

The method by Laplace to find a preliminary orbit uses the approximation

$$\boldsymbol{q} = \boldsymbol{q}_\oplus$$

that is, it supposes that the observations are taken from the center of the Earth. We start from three or more observations (α_i, δ_i) at times $t_i, i = 1, 2, 3 \ldots$. From them we can interpolate to get $(\alpha, \delta, \dot{\alpha}, \dot{\delta}, \ddot{\alpha}, \ddot{\delta})$ at a mean time $\bar{t}$. Generally, a 2-deg polynomial fit is enough for this purpose.

With the notation used so far, we can define

$$\mathbf{v} = \frac{\mathrm{d}\hat{\boldsymbol{\rho}}}{\mathrm{d}t} = \eta \hat{\mathbf{v}}$$

where $\eta = ||\mathbf{v}||$. Using the arc length parameter s, defined by $\mathrm{d}s/\mathrm{d}t = \eta$, we have $\mathrm{d}\hat{\boldsymbol{\rho}}/\mathrm{d}s = \hat{\mathbf{v}}$. And so we consider the moving orthogonal frame $\{\hat{\boldsymbol{\rho}}, \hat{\mathbf{v}}, \hat{\mathbf{n}}\}$ given by

$$\hat{\boldsymbol{\rho}}, \quad \hat{\mathbf{v}} = \frac{\mathrm{d}\hat{\boldsymbol{\rho}}}{\mathrm{d}s}, \quad \hat{\mathbf{n}} = \hat{\boldsymbol{\rho}} \times \hat{\mathbf{v}}$$

The geodesic curvature κ is introduced by the relation

$$\frac{\mathrm{d}\hat{\mathbf{v}}}{\mathrm{d}s} = -\hat{\boldsymbol{\rho}} + \kappa \hat{\mathbf{n}}$$

and it measures the deviation of the path from a great circle (i.e., a geodesic on the sphere). The value of κ depends on the whole vector $(\alpha, \delta, \dot{\alpha}, \dot{\delta}, \ddot{\alpha}, \ddot{\delta})$. The acceleration of $\boldsymbol{\rho}$ is given by

$$\frac{d^2}{dt^2}\boldsymbol{\rho} = (\ddot{\rho} - \rho\eta^2)\hat{\boldsymbol{\rho}} + (\rho\dot{\eta} + 2\dot{\rho}\eta)\hat{\mathbf{v}} + (\rho\eta^2\kappa)\hat{\mathbf{n}}$$

On the other hand we have

$$\frac{d^2}{dt^2}\boldsymbol{\rho} = \frac{d^2}{dt^2}(\mathbf{r} - \mathbf{q})$$

where, according to the equations of the two-body motion,

$$\frac{d^2}{dt^2}\mathbf{r} = -\frac{\mu}{r^3}\mathbf{r}, \quad \frac{d^2}{dt^2}\mathbf{q} = -\frac{\mu + \mu_\oplus}{q^3}\mathbf{q}$$

with $r = \|\mathbf{r}\|$ and μ, $\mu_\oplus$ the masses of the Sun and of the Earth, respectively.

Neglecting the mass of the Earth and projecting the equation of motion on $\hat{\mathbf{n}}$ at time $\bar{t}$ we obtain

$$C\frac{\rho}{q} = 1 - \frac{q^3}{r^3} \quad \text{where} \quad C = \frac{\eta^2\kappa q^3}{\mu(\hat{\mathbf{q}} \cdot \hat{\mathbf{n}})} \tag{2}$$

where $\rho, q, r, \eta, \hat{\mathbf{q}}, \hat{\mathbf{n}}, C$ are the values at time $\bar{t}$.

Equation (2) is called the dynamical equation of Laplace's method.

Here ρ and r are unknowns, whereas the other quantities are obtained by interpolation.

Using Eq. (2) and the geometric equation

$$r^2 = q^2 + \rho^2 + 2q\rho \cos\epsilon$$

we can write a polynomial equation of degree 8 for r by eliminating the geocentric distance ρ:

$$C^2r^8 - q^2(C^2 + 2C\cos\epsilon + 1)r^6 + 2q^5(C\cos\epsilon + 1)r^3 - q^8 = 0$$

Projecting the equation of motion on $\hat{\mathbf{v}}$ yields

$$\rho\dot{\eta} + 2\dot{\rho}\eta = \mu(\mathbf{q} \cdot \hat{\mathbf{v}})\left(\frac{1}{q^3} - \frac{1}{r^3}\right) \tag{3}$$

Now we can use Eq. (3) to compute $\dot{\rho}$ from the values of r, ρ found by the geometric and dynamical equations. This is enough to compute a preliminary orbit.

B. GAUSS'S METHOD

Gauss's method does not make the approximation introduced by Laplace and naturally deals with topocentric observations. This method uses three observations (α_i, δ_i), $i = 1, 2, 3$, related to the heliocentric position of the observed body

$$\mathbf{r}_i = \boldsymbol{\rho}_i + \mathbf{q}_i$$

at times t_i, with $t_1 < t_2 < t_3$.

Here $\boldsymbol{\rho}_i$ denotes the topocentric position of the observed body, and $\mathbf{q}_i$ is the heliocentric position of the observer at time t_i. The basic assumption of this method is that $t_i - t_j$ is much smaller than the period of the orbit.

We consider a two-body model so that we can assume the coplanarity condition

$$\lambda_1 \mathbf{r}_1 - \mathbf{r}_2 + \lambda_3 \mathbf{r}_3 = 0, \quad \lambda_1, \lambda_3 \in \mathbb{R} \tag{4}$$

The vector product of both members of Eq. (4) with $\mathbf{r}_i$, $i = 1, 3$, together with the projection along the direction $\hat{\mathbf{c}}$ of the angular momentum yields

$$\lambda_1 = \frac{\mathbf{r}_2 \times \mathbf{r}_3 \cdot \hat{\mathbf{c}}}{\mathbf{r}_1 \times \mathbf{r}_3 \cdot \hat{\mathbf{c}}}, \qquad \lambda_3 = \frac{\mathbf{r}_1 \times \mathbf{r}_2 \cdot \hat{\mathbf{c}}}{\mathbf{r}_1 \times \mathbf{r}_3 \cdot \hat{\mathbf{c}}}$$

Now dynamics come into play. Following Refs. [8, 9] the differences $\mathbf{r}_1 - \mathbf{r}_2$, $\mathbf{r}_3 - \mathbf{r}_2$ can be expanded in powers of $t_{i2} = t_i - t_2 = \mathcal{O}(\Delta t)$, $i = 1, 3$.

We can leave only $\mathbf{r}_2$, $\dot{\mathbf{r}}_2$ in the expansion by replacing the second derivative $\ddot{\mathbf{r}}_2$ with $-\mu \mathbf{r}_2 / r_2^3$. Finally, we get the so-called expansion in f, g series

$$\mathbf{r}_i = f_i \mathbf{r}_2 + g_i \dot{\mathbf{r}}_2$$

where

$$f_i = 1 - \frac{\mu}{2} \frac{t_{i2}^2}{r_2^3} + \mathcal{O}(\Delta t^3), \quad g_i = t_{i2} \left(1 - \frac{\mu}{6} \frac{t_{i2}^2}{r_2^3} \right) + \mathcal{O}(\Delta t^4) \tag{5}$$

Using the f, g series we have

$$\mathbf{r}_i \times \mathbf{r}_2 = -g_i \mathbf{c}, \quad i = 1, 3$$
$$\mathbf{r}_1 \times \mathbf{r}_3 = (f_1 g_3 - f_3 g_1) \mathbf{c}$$

so that

$$\lambda_1 = \frac{g_3}{f_1 g_3 - f_3 g_1}, \qquad \lambda_3 = \frac{-g_1}{f_1 g_3 - f_3 g_1} \tag{6}$$

Inserting the expressions of f_i, g_i into Eq. (6) we obtain

$$\lambda_1 = \frac{t_{32}}{t_{31}}\left[1 + \frac{\mu}{6r_2^3}(t_{31}^2 - t_{32}^2)\right] + \mathcal{O}(\Delta t^3)$$
$$\lambda_3 = \frac{t_{21}}{t_{31}}\left[1 + \frac{\mu}{6r_2^3}(t_{31}^2 - t_{21}^2)\right] + \mathcal{O}(\Delta t^3)$$

From the scalar product of both members of Eq. (4) with $\hat{\boldsymbol{\rho}}_1 \times \hat{\boldsymbol{\rho}}_3$, using the relations $\mathbf{r}_i = \boldsymbol{\rho}_i + \mathbf{q}_i$, we obtain

$$\rho_2(\hat{\boldsymbol{\rho}}_1 \times \hat{\boldsymbol{\rho}}_3 \cdot \hat{\boldsymbol{\rho}}_2) = \hat{\boldsymbol{\rho}}_1 \times \hat{\boldsymbol{\rho}}_3 \cdot (\lambda_1 \mathbf{q}_1 - \mathbf{q}_2 + \lambda_3 \mathbf{q}_3) \tag{7}$$

We define

$$V = \hat{\boldsymbol{\rho}}_1 \times \hat{\boldsymbol{\rho}}_2 \cdot \hat{\boldsymbol{\rho}}_3$$

By substituting the expressions of λ_1, λ_3 into Eq. (7), using relations

$$t_{31}^2 - t_{32}^2 = t_{21}(t_{31} + t_{32})$$
$$t_{31}^2 - t_{21}^2 = t_{32}(t_{31} + t_{21})$$

we can write

$$-V\rho_2 t_{31} = \hat{\boldsymbol{\rho}}_1 \times \hat{\boldsymbol{\rho}}_3 \cdot \big(t_{32}\mathbf{q}_1 - t_{31}\mathbf{q}_2 + t_{21}\mathbf{q}_3\big)$$
$$+ \hat{\boldsymbol{\rho}}_1 \times \hat{\boldsymbol{\rho}}_3 \cdot \left[\frac{\mu}{6r_2^3}[t_{32}t_{21}(t_{31} + t_{32})\mathbf{q}_1 + t_{32}t_{21}(t_{31} + t_{21})\mathbf{q}_3]\right] + \mathcal{O}(\Delta t^4)$$

By the smallness of $t_i - t_j$ we can neglect the $\mathcal{O}(\Delta t^4)$ terms and set

$$A(\mathbf{q}_1, \mathbf{q}_2, \mathbf{q}_3) = q_2^3\, \hat{\boldsymbol{\rho}}_1 \times \hat{\boldsymbol{\rho}}_3 \cdot [t_{32}\mathbf{q}_1 - t_{31}\mathbf{q}_2 + t_{21}\mathbf{q}_3]$$
$$B(\mathbf{q}_1, \mathbf{q}_3) = \frac{\mu}{6} t_{32}t_{21}\hat{\boldsymbol{\rho}}_1 \times \hat{\boldsymbol{\rho}}_3 \cdot [(t_{31} + t_{32})\mathbf{q}_1 + (t_{31} + t_{21})\mathbf{q}_3]$$

In this way the last equation becomes

$$-\frac{V\,\rho_2\,t_{31}}{B(\mathbf{q}_1, \mathbf{q}_3)}\,q_2^3 = \frac{q_2^3}{r_2^3} + \frac{A(\mathbf{q}_1, \mathbf{q}_2, \mathbf{q}_3)}{B(\mathbf{q}_1, \mathbf{q}_3)}$$

Finally, letting

$$C = \frac{V\,t_{31}\,q_2^4}{B(\mathbf{q}_1, \mathbf{q}_3)}, \quad \gamma = -\frac{A(\mathbf{q}_1, \mathbf{q}_2, \mathbf{q}_3)}{B(\mathbf{q}_1, \mathbf{q}_3)}$$

we obtain the dynamical equation of Gauss' method:

$$C\,\frac{\rho_2}{q_2} = \gamma - \frac{q_2^3}{r_2^3} \tag{8}$$

We can find the possible values for r_2 by coupling Eq. (8) with the geometric equation

$$r_2^2 = \rho_2^2 + q_2^2 + 2\rho_2 q_2 \cos\epsilon_2$$

This yields the polynomial equation

$$P(r) = C^2 r^8 - q^2(C^2 + 2C\gamma\cos\epsilon + \gamma^2)r^6 + 2q^5(C\cos\epsilon + \gamma)r^3 - q^8 = 0$$

The roots of this polynomial allow to compute the value of $\mathbf{r}_2$. Using Gibbs' formulas [8, Chapter 8] the velocity vector $\dot{\mathbf{r}}_2$ can be computed and we get a preliminary orbit.

III. LEAST-SQUARES SOLUTIONS

The preliminary orbit can be improved, obtaining smaller residuals with respect to the observations. This can be done in several ways and here we present the least squares method. This technique was first introduced by Gauss. Consider the differential equation

$$\frac{\mathrm{d}\mathbf{y}}{\mathrm{d}t} = \mathbf{f}(\mathbf{y}, t, \boldsymbol{\mu}) \tag{9}$$

giving the state $\mathbf{y} \in \mathbb{R}^p$ of the system at time t. For example $p = 6$ if $\mathbf{y}$ is a vector of orbital elements. Moreover, the vector $\boldsymbol{\mu} \in \mathbb{R}^{p'}$ represents the dynamical parameters. The integral flow, solution of Eq. (9) for initial data $\mathbf{y}_0$ at time t_0, is denoted by $\Phi_{t_0}^t(\mathbf{y}_0, \boldsymbol{\mu})$.

We also introduce the observation function

$$\mathbf{R} = (R_1, \ldots, R_k), \quad R_j = R_j(\mathbf{y}, t, \boldsymbol{\nu}), \quad j = 1 \ldots k$$

depending on the state $\mathbf{y}$ of the system at time t. The vector $\boldsymbol{\nu} \in \mathbb{R}^{p''}$ represents the kinematic parameters.

Moreover, we define the prediction function as the composition of the integral flow with the observation function:

$$\mathbf{r} = (r_1, \ldots, r_k), \quad \mathbf{r}(t) = \mathbf{R}(\Phi_{t_0}^t(\mathbf{y}_0, \boldsymbol{\mu}), t, \boldsymbol{\nu})$$

The components r_i give a prediction for a specific observation at time t, for example, the right ascension $\alpha(t)$, or the declination $\delta(t)$. For example, assume the available observations at time τ_j are the right ascension α_j and the declination δ_j, for $j = 1, \dots, h$. Then $k = 2$ and we let $m = hk$. We chose the times such that $t_{2j-1} = t_{2j} = \tau_j$. And so the observation functions become

$$\begin{cases} r_{2j-1} = \alpha_j \\ r_{2j} = \delta_j \end{cases}$$

We can group the multidimensional data and predictions into two arrays, with components

$$r_i, \quad r(t_i), \quad i = 1 \dots m$$

respectively, and define the vector of the residuals as the difference between the observations and the predictions

$$\boldsymbol{\xi} = (\xi_1, \dots, \xi_m), \quad \xi_i = r_i - r(t_i), \quad i = 1 \dots m$$

The least-squares principle asserts that the solution of the orbit determination problem makes the target function

$$\mathcal{Q}(\boldsymbol{\xi}) = \frac{1}{m}\, \boldsymbol{\xi}^T \boldsymbol{\xi} \tag{10}$$

attain its minimal value.

We observe that

$$\boldsymbol{\xi} = \boldsymbol{\xi}(\mathbf{y}_0, \boldsymbol{\mu}, \boldsymbol{\nu})$$

and select part of the components of $(\mathbf{y}_0, \boldsymbol{\mu}, \boldsymbol{\nu}) \in \mathbb{R}^{p+p'+p''}$ to form the vector $\mathbf{x} \in \mathbb{R}^N$ of the fit parameters, that is, the parameters to be determined by fitting them to the data.

Let us define

$$Q(\mathbf{x}) = \mathcal{Q}(\boldsymbol{\xi}(\mathbf{x}; \mathbf{z}))$$

with $\mathbf{z}$ the vector of consider parameters, that is, the remaining components of $(\mathbf{y}_0, \boldsymbol{\mu}, \boldsymbol{\nu})$ fixed at some assumed value.

An important requirement is that $m \geqslant N$.

We introduce the $m \times N$ design matrix

$$B = \frac{\partial \boldsymbol{\xi}}{\partial \mathbf{x}}(\mathbf{x})$$

and search for the minimum of $Q(\mathbf{x})$ by looking for solutions of

$$\frac{\partial Q}{\partial \mathbf{x}} = \frac{2}{m}\, \boldsymbol{\xi}^T B = \mathbf{0} \tag{11}$$

To search for solutions of Eq. (11) we can use Newton's method. It involves the computation of the second derivatives of the target function

$$\frac{\partial^2 Q}{\partial \mathbf{x}^2} = \frac{2}{m} C_{\text{new}}, \quad C_{\text{new}} = B^T B + \boldsymbol{\xi}^T H \tag{12}$$

where

$$H = \frac{\partial^2 \boldsymbol{\xi}}{\partial \mathbf{x}^2}(\mathbf{x})$$

is a three-index array of shape $m \times N \times N$. Moreover, by $\boldsymbol{\xi}^T H$ we mean the matrix with components $\sum_i \xi_i \, (\partial^2 \xi_i / \partial x_j \partial x_k)$. The computation of the matrix C_{new} may be cumbersome so that one variant of Newton's method is commonly used. An iterative method, known as differential corrections, is often used to minimize the target function $Q(\mathbf{x})$. It consists of taking the so-called normal matrix $C = B^T B$ in place of C_{new}.

Given an initial guess (i.e., the preliminary orbit), at each iteration we have

$$\mathbf{x}_{k+1} = \mathbf{x}_k - C^{-1} \, B^T \boldsymbol{\xi} \tag{13}$$

where B is computed at $\mathbf{x}_k$.

In this way we are neglecting the term $\boldsymbol{\xi}^T H(\mathbf{x}^* - \mathbf{x}_k)$ in Eq. (13). Notice that this approximation works if the residuals are small enough.
One iteration of differential corrections is just the solution of a linear least squares problem

$$C(\mathbf{x}_{k+1} - \mathbf{x}_k) = -B^T \, \boldsymbol{\xi} \tag{14}$$

Equation (14) is called normal equation: it can be obtained by truncating the target function

$$Q(\mathbf{x}) \simeq Q(\mathbf{x}_k) + \frac{2}{m} \, \boldsymbol{\xi}^T \, B \, (\mathbf{x} - \mathbf{x}_k) + \frac{1}{m} \, (\mathbf{x} - \mathbf{x}_k)^T \, C \, (\mathbf{x} - \mathbf{x}_k)$$

Notice that it is not Taylor's expansion of order 2, because C_{new} is replaced by C.

Let $\mathbf{x}_*$ be the value of $\mathbf{x}$ at convergence. The inverse of the normal matrix

$$\Gamma = C^{-1} \tag{15}$$

is called covariance matrix and its value in $\mathbf{x}_*$ can be used to estimate the uncertainty of the solution of the differential corrections algorithm. In fact, the eigenvalues of Γ are proportional to the length of the axes of the confidence ellipsoids

$$\frac{1}{m}(\mathbf{x} - \mathbf{x}^*)^T C (\mathbf{x} - \mathbf{x}^*) \leqslant \sigma^2 \tag{16}$$

where $\sigma \in \mathbb{R}$ can be selected within a probabilistic interpretation of the observational errors.

IV. MULTIPLE SOLUTIONS

From the two-body dynamics, both Laplace's and Gauss' methods lead to an equation of the form

$$C\,\frac{\rho}{q} = \gamma - \frac{q^3}{r^3} \quad \text{(dynamical equation)} \tag{17}$$

with C, γ real parameters depending on the observations. This equation gives the possible values for the heliocentric and topocentric (or geocentric) distance of the asteroid (respectively r and ρ). It can be solved taking into account the geometric equation

$$r^2 = \rho^2 + 2q\rho\cos\epsilon + q^2 \tag{18}$$

Note that Eqs. (17) and (18) define two surfaces of rotation around the axis $\hat{\mathbf{q}}$ passing through the center of the Sun and the observer. The observation line and the axis $\hat{\mathbf{q}}$ define a reference plane where we can study the intersection problem. The problem takes the form

$$\begin{cases} (q\gamma - C\rho)r^3 - q^4 = 0 \\ r^2 - q^2 - \rho^2 - 2q\rho\cos\epsilon = 0 \\ r, \rho > 0 \end{cases} \tag{19}$$

We can eliminate the variable ρ by substitution and get the reduced problem

$$P(r) = 0, \quad r > 0 \tag{20}$$

with

$$P(r) = C^2 r^8 - q^2(C^2 + 2C\gamma\cos\epsilon + \gamma^2)r^6 + 2q^5(C\cos\epsilon + \gamma)r^3 - q^8 \tag{21}$$

The latter reduced version is useful to give an upper bound for the number of solutions of Eq. (19). In fact, the polynomial $P(r)$ has only four monomials and Descartes' rule of signs allows the presence of at most three positive solutions. Moreover, from a root $\bar{r}$ of $P(r)$ we can get one and only one corresponding value of $\bar{\rho}$ from the geometric equation. Passing to the reduced problem one can introduce some solutions $\bar{r}$, roots of $P(r)$, that actually are not solutions of the intersection problem. These are called spurious solutions and give a non-positive value of ρ through the dynamical equation.

Multiple solutions of the intersection problem correspond to multiple solutions of the preliminary orbit problem, therefore, it is worth counting the maximum number of such solutions.

A. CHARLIER'S THEORY

In 1910, Charlier gave a geometric interpretation of the occurrence of multiple solutions in preliminary orbit determination with Laplace's method, assuming geocentric observations ($\gamma = 1$). By the previous discussion, Charlier's theory corresponds to the study of the solutions of the intersection problem (19) with $\gamma = 1$. The main result of this theory is that *the condition for the appearance of another solution depends in a simple way on the position of the observed body (MNRAS, 1910)*. The method assumes that *the parameters C, ϵ are such that a solution of the corresponding intersection problem with $\gamma = 1$ always exists.* This condition is accomplished, for example, when the model is exact and the parameters are free of errors.

To prove Charlier's result we first notice that $r = q$ is a root of $P(r)$. However, this is not a solution of Eq. (19) because it corresponds to the position of the center of the Earth and must be discarded. Moreover, there are exactly three changes of sign in the sequence

$$\{C^2, -q^2(C^2 + 2C\cos\epsilon + 1), 2q^5(C\cos\epsilon + 1), -q^8\}$$

of the coefficients of $P(r)$.

Using Descartes' rule of signs applied to $P(r)$ and $P(-r)$, we deduce the existence of three positive and one negative real roots. The number of positive solutions is odd, and one is discarded. Hence, by Charlier's hypothesis there must exist two.

Then the problem is reduced to decide how many of the two positive roots different from $r = q$ are not spurious. Let $P(r) = (r - q)P_1(r)$ where $P_1(r)$ is a 7-deg polynomial. We have

$$P_1(q) = 2q^7C[C - 3\cos\epsilon]$$

Noting that $P_1(0) = q^7 > 0$ and $P_1(r) \to +\infty$ as $r \to +\infty$ we have that if $P_1(q) < 0$ there are two roots of $P(r)$: $r_1 < q$ and $r_2 > q$. From Charlier's hypothesis one of them is spurious. Analogously, if $P_1(q) > 0$ both roots are either greater or smaller than q. In both cases we have 2 different solutions of Eq. (19). To represent these two possibilities let us express, from the dynamical equation, the parameter C as an algebraic function in geocentric polar coordinates (ρ, ϵ) where

$$C(\rho, \epsilon) = \frac{q}{\rho}\left[1 - \frac{q^3}{r^3}\right], \quad r = \sqrt{\rho^2 + q^2 + 2q\rho\cos\epsilon}$$

With reference to Fig. 1, we introduce the curves

$$C = 0, \quad C - 3\cos\epsilon = 0$$

that we call zero circle and limiting curve, respectively.

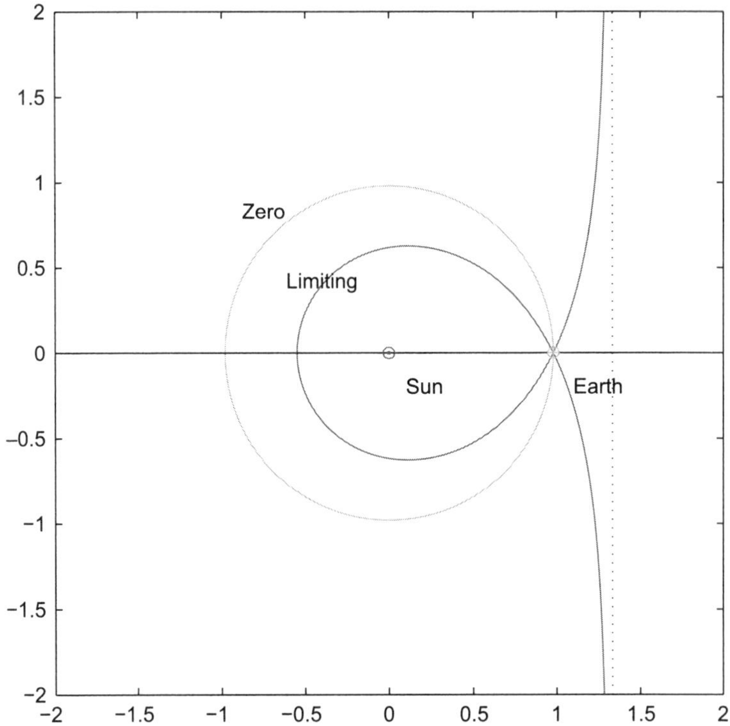

Fig. 1 The zero circle and the limiting curve, whose equation in heliocentric rectangular coordinates (x, y) is $4 - 3(x/q) = (q^3/r^3)$.

The position of the observed body corresponds to the intersection of the level curve $\mathscr{C}^{(1)}(x, y) = \mathcal{C}$ with the observation line (defined by ϵ), where

$$\mathscr{C}^{(1)} = \mathcal{C}^{(1)} \circ \Psi$$

Here

$$\mathcal{C}^{(1)}(r, \rho) = \frac{q}{\rho}\left[1 - \frac{q^3}{r^3}\right]$$

and $(x, y) \mapsto \Psi(x, y) = (r, \rho)$ is the map from rectangular to bipolar coordinates (see Fig. 2).

The set of tangency points between an observation line and a level curve of $\mathscr{C}^{(1)}$ is called the singular curve. This curve is defined as the set of points (ρ, ϵ) such that

$$F(\mathcal{C}, \rho, \epsilon) = F_\rho(\mathcal{C}, \rho, \epsilon) = 0 \tag{22}$$

where F_ρ is the derivative of F with respect to ρ. Here

$$F(\mathcal{C}, \rho, \epsilon) = \mathcal{C}\frac{\rho}{q} - 1 + \frac{q^3}{r^3(\rho, \epsilon)}$$

and

$$F_\rho(\mathcal{C}, \rho, \epsilon) = \frac{\mathcal{C}}{q} - 3\frac{q^3}{r^5}(\rho + q\cos\epsilon)$$

We can eliminate the dependence on $\mathcal{C}$ by taking the difference

$$F(\mathcal{C}, \rho, \epsilon) - \rho F_\rho(\mathcal{C}, \rho, \epsilon)$$

Passing to heliocentric rectangular coordinate, the function $r^5(F - \rho F_\rho)$ can be written as

$$\mathcal{G}(x, y) = -r^5 + q^3(4r^2 - 3qx), \quad r = \sqrt{x^2 + y^2}$$

Hence, we can define the singular curve as the set

$$\mathcal{S} = \{(x, y) : \mathcal{G}(x, y) = 0\}$$

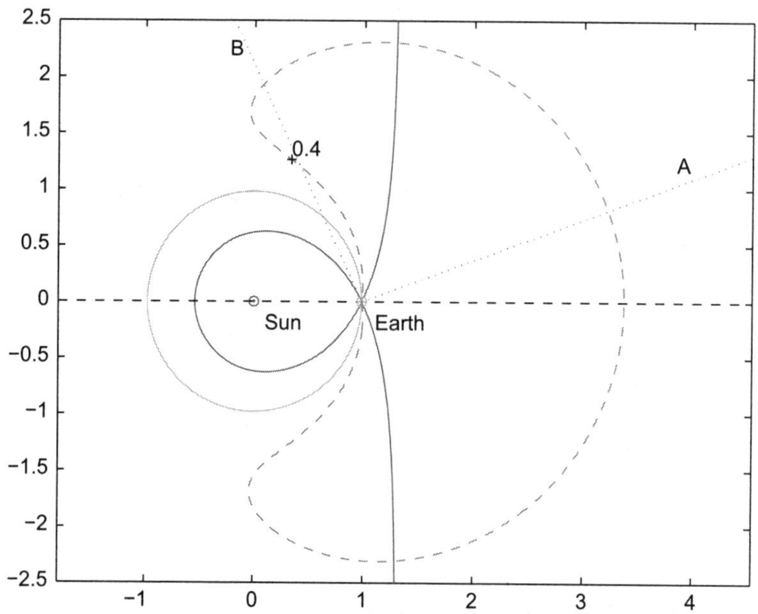

Fig. 2 The intersection of the same level curve with two different lines of sight. Observing in the direction *A* we find one solution only, whereas in direction *B* we have two solutions.

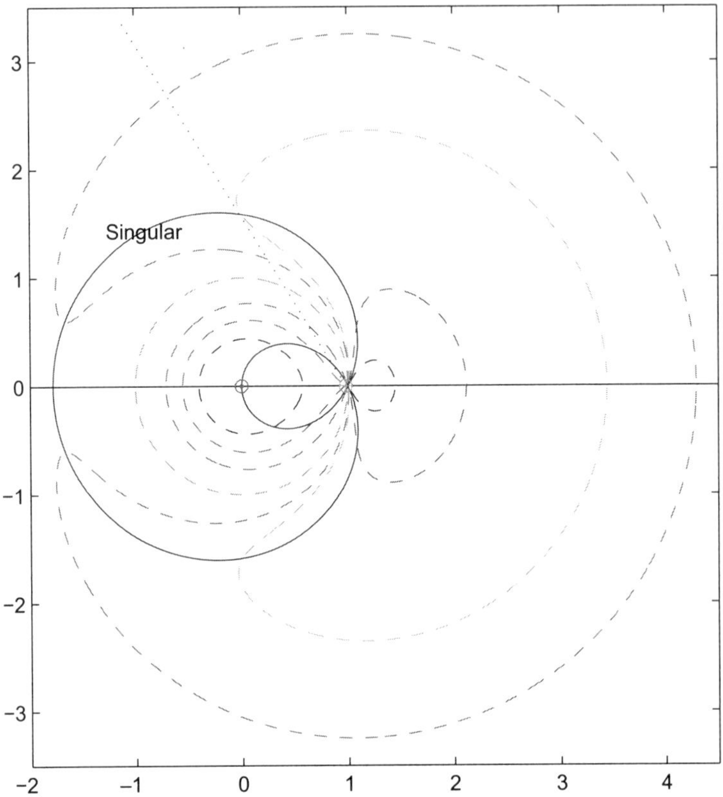

Fig. 3 The singular curve, consisting of the tangency points between the level curves and the observation lines.

The curve $\mathcal{S}$ contains all the points satisfying Eq. (22) plus the origin and is plotted in Fig. 3.

In Fig. 4 we summarize the results. Alternative solutions occur in two regions: the interior of the limiting curve loop and outside the zero circle, on the left of the unbounded branches of the limiting curve. For example, if the observations are taken at opposition then there is only one of the solution, whereas for close to quadrature multiple solutions appear.

B. GENERALIZATION OF CHARLIER'S THEORY

Charlier's theory only deals with Laplace's method, that corresponds to the case $\gamma = 1$ in the intersection problem (19). It is natural to ask for a generalization of his results to the case of a generic $\gamma \in \mathbb{R}$ that occurs in Gauss' method. Moreover, this approach is also useful when dealing with space-based observations. As before, we introduce topocentric radial coordinates (ρ, ϵ) in the reference plane and assume, as a generalized Charlier's hypothesis, that *the parameters* $\gamma, \mathcal{C}, \epsilon$ *are such that the corresponding intersection problem admits at least one solution.* Inspired by Charlier's theory, we define

$$\mathscr{C}^{(\gamma)} = \mathcal{C}^{(\gamma)} \circ \Psi, \quad \mathcal{C}^{(\gamma)}(r, \rho) = \frac{q}{\rho}\left[\gamma - \frac{q^3}{r^3}\right]$$

where $\gamma \in \mathbb{R}$, $\gamma \neq 1$ and Ψ represents the passage from rectangular to bipolar coordinates. The zero circle is defined by the relation

$$\mathcal{C} = 0$$

and corresponds to the heliocentric circle with radius

$$r_0 = q/\sqrt[3]{\gamma}, \quad \text{for } \gamma > 0$$

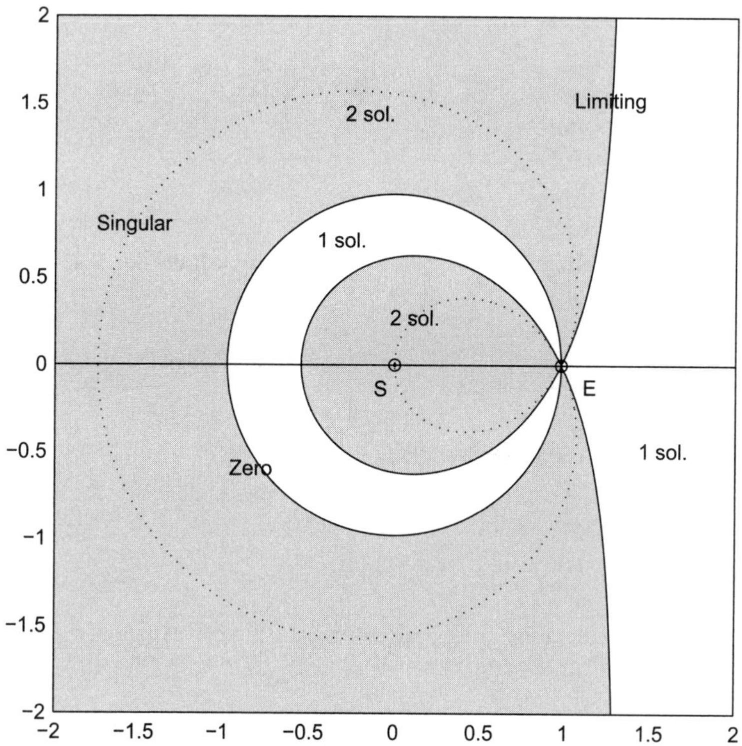

Fig. 4 Regions of multiple solutions. Reprinted from [7] with permission from Springer.

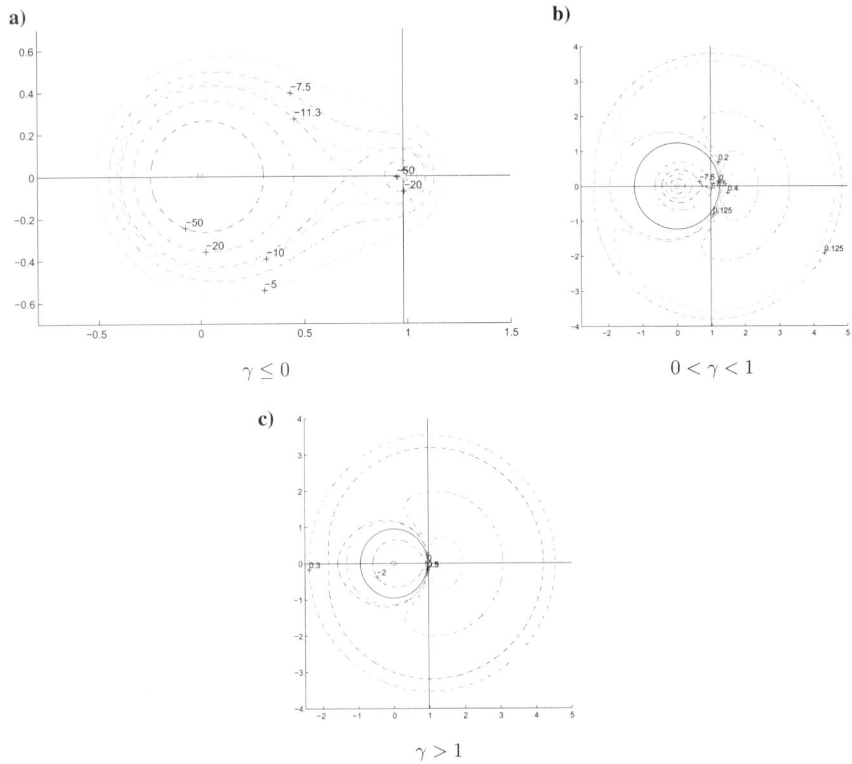

Fig. 5 Levels curves of $\mathcal{C}^{(\gamma)}$: a) $\gamma \leqslant 0$, b) $0 < \gamma < 1$, and c) $\gamma > 1$.

Moreover, a simple study of the level curves of $\mathcal{C}^{(\gamma)}$ gives the portraits of Fig. 5.

We have the following cases:

1. For $\gamma \leqslant 0$ there is only one saddle point
2. For $0 < \gamma < 1$ there are three critical points: one saddle inside the zero circle and one saddle and one maximum outside
3. For $\gamma \geqslant 1$ there is only one saddle point

Concerning the singular curve, we can repeat the construction made in the case $\gamma = 1$: it can be defined as the set

$$\mathcal{S} = \{(x, y) : \mathcal{G}(x, y) = 0\}, \quad \mathcal{G}(x, y) = -\gamma r^5 + q^3(4r^2 - 3qx)$$

Depending on the value of γ we get the portraits in Fig. 6.

Notice that we have the following situations:

1. For $\gamma \leqslant 0$ it has a unique component, which is convex
2. For $0 < \gamma < 1$ it has two components, one inside and one outside the zero circle
3. For $\gamma \geqslant 1$ it has a unique non convex component intersecting the zero circle

From Descartes' rule of signs the number of solutions of the intersection problem (19) can not be more than three. In particular, for (γ, C, ϵ) fulfilling the

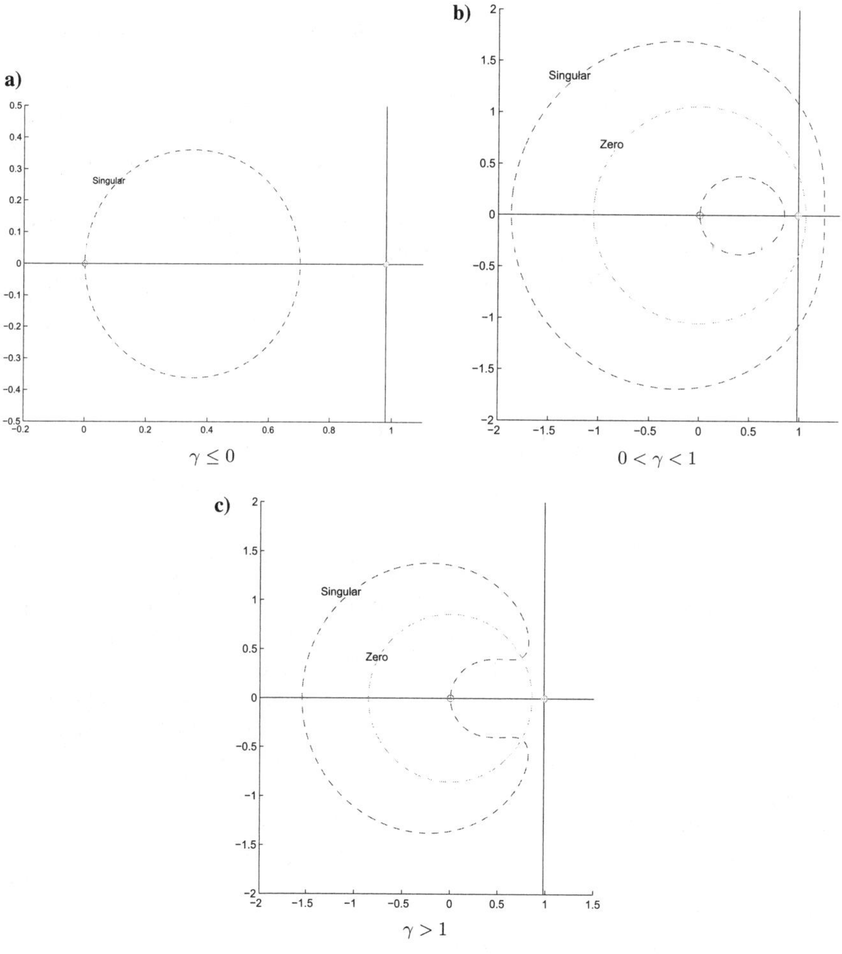

Fig. 6 Shape of the singular curves in the three different cases: a) $\gamma \leqslant 0$, b) $0 < \gamma < 1$, and c) $\gamma > 1$.

generalized Charlier's hypothesis with $\gamma \neq 1$, if the number of solutions is even they are two, if it is odd they are either one or three. More precisely, one can prove that for $(\gamma, \mathcal{C}, \epsilon) \in \mathbb{R}^2 \times [0, \pi]$, with $\gamma \neq 1$, the following assertions hold:

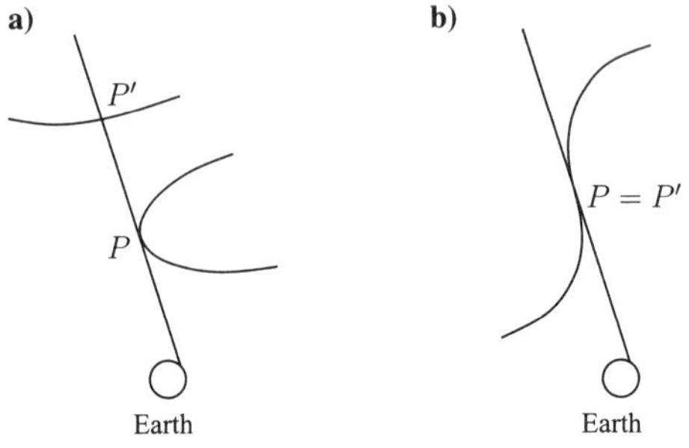

Fig. 7 Residual point and self-residual point.

1. If $\gamma \leqslant 0$ the number of solutions of Eq. (19) is odd, and we have solutions only for $\mathcal{C} < 0$.
2. If $0 < \gamma < 1$ the number of solutions of Eq. (19) is even for $\mathcal{C} > 0$, odd for $\mathcal{C} \leqslant 0$.
3. If $\gamma > 1$ the number of solutions of Eq. (19) is even for $\mathcal{C} \leqslant 0$, odd for $\mathcal{C} > 0$.

In the general case, for $\gamma \in \mathbb{R}$ we cannot define the limiting curve following Charlier's approach, because $r = q$ is no more a root of $P(r)$.

For $\gamma \neq 1$ and positive the radius of the zero circle is $r_0 = q/\sqrt[3]{\gamma}$. Hence, for every $\gamma \neq 1$ we define the sets

$$\mathcal{D}_2(\gamma) = \begin{cases} \emptyset & \text{if } \gamma \leqslant 0 \\ \{(x, y) : r > r_0\} & \text{if } 0 < \gamma < 1 \\ \{(x, y) : r \leqslant r_0\} & \text{if } \gamma > 1 \end{cases}$$

and

$$\mathcal{D}(\gamma) = \mathbb{R}^2 \setminus \big(\mathcal{D}_2(\gamma) \cup \{(q, 0)\}\big)$$

It comes from the result stated above that points in $\mathcal{D}_2(\gamma)$ correspond to an intersection problem with two solutions, points in $\mathcal{D}(\gamma)$ to a problem with one or three solutions. Our aim is to divide the set $\mathcal{D}(\gamma)$ into two connected regions, each of them with the same number of solutions of Eq. (19). To do that let us concentrate on the portion of the singular curve contained in $\mathcal{D}(\gamma)$ and call it $\mathfrak{S} = \mathcal{S} \bigcap \mathcal{D}$ (from now on, we drop the dependence on γ). Note that in $\mathcal{D}$ the solutions are one or three and the solutions on the singular curve have multiplicity at least two. Therefore, the points in $\mathfrak{S}$ correspond to an intersection problem with three solutions. Two situations can occur and are represented in Fig. 7.

To explain it, fix $\gamma \neq 1$ and suppose that a point $P \in \mathfrak{S} = \mathcal{S} \bigcap \mathcal{D}$ is represented by $(\bar{\rho}, \overline{\psi})$. We recall the definition

$$F(\mathcal{C}, \rho, \psi) = \mathcal{C}\frac{\rho}{q} - \gamma + \frac{q^3}{r^3}$$

The first case occurs when $F_{\rho\rho}(\mathcal{C}, \bar{\rho}, \overline{\psi}) \neq 0$, for a value of $\mathcal{C}$ such that $F(\mathcal{C}, \bar{\rho}, \overline{\psi}) = 0$. In this case, P gives a solution with multiplicity 2 and there exists another point P' on the same level curve and observation line corresponding to the third solution. In this case we say that P' is the residual point related to P (see Fig. 7a). The other case occurs when $F_{\rho\rho}(\mathcal{C}, \bar{\rho}, \overline{\psi}) = 0$. In this case the observation line is tangent in P to both the singular curve and the level curve. Then P gives a solution with multiplicity three and we call it self-residual point (Fig. 7b). Now we are ready to define the limiting curve in the case $\gamma \neq 1$. The limiting curve $\mathcal{L}$ is the set composed by all the residual points related to the points in $\mathfrak{S}$. We agree that the point $(x, y) = (q, 0)$ corresponding to the observer position is the residual point related to the origin, in case the latter belongs to $\mathfrak{S}$. In Fig. 8, we illustrate the construction of $\mathcal{L}$: we mark with "x" a point of $\mathfrak{S}$, and with "+" the associated residual point on the same observation line (dashed).

The curve defined in this way has the following properties:

1. *Separating property*: the limiting curve $\mathcal{L}$ separates $\mathcal{D}$ into two connected regions $\mathcal{D}_1, \mathcal{D}_3$: $\mathcal{D}_3$ contains the whole portion $\mathfrak{S}$ of the singular curve. If $\gamma < 1$ then $\mathcal{L}$ is a closed curve, if $\gamma > 1$ it is unbounded.

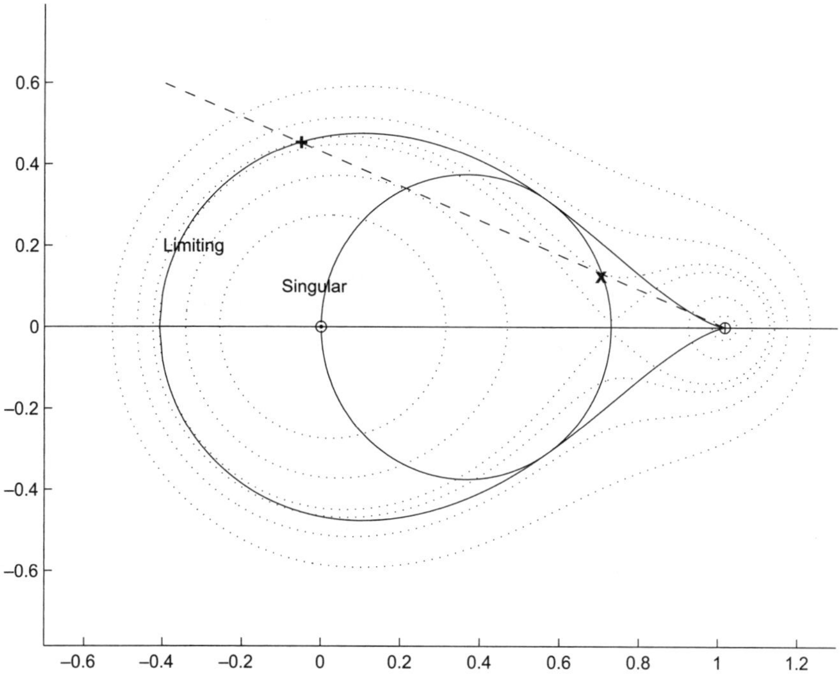

Fig. 8 Sketch of the construction of the limiting curve.

2. *Transversality*: the level curves of $\mathscr{C}^{(\gamma)}(x,y)$ cross $\mathcal{L}$ transversely, except for at most the two self-residual points and for the points where $\mathcal{L}$ meets the x-axis.
3. *Limiting property*: For $\gamma \neq 1$ the limiting curve $\mathcal{L}$ divides the set $\mathcal{D}$ into two connected regions $\mathcal{D}_1, \mathcal{D}_3$: the points of $\mathcal{D}_1$ are the unique solutions of the corresponding intersection problem; the points of $\mathcal{D}_3$ are solutions of an intersection problem with three solutions.

A summary of all the different cases is given in Fig. 9. Notice that the scenario is very different from the one of Charlier's case: actually one can also have the case of three solutions.

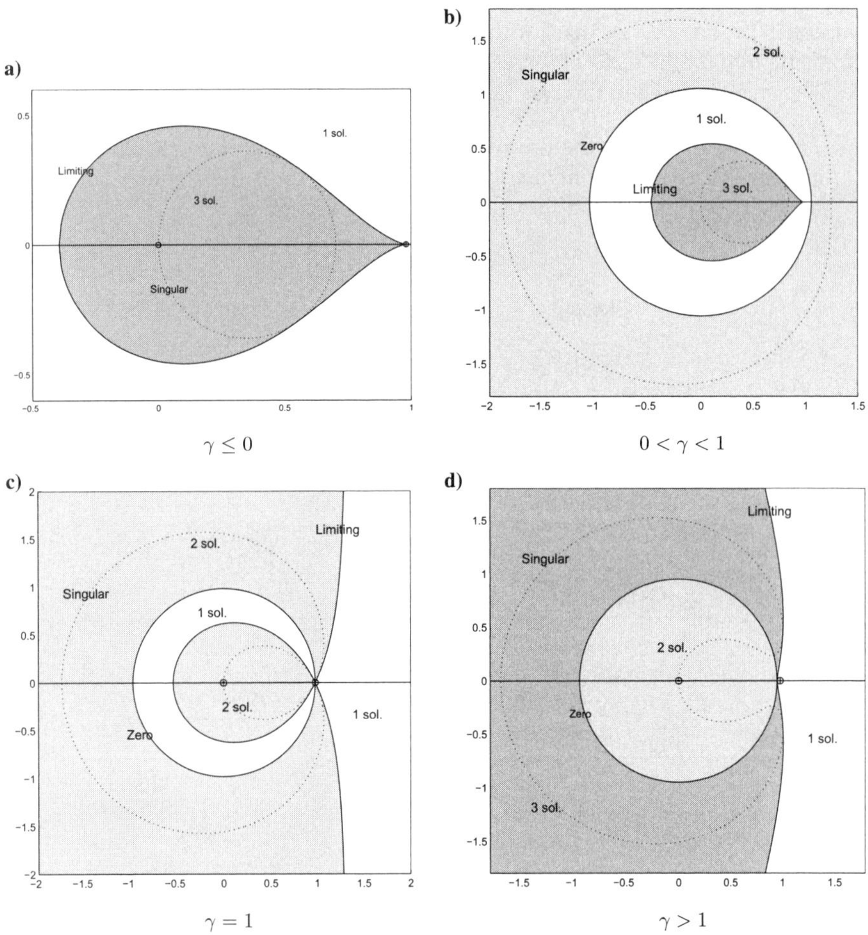

Fig. 9 Summary of the results on alternative solutions for all the qualitative different cases: a) $\gamma \leqslant 0$, b) $0 < \gamma < 1$, c) $\gamma = 1$, and d) $\gamma > 1$. Reprinted from [7] with permission from Springer.

V. CONCLUSIONS

The recent improvements in the observational techniques are stimulating the research about the determination of orbits. Moreover, the huge amount of astrometric data that are produced needs to be processed by efficient methods. In some situations classical methods are efficient and are still used nowadays. Thus, they are worth to be investigated and they can still hide some unsolved features.

REFERENCES

[1] Laplace, P. S., "Memories on the Determination of Comets Orbits," *Memoirs of the Academy of Royal Science Paris in Laplace's Collected Works*, Vol. 10, 1780, pp. 93–146.

[2] Gauss, C. F., *Theoria motus corporum coelestium in sectionibus conicis solem ambientum*, Hamburg 1809, English translation by C. H. Davis, Dover, 1963, reprinted 2004.

[3] Celletti, A., and Pinzari, G., "Four Classical Methods for Determining Planetary Elliptic Elements: A Comparison," *Celestial Mechanics and Dynamical Astronomy*, Vol. 93, No. 1–4, 2005, pp. 1–52.

[4] Milani, A., Gronchi, G. F., Farnocchia, D., Knezevic, Z., Jedicke, R., Dennau, L., and Pierfederici, F., "Topocentric Orbit Determination: Algorithms for the Next Generation Surveys," *Icarus*, Vol. 195, 2008, pp. 474–492.

[5] Charlier, C. V. L., "On Multiple Solutions in the Determination of Orbits from Three Observations," *Monthly Notices of the Royal Astronomical Society*, Vol. 71, 1910, pp. 120–124.

[6] Charlier, C. V. L., "Second Note on Multiple Solutions in the Determination of Orbits from Three Observations," *Monthly Notices of the Royal Astronomical Society* , Vol. 71, 1910, pp. 454–459.

[7] Gronchi, G. F., "Multiple Solutions in Preliminary Orbit Determination from Three Observations," *Celestial Mechanics and Dynamical Astronomy*, Vol. 103, No. 4, 2009, pp. 301–326.

[8] Herrick, S., "Orbit Determination, Space Navigation, Celestial Mechanics," *Astrodynamics*, With a foreword by Charles Stark Draper, Vol. 1, Van Nostrand Reinhold Co., London, 1971.

[9] Everhart, E., and Pitkin, E. T., "Universal Variables in the Two-Body Problem," *American Journal of Physics* , Vol. 51/8, 1983, pp. 712–717.

CHAPTER 11

An Introduction to Optimal Control Problem and Space Trajectory Optimization with Some Applications

Francesco Topputo* **and Chiara Tardioli***
University of Strathclyde, Glasgow, United Kingdom

Pierluigi Di Lizia†
Politecnico di Milano, Milan, Italy

I. INTRODUCTION

The aim of this chapter is to introduce the optimal control theory and specialize it to the case of space trajectory optimization. First, the basic notions of optimal control theory and nonlinear programming (NLP) problems are introduced. Some numerical examples are also shown. Next, the space trajectory optimization problem is presented, and practical problems are given. Both impulsive and low-thrust propulsions are treated.

Optimal control theory aims to find a control law for a given system such that certain optimality conditions are satisfied and a performance index (also called objective or cost function) is minimized. The optimal control can be seen as a special case of *Pontryagin's maximum principle* (a necessary condition also known as Pontryagin's minimum principle or simply Pontryagin's principle), formulated in 1956 by the Russian mathematician Lev Semenovich Pontryagin and his collaborators [1].

A crucial question when it comes to optimize a function on a given set, is whether a minimum or a maximum solution actually exists for that function in that set. For example, suppose f is a real function, then the cases where a minimum does not exist are $f = (x) = x$ defined in the open interval $(0, 1)$; $f = (x) = -x$ defined in the open interval $(0, +\infty)$; and f has a minimizer at a finite value but there the function is not defined because it is discontinuous. Conversely, if the domain of f is nonempty, closed, and bounded, and if f is continuous on it, then a minimum exists. This is what the well-known Weierstrass' theorem states. A generalization of the theorem holds for vector functions of several variables. In

*Mechanical and Aerospace Engineering.
†Department of Aerospace Science and Technology.

the rest of the chapter, we deal with continuously differentiable functions defined in closed-bounded sets; thus, a minimum, though difficult to compute, always exists.

Classical methods to solve the optimal control problem are divided into two groups: indirect methods and direct methods. Indirect methods use the calculus of variation to construct an analytical system of differential-algebraic equations (DAEs) for the state and the controls, whose boundary conditions are given both at the initial and final time. The optimal problem is so reduced to a two-point boundary value problem (TPBVP). Direct methods transform the optimal control problem into a sequence of NLP problems, related to the necessary conditions for optimality, to be solved iteratively until a certain convergence condition is satisfied. An important reference for indirect methods is Ref. [2], while for direct methods see Ref. [3].

Both direct and indirect methods are based on the use of Newton's method. This implies that both of them depend on a starting guess, which must be close enough to the optimal solution to converge quickly. However, this request is difficult to satisfy and in the past few decades, new approximate methods have been developed to define new procedures that are quasi-independent or independent from an initial guess, even if only suboptimal solutions can be guaranteed. Examples are the state-dependent Riccati equations (SDREs) and approximating sequence of Riccati equation (ASRE) methods. By using the well-known theory of linear optimal control, these methods rewrite the nonlinear problem in a pseudolinear form. Then, they apply the common procedure to solve linear optimal control problems over an infinite time horizon (DSRE) or a finite time horizon (ASRE). In this chapter, we will present a modification of the ASRE technique which uses a factorization of the state-transition matrix (STM), as introduced in Refs. [4, 5]. Whereas, explanations on SDRE methods are left to Refs. [6,7].

The chapter is organized as follows. The optimal control problem is stated in Sec. II. The associated necessary conditions for optimality are derived in Sec. III. Section IV recalls Newton's method and explains how to solve an NLP problem. Direct methods are analyzed in Sec. V. Finally, the modified ASRE and some possible applications in astrodynamics are treated in Sec. VI.

II. ANALYTICAL FORMULATION OF THE OPTIMAL CONTROL PROBLEM

Consider the dynamical system in the form

$$\dot{\mathbf{x}}(t) = \mathbf{f}(\mathbf{x}(t), \mathbf{u}(t), t) \tag{1}$$

where $\mathbf{x}(t)$ represents an n-dimensional state vector and $\mathbf{u}(t)$ is the m-dimensional control vector, as continuously differentiable functions of time.

The objective function can be written in the Bolza form as

$$J = \varphi(\mathbf{x}(t_f), t_f) + \int_{t_0}^{t_f} L(\mathbf{x}(t), \mathbf{u}(t), t)\, dt \tag{2}$$

where φ is a terminal cost function depending on the final time t_f and the integral expresses a cost incurred during the entire trajectory. The problem may have some fixed initial conditions $x_i(t_0)$, for $i = 1, \ldots, k$, with $k \leqslant n$, as well as some fixed terminal conditions $x_i(t_f)$ or fixed final time t_f. We impose the following conditions, called constraints of the optimal problem:

$$\Psi(\mathbf{x}(t_f), t_f) = 0, \quad \Psi \in \mathbb{R}^p, \ p \leqslant n \tag{3}$$

$$\mathbf{C}(\mathbf{u}(t), t) \leqslant 0, \quad \mathbf{C} \in \mathbb{R}^q, \ q \leqslant m \tag{4}$$

The equalities are some boundary conditions at the final time t_f, while the inequalities are path constraints on the control variables.

The *optimal control problem* is formulated as follows. Given the dynamical system (1), determine the m control function $u_1, \ldots, u_m$ such that the performance index J given by Eq. (2) is minimized and the systems of constraints (3) and (4) are satisfied.

The necessary condition for optimality may be derived using the calculus of variation. Unfortunately, solving the resulting system of equations and constraints in a closed form is either difficult or impossible. Therefore, numerical optimization methods are used to solve the problem. They are generally divided into two groups. Indirect methods are based on reducing the optimal control problem to a boundary value problem using the calculus of variation. Direct methods are based on reducing the optimal control problem to an NLP problem. The problem is discretized and the system of equations are verified by integrating them, using either implicit or explicit methods. A comprehensive survey about direct and indirect methods can be found in Refs. [8,9].

Example 1

For low-thrust Earth–Mars transfer the dynamics of the controlled Keplerian motion of a spacecraft, in the gravitational field of the Sun, is given by

$$\ddot{\mathbf{r}} = -\frac{\mu}{r^3}\mathbf{r} + \mathbf{u}, \quad \mathbf{r} \in \mathbb{R}^3, \quad \mathbf{u} \in \mathbb{R}^3 \tag{5}$$

with μ the standard gravitational parameter $\mu = GM_\odot$, where G is the gravitational constant and $M_\odot$ is the mass of the Sun, and $r = \| \mathbf{r} \|$, where $\| \cdot \|$ is the Euclidean norm in $\mathbb{R}^3$. An optimal control problem consists of minimizing the objective function

$$J = \int_{t_0}^{t_f} L(\mathbf{x}, \mathbf{u}, t)\, dt = \frac{1}{2}\int_{t_0}^{t_1} \mathbf{u}^T\mathbf{u}\, dt$$

subject to the equality constraints

$$\mathbf{r}(t_o) = \mathbf{r}_{\oplus}(t_0), \quad \mathbf{v}(t_o) = \mathbf{v}_{\oplus}(t_0)$$
$$\mathbf{r}(t_f) = \mathbf{r}_{\male}(t_f), \quad \mathbf{v}(t_f) = \mathbf{v}_{\male}(t_f)$$

and to the inequality constraints

$$\| \mathbf{u} \| \leqslant u^{\max}$$

where t_0, t_f are the initial and final time, respectively, $\mathbf{r}_{\oplus}$, $\mathbf{r}_{\male}$ are the Earth's and Mars' radius, respectively (we assume that the orbits of the planets are circular coplanar), and $u^{\max}$ is a constant value.

In astrodynamics, optimal control theory is a powerful tool used for space mission design. For example, it was used in NEAR and Deep Space 1 missions. However, a classical optimal trajectory design problem is stated in a slightly different manner than the classical optimal control problem illustrated previously. The formulation proposed by Betts [3] better fits the numerical methods used to solve the trajectory design problem. Typically, the dynamical system is defined by a set of ordinary differential equations (ODEs) written in explicit form, referred to as the *state equations*

$$\dot{\mathbf{x}} = \mathbf{f}(\mathbf{x}(t), \mathbf{u}(t), \mathbf{p}, t) \tag{6}$$

where $\mathbf{x} : [t_0, t_f] \rightarrow \mathbb{R}^n$ is the state vector, $\mathbf{u} : [t_0, t_f] \rightarrow \mathbb{R}^m$ is the control vector, $\mathbf{p} \in \mathbb{R}^{n_p}$ is a constant parameter vector, and $t_0 \leqslant t \leqslant t_f$ is the time. The boundary conditions are defined as

$$\begin{aligned} \boldsymbol{\psi}_{0,l} &\leqslant \boldsymbol{\psi}_0(\mathbf{x}(t_0), \mathbf{u}(t_0), \mathbf{p}, t_0) \leqslant \boldsymbol{\psi}_{0,u} \\ \boldsymbol{\psi}_{1,l} &\leqslant \boldsymbol{\psi}_1(\mathbf{x}(t_0), \mathbf{u}(t_0), \mathbf{p}, t_0) \leqslant \boldsymbol{\psi}_{1,u} \end{aligned} \tag{7}$$

for some fixed lower- and upper-bound vectors $\boldsymbol{\psi}_{0,l}, \boldsymbol{\psi}_{0,u}, \boldsymbol{\psi}_{1,l}, \boldsymbol{\psi}_{1,u}$. In addition, the solution must satisfy algebraic path constraints of the form

$$\mathbf{g}_l \leqslant \mathbf{g}(\mathbf{x}(t), \mathbf{u}(t), \mathbf{p}, t) \leqslant \mathbf{g}_u \tag{8}$$

as well as simple bounds on the state and control variables

$$\mathbf{x}_l \leqslant \mathbf{x}(t) \leqslant \mathbf{x}_u, \quad \mathbf{u}_l \leqslant \mathbf{u}(t) \leqslant \mathbf{u}_u$$

The basic problem is to determine the controls $\mathbf{u}(t)$ and parameter $\mathbf{p}$ to minimize the performance index, here written in Mayer form for notational simplicity

$$J = \phi(\mathbf{x}(t_f), t_f) \tag{9}$$

III. INDIRECT METHODS

In this section, we derive the necessary condition to solve the optimal control problem using the calculus of variation. This requires the addition of the Lagrange multipliers (or adjoint variables), equal in number to the state variables, and their equations. The size of the dynamical system is now doubled. This approach is called an *indirect method.*

A. EULER–LAGRANGE EQUATIONS

Consider the optimal control problem with the dynamical system (1) with performance index (2) and constraints (3), (4). The inequality constraints (4) may be regarded as "continuous," since they must be satisfied over the entire interval $[t_0, t_f]$. While the equality constraints (3) may be considered as "discrete," since they are imposed at a fixed time t_f. A solution can be found by the calculus of variations.

We introduce two kinds of multipliers: the Lagrangian multipliers $\boldsymbol{\nu} \in \mathbb{R}^p, p \leqslant n$ for the final boundary constraints, and the adjoint or costate variables $\boldsymbol{\lambda}(t) \in \mathbb{R}^n$, $\boldsymbol{\mu}(t) \in \mathbb{R}^q$, $q \leqslant m$, for the dynamics and path constraints. The *augmented performance index* is given by

$$
\begin{aligned}
\bar{J} &= \varphi(\mathbf{x}(t_f), t_f) + \boldsymbol{\nu}^T \boldsymbol{\Psi}(\mathbf{x}(t_f), t_f) \\
&\quad + \int_{t_0}^{t_f} \Big(L(\mathbf{x}(t), \mathbf{u}(t), t) + \boldsymbol{\lambda}^T(t)(\mathbf{f}(\mathbf{x}(t), \mathbf{u}(t), t) - \dot{\mathbf{x}}(t)) + \boldsymbol{\mu}^T(t)\mathbf{C}(\mathbf{u}(t), t) \Big) dt
\end{aligned}
$$

The dynamics is included in the formulation as a set of constraints, that is, the optimal solution must both minimize the objective function and satisfy the dynamics. Moreover, the generic component of vector $\boldsymbol{\mu}$ must satisfy the following relations:

$$
\begin{aligned}
C_k(\mathbf{u}(t), t) < 0 &\Rightarrow \mu_k(t) = 0 \\
C_k(\mathbf{u}(t), t) = 0 &\Rightarrow \mu_k(t) \geqslant 0
\end{aligned}
$$

Integrating the term $\boldsymbol{\lambda}^T \dot{\mathbf{x}}$ by parts, and dropping the dependency from t for notational simplicity, we obtain

$$
\begin{aligned}
\bar{J} &= \varphi(\mathbf{x}_f, t_f) + \boldsymbol{\nu}^T \boldsymbol{\Psi}(\mathbf{x}_f, t_f) - \boldsymbol{\lambda}_f^T \mathbf{x}_f + \boldsymbol{\lambda}_0^T \mathbf{x}_0 \\
&\quad + \int_{t_0}^{t_f} \Big(L(\mathbf{x}, \mathbf{u}, t) + \boldsymbol{\lambda}^T(\mathbf{f}(\mathbf{x}, \mathbf{u}, t) + \boldsymbol{\mu}^T \mathbf{C}(\mathbf{u}, t) \Big) \, dt
\end{aligned}
$$

where $\mathbf{x}_0 = \mathbf{x}(t_0)$, $\mathbf{x}_f = \mathbf{x}(t_f)$, and $\boldsymbol{\lambda}_0 = \boldsymbol{\lambda}(t_0)$. The problem is reduced to identify a stationary point of $\bar{J}$. This can be achieved by imposing the gradient to be zero.

The optimization variables are the state vector $\mathbf{x}$ and the control vector $\mathbf{u}$; the Lagrange multipliers and costate variables $\boldsymbol{\nu}$, $\boldsymbol{\lambda}$, and $\boldsymbol{\mu}$; some unknown components of the initial state $\mathbf{x}_0$, named $x_{i,0}, i = \bar{k}+1, \ldots, n$, for some $\bar{k} \in \mathbb{N}$; the final state $\mathbf{x}_f$; and the final time t_f. In explicit form, the zero-gradient conditions are

$$
\begin{aligned}
&\frac{\partial \bar{J}}{\partial \boldsymbol{\lambda}} = 0 \Rightarrow \dot{\mathbf{x}}(t) = \mathbf{f}(\mathbf{x}, \mathbf{u}, t) \\
&\frac{\partial \bar{J}}{\partial \mathbf{x}} = 0 \Rightarrow \dot{\boldsymbol{\lambda}} = -\left(\frac{\partial \mathbf{f}}{\partial \mathbf{x}}\right)^T - \boldsymbol{\lambda}\left(\frac{\partial L}{\partial \mathbf{x}}\right)^T \\
&\frac{\partial \bar{J}}{\partial \mathbf{u}} = 0 \Rightarrow \left(\frac{\partial L}{\partial \mathbf{u}}\right)^T + \left(\frac{\partial \mathbf{f}}{\partial \mathbf{u}}\right)^T \boldsymbol{\lambda} + \left(\frac{\partial \mathbf{C}}{\partial \mathbf{u}}\right)^T \boldsymbol{\mu} = 0 \\
&\frac{\partial \bar{J}}{\partial x_{i,0}} = 0 \Rightarrow \lambda_{i,0} = 0 \quad \text{for } i = \bar{k}+1, \ldots, n \\
&\frac{\partial \bar{J}}{\partial \mathbf{x}_f} = 0 \Rightarrow \boldsymbol{\lambda}_f = \left(\frac{\partial \varphi}{\partial \mathbf{x}_f}\right)^T + \left(\frac{\partial \Psi}{\partial \mathbf{x}_f}\right)^T \boldsymbol{\nu} \\
&\frac{\partial \bar{J}}{\partial \boldsymbol{\nu}} = 0 \Rightarrow \Psi(\mathbf{x}_f, t_f) = 0 \\
&\frac{\partial \bar{J}}{\partial \boldsymbol{\mu}} = 0 \Rightarrow \mathbf{C}(\mathbf{u}, t) \leqslant 0 \\
&\frac{\partial \bar{J}}{\partial t_f} = 0 \Rightarrow \frac{\partial \varphi}{\partial t_f} + \boldsymbol{\nu}^T \frac{\partial \Psi}{\partial t_f} + \left(\frac{\partial \varphi}{\partial \mathbf{x}_f} + \boldsymbol{\nu}^T \frac{\partial \Psi}{\partial \mathbf{x}_f}\right) \mathbf{f}(\mathbf{x}, \mathbf{u}, t) \\
&\qquad\qquad\qquad + L(\mathbf{x}_f, \mathbf{u}_f, t_f) + \boldsymbol{\mu}(t_f)^T \mathbf{C}(\mathbf{x}_f, t_f) = 0
\end{aligned}
$$

To write the necessary conditions in a compact form, it is convenient to define the *Hamiltonian*[‡]

$$H(\mathbf{x}, \mathbf{u}, \boldsymbol{\lambda}, t) = L(\mathbf{x}(t), \mathbf{u}(t), t) + \boldsymbol{\lambda}(t)^T \mathbf{f}(\mathbf{x}(t), \mathbf{u}(t), t) + \boldsymbol{\mu}(t)^T \mathbf{C}(\mathbf{u}(t), t)$$

and the auxiliary function

$$\phi = \varphi(\mathbf{x}_f, t_f) + \boldsymbol{\nu}^T \Psi(\mathbf{x}_f, t_f) - \boldsymbol{\lambda}_f^T \mathbf{x}_f + \boldsymbol{\lambda}_0^T \mathbf{x}_0$$

[‡]It has the same name but it is different from the Hamiltonian of classical mechanics, for example, see Ref. [10].

The necessary conditions for optimality, referred to as the *Euler–Lagrange equations*, are

$$\dot{\mathbf{x}} = H_{\boldsymbol{\lambda}}^T \quad \text{(dynamical equations)} \tag{10}$$

$$\dot{\boldsymbol{\lambda}} = -H_{\mathbf{x}}^T \quad \text{(adjoint equations)} \tag{11}$$

$$\mathbf{0} = H_{\mathbf{u}}^T \quad \text{(control equations)} \tag{12}$$

The partial derivatives $H_{\boldsymbol{\lambda}}, H_{\mathbf{x}}$, and $H_{\mathbf{u}}$, are considered as row vectors; so, in the equations, their transposes are taken. The control equation (12) can be thought as a simplified statement of *Pontryagin's minimum principle*, stating that at any time on the optimal trajectory, the control variables are chosen to minimize the Hamiltonian.

The differential-algebraic equations (10–12) must be solved together with the $2n$ boundary conditions, also called *transversality conditions*

$$\lambda_{i,0} = 0, \quad i = \bar{k}+1,\ldots,n, \quad \bar{k} \geqslant 0, \quad \boldsymbol{\lambda}(t_f) - \Phi_{\mathbf{x}}^T\Big|_{t=t_f} \tag{13}$$

with the $p+q+1$ additional constraints (3), (4), and with the condition at the final time

$$(\Phi_f + H)\Big|_{t=t_f} = 0 \tag{14}$$

where Φ_f is the partial derivative of Φ with respect to the final time t_f. This is often referred to as a *TPBVP*. More details can be found in Ref. [2].

Example 2

This example is taken from Ref. [11]. Consider the simplified optimal control problem with dynamics

$$\dot{x}_1 = \frac{x_1}{2} + u, \quad \dot{x}_2 = u^2 + x_1 u + \frac{5}{4}x_1^2$$

and boundary conditions $t_0 = 0$, $t_f = 1$ and $x_1(0) = 1$, $x_2(0) = 0$. The objective function is $J = x_2(1)$. The analytical solution of the problem is found by solving the necessary conditions for optimality (10–12):

$$x_1(t) = \frac{\cosh(1-t)}{\cosh(1)}$$

$$u(t) = -\frac{(2\tanh(1-t)+1)\cosh(1-t)}{2\cosh(1)}$$

B. LOW-THRUST TRANSFER TO HALO ORBIT

This example is taken from Ref. [12]. We consider a dynamical model Earth–Moon–spacecraft, with masses m_1, m_2, and m_3, respectively. We suppose that the orbit of the two celestial bodies is circular and coplanar, and that the artificial satellite has negligible mass, that is, its gravitational field does not influence the motion of the other bodies, and $m_1 > m_2 \gg m_3$. The problem is known as the circular-restricted three-body problem (CRTBP).

The goal is to find an optimal, fuel-minimizing trajectory from an initial geostationary transfer orbit (GTO) about the Earth to a final Halo orbit around the Lagrange point L_1, through low-thrust and stable manifolds.

To simplify the dynamical equations, we set the universal gravitational constant G to unity and define the mass ratio $\mu = m_2/(m_1 + m_2)$. By hypothesis, $m_1 \geqslant m_2$, $\mu \in [0, 0.5]$, then the normalized masses of the two principal bodies are $1 - \mu$ and μ, respectively.

The controlled equations of motion for the spacecraft are

$$\ddot{x} - 2\dot{y} = \Omega_x + u_x, \quad \ddot{y} - 2\dot{x} = \Omega_y + u_y, \quad \ddot{z} = \Omega_z + u_z \tag{15}$$

where $\Omega_x, \Omega_y, \Omega_z$ are the partial derivatives w.r.t. x, y, z, respectively, of the function

$$\Omega(x, y, z) = \frac{1}{2}(x^2 + y^2) + \frac{1 - \mu}{r_1} + \frac{\mu}{r_2} + \frac{1}{2}\mu(1 - \mu)$$

with

$$r_1 = \sqrt{(x + \mu)^2 + y^2 + z^2}, \quad r_2 = \sqrt{(x + \mu - 1)^2 + y^2 + z^2}$$

The vector $\mathbf{u} = (u_x, u_y, u_z)^T$ denotes the control function, that is, in the case of a low-thrust trajectory, the acceleration given by the engine, and $\mathbf{x} = (x, y, z, \dot{x}, \dot{y}, \dot{z})$ represents the state vector.

To design optimal low-thrust transfers to the Earth–Moon halo orbit, the optimal control problem should be stated in the formalism of the previous section. We start rewriting the controlled-restricted three-body equations (15) as

$$\dot{x} = v_x, \quad \dot{y} = v_y, \quad \dot{z} = v_z$$
$$\dot{v}_x = 2v_y + \Omega_x + u_x, \quad \dot{v}_y = -2v_x + \Omega_x + u_y, \quad \dot{v}_z = \Omega_x + u_z$$

In a compact form, the previous system of equations corresponds to $\dot{\mathbf{x}} = \mathbf{f}(\mathbf{x}, \mathbf{u})$, with state $\mathbf{x} = (x, y, z, v_x, v_y, v_z)$ and control $\mathbf{u} = (u_x, u_y, u_z)$. We want to find the control vector history $\mathbf{u} = \mathbf{u}(t), t_0 \leqslant t \leqslant t_f$ that minimizes the following scalar performance index:

$$J = \int_{t_0}^{t_f} L(\mathbf{x}, \mathbf{u}, t)\,\mathrm{d}t = \frac{1}{2}\int_{t_0}^{t_f} \mathbf{u}^T\mathbf{u}\,\mathrm{d}t \tag{16}$$

while satisfying certain constraints. These constraints may be represented by two boundary conditions, defined at the end points of the optimal control interval, or/and by an inequality condition defined along the whole low-thrust arc. In this simplified case, we assume to fix the final time t_f and the two boundary states $\mathbf{x}_0$, $\mathbf{x}_f$. The Euler–Lagrange equation (12) leads to

$$0 = \frac{\partial L}{\partial u_i} + \sum_{j=1}^{6} \frac{\partial f_j}{\partial u_i} \lambda_j \Rightarrow u_i = -\lambda_{3+i}, \quad i = 1, \ldots, 3 \tag{17}$$

where f_j is a component of the vector field. The algebraic equations (17) can be inserted into the ordinary differential equations (10) and (11). Adding the constraint equations

$$\mathbf{x}(t_0) - \mathbf{x}_0 = 0, \quad \mathbf{x}(t_f) - \mathbf{x}_f = 0$$

the original problem is finally reduced into a simplified TPBVP. The solution can be computed numerically using a Newton method starting from a suitable initial condition. The satisfaction of constraints is checked by evaluating the control parameters (u_x, u_y, u_z).

To find the fuel-minimizing trajectory, we put $\mathbf{u}(t) = \mathbf{T}(t)/m(t)$, where $\mathbf{T}$ is the thrust force and m is the low-thrust propulsion defined by its variation

$$\dot{m} = \frac{T(\tau)}{I_{sp}\, g_0}$$

where I_{sp} and g_0 are, respectively, the specific impulse of the engine and the gravitational acceleration at sea level. The problem is to find $T(t), t \in [t_0, t_f]$ that minimizes the performance index

$$J = \int_{t_0}^{t_f} m_p$$

where m_p stands for the propellant mass needed to carry out the transfer, m_i is the wet mass placed in the translunar orbit, and $m(t_f)$ is the spacecraft mass. To make the whole optimization robust, the first guesses are processed to minimize the performance index (16). A comprehensive study on Earth-to-Halo transfers, together with some numerical examples, can been found in Ref. [13].

IV. INTRODUCTION TO NONLINEAR PROGRAMMING

This section presents numerical methods for solving the optimal control problem. The main expedient to effectively solve all numerical optimization methods is to reduce them to a sequence of simpler subproblems. In particular, the solution of an optimal control problem will require the solution of one or more finite-dimensional subproblems. This section focuses on the NLP problem. The NLP problem requires finding a finite set of variables such that an objective function or performance index is optimized without violating a set of constraints. As opposite to the optimal control problem, no dynamics is involved into an NLP problem.

Special cases of the NLP problem include linear programming (LP), quadratic programming (QP), and least-squares problem. LP is very attractive and guarantees a global optimal solution, although the formulation involves simplifying assumptions and linearization. In QP, the objective function is quadratic of the form $(1/2)x^TQx + q^T$ and all constraints are linear. Least squares are widely used in orbit determination to compute a preliminary orbit minimizing the errors due to the observations [14].

A. NEWTON'S METHODS

The fundamental approach to most iterative schemes was suggested over 300 years ago by Newton. We analyze the simplest form of Newton's method in one variable. Given a function $f \in C^2$, that is, f is continuous and its derivative f' exists and it is continuous, find the value of the real variable x such that $f(x) = 0$. Let us denote the solution by x^* and let us assume $\hat{x}$ is a first guess for the solution. The basic idea of Newton's method is to approximate $f(x)$ up to the first order in a Taylor series expansion about the point $\hat{x}$. This yields a linear approximation for f at the point $\hat{x}$:

$$f(x) = f(\hat{x}) + f'(\hat{x})(x - \hat{x})$$

It is reasonable to consider x as a new estimate for the root and compute it by solving the previous equation with $f(x) = 0$:

$$x = \hat{x} + \Delta x \quad \text{with} \quad \Delta x = -[f'(\hat{x})]^{-1} f(\hat{x}) \tag{18}$$

and Δx is usually called *differential correction*. Of course, in general, $f(x)$ is not a linear function of x, and consequently we cannot expect that $f(x) = 0$. However, we might expect that x is a better estimate for the root x^* than the original guess $\hat{x}$, that is, in formulæ

$$|x - x^*| \leqslant |\hat{x} - x^*| \quad \text{and} \quad |f(x)| \leqslant |f(\hat{x})|$$

If the new point is an improvement, then it makes sense to repeat the process, defining a sequence of points $x^{(0)}, x^{(1)}, x^{(2)}, \ldots$ such that the generic point $x^{(k)}$ is given by

$$x^{(0)} = \hat{x}, \quad x^{(k)} = x^{(k-1)} + \Delta x^{(k)}, \quad k \geqslant 1$$

with $\hat{x}$ the initial guess and $\Delta x^{(k)} = -[f'(x^{(k-1)})]^{-1} f(x^{(k-1)})$. The sequence is said to be *convergent* to x^* if

$$\lim_{k \to \infty} |x^{(k)} - x^*| = 0$$

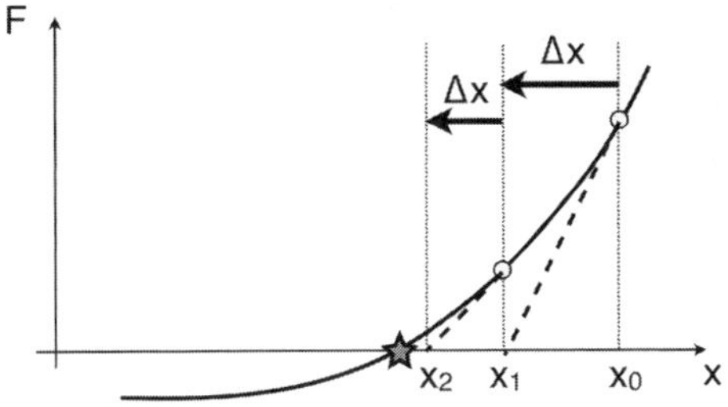

Fig. 1 Newton's method to find a root of one-variable real function.

In practice, the sequence is ended when the computed zero is close enough to the real solution. Indeed, fixed a threshold $\epsilon > 0$, if the condition $|x^{(\bar{k})} - x^*| \leqslant \epsilon$ is satisfied for a certain $\bar{k} \in \mathbb{N}$, then $|x^{(k)} - x^*| \leqslant \epsilon$ for every $k \geqslant \bar{k}$. The rate of convergence for Newton's method is said to be quadratic.

Geometrically, to find a zero of a function we can think of choosing a point $\hat{x}$ close enough to the real solution x^*, compute the tangent line to f at $\hat{x}$, and take the intersection between this line and the x-axis as the new point (Fig. 1).

The solution of Newton's method depends on the choice of the first guess. Indeed, since it is based on a first-order approximation of the function f, different initial conditions may lead to different zeros. Although, requiring further properties to f, for example, convexity, we may assure that the method is global and does not depend on the first guesses.

Methods for correcting the difficulties with Newton's method are referred to as *globalization strategies*. Among them we mention the two classical ones: the line-search method, which alters the magnitude of the step $\Delta x^{(k)}$, and the thrust region method, which changes both the magnitude and the direction of the step. All globalization strategies do not change the convergence, they just detect a problem with the (unmodified) Newton's method, and then, correct the deficiency.

Another downside of Newton's method is the computation of the slope $f'(x)$ at each iteration. This can be difficult and/or time consuming, especially when $f(x)$ is highly nonlinear. To eliminate the explicit calculation of the slope, various modifications of Newton's method have been proposed and they are called quasi-Newton method. The simplest one is the secant method: at the k-iteration the correction is

$$\Delta x^{(k)} = -B^{-1} f(x^{(k)}) \quad \text{with} \quad B = \frac{f(x^{(k)}) - f(x^{(k-1)})}{x^{(k)} - x^{(k-1)}}, \quad k \geqslant 1$$

The slope is approximated using information from previous iterates instead of a direct evaluation. However, the secant method is superlinearly convergent, which, in general, is not as fast as the quadratical convergence of Newton's algorithm. Thus, we can expect that more iterations are needed for convergence, even though the cost per iteration is lower.

Newton's method can be easily generalized to the multidimensional case. Given $\mathbf{f} = (f_1, \ldots, f_m)$, vectors of continuously differentiable functions, find $\mathbf{x}^* = (x_1^*, \ldots, x_n^*) \in R^n$ such that $\mathbf{f}(\mathbf{x}^*) = \mathbf{0}$. As in one variable, a linear approximation of $\mathbf{f}$ at the starting point $\hat{\mathbf{x}}$ leads to

$$\mathbf{f}(\mathbf{x}) = \mathbf{f}(\hat{\mathbf{x}}) + \mathrm{J}(\hat{\mathbf{x}})(\mathbf{x} - \hat{\mathbf{x}})$$

where

$$\mathrm{J}(\hat{\mathbf{x}}) = \frac{\partial \mathbf{f}}{\partial \mathbf{x}}(\hat{\mathbf{x}}) = \left(\frac{\partial f_j}{\partial x_i}(\hat{\mathbf{x}}) \right)_{i=1,\ldots,n;\, j=1,\ldots,m}$$

is the Jacobian matrix at point $\hat{\mathbf{x}}$. Then, requiring $\mathbf{f}(\mathbf{x}) = \mathbf{0}$, we obtain the correction

$$\Delta\mathbf{x} = \mathbf{x} - \hat{\mathbf{x}} = -\left[\mathrm{J}(\hat{\mathbf{x}})\right]^{-1} \mathbf{f}(\hat{\mathbf{x}}) \tag{19}$$

and the new starting point is $\mathbf{x} = \hat{\mathbf{x}} + \Delta\mathbf{x}$.

The multidimensional version of Newton's method is again quadratically convergent, provided it is within a region of convergence, and it may diverge unless appropriate globalization strategies are used. Furthermore, in Eq. (19) we assumed that the Jacobian matrix J is nonsingular, that is, $\det \mathrm{J} \neq 0$, which is analogous to requiring that in the univariate case $f'(x) \neq 0$. Finally, the method requires the explicit computation of the matrix J, which can be costly.

B. UNCONSTRAINED OPTIMIZATION

This subsection describes how Newton's methods can be applied to the multidimensional unconstrained minimization problem. The goal is to find $\mathbf{x} = (x_1, \ldots, x_n) \in R^n$ such that the real function $f(\mathbf{x})$ is a minimum. For this purpose, we approximate f using a Taylor series expansion up to order 2 about the point $\hat{\mathbf{x}}$:

$$f(\mathbf{x}) = f(\hat{\mathbf{x}}) + \operatorname{grad} f(\hat{\mathbf{x}})(\mathbf{x} - \hat{\mathbf{x}}) + \frac{1}{2}(\mathbf{x} - \hat{\mathbf{x}})^T \mathrm{H}(\hat{\mathbf{x}})(\mathbf{x} - \hat{\mathbf{x}})$$

where

$$\operatorname{grad} f = (\nabla_{\mathbf{x}} f)^T = \left(\frac{\partial f}{\partial x_1}, \ldots, \frac{\partial f}{\partial x_n} \right), \quad \mathrm{H} = \left(\frac{\partial f}{\partial x_i \partial x_j} \right)_{i=1,\ldots,n; j=1,\ldots,m}$$

are the gradient vector and the Hessian matrix, respectively. The correction $\Delta\mathbf{x} = \mathbf{x} - \hat{\mathbf{x}}$ is also called the search direction, $\operatorname{grad} f(\hat{\mathbf{x}})\Delta\mathbf{x}$ is the directional derivative along $\Delta\mathbf{x}$, and $(\Delta\mathbf{x})^T\mathrm{H}(\hat{\mathbf{x}})\Delta\mathbf{x}$ is called the curvature or second directional derivative in the direction $\Delta\mathbf{x}$.

If $\hat{\mathbf{x}} = \mathbf{x}^*$ is the minimum point, then it is immediate to show that *sufficient conditions* are

$$\nabla_{\mathbf{x}} f(\mathbf{x}^*) = \mathbf{0}, \quad (\mathbf{x} - \mathbf{x}^*)^T \mathrm{H}(\mathbf{x}^*)(\mathbf{x} - \mathbf{x}^*) > 0$$

and *necessary conditions* are

$$\nabla_{\mathbf{x}} f(\mathbf{x}^*) = \mathbf{0}, \quad (\mathbf{x} - \mathbf{x}^*)^T \mathrm{H}(\mathbf{x}^*)(\mathbf{x} - \mathbf{x}^*) \geqslant 0$$

for every $\mathbf{x}$ in a neighborhood of $\mathbf{x}^*$. In the first case $\mathbf{x}^*$ is called a strong local minimum and the Hessian matrix $\mathrm{H}(\mathbf{x}^*)$ is said to be positive definite. In the second case, there are some directions in which $\mathbf{x}^*$ is not the only minimum and $\mathrm{H}(\mathbf{x}^*)$ is said to be positive semidefinite. For the sake of completeness, if there are directions with both positive and negative, the matrix is called indefinite.

If $\hat{\mathbf{x}}$ is not a minimum point, it is reasonable to choose a new point $\mathbf{x}$ such that the gradient at $\mathbf{x}$ is null. Thus, for every $\hat{\mathbf{x}}$ in a neighborhood of $\mathbf{x}$, the linear approximation applied to the gradient leads to

$$\nabla_{\mathbf{x}} f(\mathbf{x}) = \mathbf{0} = \nabla_{\mathbf{x}} f(\hat{\mathbf{x}}) + \mathrm{H}(\hat{\mathbf{x}})(\mathbf{x} - \hat{\mathbf{x}})$$

which can be solved using Newton's method with the first guess $\hat{\mathbf{x}}$ and the search direction is

$$\Delta\mathbf{x} = \mathbf{x} - \hat{\mathbf{x}} = -\mathrm{H}(\hat{\mathbf{x}})^{-1} \nabla_{\mathbf{x}} f(\hat{\mathbf{x}}) \tag{20}$$

Then the new point is as usual $\mathbf{x} = \hat{\mathbf{x}} + \Delta\mathbf{x}$. To guarantee that we are moving towards a minimum and not only a stationary point, we require the so-called descent condition

$$\mathrm{grad} f(\mathbf{x}) \Delta\mathbf{x} < 0 \tag{21}$$

Indeed, substituting Eq. (20) into Eq. (21), we obtain $-(\nabla_{\mathbf{x}} f)^T \mathrm{H}^{-1} \nabla_{\mathbf{x}} f$, which holds if the Hessian matrix is positive definite.

Although the derivation of the system (20) was motivated by a Taylor series expansion, an alternative approach is to choose $\Delta\mathbf{x}$ to minimize the quadratic objective function

$$(\nabla_{\mathbf{x}} f)^T \Delta\mathbf{x} + \frac{1}{2} \Delta\mathbf{x}^T \mathrm{H} \Delta\mathbf{x}$$

Indeed, the necessary optimality condition is $\mathrm{H}\Delta\mathbf{x} + \nabla_{\mathbf{x}} f = 0$, which is equivalent to Eq. (20).

To sum up, given the function to be minimized, $f(\mathbf{x})$, each iteration of the Newton method is equivalent to approximate f around the current solution $\mathbf{x}$ with a quadratic form, then find the offset, $\Delta\mathbf{x}$, to the zero-gradient point of the quadratic form, and finally use $\Delta\mathbf{x}$ as a correction in the original optimization problem (Fig. 2).

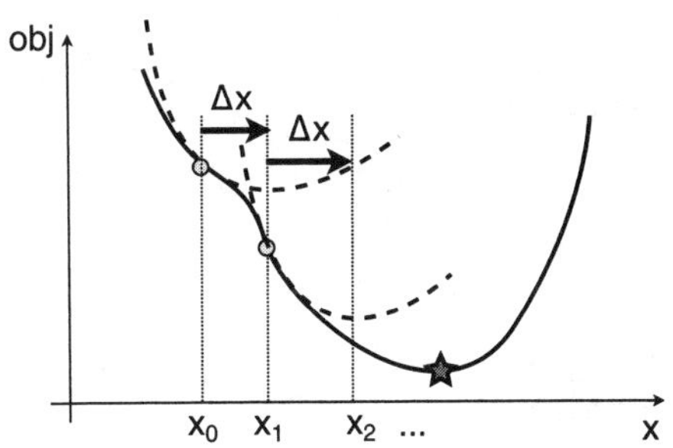

Fig. 2 Unconstrained optimization of one-variable real function.

TABLE 1 NUMERICAL RESULTS USING THE MATLAB FUNCTION `fmincon`

	Minimum Value of f	Minimum Points (x, y)	Number of Iterations
Problem A	2.97×10^{-13}	$(-0.54, -0.54) \times 10^{-6}$	2
Problem B	-28.05	$(-6.61, 0)$	10
Problem C	-321.92	$(-7.31, 0)$	6

Example 3

Consider the three unconstrained optimizations problems:

$$\text{(A)} \quad f(x,y) = \frac{1}{2}(x^2 + y^2)$$

$$\text{(B)} \quad f(x,y) = 5x + e^{-(x+5)} + y^2$$

$$\text{(C)} \quad f(x,y) = \frac{x^4}{10} - 10x^2 + 10x + y^2$$

A numerical solution can be computed using the Matlab built-in function `fmincon` and coding an SQP solver. The results for the three problems are shown in Table 1.

Example 4

The goal for Earth–Mars two-impulse transfer is to minimize $\Delta v = \Delta v_1 + \Delta v_2$ necessary for a two-impulsive transfer from the Earth to Mars. The optimal variables are the departure time t_0 and the time of flight t_{tof}. Using the same notation of Example 1, we compute the ephemeris of the Earth at epoch t_0, $(\mathbf{r}_{\oplus}, \mathbf{v}_{\oplus})$, and that of Mars at epoch $t_0 + t_{tof}$, $(\mathbf{r}_{\male}, \mathbf{v}_{\male})$. The escape velocity v_1 and the arrival velocity v_2 are evaluated by solving Lambert's problem (Fig. 3).

The objective function is $f(\mathbf{x}) = \Delta v(\mathbf{x}) = \Delta v(t_0, t_{tof})$, with t_0 varying from 0 to 1460 days (about 4 years), and t_{tof} from 100 to 600 days, starting from the date 1 January 2000 (Fig. 4).

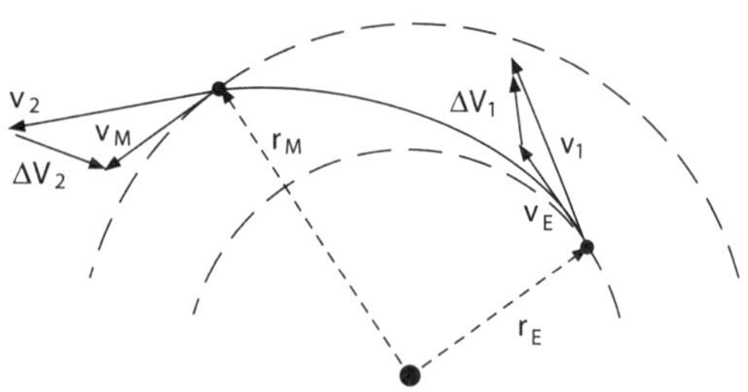

Fig. 3 Lambert's problem to compute the escape velocities v_1, v_2.

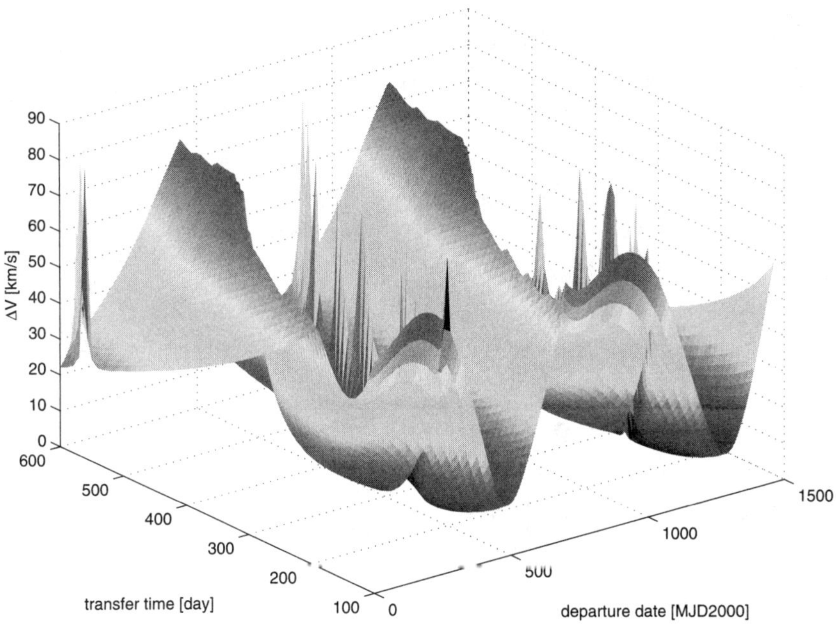

Fig. 4 Objective function of the two-impulse transfer from the Earth to Mars.

Necessary conditions for optimality are

$$\frac{\partial \Delta v}{\partial t_0} = 0, \quad \frac{\partial \Delta v}{\partial t_{\text{tof}}} = 0$$

In a generic iteration, given the current estimate $\mathbf{x} = (t_0, t_{\text{tof}})$, we need to compute the corrections $\Delta\mathbf{x} = (\Delta t_0, \Delta t_{\text{tof}})$ by using Eq. (20) with Hessian matrix

$$\mathrm{H} = \begin{bmatrix} \dfrac{\partial^2 \Delta v}{\partial^2 t_0} & \dfrac{\partial^2 \Delta v}{\partial^2 t_0 \partial t_{\text{tof}}} \\[2ex] \dfrac{\partial^2 \Delta v}{\partial t_{\text{tof}} \partial t_0} & \dfrac{\partial^2 \Delta v}{\partial^2 t_{\text{tof}}} \end{bmatrix}$$

Using $t_0 = 700$ days and $t_{\text{tof}} = 550$ days as an initial guess, Matlab built-in funtion `fmincon` finds the solution $t_0 = 413.6$ days and $t_{\text{tof}} = 271.4$ days, corresponding to $J = 7.437\ \text{km/s}$ (Fig. 5).

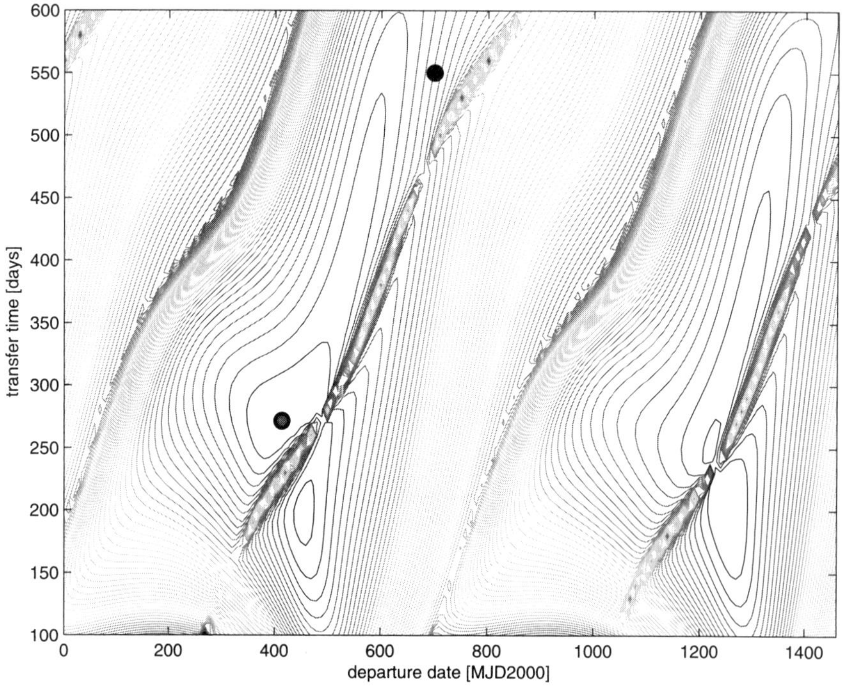

Fig. 5 Contour lines of the cost function represented in Fig. 4, initial guess (black point), and the solution found by `fmincon` (dark grey point).

C. EQUALITY-CONSTRAINED OPTIMIZATION

This subsection describes how Newton's method can be applied to optimize an objective function while satisfying a set of equality constraints. The problem consists of finding a point $\mathbf{x} \in \mathbb{R}^n$ to minimize the function $f(\mathbf{x})$ subject to $m \leqslant n$ constraints $c_i(\mathbf{x}) = 0,\ i = 1, \ldots, m$. The classical approach is to define the *Lagrangian*

$$L(\mathbf{x}, \boldsymbol{\lambda}) = f(\mathbf{x}) - \boldsymbol{\lambda}^T \mathbf{c}(\mathbf{x}) = f(\mathbf{x}) - \sum_{i=1}^{m} \boldsymbol{\lambda}_i c_i(\mathbf{x}) \tag{22}$$

where $\boldsymbol{\lambda}$ is an vector in $\mathbb{R}^m$ of Lagrange multipliers and $\mathbf{c} = (c_1, \ldots, c_m)$.

Analogously to the unconstrained case, the necessary conditions for the point $(\mathbf{x}, \boldsymbol{\lambda})$ to be an optimum are

$$\mathbf{0} = \nabla_{\mathbf{x}} L(\mathbf{x}, \boldsymbol{\lambda}) = \nabla f(\mathbf{x}) - \mathrm{J}(\mathbf{x})^T \boldsymbol{\lambda} = \nabla f(\mathbf{x}) - \sum_{i=1}^{m} \boldsymbol{\lambda}_i \nabla_{\mathbf{x}} c_i(\mathbf{x}) \tag{23}$$

$$\mathbf{0} = \nabla_{\boldsymbol{\lambda}} L(\mathbf{x}, \boldsymbol{\lambda}) = -\mathbf{c}(\mathbf{x}) \tag{24}$$

where $\mathrm{J}(\mathbf{x})$ is the Jacobian matrix of $\mathbf{c}(\mathbf{x})$. Since these conditions do not distinguish between a point that is a minimum, a maximum, or another kind of stationary point, we require further conditions on the curvature of f. Let

$$H_L = \nabla^2_{\mathbf{xx}} L = \nabla^2_{\mathbf{xx}} f - \sum_{i=1}^{m} \lambda_i \nabla^2_{\mathbf{xx}} c_i$$

be the Hessian of the Lagrangian, then a sufficient condition is that $\mathbf{v}^T H_L \mathbf{v} > 0$ for any vector $\mathbf{v}$ in the constraint tangent space.

The necessary conditions (23) and (24) define a system of $n+m$ equation in $n+m$ variables, which can be solved using Newton's method. Let $(\hat{\mathbf{x}}, \hat{\boldsymbol{\lambda}})$ be the initial guess, we proceed by expanding the equations in a Taylor series about this point up to the next order:

$$\begin{aligned} \mathbf{0} &= \nabla f(\hat{\mathbf{x}}) - \mathrm{J}(\hat{\mathbf{x}})^T \hat{\boldsymbol{\lambda}} + H_L(\hat{\mathbf{x}})(\mathbf{x} - \hat{\mathbf{x}}) - \mathrm{J}(\hat{\mathbf{x}})^T(\boldsymbol{\lambda} - \hat{\boldsymbol{\lambda}}) \\ \mathbf{0} &= -\mathbf{c}(\hat{\mathbf{x}}) - \mathrm{J}(\hat{\mathbf{x}})(\mathbf{x} - \hat{\mathbf{x}}) \end{aligned}$$

After simplification, we end up in a linear system, referred to as Karush–Kuhn–Tucker (KKT) system:

$$\begin{bmatrix} H_L(\hat{\mathbf{x}}) & -(\mathrm{J}(\hat{\mathbf{x}})^T \\ \mathrm{J}(\hat{\mathbf{x}}) & \mathbf{0} \end{bmatrix} \begin{bmatrix} \Delta\mathbf{x} \\ \boldsymbol{\lambda} \end{bmatrix} = \begin{bmatrix} -\mathrm{J}(\hat{\mathbf{x}}) \\ -\mathbf{c}(\hat{\mathbf{x}}) \end{bmatrix} \tag{25}$$

where the unknown variables are the search direction $\Delta\mathbf{x} = \mathbf{x} - \hat{\mathbf{x}}$ and the Lagrange multipliers vector $\boldsymbol{\lambda}$ at the new point. Then, we update $(\mathbf{x}, \boldsymbol{\lambda}) = (\hat{\mathbf{x}} + \Delta\mathbf{x}, \boldsymbol{\lambda})$ and we again apply Newton's method with the starting point $(\hat{\mathbf{x}}, \hat{\boldsymbol{\lambda}}) = (\mathbf{x}, \boldsymbol{\lambda})$ until convergence.

As in the unconstraint case, instead of using a Taylor series expansion, we can derive the KKT system (25) by choosing $\Delta\mathbf{x}$ to minimize the quadratic objective

$$(\nabla f)^T \Delta\mathbf{x} + \frac{1}{2} \Delta\mathbf{x}^T H_L \Delta\mathbf{x}$$

subject to the linear constraints $\mathrm{J}\Delta\mathbf{x} = -\mathbf{c}$. Indeed, using Lagrangian multipliers we define the Lagrangian

$$\tilde{L} = (\nabla f)^T \Delta\mathbf{x} + \frac{1}{2} \Delta\mathbf{x}^T H_L \Delta\mathbf{x} - \boldsymbol{\lambda}^T(\mathrm{J}\Delta\mathbf{x} + \mathbf{c})$$

and the necessary optimality conditions are

$$\frac{\partial \tilde{L}}{\partial \Delta\mathbf{x}} = 0, \quad \frac{\partial \tilde{L}}{\partial \boldsymbol{\lambda}} = 0$$

which are equivalent to the KKT system (25). In other words, a linear approximation to the gradient, given by Taylor series expansion, is equivalent to making a quadratic model for the quantity being optimized, $\Delta\mathbf{x}$ in this case. Note that the quadratic approximation is made to the Lagrangian (22) and not just the objective function f. Since we end up with a quadratic–linear model, Newton's method will converge in one step for a quadratic objective with linear equality constraints.

D. INEQUALITY-CONSTRAINED OPTIMIZATION

In this subsection we consider a problem with inequality constraints. The goal is to choose $\mathbf{x} \in \mathbb{R}^n$ to minimize $f(\mathbf{x})$ subject to the m inequality constraints $\mathbf{c}(\mathbf{x}) \geqslant \mathbf{0}$.

In contrast to equality-constrained problems, introduced in Sec. C, which require $m \leqslant n$, the number of inequality constraints can exceed the number of variables. A point that satisfies the constraints is said to be *feasible* and the collection of all feasible points is called the *feasible region*. Conversely, points in the infeasible region violate one or more of the constraints.

The total set of constraints is partitioned into two subsets, namely the active and inactive constraints. The active constraints are satisfied as equalities and the inactive constraints are satisfied as inequalities. Fortunately, to identify the active constraints, there is an additional necessary condition: the algebraic sign of the Lagrange multipliers that corresponds to the active constraints must be nonnegative.

E. GENERAL FORM OF A NONLINEAR-PROGRAMMING PROBLEM

The general NLP problem can be stated as follows. Given a continuously differentiable function $f: \mathbb{R}^n \to \mathbb{R}$, find a point $\mathbf{x}^*$ to minimize the scalar objective function $f(\mathbf{x})$ subject to the m constraints $\mathbf{c}_1 \leqslant \mathbf{c}(\mathbf{x}) \leqslant \mathbf{c}_2$ and the boundary conditions $\mathbf{x}_1 \leqslant \mathbf{x} \leqslant \mathbf{x}_2$. Equality constraints can be imposed by setting $\mathbf{c}_1 = \mathbf{c}_2$.

Different methods were developed for solving the general NLP problem since Kuhn and Tucker published the first paper about this subject in 1950. Among them, we cite the sequential quadratic programming (SQP), which consists of solving a series of QP subproblems that converge in to the optimal solution of the original problem, and the interior-point methods (IPMs) [3, 15], which constrain the search to the feasible region by introducing barrier terms to the objective function. Suppose we want to minimize $f(\mathbf{x})$ subject to inequalities $g_i(\mathbf{x}) \geqslant 0$, $i = 1, \ldots, m$, then the objective function becomes

$$\tilde{f}(\mathbf{x}) = f(\mathbf{x}) + \sum_{i=1}^{m} \mu_i e^{-g_i(\mathbf{x})}$$

where the sum is the *penalty* term, μ is the m-vector of Lagrange multipliers, and $e^{-g_i(\mathbf{x})}$ is the *barrier* function.

F. WHAT CAN GO WRONG

So far, we presented how Newton's method works. However, in practice, a series of problems may occur. In the following, we listed the most common difficulties which can be encountered [3]:

1. *Infeasible constraints*—One of the most common difficulties occurs when the NLP problem has infeasible constraints, that is, the constraints taken all together have no solution.
2. *Rank-deficient constraints*—The constraints are consistent, that is, they do have a solution. However, at the solution, the Jacobian matrix may be either ill-conditioned or rank deficient.
3. *Constraint redundancy*—If the problem formulation contains constraints that are redundant, we may have Lagrange multipliers near zero or difficulty detecting the active set.
4. *Discontinuities*—Perhaps the biggest obstacle encountered in the practical application of NLP methods is the presence of discontinuous behavior. All the numerical methods described assume that the objective function and constraints are continuous and differentiable. When discontinuities are present in the problem functions, the standard quadratic/linear models that form the basis for most NLP methods are no longer appropriate.
5. *Scaling*—Poor scaling can make a good algorithm bad. Scaling changes the convergence rate, termination tests, and numerical conditioning.
6. *Nonunique solution*—If the problem formulation has more than one solution, then we can require that Hessian approximation remains positive definite at each step of Newton's method. However, if we repeated the procedure with a different initial guess, it is quite likely that a different solution will be computed.

Example 5

Let $\mathbf{x} = (x, y, z) \in \mathbb{R}^3$ and consider the NLP problem

$$\min_{\mathbf{x}\in\mathbb{R}^3} f(\mathbf{x}) = x^2 + xy + 2y^2 - 6x - 2y - 12z$$

$$\text{s.t. } g_1(\mathbf{x}) = 2x^2 + y^2 \leqslant 15$$

$$g_2(\mathbf{x}) = x - 2y - z \geqslant -3$$

$$x, y, z \geqslant 0$$

We solved this problem using the function `fmincon` of the Optimization Toolbox in Matlab, following the example in Ref. [16]. The initial guess $\mathbf{x}_0 = (1, 1, 1)$ leads to the optimal value $f(\mathbf{x}) = -77.80$, corresponding to $x = 2.74$, $y = 0$, and $z = 5.74$. The M-files are listed in Table 2.

TABLE 2 MATLAB CODE USED TO SOLVE EXAMPLE 5

```
%%%%%%%%%%%%%%%%%%%%%%%%%%%%%%%%%%%%%%%%%%%%%%%%%%%%%%%%%%%%%%
%                Main program
%%%%%%%%%%%%%%%%%%%%%%%%%%%%%%%%%%%%%%%%%%%%%%%%%%%%%%%%%%%%%%
x0 = [0; 8; 1];    % inital guess
xL = [0; 0 ; 0];  % lower bounds
xU = [];  % upper bounds
% Set option for fmincon
options = optimset('Display', 'iter', 'GradObj', 'on', 'GradConstr', 'on', ...
            'DerivativeCheck', 'on', 'LargeScale', 'off', 'HessUpdate',
            'bfgs', ...
            'Diagnostics', 'on', 'TolX', 1e-7, 'TolFun', 1e-7, 'TolCon',
            1e-7, ...
            'MaxFunEval', 100, 'MaxIter', 100)
% Call to the optimiser
[xopt, fopt, iout] = fmincon( @SQP_fun, x0, [], [], [], [], xL, xU, ...
                  @SQP_ctr, options );
```

```
%%%%%%%%%%%%%%% FUNCTION TO BE MINIMIZED %%%%%%%%%%%%%%
% Objective: f(x,y) := x^2+xy+y^2-6y^2-6x-2y-12z
%%%%%%%%%%%%%%%%%%%%%%%%%%%%%%%%%%%%%%%%%%%%%%%%%%%%%%%%%%%%%%
function [F,dF] = SQP_fun(x)
    F = x(1)^2 + x(1)*x(2) + 2*x(2)^2 - 6*x(1) - 2*x(2) - 12*x(3);
    if nargout >1
        dF = [2*x(1)+x(2)-6 x(1)+4*x(2)-2 -12];
    end
end
```

```
%%%%%%%%%%%%%%%%%%%%%%% CONSTRAINTS %%%%%%%%%%%%%%%%%%%%%
% inequality constraint: g_1(x,y,z) := 2x^2+y^2 <= 15
% inequality constraint: g_2(x,y,z) := x-2y-z   >= -3
% inequality constraint: x,y,z  >= 0
%%%%%%%%%%%%%%%%%%%%%%%%%%%%%%%%%%%%%%%%%%%%%%%%%%%%%%%%%%%%%%
function [g,h,dg,dh] = SQP_ctr(x)
    g = [ 2*x(1)^2+x(2)^2-1  -3-x(1)+2*x(2)+x(3) ]; % inequality constraints
    h = [];% equality constraints
    if nargout > 2
        dg = [ 4*x(1) -1 ; 2*x(2) 2; 0 1];% gradient of inequality constraints
        dh = []; % gradient of equality constraints
    end
end
```

V. DIRECT METHODS

In Sec. III we showed how to solve the optimal control problem with the Euler–Lagrange equations (10–12) (indirect method). Another way is to translate the optimal control problem into an NLP problem. The difference between them is that an NLP problem is characterized by a finite set of variables **x** and constraints **c**. Instead, optimal control problems can involve continuous functions for the state $\mathbf{x}(t)$ and control $\mathbf{u}(t)$ vectors. It will be convenient to view the optimal control problem as an infinite-dimensional extension of an NLP problem. However, practical methods for solving optimal control problems require Newton-based iterations with a finite set of variables and constraints. This goal can be achieved by transcribing or converting the infinite-dimensional problem into a finite-dimensional approximation. This procedure is called *direct transcription* and the approach is said to be a *direct method.*

The solution of the optimal control problem given by a direct approach is strictly connected to the numerical integration of the differential equations. The core of this method consists in fact of the way of dealing with the dynamical system: the set of differential equations can be transcribed into a finite set of equality constraints. If the solution to the NLP problem satisfies these constraints, then the original optimal control problem is solved within the degree of accuracy of the numerical scheme used [3, 8, 17, 18]. Classical transcription methods are collocation and multiple shooting.

A. DIRECT TRANSCRIPTION AND COLLOCATION

Consider the optimal control problem formulated through Eqs. (6) and (9). All direct approaches start by dividing the time domain $[t_0, t_f]$ into s subintervals with nodes (or mesh or grid points)

$$t_0 = t_1 < t_2 < \cdots < t_N = t_f$$

where $N = s + 1$ and $h_k = t_{k+1} - t_k$ is the difference between two consecutive nodes. If the intervals have the same length, then the step size of the discretization is $h = (t_f - t_0)/(N - 1)$. For each node t_k, the states and the controls can be discretized by defining $\mathbf{x}_k = \mathbf{x}(t_k) \in \mathbb{R}^n$ and $\mathbf{u}_k = \mathbf{u}(t_k) \in \mathbb{R}^m$. Consequently, a new variable vector of the problem is

$$\mathbf{y} = (t_f, \mathbf{x}_1, \mathbf{u}_1, \ldots, \mathbf{x}_N, \mathbf{u}_N)^T$$

As a result of the transcription, the optimal control constraints (7) and (8) are replaced by the NLP constraints

$$\mathbf{c}_l \leqslant \mathbf{c}(\mathbf{y}) \leqslant \mathbf{c}_u \tag{26}$$

with

$$\begin{aligned}
\mathbf{c}(\mathbf{y}) &= (\boldsymbol{\zeta}_1, \ldots, \boldsymbol{\zeta}_{N-1}, \boldsymbol{\psi}_0, \boldsymbol{\psi}_f, \mathbf{g}_1, \ldots, \mathbf{g}_N)^T \\
\mathbf{c}_l &= (\mathbf{0}, \ldots, \mathbf{0}, \mathbf{g}_l, \ldots, \mathbf{g}_l)^T \\
\mathbf{c}_u &= (\mathbf{0}, \ldots, \mathbf{0}, \mathbf{g}_u, \ldots, \mathbf{g}_u)^T
\end{aligned}$$

where $\boldsymbol{\zeta}_k, k = 1, \ldots, N-1$ are the defect constraints derived by the numerical integration scheme and replace the differential Eqs. (6) (collocation). The first $n(N-1)$ equality constraints in Eq. (26) require each defect vector to be zero, so that the differential equations (6) are approximately satisfied. The boundary conditions (7) are directly enforced by the equality constraints on $\boldsymbol{\psi}_0$ and $\boldsymbol{\psi}_f$, and the nonlinear path constraints (8) are imposed at the nodes. In a similar way, the objective function, either in form (2) or (9), can be written in terms of the new vector $\mathbf{y}$. Thus, the optimal control problem has been reformulated into the form of a general NLP problem and it can be solved with the techniques described in Sec. IV.

It remains to show how to compute the defects $\boldsymbol{\zeta}_k, k = 1, \ldots, N-1$. The simplest of the collocation methods uses Euler's scheme. The solution is replaced with its linear approximation

$$\mathbf{x}_k = \mathbf{x}_k + \dot{\mathbf{x}}(t_k)(t_k + h - t_k) = \mathbf{x}_k + \dot{\mathbf{x}}(t_k)h = \mathbf{x}_k + \mathbf{f}_k h$$

where $\mathbf{f}_k = \mathbf{f}(\mathbf{x}_k, \mathbf{u}_k, \mathbf{p}, t_k)$ and the last equality follows from Eq. (6). So, the defect constraints are

$$\boldsymbol{\zeta}_k = \mathbf{x}_{k+1} - \mathbf{x}_k - h\mathbf{f}_k = 0$$

Another collocation method is the classical Runge–Kutta scheme, where the defects are of the form

$$\boldsymbol{\zeta}_k = \mathbf{x}_{k+1} - \mathbf{x}_k - h\sum_{j=1}^{4} \beta_j \mathbf{f}_{kj}$$

with appropriate definitions of $\mathbf{f}_{kj}$ [3].

B. MULTIPLE SHOOTING

Another method for solving an optimal control problem via NLP algorithm is the multiple-shooting method [3, 19]. The idea is to divide the trajectory into shorter pieces or segments [18]. Thus, we divide the time domain into smaller intervals of the form

$$t_0 = t_1 < t_2 < \cdots < t_N = t_f$$

On each time subinterval $[t_k, t_{k+1}]$, for $0 \leqslant k \leqslant N-1$, we define M nodes for the controls $\mathbf{u}_k^{(1)}, \ldots, \mathbf{u}_k^{(M-1)}$, $\mathbf{u}_k^{(M)} = \mathbf{u}_{k+1}^{(1)}$ and on the generic interval k, the variable vector is

$$\mathbf{y}_k = (\mathbf{x}_k, \mathbf{u}_k^{(1)} \ldots, \mathbf{u}_k^{(M-1)})^T, \quad k = 1, \ldots, N-1$$

$$\mathbf{y}_N = (t_f, \mathbf{x}_N, \mathbf{u}_N^{(1)})$$

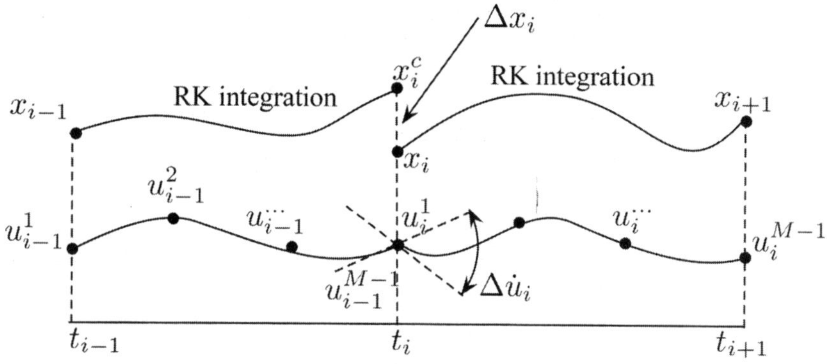

Fig. 6 Multiple-shooting method.

By interpolation, we can easily compute the control history corresponding to each leg and patch them together as $\mathbf{u} = \mathbf{u}(t),\ t_1 \leqslant t \leqslant t_N$. Alternatively, we can propagate (integrate) the differential equations from t_{k-1} to the end of the segment at t_k obtaining a value $\mathbf{x}_k^c$ at node t_k.

It is reasonable to require that the segments join at the boundaries so that we impose a system of constraints on the defects

$$\zeta_k = \mathbf{x}_k^c - \mathbf{x}_k = \mathbf{0}, \quad k = 1, \ldots, N \tag{27}$$

and the vector of variables for the NLP problem is

$$\mathbf{y} = (t_f, \mathbf{y}_1 \ldots, \mathbf{y}_N)^T$$

The multiple-shooting method is illustrated in Fig. 6.

Both multiple shooting and collocation can be considered as robust methods, even if highly nonlinear dynamics is involved. Collocation can better manage discontinuities of the control functions, however, a higher number of variables are added to the original problem.

Example 6

The TPBVP associated to the problem given in Example 2 can be solved analytically or using direct transcription and collocation. Figure 7, taken from Ref. [11], shows the comparison between the two methods.

Example 7

The optimal control problem given in Example 1 for low-thrust Earth–Mars transfer can be solved using the simple-shooting method, that is, multiple shooting with $N = 1$. A cubic spline on $M = 4$ points is used to

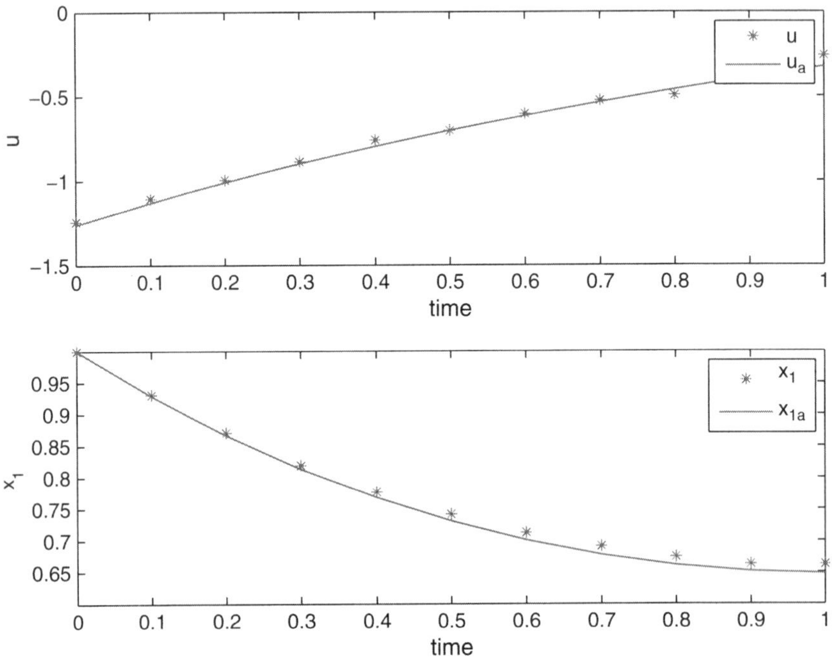

Fig. 7 Comparison of the state $x(t)$ and control $u(t)$ histories obtained using direct transcription and collocation with the corresponding analytical solutions (taken from Ref. [11]).

compute $\mathbf{u}(t)$. Earth's ephemerides are used to set the initial condition for the integration of the shooting method. In this way, the constraints on $\mathbf{x}_0$ are automatically satisfied. Finally, the NLP problem with variable vector $(t_0, t_f, \mathbf{u}^{(1)}, \ldots, \mathbf{u}^{(4)}) \in \mathbb{R}^{14}$ is solved using the ballistic Lambert's arc as first guess (Fig. 8). All the details can be found in Ref. [20].

C. EXAMPLES

In the following section, three examples will be described in detail. The first one is the controlled trajectory in relative dynamics, the second one is Mars aero-gravity assist (AGA), and the last one is an optimization of spacecraft trajectories with a multiple-asteroid rendezvous using a low-thrust propulsion system.

1. CONTROLLED TRAJECTORY IN RELATIVE DYNAMICS

Given the following equations for the relative dynamics

$$\ddot{x} - 2n\dot{y} - 3n^2x = 0, \quad \ddot{y} + 2n\dot{x} = 0, \quad \ddot{z} + n^2z = 0$$

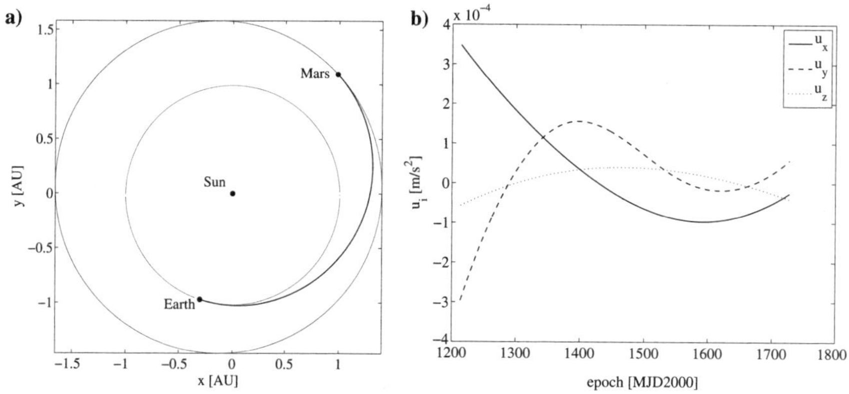

Fig. 8 Low-thrust Earth–Mars transfer using a multiple-shooting technique: a) first guess and b) solution (taken from Ref. [20]).

we want to minimize the performance index

$$J = \int_{t_0}^{t_f} L(\mathbf{x}, \mathbf{u}, t)\, \mathrm{d}t = \frac{1}{2}\int_{t_0}^{t_1} \mathbf{u}^T \mathbf{u}\, \mathrm{d}t$$

where t_0 is the initial time, t_f is the final time, and $\mathbf{u} = \mathbf{u}(t)$ are the controls, subject to

$$\mathbf{r}(t_o) = \mathbf{r}_0, \quad \mathbf{v}(t_o) = \mathbf{v}_0$$
$$\mathbf{r}(t_f) = \mathbf{r}_f, \quad \mathbf{v}(t_f) = \mathbf{v}_f$$

To solve the optimal control problem, a simple shooting-as-transcript technique is used. The interval $[t_0, t_f]$ is divided into seven subintervals and the control $\mathbf{u}(t)$ is computed interpolating over $M = 6$ points. Finally, $\mathbf{u}(t) = \mathbf{0}$ is chosen as the first guess for the NLP problem. An example of the application of the above procedure to reconfigure a four-satellite formation is reported in Fig. 9 (see Ref. [21]).

2. MARS AGA

An AGA is a spacecraft maneuver designed to change the spacecraft velocity when arriving at a celestial body with an atmosphere. Whereas, a gravity assist (GA) uses only the gravity of a celestial body to change the direction of the spacecraft trajectory (Fig. 10).

The AGA can be formalized as an optimal control problem: the dynamics can be transcribed into a set of nonlinear constraints and the arising NLP problem can be solved with an SQP solver. The problem is analyzed in detail in Ref. [22].

We consider a local noninertial reference frame (x, y, z) with origin in the center mass of the spacecraft and x-axis from the Earth to the spacecraft. We also

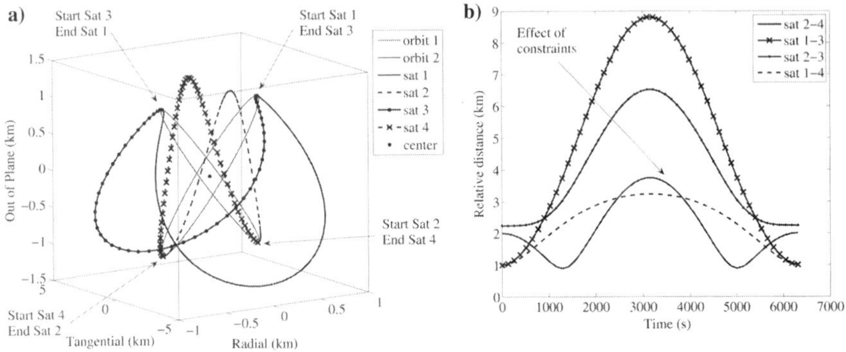

Fig. 9 Reconfiguration of two pairs of satellites orbiting on two different bounded relative orbits that switch the relative orbit and simultaneously perform a rephasing: resulting trajectory and relative distances (taken from Ref. [21]).

consider an inertial planetocentric ecliptic reference frame (X, Y, Z). The relative orientation between the two frames is defined by the longitude θ and the latitude φ, whereas the flight path angle γ and the heading angle ψ are introduced to identify the vehicle velocity vector $\mathbf{v}$ in the local reference frame (Fig. 11).

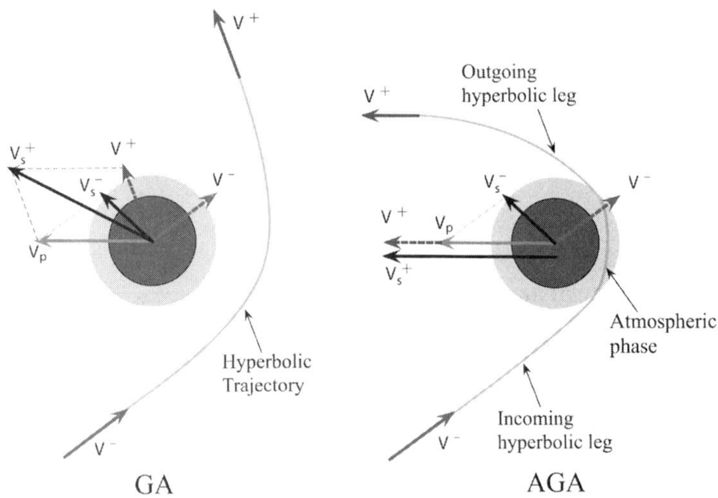

Fig. 10 AGA versus GA with a planet (taken from Ref. [23]).

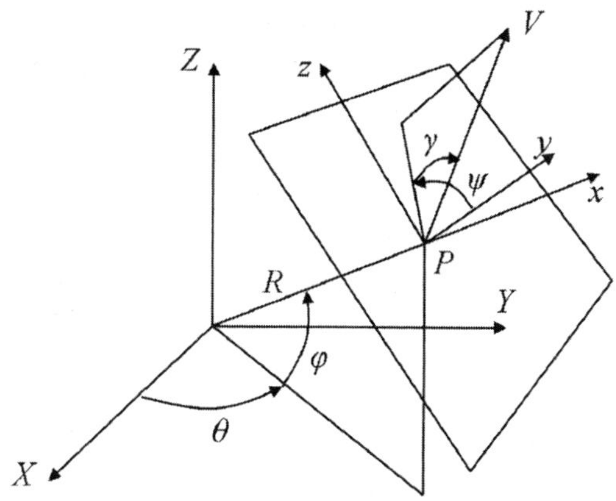

Fig. 11 Local reference frame and state vector variables $\mathbf{x} = (R, \theta, \varphi, v, v_\gamma, v_\varphi)$.

The dynamical equations of the vehicle center of mass are

$$\begin{aligned}
\dot{R} &= v \sin\gamma \\
\dot{\theta} &= \frac{v\cos\gamma\cos\psi}{R\cos\varphi} \\
\dot{\varphi} &= \frac{v\cos\gamma\cos\psi}{R} \\
\dot{v} &= -\frac{D}{m} - G\sin\gamma \\
\dot{v}_\gamma &= \frac{L\cos\sigma}{m} - G\cos\gamma + \frac{v^2\cos\gamma}{R} \\
\dot{v}_\varphi &= \frac{L\sin\sigma}{m\cos\gamma} - v^2\frac{\tan\varphi\cos\gamma\cos\psi}{R}
\end{aligned} \tag{28}$$

where G is the gravitational constant, L and D are the lift and drag force, respectively, and σ is the bank angle, defined as the angle between the lift vector and the plane described by the local vertical and the velocity vector. To simplify the equations, we assume that only planar maneuvers can be performed, consequently $\sigma = 0$, and the control is limited to the lift coefficient ratio λ.

The associated optimal control problem is formulated as follows. Consider the dynamics given by Eq. (28), with state vector $\mathbf{x} = (R, \theta, \phi, v, v_\gamma, v_\varphi)$ and control $\mathbf{u} = \lambda$. Find the optimal control law $\lambda(t)$ to maximize the absolute value of the final heliocentric velocity vector, that is, $J = \|v_s^+\|$ subject to the initial conditions

$$R(t_0) = R_a, \quad v_\infty^-(\mathbf{x}_0) = v_\infty^-$$

and final constraints $R(t_f) = R_a$, where R_a is the planetary-sensible atmosphere radius and v_∞^- is the spacecraft velocity going into the planet sphere of influence.

An additional constraint is imposed on the convective-heating variable at a stagnation point [24]:

$$q = q_c(\mathbf{x}(t), \mathbf{u}(t)) + q_r(\mathbf{x}(t), \mathbf{u}(t)) \leqslant q_{\max}$$

Since the dynamics is highly sensitive to control variables during the bulk of the atmospheric path, a multiple-shooting technique has been used to solve the Mars AGA. We divide the time span $[t_0, t_f]$ in N uniform subintervals. In each subinterval, the control profile is parameterized by means of a cubic spline defined by M points. Then, on a generic node k the variable vector is represented by

$$\mathbf{y}_k = (\mathbf{x}_k, \lambda_k^{(1)}, \ldots, \lambda_k^{(M-1)}) \in \mathbb{R}^6 \times \mathbb{R}^{M-1}, \quad k = 1, \ldots, N-1$$
$$\mathbf{y}_N = (\mathbf{x}_N, \lambda_N) \in \mathbb{R}^6 \times \mathbb{R}$$

and the complete variable vector is given by $\mathbf{y} = (t_f, \mathbf{y}_1, \ldots, \mathbf{y}_N)$.

In each interval we interpolate the control variables defined as $\boldsymbol{\lambda}_k = (\lambda_k^{(1)}, \ldots, \lambda_k^{(M-1)}, \lambda^{(1)})$ for $k = 1, \ldots, N-1$ by means of a cubic spline such that, if $\hat{\lambda}_k(t)$ is the result of the interpolation on the single interval $I_k = [t_k, t_{k+1}]$, then $\lambda(t) = \lambda_k(t)$ for every $t \in I_k$. In each interval I_k, $k = 1, \ldots, N-1$ we also perform a numerical integration of the ODEs defining the dynamics by a eighth-order fixed-step Runge–Kutta scheme, that allows to compute the value $\mathbf{x}_{k+1}^c$. Finally, the dynamics constraints are transcribed into a set of $6(N-1)$, algebraic constraints given by Eq. (27).

The optimal control problem is then transcribed into a discrete form and solved using an SQP solver. The required initial estimate for the first guess solution in terms of both state and control variables is computed by solving the same problem using a single-shooting method with a simpler representation of the controls and a stochastic optimizer based on the genetic algorithms, to enhance convergence to the global minimum.

In this example, the transcription parameters are $N = 11$, $M = 4$; so, $\mathbf{y} \in \mathbb{R}^{97}$ and the NLP problem contains 60. A complete description of the results can be found in Ref. [22].

3. LOW-THRUST TRAJECTORY OPTIMIZATION

This example is inspired by the second Global Trajectory Optimization Competition.[§] The problem concerns a multiple-asteroid rendezvous. A trajectory must be designed for a low-thrust spacecraft which is launched from Earth and subsequently performs a rendezvous with four asteroids, whose ephemeris is given as

[§]Website: http://sophia.estec.esa.int/gtoc_portal

problem data. The motion of the spacecraft is given by

$$\ddot{\mathbf{r}} + \frac{\mu}{r^3}\mathbf{r} = \frac{\mathbf{T}}{m}, \quad \dot{m} = -\frac{T}{I_{sp}\, g}, \quad \mathbf{r} \in \mathbb{R}^3, \quad r = ||\mathbf{r}||$$

where $\mathbf{T}$ is the thrust acceleration vector and T is its magnitude.

The objective of the optimization is to maximize the quantity

$$J = \frac{m_f}{t_f - t_0}, \quad m_f = m_0 \exp\left(-\frac{1}{I_{sp}\, g}\int_{t_0}^{t_f} \parallel \mathbf{u} \parallel \, \mathrm{d}\tau\right)$$

where m_0 and m_f are the initial and final spacecraft mass, subject to the following constraints:

$$\begin{aligned}
\mathbf{r}(t_{\text{dep},P}) &= \mathbf{r}_P(t_{\text{dep},P}), \quad \mathbf{v}(t_{\text{dep},P}) = \mathbf{v}_P(t_{\text{dep},P}) \\
\mathbf{r}(t_{\text{arr},P}) &= \mathbf{r}_P(t_{\text{arr},P}), \quad \mathbf{v}(t_{\text{arr},P}) = \mathbf{v}_P(t_{\text{arr},P}) \\
\parallel \mathbf{u} \parallel &\leqslant u^{\max} \\
t_{\text{dep},P_i} - t_{\text{arr},P_{i-1}} &\leqslant 90 \text{ days}
\end{aligned}$$

The problem has been implemented in Ref. [25] using a transcription technique. The optimization variables were the four departure epochs (from the Earth and from three asteroids), the four transfer times, and the control parameters deriving from transcription. The variable vector had a dimension of about 1000. The solution is illustrated in Fig. 12.

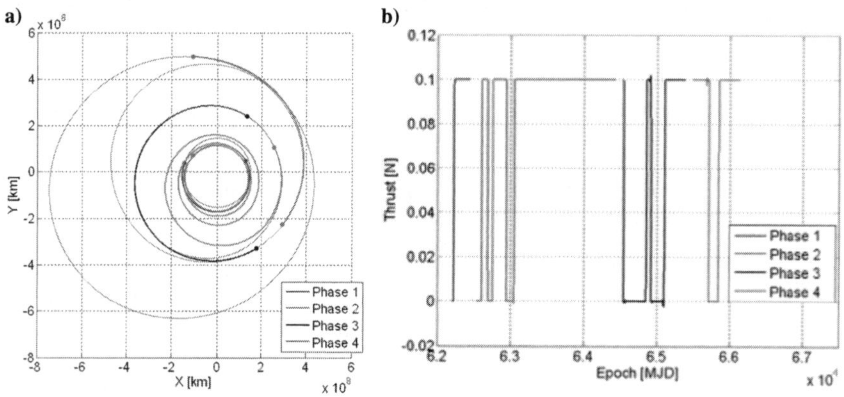

Fig. 12 A solution for the low-thrust optimization trajectory problem (taken from Ref. [25]): a) Low-thrust optimization trajectory and b) low-thrust magnitude over time.

VI. AN APPROXIMATE METHOD USING THE STM

Both indirect and direct methods use an iterative procedure that implements some form of Newton's method to find the zeros of a nonlinear function. This implies the need or an initial guess solution. However, guessing an appropriate initial solution is not trivial and requires a deep knowledge of the problem. Indirect methods need to provide initial guess values for the Lagrange multipliers, which, typically, are not physical quantities, while for direct methods, the initial trajectory and control have to be guessed at discrete points over the whole time interval.

In this section, we will present a modification of the method known as ASRE. By using the well-known theory of linear optimal control, this method rewrites the nonlinear problem in a pseudolinear form, and the objective function into a linear–quadratic function. At each iteration, these functions are evaluated by using the solutions at the previous iteration, and therefore, the problem turns out to be a series of time-varying linear quadratic regulators (TVLQRs). The new method, introduced in Refs. [4,5], uses a factorization of the STM. Three different final conditions are studied: the final state fully specified [hard-constrained problem (HCP)], final state not specified [soft-constrained problem (SCP)], and final state not completely specified [mixed-constrained problems (MCPs)].

The main advantage of this approximate method is that it does not require guessing any initial solution or Lagrange multiplier. In fact, iterations start by evaluating the state and control-dependent functions using the initial condition and zero control, respectively. As the method is approximated, only suboptimal solutions are guaranteed. However, they could be used as first guess solutions for either indirect or direct methods.

A. APPROXIMATING SEQUENCE OF RICCATI EQUATION

Optimal control of general nonlinear dynamic systems of the form (1) and performance index (2) can be transformed into a nonlinear affine system of the form

$$\dot{\mathbf{x}}(t) = A(\mathbf{x}, t)\mathbf{x}(t) + B(\mathbf{x}, \mathbf{u}, t)\mathbf{u}(t) \tag{29}$$

with boundary conditions $\mathbf{x}(t_i) = \mathbf{x}_i$, and linear–quadratic objective function

$$J = \frac{1}{2}\mathbf{x}^T(t_f)S(\mathbf{x}(t_f), t_f)\mathbf{x}(t_f) + \frac{1}{2}\int_{t_0}^{t_f} [\mathbf{x}^T Q(\mathbf{x}, t)\mathbf{x} + \mathbf{u}^T R(\mathbf{x}, t)\mathbf{u}]\,\mathrm{d}t \tag{30}$$

where $\mathbf{x} \in \mathbb{R}^n$ is the state vector, $\mathbf{u} \in \mathbb{R}^m$ is the control vector, unconstrained, $A(t) \in \mathbb{R}^{n\times n}, B(t) \in \mathbb{R}^{n\times m}$ are matrix functions of time, in general with continuous entries, $S, Q \in \mathbb{R}^{n\times n}$ are positive semidefinite, and $R \in \mathbb{R}^{m\times m}$ is positive definite.

In finite-horizon optimal control problems, one is interested in finding the optimal control history $\mathbf{u}(t)$, with $t_i \leqslant t \leqslant t_f$ which minimizes the cost function (30). The solution to this problem is found by solving algebraic Riccati equations (ASRE). This method reduces problem (29) and (30) to iteratively solve a sequence of TVLQR problems based on a pseudofactorization of the dynamics. The idea of the ASRE is to use state-dependent matrices $A(\mathbf{x}, t)$, $B(\mathbf{x}, \mathbf{u}, t)$, $Q(\mathbf{x}, t)$, and $R(\mathbf{x}, t)$ such that for given arguments $\bar{\mathbf{x}}(t)$, $\bar{\mathbf{u}}(t)$ they depending on time.

At iteration 0, the problem consists of finding $\mathbf{x}^{(0)}(t)$ and $\mathbf{u}^{(0)}(t)$ given the initial conditions $\mathbf{x}_i$ and zero control. *Problem 0* is stated as follows: the dynamics and the objective function are

$$\dot{\mathbf{x}}^{(0)} = A(\mathbf{x}_i, t)\mathbf{x}^{(0)} + B(\mathbf{x}_i, \mathbf{0}, t)\mathbf{u}^{(0)}$$

$$J = \frac{1}{2}[\mathbf{x}^{(0)}(t_f)]^T S(\mathbf{x}_i, t_f)\mathbf{x}^{(0)}(t_f) + \frac{1}{2}\int_{t_0}^{t_f} [[\mathbf{x}^{(0)}]^T Q(\mathbf{x}_i, t)\mathbf{x}^{(0)} + [\mathbf{u}^{(0)}]^T R(\mathbf{x}_i, t)\mathbf{u}^{(0)}]\, dt$$

All the matrices A, B, S, Q, and R depend only on time and so problem 0 is a standard TVLQR. It can be solved for $\mathbf{x}^{(0)}(t)$ and $\mathbf{u}^{(0)}(t)$ for every $t \in [t_i, t_f]$.

At a generic iteration k, the problem consists of finding $\mathbf{x}^{(k)}(t)$ and $\mathbf{u}^{(k)}(t)$ given state $\mathbf{x}^{(k-1)}(t)$ and control $\mathbf{u}^{(k-1)}$. *Problem k* is defined as follows:

$$\dot{\mathbf{x}}^{(k)} = A(\mathbf{x}^{(k-1)}, t)\mathbf{x}^{(k)} + B(\mathbf{x}^{(k-1)}, \mathbf{u}^{(k-1)}, t)\mathbf{u}^{(k)}$$

and the objective function is

$$J = \frac{1}{2}[\mathbf{x}^{(k)}(t_f)]^T S(\mathbf{x}^{(k-1)}, t_f)\mathbf{x}^{(k)}(t_f) + \frac{1}{2}\int_{t_0}^{t_f} [[\mathbf{x}^{(k)}]^T Q(\mathbf{x}^{(k-1)}, t)\mathbf{x}^{(k)} + [\mathbf{u}^{(k)}]^T R(\mathbf{x}^{(k-1)}, t)\mathbf{u}^{(k)}]\, dt$$

All the matrices A, B, S, Q, and R depend only on time, since $\mathbf{x}^{(k-1)}$ and $\mathbf{u}^{(k-1)}$ have been computed at the previous iteration. Problem k is again a TVLQR that can be solved for $\mathbf{x}^{(k)}(t)$ and $\mathbf{u}^{(k)}(t)$ for every $t \in [t_i, t_f]$.

Iterations are terminated when a certain convergence criterion is satisfied. In Ref. [4] the condition

$$\| \mathbf{x}^{(k)} - \mathbf{x}^{(k-1)} \|_\infty = \max_{t\in[t_i,t_f]} \{|x_j^{(k)}(t) - x_j^{(k-1)}(t)|\} \leqslant \epsilon$$

is used with a fixed threshold $\epsilon > 0$. The procedure ends when the difference between each component of the state, evaluated for all times, changes by less than ϵ between two successive iterations.

In the following subsection we describe the modified ASRE as proposed by Topputo and Bernelli-Zazzera [4]. For the interested reader, a reference on ASRE method is shown in Cimen and Banks [26], whereas a classical reference about LQR is shown in Anderson and Moore [27].

B. SOLUTION OF THE TIME-VARYING LINEAR QUADRATIC REGULATOR BY THE STM

Suppose the dynamics can be factorized as

$$\dot{\mathbf{x}}(t) = A(t)\mathbf{x}(t) + B(t)\mathbf{u}(t) \tag{31}$$

and the objective function is factorized as well in

$$J = \frac{1}{2}\mathbf{x}^T(t_f)S(t_f)\mathbf{x}(t_f) + \frac{1}{2}\int_{t_0}^{t_f}[\mathbf{x}^T Q(t)\mathbf{x} + \mathbf{u}^T R(t)\mathbf{u}]\,\mathrm{d}t \tag{32}$$

where Q, R, and S are, respectively, positive semidefinite and positive-definite time-varying matrices with appropriate dimensions.

The Hamiltonian of this optimal problem is

$$H = \mathbf{x}^T Q(t)\mathbf{x} + \mathbf{u}^T R(t)\mathbf{u} + \boldsymbol{\lambda}^T A(t)\mathbf{x} + \boldsymbol{\lambda}^T B(t)\mathbf{u} \tag{33}$$

where $\boldsymbol{\lambda} \in \mathbb{R}^n$ are the Lagrange multipliers. Using the Euler–Lagrange equations (10–12), the necessary conditions for optimality are

$$\begin{aligned}
\dot{\mathbf{x}} &= A(t)\mathbf{x} + B(t)\mathbf{u} \\
\boldsymbol{\lambda} &= -Q(t)\mathbf{x}(t) - A^T(t)\boldsymbol{\lambda} \\
\mathbf{0} &= R(t)\mathbf{u} + B^T(t)\boldsymbol{\lambda}
\end{aligned}$$

From the last equation, we compute $\mathbf{u}$ and a linear system of $2n$ differential equations in the variables $\mathbf{x}, \boldsymbol{\lambda}$ is obtained

$$\begin{aligned}
\dot{\mathbf{x}} &= A(t)\mathbf{x} - B(t)R^{-1}B(t)^T\lambda \\
\mathbf{u} &= -Q(t)\mathbf{x} - R(t)B^T(t)\boldsymbol{\lambda}
\end{aligned} \tag{34}$$

with the appropriate boundary conditions. Instead of solving the problem via the Riccati equation, it is possible to employ the STM. In general the STM is the matrix

$$\boldsymbol{\Phi} = \begin{bmatrix} \phi_{\mathbf{xx}}(t_i, t) & \phi_{\mathbf{x}\boldsymbol{\lambda}}(t_i, t) \\ \phi_{\boldsymbol{\lambda}\mathbf{x}}(t_i, t) & \phi_{\boldsymbol{\lambda\lambda}}(t_i, t) \end{bmatrix}$$

that solves the system

$$\begin{bmatrix} \mathbf{x}(t) \\ \boldsymbol{\lambda}(t) \end{bmatrix} = \boldsymbol{\Phi}(t_i, t) \begin{bmatrix} \mathbf{x}(t_i) \\ \boldsymbol{\lambda}(t_i) \end{bmatrix} \tag{35}$$

which related the initial conditions at time t_i and conditions at a generic time $t \in [t_i, t_f]$.

The STM can be computed by solving the variational equation

$$\frac{\partial \mathbf{\Phi}(t_i, t)}{\partial t} = \begin{bmatrix} A(t) & -B(t)R(t)^{-1}B(t) \\ -Q(t) & -A(t)^T \end{bmatrix} \mathbf{\Phi}(t_i, t)$$

with an initial condition $\mathbf{\Phi}(t_i, t) = \mathbf{I}$, since the transition from a condition to itself is performed by an identity matrix.

If the multipliers $\boldsymbol{\lambda}_i = \boldsymbol{\lambda}(t_i)$ were known, then, using Eq. (34), it would be possible to compute $\mathbf{x}(t), \boldsymbol{\lambda}(t)$ through Eq. (35). However, in general, $\boldsymbol{\lambda}_i$ are unknown and they must be evaluated using Eq. (35) and the boundary conditions. There are three possible cases: hard, soft, and MCPs.

1. HCP

In the HCP both initial and final conditions are prescribed for all the n states. The HCP is stated as follows. Given the dynamics (31), minimize the performance index

$$J = \frac{1}{2}\int_{t_0}^{t_f} [\mathbf{x}^T(t)Q(t)\mathbf{x}(t) + \mathbf{u}^T(t)R(t)\mathbf{u}(t)]\, \mathrm{d}t$$

subject to the boundary constraints $\mathbf{x}(t_i) = \mathbf{x}_i$, $\mathbf{x}(t_f) = \mathbf{x}_f$.

To solve the HCP, we write the first equation of system (35) at $t = t_f$

$$\mathbf{x}_f = \phi_{\mathbf{xx}}(t_i, t_f)\mathbf{x}_i + \phi_{\mathbf{x}\boldsymbol{\lambda}}(t_i, t_f)\boldsymbol{\lambda}_i$$

If the $n \times n$ matrix $\phi_{\mathbf{x}\boldsymbol{\lambda}}(t_i, t_f)$ is invertible, that is $\det \phi_{\mathbf{x}\boldsymbol{\lambda}}(t_i, t_f) \neq 0$, then we can solve it for $\boldsymbol{\lambda}_i$ and we obtain

$$\boldsymbol{\lambda}_i(\mathbf{x}_i, \mathbf{x}_f, t_i, t_f) = [\phi_{\mathbf{x}\boldsymbol{\lambda}}(t_i, t_f)]^{-1}[\mathbf{x}_f - \phi_{\mathbf{xx}}(t_i, t_f)\mathbf{x}_i]$$

The value of $\boldsymbol{\lambda}_i$ can now be substituted into Eqs. (35) and (34) to compute $\mathbf{x}(t), \boldsymbol{\lambda}(t)$, and $\mathbf{u}(t)$.

2. SCP

In the SCP only initial conditions are prescribed, and the final state will appear inside the objective function that takes form (32). The SCP is stated as follows. Given the dynamics (31), minimize the performance index (32) subject to the boundary constraint $\mathbf{x}(t_i) = \mathbf{x}_i$.

The dynamical systems contain $2n$ equations but we have only n boundary conditions. We find the other n conditions using the transversality condition (13)

$$\boldsymbol{\lambda}_f = \left(\frac{\partial H}{\partial \mathbf{x}_f}\right)^T = S(t_f)\mathbf{x}(t_f)$$

where H is the Hamiltonian (33) associated to SCP. Writing the system (35) at $t = t_f$ and using the previous equation, we get the system of equations

$$\begin{aligned} \mathbf{x}(t_f) &= \phi_{\mathbf{xx}}(t_i, t_f)\mathbf{x}_i + \phi_{\mathbf{x}\boldsymbol{\lambda}}(t_i, t_f)\boldsymbol{\lambda}_i \\ S(t_f)\mathbf{x}(t_f) &= \phi_{\boldsymbol{\lambda}\mathbf{x}}(t_i, t_f)\mathbf{x}_i + \phi_{\mathbf{xx}}(t_i, t_f)\boldsymbol{\lambda}_i \end{aligned}$$

that can be solved for $\boldsymbol{\lambda}_i$

$$\boldsymbol{\lambda}_i(\mathbf{x}_i, t_i, t_f) = [\phi_{\boldsymbol{\lambda}\boldsymbol{\lambda}}(t_i, t_f) - S(t_f)\phi_{\mathbf{x}\boldsymbol{\lambda}}(t_i, t_f)]^{-1}[S(t_f)\phi_{\mathbf{xx}}(t_i, t_f) - \phi_{\mathbf{xx}}]$$

provided that the $n \times n$ matrix $\phi_{\boldsymbol{\lambda}\boldsymbol{\lambda}}(t_i, t_f) - S(t_f)\phi_{\mathbf{x}\boldsymbol{\lambda}}(t_i, t_f)$ is invertible. Finally, substituting the value of $\boldsymbol{\lambda}_i$ in Eqs. (35) and (34), we obtain a solution for $\mathbf{x}(t)$, $\boldsymbol{\lambda}(t)$, and $\mathbf{u}(t)$.

3. MCP

In the MCP some components of the final state are given and some are unknown, that is, $x_k(t_f) = x_{k,f}$ for some $k \in \{1, \ldots, n\}$. Suppose that $x_k(t_f) = x_{k,f}$ for $k = 1, \ldots, m$ with $m < n$ are the final state given. Then, we can partition the state vector as $\mathbf{x} = (\mathbf{x}_1, \mathbf{x}_2)$, with $\mathbf{x}_1 \in \mathbb{R}^m$ and $\mathbf{x}_2 \in \mathbb{R}^{n-m}$. The MCP is stated as follows. Given the dynamics (31), minimize the performance index (32) subject to the boundary constraint $\mathbf{x}(t_i) = \mathbf{x}_i$ and $\mathbf{x}_2(t_f) = \mathbf{x}_{2,f}$. Note that $\mathbf{x}_1(t_i) = \mathbf{x}_{1,i}, \mathbf{x}_2(t_i) = \mathbf{x}_{2,i}$ and that $S(t_f)$ is an $(n - m) \times (n - m)$ matrix.

The costate vector can be partitioned as well: $\boldsymbol{\lambda} = (\boldsymbol{\lambda}_1, \boldsymbol{\lambda}_2)$, with $\boldsymbol{\lambda}_1 \in \mathbb{R}^m$ and $\boldsymbol{\lambda}_2 \in \mathbb{R}^{n-m}$. Also the STM can be partitioned in 16 subblocks such that

$$\begin{bmatrix} \mathbf{x}_1(t) \\ \mathbf{x}_2(t) \\ \boldsymbol{\lambda}_1(t) \\ \boldsymbol{\lambda}_2(t) \end{bmatrix} = \boldsymbol{\Phi}(t_i, t) \begin{bmatrix} \mathbf{x}_1(t_i) \\ \mathbf{x}_2(t_i) \\ \boldsymbol{\lambda}_1(t_i) \\ \boldsymbol{\lambda}_2(t_i) \end{bmatrix}$$

Thus, evaluating the previous system at $t = t_f$, we can solve it for $\boldsymbol{\lambda}_{1,i}$ using $\mathbf{x}_{1,f}$ (as in the HCP) and for $\boldsymbol{\lambda}_{2,i}$ using the transversality condition $\boldsymbol{\lambda}_2 = S(t_f)\mathbf{x}_2$ (as in the SCP). A complete description of the procedure can be found in Ref. [4].

C. EXAMPLES

In this subsection, we will present three numerical examples taken from Ref. [4]: the first one corresponds to the controlled relative spacecraft motion, the second one to the controlled two-body dynamics for low-thrust transfers, and the last one to the controlled station-keeping of GEO satellites, is also shown. For further details see also Ref. [28].

1. LOW-THRUST RENDEZVOUS

The model is that of a particle in a central gravity field, with constant angular momentum, that is, the motion is constrained to stay in a plane. The problem is simplified by using normalized units: the unit length is the orbital radius, the unit time is such that the orbital period is 2π, and the gravitational parameter μ is unitary. The equations of motion in a rotating frame are given by

$$\dot{x}_1 = x_3, \quad \dot{x}_3 = 2x_4 - (1 + x_1)\left(\frac{1}{r^3} - 1\right) + u_1$$

$$\dot{x}_2 = x_4, \quad \dot{x}_4 = -2x_3 - x_2\left(\frac{1}{r^3} - 1\right) + u_2$$

where $r = \sqrt{(x_1 + 1)^2 + x_2^2}$. The origin of the frame corresponds to a circular orbit. The coordinates x_1, x_2, x_3, x_4 represent the radial and transversal displacement and the radial and transversal velocity deviation from the circular orbit, respectively. Whereas, u_1, u_2 are the radial and transversal control.

We use the modified ASRE to solve the simplified rendezvous problem. We consider the initial condition $\mathbf{x}_i = (0.2, 0.2, 0.1, 0.1)$, with $\mathbf{x}_i = (x_{1,i}, x_{2,i}, x_{3,i}, x_{4,i})$.

The dynamical system can be factorized as

$$\dot{\mathbf{x}} = A(\mathbf{x})\mathbf{x} + B(\mathbf{x})\mathbf{u}$$

where $\mathbf{x} = (x_1, x_2, x_3, x_4)$, $\mathbf{u} = (u_1, u_2)$, and

$$A(\mathbf{x}) = \begin{bmatrix} 0 & 0 & 1 & 0 \\ 0 & 0 & 0 & 1 \\ f(x_1, x_2)\left(1 + \dfrac{1}{x_1}\right) & 0 & 0 & 2 \\ 0 & f(x_1, x_2) & -2 & 0 \end{bmatrix}, \quad B(\mathbf{x}) = \begin{bmatrix} 0 & 0 \\ 0 & 0 \\ 1 & 0 \\ 0 & 1 \end{bmatrix} \tag{36}$$

with

$$f(x_1, x_2) = 1 - \frac{1}{r^3}$$

For the HCP, the terminal boundary condition is located at the origin, that is, $\mathbf{x}_f = (0, 0, 0, 0)$. For the SCP, the final condition is not fixed by definition, so $\mathbf{x}(t_f)$ is to determine. The objective functions are

$$J^{(\mathrm{HPC})} = \frac{1}{2}\int_{t_i}^{t_f} \mathbf{u}^T\mathbf{u}\,\mathrm{d}t, \quad J^{(\mathrm{SCP})} = \frac{1}{2}\mathbf{x}^T S(t_f)\mathbf{x} + \frac{1}{2}\int_{t_i}^{t_f} \mathbf{u}^T\mathbf{u}\,\mathrm{d}t$$

for the SCP and HCP, respectively, with $t_i = 0$ and $t_f = 0$, and with $S = \mathrm{diag}(25, 15, 10, 10)$ for the SCP. In both cases, HCP and SCP, the rendezvous problem is reduced to minimize a quadratic continuous-time cost functional of the form (30) subject to a linear first-order dynamics of the form (29), with $A(\mathbf{x})$ and $B(\mathbf{x})$ defined in Eq. (36) and by choosing Q, the null matrix of order 4 and R is the identity matrix of order 2.

The modified ASRE leads to $J^{(\mathrm{HPC})} = 0.9586$ and $J^{(\mathrm{SCP})} = 0.5660$, after five iterations in 0.447 s of central processing unit (CPU) time and six iterations in 0.492 s of CPU time, respectively, for a tolerance of $\epsilon = 10^{-9}$. The CPU time referred to an Intel Core 2 Duo 2 GHz with 4-GB RAM running Mac OS X 10.6. The results are shown in Fig. 13.

2. LOW-THRUST ORBITAL TRANSFER

The motion of a spacecraft in the gravitational field of the Sun is given by Eq. (5). In the following we show how to transfer a spacecraft from a given orbit to another using the STM. This example is taken from Refs. [4, 5].

Using normalized units as in the previous section, and the polar coordinates $x = r\cos\theta$, $y = r\sin\theta$, the state $\mathbf{x} = (x_1, x_2, x_3, x_4)$ is given by the radial distance from the attractor ($x_1 = r$), the angular phase ($x_2 = \theta$), the radial velocity ($x_3 = \dot{r}$), and the transversal velocity ($x_4 = \dot{\theta}$). Whereas, the components of the control $\mathbf{u} = (u_1, u_2)$ are the radial and transversal accelerations, respectively. The

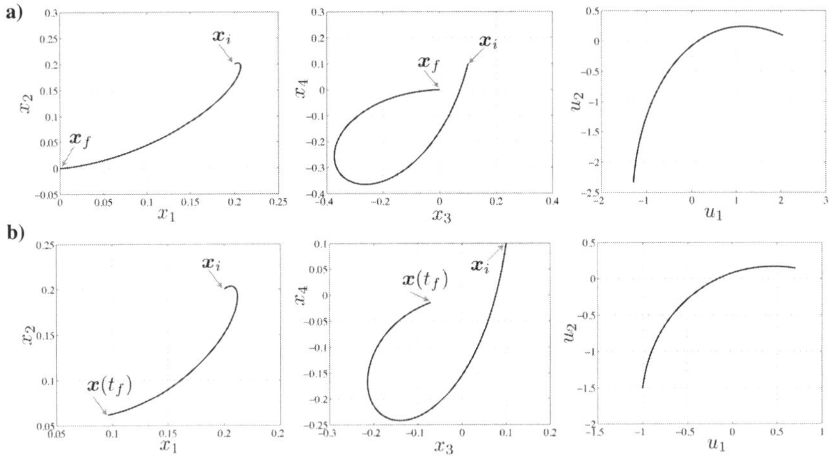

Fig. 13 Low-thrust rendezvous using the modified ASRE: a) HCP and b) SCP (taken from Ref. [4]).

differential equations of motion in polar coordinates are

$$
\begin{aligned}
\dot{x}_1 &= x_3, \quad \dot{x}_3 = x_1 x_4^2 - \frac{1}{x_1^2} + u_1 \\
\dot{x}_2 &= x_4, \quad \dot{x}_4 = -\frac{2x_3 x_4}{x_1} + \frac{u_2}{x_1}
\end{aligned}
\tag{37}
$$

and the objective function is given by

$$
J = \frac{1}{2}\int_{t_i}^{t_f} \mathbf{u}^T \mathbf{u}\, \mathrm{d}t
$$

with $t_i = 0$ and $t_f = \pi$.

Suppose we want to perform a low-thrust transfer from the Earth to Mars. The two planets are considered to move on circular coplanar orbits. The initial condition is $\mathbf{x}_i = (1, 0, 0, 1)$, where the initial angle θ is arbitrarily chosen to be zero at initial time. Indeed, $r(t_i) = 1$ (the Earth's radius), $\dot{r}_1(t_i) = 0$, and $\dot{\theta}(t_i) = 1$, thanks to the third Kepler's law. The final boundary condition must satisfy $r(t_f) = 1.52$ (the radius of Mars), $\dot{r}(t_f) = 0$, and $\dot{\theta}(t_i) = 1.52^{-3/2}$. The final angle can be chosen arbitrarily, or even left as a free parameter, to be determined by the optimal control. In Ref. [4] it is chosen to solve two HCP problems, with the final state $\mathbf{x}_f = (1.52, \pi, 0, 1.52^{-3/2})$ (HCP-1) and $\mathbf{x}_f = (1.52, 3\pi/2, 0, 1.52^{-3/2})$ (HCP-2).

The dynamical system (37) is factorized as Eq. (29), where

$$
A(\mathbf{x}) = \begin{bmatrix} 0 & 0 & 1 & 0 \\ 0 & 0 & 0 & 1 \\ -\frac{1}{x_1^3} & 0 & 0 & x_1 x_4 \\ 0 & 0 & -\frac{2x_4}{x_1} & 0 \end{bmatrix}, \quad B(\mathbf{x}) = \begin{bmatrix} 0 & 0 \\ 0 & 0 \\ 1 & 0 \\ 0 & \frac{1}{x_1} \end{bmatrix}
$$

and the performance index is put in the form (30) by choosing Q, the null matrix of order 4 and R, the identity matrix of order 2.

Using the same tolerance as in the previous example, and performing the computation with the same processor, the number of iterations required for convergence increases, together with the computational time. For the HCP-1, the modified ASRE method finds the value $J = 0.5298$ in 22 iterations and about 6 s, while for the HCP-2, it finds the value $J = 4.8665$ in 123 iterations in almost 1 min. The results are illustrated in Fig. 14.

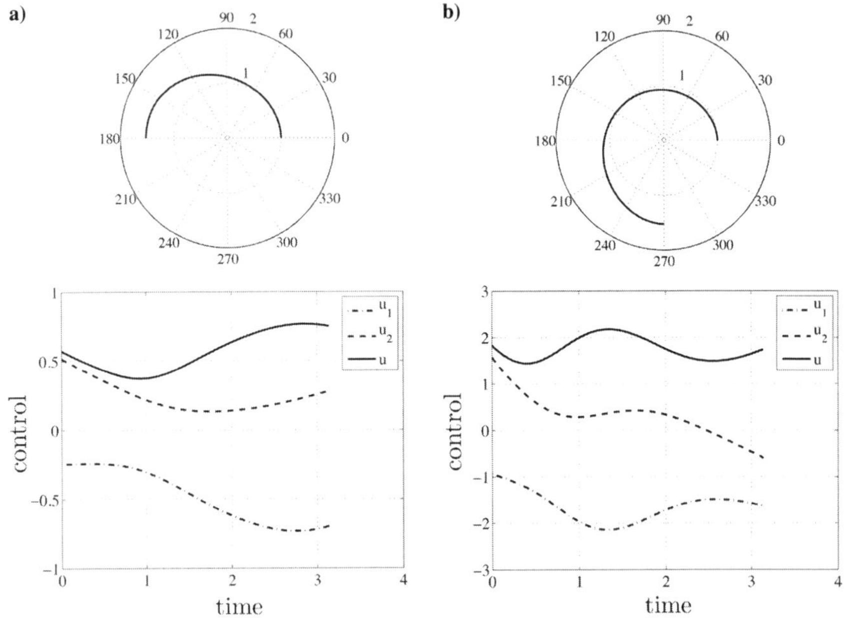

Fig. 14 Low-thrust Earth–Mars orbital transfer using the STM (taken from Ref. [4]): a) HCP-1 and b) HCP-2.

3. LOW-THRUST STATION-KEEPING OF GEO SATELLITES

In astrodynamics orbital station-keeping is the orbital maneuvers needed to keep a spacecraft in a particular assigned orbit.

The model is a GEO satellite in orbit around the Earth. The reference system is a rotating frame jointly with the planet, with origin in the centre of the Earth. The x-axis and the y-axis lie in the equatorial plane, while the z-axis points toward the North pole. We consider normalized units: the length unit is the nominal GEO radius, the gravitational parameter is $\mu = 1$ and the angular velocity is $\omega = 1$. The nominal satellite location is fixed at 60 deg longitude. Then, in spherical coordinates $x = r\cos\theta\cos\varphi$, $y = r\sin\theta\cos\varphi$, $z = r\sin\varphi$, with $r > 0$, $0 \leqslant \theta \leqslant 2\pi$ and $0 \leqslant \varphi \leqslant \pi$, the dynamic equations are

$$
\begin{aligned}
\dot{x}_1 &= x_4, \quad \dot{x}_2 = x_5, \quad \dot{x}_3 = x_6 \\
\dot{x}_4 &= -\frac{1}{x_1} + x_1 x_6^2 + x_1(x_5+1)\cos^2 x_3 + a_1 + u_1 \\
\dot{x}_5 &= 2x_6(x_5+1)\tan x_3 - 2\frac{x_4}{x_1}(x_5+1) + \frac{a_2}{x_1\cos x_3} + \frac{u_2}{x_1\cos x_3}
\end{aligned} \tag{38}
$$

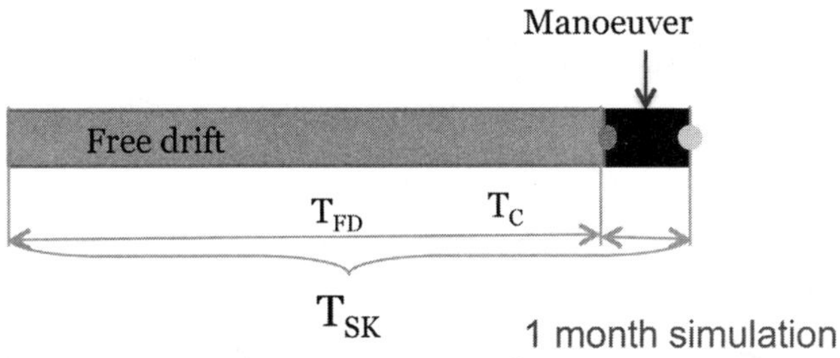

Fig. 15 Control cycle (taken from Ref. [29]).

$$\dot{x}_6 = -2\frac{x_4}{x_1}x_6 - (x_5+1)^2 \sin x_3 \cos x_3 + \frac{a_3}{x_1} + \frac{u_3}{x_1}$$

where $x_1 = r$, $x_2 = \varphi$, and $x_3 = \theta$, a_1, a_2, and a_3 are the components of the perturbations, and u_1, u_2, and u_3 are the components of controls in spherical coordinates to be determined to stabilize the GEO satellite. The state vector is $\mathbf{x} = (x_1, x_2, x_3, x_4, x_5, x_6)$.

The optimal control problem can be reformulated as an MCP, with $\mathbf{x}_{f,1} = 1$ and $\mathbf{x}_{f,j}$ are free for $j = 2, \ldots, 6$. The initial condition is

$$\mathbf{x}_i = (1, 0.05 \cdot 180/\pi, 0.05 \cdot 180/\pi, 0, 0, 0)$$

that is, the nominal satellite state is perturbed in longitude and latitude by 0.05 deg. The objective function can be put in the form (30) with

$$Q = \text{diag}(0,1,1,1,1,1), \quad R = \text{diag}(1,1,1), \quad S = 100\,\text{diag}(1,1,1,1,1)$$

The initial and the final time are $t_i = 0$ and $t_f = \pi$, respectively. The dynamics (38) can be factorized and put in form (29). The numerical experiments

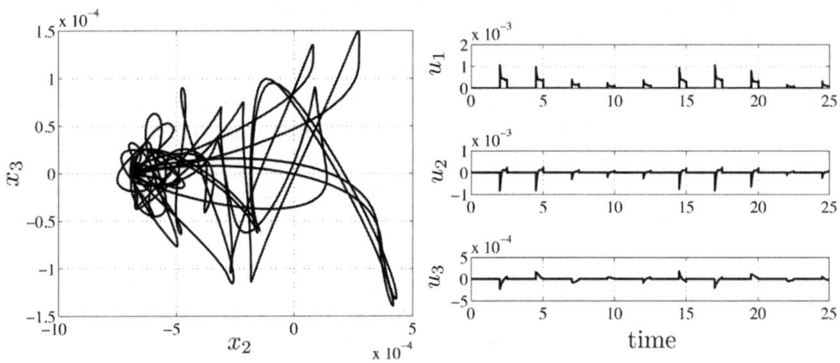

Fig. 16 Station-keeping simulation for one month (taken from Ref. [29]).

have been performed considering a control cycle made up of the combination of free-drift (lasting about 2.5 days) and station-keeping maneuvers (lasting about 0.5 days), see Fig. 15. The results of 1-month simulation are illustrated in Fig. 16. One-year station-keeping simulation has been described in Ref. [29].

VII. CONCLUSIONS

In this chapter we have seen how the nonlinear optimal control problem is formulated and solved. Classical methods are divided in two groups: indirect and direct methods.

Indirect methods use the calculus of variation to reduce the problem to the so-called TPBVP, that is, a system of differential equations for the state and the controls [Euler–Lagrange equations (10–12)], whose boundary conditions are given both at the initial and final time [transversality conditions (13) and (14)]. Unfortunately, it is generically difficult to construct these expressions and solve the problem analytically, because the differential equations can be nonlinear and nontrivial and because the system can be very large. Besides, the approach is not flexible, in the sense that if we pose a new problem, we need to compute new derivatives.

Direct methods transform the continuous optimization problem into a discrete parameter optimization NLP problem. The main advantages of direct methods are that there is no need of deriving the equations related to the necessary conditions for optimality, and that they are more versatile and easier to implement as a stand-alone software. Besides, control constraints are included trivially and constraints such as dynamic pressure constraints, which are problematic for indirect methods, are simply included. On the other hand, direct methods know states and controls only at discrete points and require numerical techniques to effectively estimate Hessian and Jacobian matrix.

Both direct and indirect methods suffer from the same great disadvantages. The main difficulty is *getting started*, that is, they need "good" first guesses to converge in to a solution. Indirect methods must not only provide a starting guess for the state, but also must guess starting values for the Lagrange multipliers $\boldsymbol{\lambda}$, which typically have no physical meaning. Direct methods use a shooting technique to find a good starting guess to converge quickly to the optimal solution.

Conversely, a starting guess that is not close enough to the optimal solution will cause the algorithms to require many iterations or even not to converge in to an optimal solution. This means that the convergence is only local and there is no guarantee for optimality. Moreover, the extremal solutions are *very sensitive* to small changes in the boundary conditions.

For a very challenging problem it is better to use an approximate numerical solution (or homotopy) to find an initial guess, before solving an NLP problem. In Ref. [29] new approximated method has been proposed. It is a modification of the ASRE: the solution of the TPBVP is avoided by means of solving time-variant linear quadratic regulators (TVLQRs) iteratively, which are simpler to solve. The improvement of the ASRE consists of using the STM and an appropriate factorization to solve the TVLQR. The modified ASRE does not need a first guess solution; however, only suboptimal solutions can be guaranteed. Indeed, factorization of a nonlinear problem is not unique.

ACKNOWLEDGMENTS

C. Tardioli would like to thank Francesco Topputo and Pierluigi Di Lizia, for their advise and for having given the talk, on which this chapter is based, during the first Stardust Open Training School. She would also like to thank Massimiliano Vasile, for having people gathered together with a different background in such a unique project, and the Marie-Curie action, for the financial support. And finally, she thanks Claudio Bogazzi for having read the first draft of this chapter.

REFERENCES

[1] Pontryagin, L. S., Boltyanskii, V. G., Gamkrelidze, R. V., and Mishchenko, E. F., *The Mathematical Theory of Optimal Processes*, John Wiley and Sons, New York, 1962.

[2] Bryson, A. E., and Ho, Y., *Applied Optimal Control: Optimization, Estimation and Control (Revised Printing)*, John Wiley and Sons, New York, 1975.

[3] Betts, J. T., *Practical Methods for Optimal Control Using Nonlinear Programming*, Advances in Design and Control, The Boeing Company Seattle, WA, 2001.

[4] Topputo, F., and Bernelli-Zazzera, F., "Approximate Solutions to Nonlinear Optimal Control Problems in Astrodynamics," *ISRN Aerospace Engineering*, Vol. 2013, 2013, pp. 1–7.

[5] Topputo, F., Miani, M., and Bernelli-Zazzera, F., "Optimal Selection of the Coefficient Matrix in State-Dependent Control Method," *Journal of Guidance, Control, and Dynamics*, Vol. 35, 2015, pp. 861–873.

[6] Cimen, T., "Systematic and Effective Design of Nonlinear Feedback Controllers via the State-Dependent Riccati Equation (SDRE) Method," *Annual Reviews in Control*, Vol. 34, 2010, pp. 40–42.

[7] Beeler, S. C., "State-Dependent Riccati Equation Regulation of Systems with State and Control Nonlinearities," Technical Report NIA Report No. 2004-08, National Institute of Aerospace, Hampton, VA, 2004.

[8] Betts, J. T., "Survey of Numerical Methods for Trajectory Optimization," *Journal of Guidance, Control, and Dynamics*, Vol. 21, 1998, pp. 193–207.

[9] Topputo, F., and Zhang, C., "Survey of Direct Transcription for Low-Thrust Space Trajectory Optimization with Applications," *Abstract and Applied Analysis*, Vol. 2014, 2014, pp. 1–15.

[10] Sussmann, H. J., and Willems, J. C., "The Brachistochrone Problem and Modern Control Theory," *Contemporary Trends in Non-Linear Geometric Control Theory and Its Applications*, edited by A. Anzaldo-Meneses, B. Bonnard, J. P. Gauthier, and F. Monroy-Perez, Word Scientific, Singapore, 2002, pp. 113–165.

[11] Becerra, V. M., "Practical Direct Collocation Methods for Computational Optimal Control," *Modeling and Optimization in Space Engineering*, edited by G. Fasano, and J. D. Pintér, Springer, New York, 2013, pp. 33–60.

[12] Armellin, R., and Topputo, F., "A Six-Order Accurate Scheme for Solving Two-Point Boundary Value Problem," *Astrodynamics, Celestial Mechanics and Dynamical Astronomy*, Vol. 96, 2006, pp. 289–309.

[13] Mingotti, G., Topputo, F., and Bernelli-Zazzera, F., "Combined Optimal Low-Thrust and Stable-Manifold Trajectories to the Earth–Moon Halo Orbits," *American Institute of Physics*, Vol. 886, 2007, pp. 100–112.

[14] Milani, A., and Gronchi, G. F., *Theory of Orbit Determination*, Cambridge University Press, New York, 2009.

[15] Fiacco, A. V., and McCormick, G. P., *Nonlinear Programming: Sequential Unconstrained Minimization Techniques*, Classics in Applied Mathematics, Society for Industrial and Applied Mathematics, Philadelphia, 1990.

[16] Chachuat, B., *Nonlinear and Dynamic Optimization: From Theory to Practice*, LA-TEACHING-2007-001, École Polytechnique Fédérale de Lausanne, 2007.

[17] Hargraves, C. R., and Paris, S. W., "Direct Trajectory Optimization Using Nonlinear Programming and Collocation," *Journal of Guidance, Control, and Dynamics*, Vol. 10, 1987, pp. 338–342.

[18] Enright, P. J., and Conway, B. A., "Discrete Approximation to Optimal Trajectories Using Direct Transcription and Nonlinear Programming," *Journal of Guidance, Control, and Dynamics*, Vol. 15, 1992, pp. 994–1001.

[19] Topputo, F., "An Optimal Two-Impulse Earth–Moon Transfers in a Four-Body Model," *Celestial Mechanics and Dynamical Astronomy*, Vol. 117, 2013, pp. 279–313.

[20] Di Lizia, P., Armellin, R., Ercoli Finzi, A., and Berz, M., "High-Order Robust Guidance of Interplanetary Trajectories Based on Differential Algebra," *Journal of Aerospace Engineering, Sciences, and Applications*, Vol. 1, 2008, pp. 43–57.

[21] Massari, M., and Bernelli-Zazzera, F., "Optimization of Low-Thrust Reconfiguration Maneuvers for Spacecraft Flying Formation," *Journal of Guidance, Control, and Dynamics*, Vol. 32, 2009, pp. 1629–1638.

[22] Armellin, R., Lavagna, M., and Ercoli-Finzi, A., "Aero-Gravity Assist Maneuvers: Controlled Dynamics Modeling and Optimization," *Celestial Mechanics and Dynamical Astronomy*, Vol. 95, 2006, pp. 391–405.

[23] Armellin, R., *Multidisciplinary Optimization of Space Vehicles for Atmospheric Maneuvers*, PhD dissertation, Politecnico di Milano, Italy, 2007.

[24] Tauber, M. E., and Sutton, K., "Stagnation-Point Radiative Heating Relations for Earth and Mars Entries," *Journal of Spacecraft Rockets*, Vol. 28, 1991, pp. 40–42.

[25] Massari, M., Armellin, R., Di Lizia, P., and Topputo, F., "Reaching NEOs: Solution for the Second Global Trajectory Optimization Competition," *XIX Congresso nazionale AIDAA*, Forlì, Italia, Sept. 17–20, 2007, pp. 1–11.

[26] Cimen, T., and Banks, S. P., "Global Optimal Feedback Control for General Nonlinear Systems with Nonquadratic Performance Criteria," *Systems and Control Letters*, Vol. 53, 2010, pp. 327–346.

[27] Anderson, B. D. O., and Moore, J. B., *Optimal Control: Linear Quadratic Methods*, Prentice-Hall, Englewood Cliffs, NJ, 1989.

[28] Miani, M. M., "A Combined Dynamic–Algebraic Method to Solve Nonlinear Optimal Control Problems with Applications," Master's thesis, Politecnico di Milano, Milan, Italy, 2013.

[29] Topputo, F., and Bernelli-Zazzera, F., "Optimal Low-Thrust Station Keeping of Geostationary Satellites," *3rd CEAS Air and Space Conference and XXI Congresso Nazionale AIDAA*, Venice, Italy, October 2011, pp. 1917–1925.

CHAPTER 12

JAXA's Approach to the Space Debris Situation

Clemens Rumpf*

University of Southampton, Southampton, United Kingdom

Helene Ma†

University of Pisa, Pisa, Italy

Seishiro Kibe‡

Japan Aerospace Exploration Agency, Chofu, Japan

I. INTRODUCTION

During the Stardust Opening Training School held in November 2013 in Glasgow, the Stardust Network was honored to receive Professor Kibe from JAXA,§ who gave two lectures about the institution and its activities related to space debris.

This chapter aims to deliver a recapitulation of the information provided by Professor Kibe. It will provide additional information and put them into context.

A general overview of JAXA's space debris-related activities and capabilities is presented at the beginning of the chapter. Subsequently, JAXA's organizational structure is described.

The space debris issue is a firm component within JAXA's spacecraft design process and the text describes how this aspect is implemented within JAXA's procedures. How the space debris issue is addressed, will determine the space environment in the future.

Possible future scenarios are presented in the last part of this chapter. Finally, a more detailed look is taken on JAXA's active debris removal efforts.

II. JAXA'S SPACE DEBRIS RELATED ACTIVITIES: MITIGATION, R&D

A. ORGANIZATIONAL SCHEME OF JAXA'S SPACE DEBRIS-RELATED ACTIVITY

The importance of space debris issues for JAXA company can be shown when in 2006, a Space Debris Coordination Committee was established in addition to the different Directorates of the company.

This committee has the responsibility to coordinate all the space debris activities among these entities, because these issues affect the activities of all of them.

*Astronautics Research Group.

†Department of Mathematics.

‡Aerospace Research and Development Directorate, Innovative Technology Research Center.

§Japanese Aerospace Exploration Agency.

It is held by a President's directive (No. 18-11) and works with domestic and international organizations such as the Government of Japan and the IADC.[¶]

Its activity is described by a system containing three main documents: one of them is dedicated for the Strategy Plan of JAXA on space debris (ref. GEH-08002) where four strategic goals are defined in the following section.

B. DEBRIS-RELATED R&D ACTIVITY IN JAXA: THE STRATEGY PLAN OF JAXA ON SPACE DEBRIS

1. FIRST GOAL

The first goal is to guarantee the success of the mission by reasonable and practical countermeasures depending on the importance of the mission. Indeed, JAXA has developed its own debris environment evolution model called LEODEEM[**] in collaboration with the Kyushu University (see Fig. 1).

Therefore, IADC takes into consideration the contribution of JAXA with the other IADC members to study the stability of LEO debris environment and concludes from all the tools provided to an instability prediction of the LEO debris population.

Besides, JAXA acts on the collision avoidance problems and provides its own ground-based debris observation by "*Kamisaibara*" radar facility and "*Bisei*" optical facility.

In addition, its optical observatory facility, "*Nyukasa*," dedicated for its researches, is used to pursue two main issues: "*Stacking method*" for GEO[††] observation and application of optical observation technology in LEO debris observation.

These two issues are described in the following paragraphs:

a. Stacking Method for GEO Observation

A new method was introduced by JAXA to detect faint objects, that are unknown and very small GEO debris and asteroids. This technique is explained in details in the Research and Development Report of JAXA in February 2008 [2].

Shown in Fig. 2, the "*Stacking method*" process takes several images with a CCD camera from where subimages are cut out in order to fit the movement of the objects.

Therefore, a median image of all the subimages is generated, removing the background stars and reducing the sky background noise efficiently. We can find this method incorporated in "*Stellar Hunter Professional*," a commercial software used to discover asteroids and comets.

[¶] Inter-agency space debris coordination committee.

[**] Low earth orbit long-term debris environment evolution model.

[††] Geostationary earth orbits.

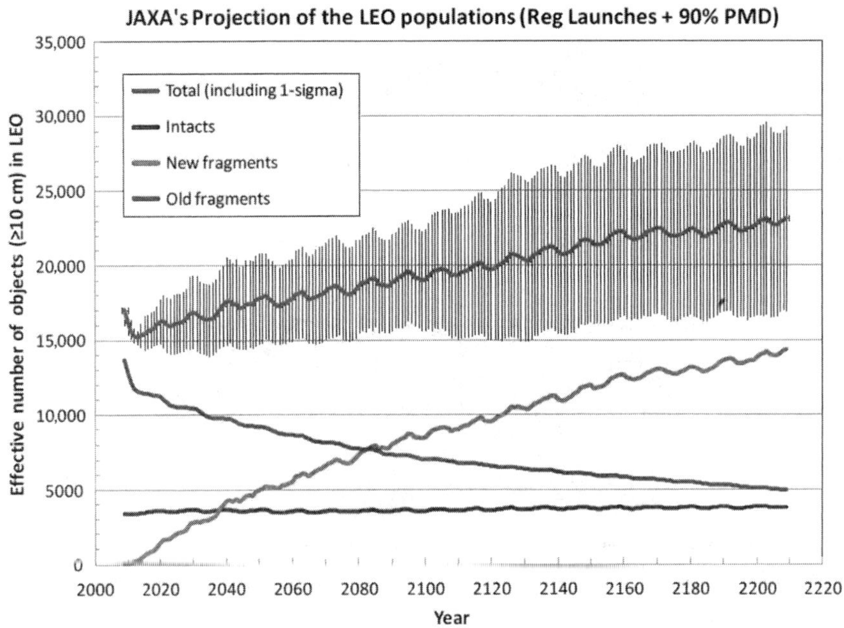

Fig. 1 Evolution model of the LEODEEM – 2013 *"Stability of the Future LEO Environment"* – AI 27.1 [1]. To identify the lines in the grey colour plot consider that the lines originate from top to bottom in the following sequence: Total, Old fragments, Intacts, New fragments.

b. Application of Optical Observation Technology in LEO Debris Observation

The LEO observation system shown in Fig. 3 with two arrays of telescopes can produce a very wide field of view virtually and can track a debris in a very long arc, determining then the orbit very accurately.

If we locate two of these systems properly, observation of consecutive passes might be possible and observation of consecutive passes enables us to do more accurate orbit determination.

For this reason, two longitudinally adjacent sites should be considered.

In each site, two narrow rectangular regions are realized using 40 optical telescopes in total (20 telescopes per each rectangular region) with the directions of sight separated by about 80–90 deg so that two consecutive passes may go through the expanded field of view (see Fig. 4).

To demonstrate the feasibility of this system concept, only two optical telescopes located longitudinally adjacent were used: one at the Ishigaki-jima Morita Observatory in Okinawa (see Fig. 5a) and another at the Rikubetsu Observatory in Hokkaido (see Fig. 5b).

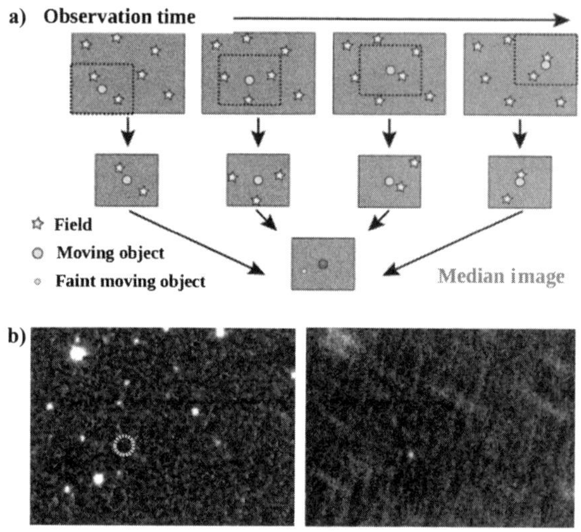

Fig. 2 Concept of the stacking method process: a) stacking method and b) a faint object detect with the stacking method: one CCD image (left) and the stacked image (right).

Some TLE-objects were observed assuming that we did not know their TLE information. After observation of a debris, the line of sight of the telescope was moved swiftly to another direction where the same debris could be observed after a while as depicted in Fig. 6. This operation provided for mimic observation data from one observation site with two telescope arrays thanks to the TLE information. The first day consecutive observation data of a debris enable us to do very accurate orbit determination, which can be verified by the second day observation data.

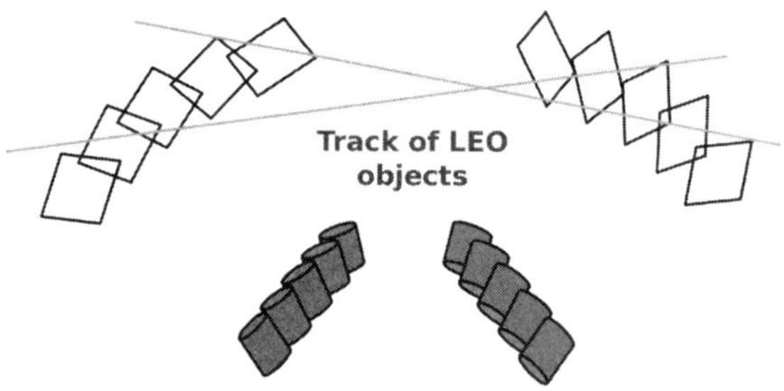

Fig. 3 Virtual field of view of the system.

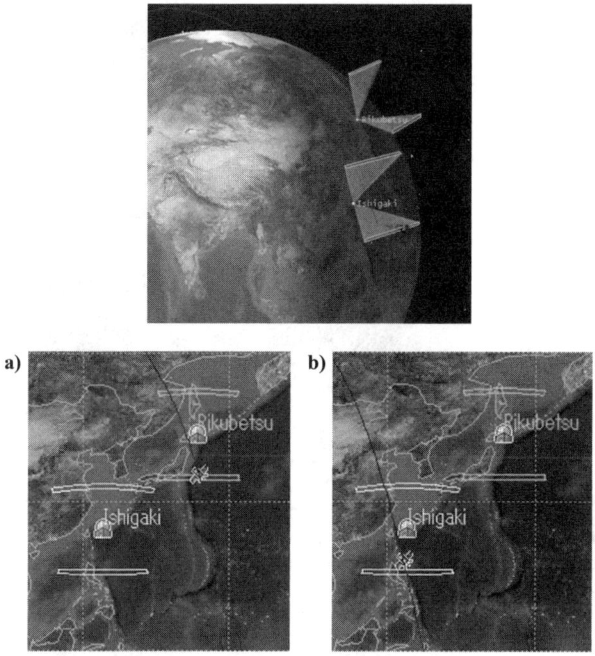

Fig. 4 Longitudinally adjacent systems for consecutive pass observation—detection of identical object at sites: a) first pass and b) second pass.

Another issue considered by JAXA for orbital debris observations is the *Space-Based Observation*, which possesses several advantages, such as no atmospheric extinction, no visible time constraint and higher sensitivity.

Thence, it enables to detect and track faint debris, which is smaller than 10 cm.

In fact, JAXA has performed a computer simulation where a telescope in a Sun-synchronous orbit aligned to the anti-solar direction was investigated in terms of the number of debris detected passing the field of view during a certain time.

It resulted in a high frequency and coverage of observation, but many of them had a rather long range and high velocity on the CCD image processor.

Consequently, this way of observation was concluded not to provide sufficient observation frequency and coverage for cataloging.

Taking into consideration the advantages and inconveniences of a space-based sensor (high sensitivity for detection/low observation frequency for cataloging), one of the solutions to compromise these pros and cons is a collaborative observation with the ground-based observation systems illustrated in Fig. 7.

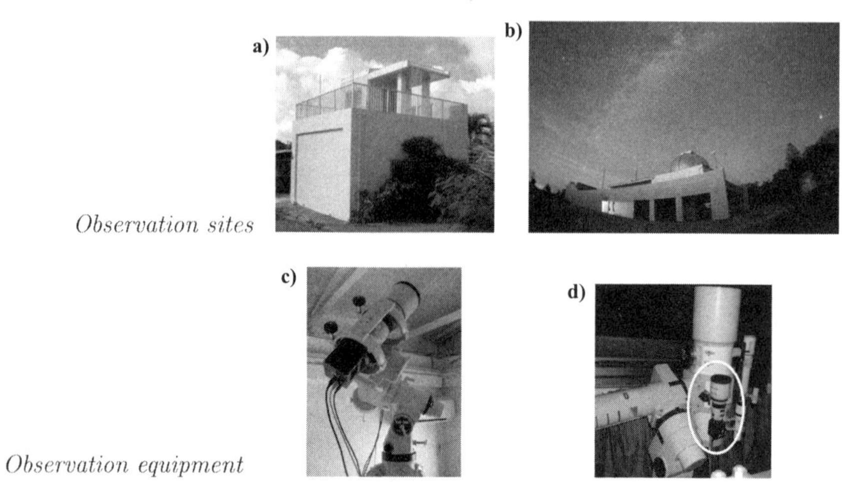

Fig. 5 Observation sites and equipment used in the feasibility demonstration experiment: *observation sites* a) Ishigaki-jima Morita Observatory (Okinawa), b) Rikubetsu Observatory (Hokkaido), *Observation equipments* c) Canon 200 mm F2+FLI ML23042, and d) Canon 300 mm F2.8+FLI ML4240.

JAXA has been compiling the hypervelocity impact test data, focusing not only on mechanical damages, such as penetration and perforation, but also on functional damages of exposed components. Data acquired are available for design purpose using a database software in Fig. 8.

In addition, JAXA has developed a collision risk analysis tool called "*TURANDOT*," which can evaluate the debris impact probability on each surface

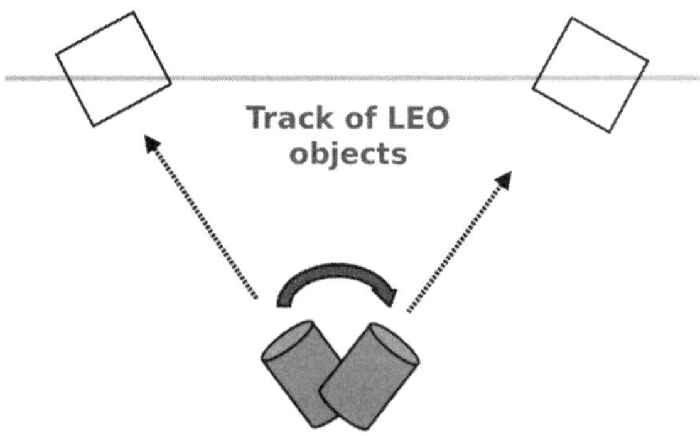

Fig. 6 Swift change of telescope direction to mimic the two arrays of the system.

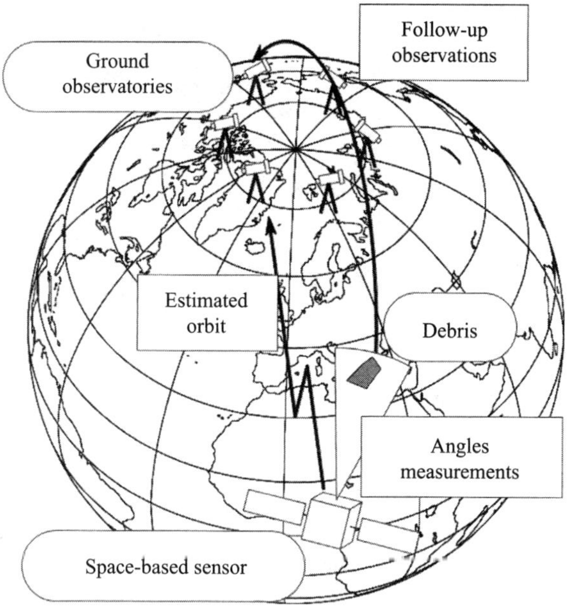

Fig. 7 Method of collaboration of ground observatories for a sensitive and frequent system from space-based observation.

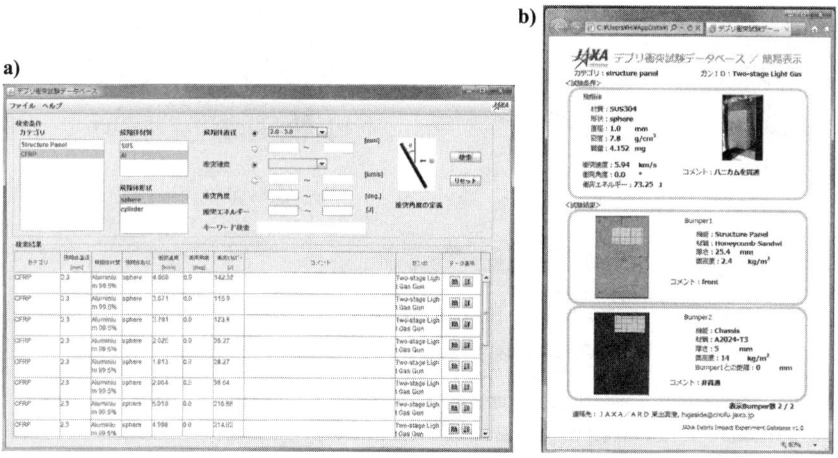

Fig. 8 Overview of the database for hyper velocity impact experimentation and hydrocode simulation: a) simple view in Japanese and b) detailed view in English.

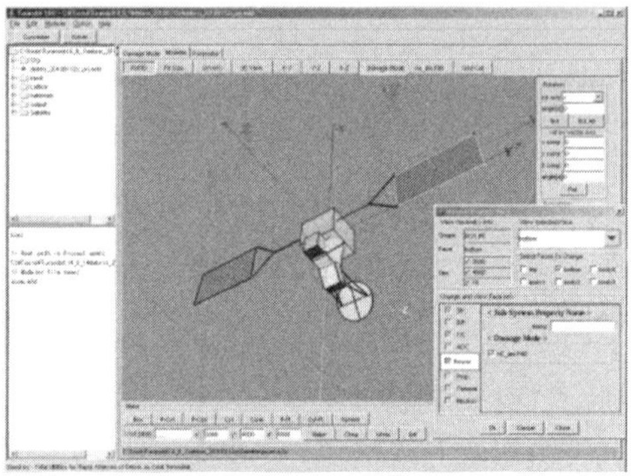

Fig. 9 Screenshot of TURANDOT tool.

of a space system using existing debris environment engineering models such as MASTER[‡‡] and ORDEM.[§§] A screenshot of this tool is shown in Fig. 9.

Moreover, JAXA has tried to set up a concept of sensor and engineering model to detect micro meteoroid and debris impact. Indeed, a large "*Flexible Printed Circuit board*" was designed with the surface featured by many thin conductive strips formed with a fine pitch on a thin polyimide film.

Once the experience done, it detects breakup signals confirmed by the apparition of perforation holes. Precisely, the number of signals can deduce the diameter

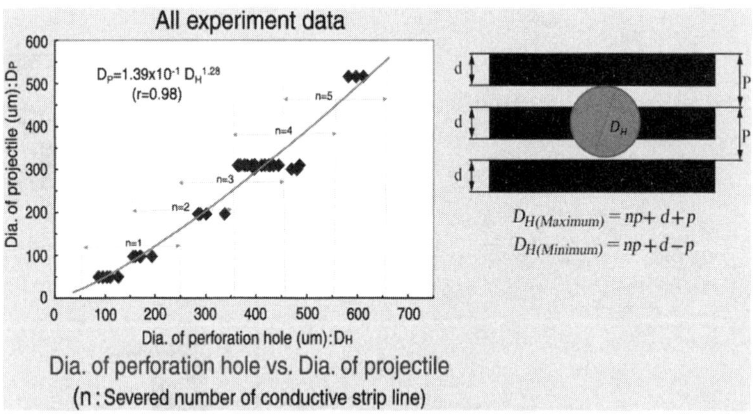

Fig. 10 Results of the experience showing the evolution of the diameter of the projectile as a function of the diameter of the perforation hole.

---

‡‡Meteoroid and space debris terrestrial environment reference.

§§Orbital debris engineering model.

of the projectile, which evolves as the diameter of the hole increases, as it is shown in Fig. 10.

This concept was implemented in KASPER[¶¶] attached to the propulsion module of HTV[***] -5 or so-called "*KOUNOTORI-5*" that was launched in August 2015 to deliver supplies to the International Space Station, and to demonstrate the sensor functional capability in space environment.

2. SECOND GOAL

The second objective according to the Strategic Plan of JAXA is to avoid a generation of new debris at some reasonable level by compromising between the cost and the reliability for sustainable space development and utilization activity.

Consequently, JAXA controls its Debris Mitigation Standard,[†††] with which the compatibility of the mission and the system design is checked and verified from the early phase of the project. JAXA contributes to the ISO standardization activity and participates actively in IADC.

One of JAXA's space debris mitigation efforts include the development and maintenance of the software tool DEMIST.[‡‡‡] Figure 11 shows this GUI-based software tool which allows engineers to check their mission and system design against JAXA's debris mitigation standard. Basic mission parameters are fed to the software which in turn provides feedback about debris collision probability, orbital lifetime of the satellite, and other mitigation-related information.

DEMIST estimates the expected number of collision avoidance maneuvers based on debris flux of cataloged objects. Based on this information, the tool is also able to provide expected total ΔV and fuel mass for these maneuvers.

Orbit lifetime is calculated considering time-variant atmospheric drag, lunisolar gravity, solar radiation pressure, and earth oblateness. If the orbit lifetime is longer than 25 years, a deorbit maneuver has to be taken into account. A deorbit maneuver places perigee at 60 km of altitude.

3. THIRD AND FOURTH GOALS

The third objective concerns about the ground safety in reentry process of space systems: JAXA improves ORSAT-J, a tool dedicated to assess the ground safety through reentry operation. Also, JAXA always participates in the IADC observation campaign of the reentering risk objects and is conducting R&D to reduce risk on ground.

[¶¶] KOUNOTORI Advanced Space Environment Research Equipment.

[***] HTV : H-II Transfer Vehicle.

[†††] Reference JMR003.

[‡‡‡] Debris mitigation standard support tools.

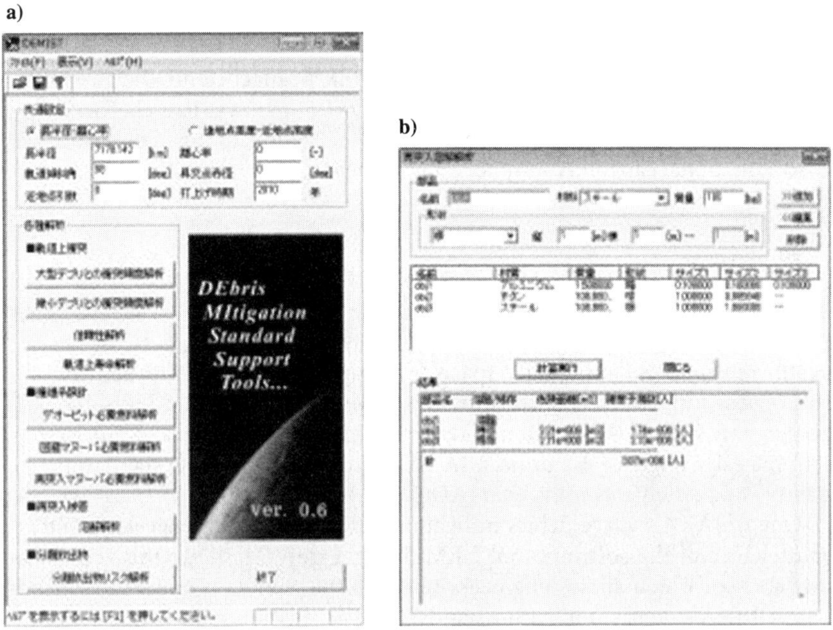

Fig. 11 Debris Mitigation Standard Support Tools *"DEMIST Ver. 0.6"*: Easily check compatibility of mission and system design with JMR003 at early stage of a project.

The fourth goal consists in developing enabling cost-effective technologies for ADR[§§§] operation and to pursue actual implementation of a practical ADR system through the framework of international cooperation.

III. JAXA'S SPACE DEBRIS MITIGATION ACTIVITY

Space debris (SD) mitigation describes the effort to avoid the generation of new SD. To this end, JAXA implemented a dedicated document system to organize and manage mitigation goals, guidelines, and activities. These documents are:

1. Space Debris Mitigation Standard (JMR003B)
2. Space Debris Mitigation Support Document (CAA-110029)
3. Space Debris Protection Manual (JERG-2-144HB001)
4. Rocket Design and Operation Manual
5. Spacecraft Design and Operation Manual

The first document (Debris Mitigation Standard) shall be described in more detail.

[§§§] Active debris removal.

A. SPACE DEBRIS MITIGATION STANDARD (JMR-003B)

This document is JAXA's reference document on the matter of SD mitigation.

1. PRINCIPLES

Seven operational principles form the basis for SD mitigation in this document.

1. Preventing on-orbit breakups of space systems
2. Transferring a spacecraft that has completed its mission in GEO into a higher orbit
3. Reducing the stay time of the orbital stage in GTO
4. Minimizing the number of objects released in orbit during operations
5. Reducing the stay time of the space system in LEO ($\leqslant$ 25 years)
6. Minimize the casualty risk on the ground due to reentry of space systems
7. Avoidance of on-orbit collisions

2. SD MITIGATION MANAGEMENT

Compliance with SD mitigation aspects is ensured through a dedicated management system. This system is also designed to achieve proper implementation of the SD mitigation guidelines into space projects. Each JAXA associated space project has to meet the following conditions:

1. Projects and contractors shall assign a responsible organization or individual responsible to study, plan, implement, and review effective measures to ensure SD mitigation.
2. Projects should develop a feasible Debris Mitigation Plan after applicability of the requirements in this standard have been reviewed in coordination with the Safety and Mission Assurance Department.
3. The mitigation plan may be incorporated into the System Safety Program Plan, which should be offered to the System Safety Review Board for review.
4. Contractors should also prepare a Space Debris Mitigation Management Plan which complies with the Debris Mitigation Plan presented by JAXA. The plan should be submitted to JAXA for approval.

3. SYSTEM SAFETY REVIEW BOARD

The debris mitigation aspects that have to be considered in each phase are presented in Fig. 12. The process starts at the mission analysis level with the operational orbit determination. Based on the orbit and the estimated ground impact, a decision whether to use a controlled reentry is made (Phase 0). As the system study advances, more detailed analysis like the debris generation during operation, which is dependent on the spacecraft design, can be made (Phase I). It

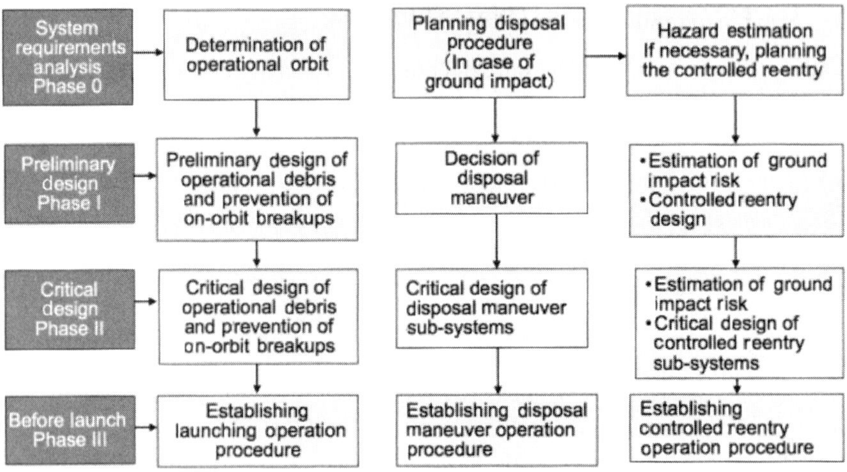

Fig. 12 SD considerations in the design process.

also becomes possible to estimate which spacecraft parts could survive reentry and pose a threat on the ground (Phase II). Shortly before launch, the procedures to limit debris generation during launch, operation, and disposal are conceived (Phase III).

The system safety review board accompanies the system design process and assesses the safety of mission operations during each phase of the system design process (Fig. 13). One of the board's purposes is to evaluate mission and system design at each phase under the aspect of space debris mitigation compliance and feasibility. This takes place at the Safety Reviews that occur at the end of each phase, I, II, III. Toward the end of the operational lifetime, a delta Phase III system safety review is also conducted to update the disposal procedure if necessary. This can be required in case any mission, system, or environmental aspects evolved differently during operations than was anticipated in the design phase.

B. COMPARISON WITH OTHER MAJOR STANDARDS

A comparison with standards of other agencies addressing space debris reveals that Japan, together with NASA, was the first to publish a standard concerning SD. This means that Japan has been one of the first nations to recognize space debris as a major concern for future space operations. The Japanese standard was conceived in 1996 and the responsible agency at that time was NASDA. The next agencies that followed in 1999 were Roscosmos, ESA, and CNES. This development led to joint agreements on the space debris matter, the IADC Space Debris Mitigation Guideline, among all agencies in 2001. In the following years, the European Code of Conduct (ECC) was devised.

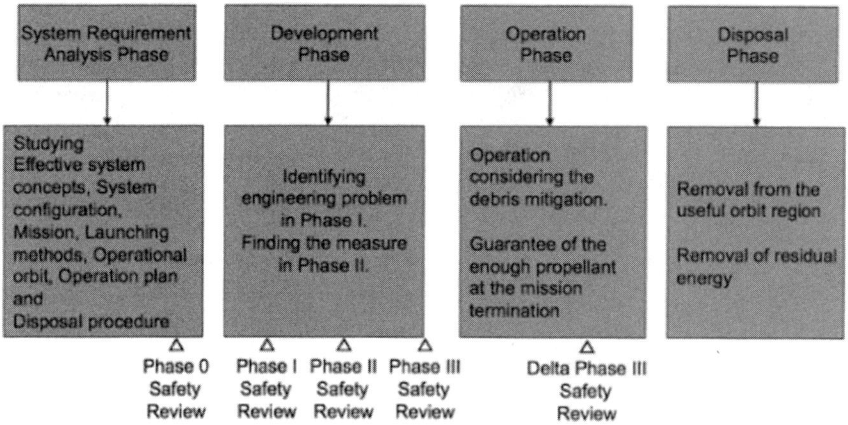

Fig. 13 System safety reviews.

The United Nations Committee on Peaceful Use of Outer Space (UNCOPUOS) devised the Space Debris Mitigation Guidelines on a global level in 2007. Following this UN document, the International Organization for Standardization (ISO) started its activity to issue their Mitigation STD document. Furthermore, the ECSS published the standard ECSS-U-AS-10C. In the same time frame, the French space act was signed into law which made space debris measures legally binding for activities operated from French soil. This step is significant because no other state has made compliance with the guidelines a law since then. More recently, UNCOPUOS is now organizing the second document providing best practices for sustainable space activities. The international development of space debris documentation is visualized in Fig. 14.

In summary, the problem of space debris is internationally recognized and thus every major space faring body has an appropriate document. The documents are listed next with their date of publication in parenthesis:

1. ESA Space Debris Mitigation for Agency Projects, Director General's Office (2008)
2. ECSS-U-AS-10C, Space Sustainability: Adoption Notice of ISO 24113 Space Systems Space Debris Mitigation Requirements (2012)
3. JAXA-JMR-003B, Space Debris Mitigation Standard (2011)
4. NPR 8715.6, NASA Procedural Requirements for Limiting Orbital Debris (2009)
5. NASA-STD-8719.14, Process for Limiting Orbital Debris (2011)
6. Space Debris Mitigation Guidelines of the COPUOS (2007)
7. IADC-02-01: IADC Space Debris Mitigation Guidelines (2007)

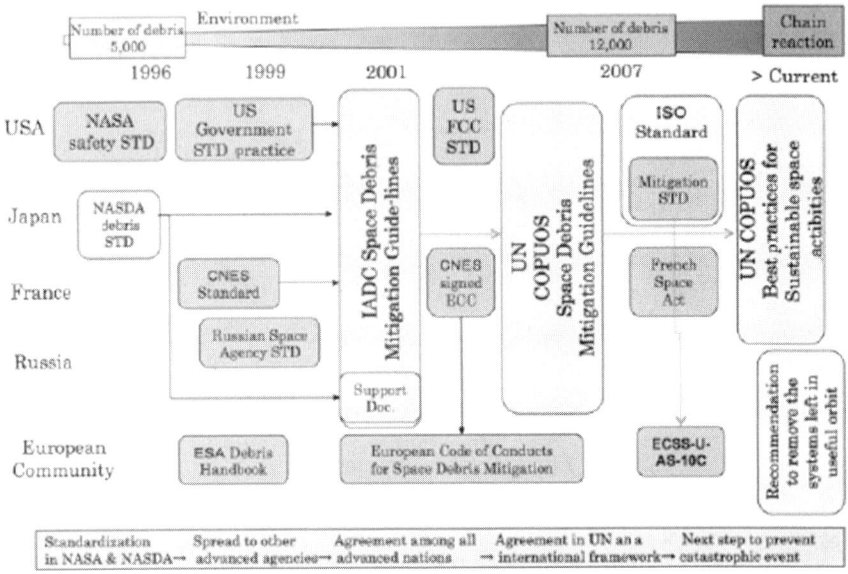

Fig. 14 Comparison with other standards.

8. ISO-24113 Space systems Space debris mitigation requirements (2011)
9. European Code of Conduct for Space Debris Mitigation (2004)

These documents offer space debris-related guidelines and requirements. The major points are summarized in Tables 1 and 2.

IV. GENERAL DISCUSSION ON ACTIVE DEBRIS REMOVAL

The increasing SD population is a threat to the sustainability of space operations. To evaluate the population evolution, the IADC conducted a joint parametric population evolution study in 2013 ("*Stability of the Future LEO Environment*"—AI 27.1). In this study, the six space debris population simulation tools by NASA, ASI, ESA, ISRO, JAXA, and UKSA are initialized with the same population. The launch traffic cycle from 2001 to 2009 is repeated over the simulation duration up to the year 2209. Satellites are assumed to have an operational lifetime of 8 years with a nominal, remaining lifetime of 25 years. The results of the different simulation tools are compared. The following results and discussions are based on Ref. [1] and other publications.

The aim of these efforts is to gain a better understanding of how the space debris population evolution can be influenced by debris mitigation and remediation actions. Furthermore, they give insight into which objects or classes thereof should be removed first. Space debris mitigation and remediation actions

TABLE 1 RECOMMENDATIONS AND REQUIREMENTS IN MAJOR WORLD DEBRIS MITIGATION STANDARDS (PART 1)

	Measures	ESA Space Debris Mitigation for Agency projects	Space Sustainability (ECSS-U-AS-10C)	JAXA (JMR-003B)	NASA (NPR 8715.6 and NPR 8719.14)
Mission-related Objects	Operational debris	Required	Required May not exceed: one for single payload launch, and two for multiple payloads	Required	1) LEO: >1 mm (decay within 25 years, and total <100 object-years) 2) GEO: >5 cm (decay -500 km within 25 years)
	Slag from Solid Motor	Slag <1 mm	Required (Slag <1 mm for GEO)	Required	–
	Pyrotechnics	Particles <1 mm	Combustion products <1 mm	Combustion products <1 mm	–
	Secondary Ejector	–	–	–	–
On-orbital breakups	Avoidance of intentional destruction	Required	Required	Required	1) <100 object-years (for >10 cm) 2) Fragments >10 mm shall be limited 1 year 3) Fragments >10 mm, collision with Operating S/C shall be limited $<10^{-6}$
	Accident During Operation	–	–	Required (Monitoring)	Probability of BU $<10^{-3}$
	Postmission breakup (Passivation, etc.)	Required (to be conducted <2 months)	Probability of BU $<10^{-3}$	Required	Required

TABLE 1 RECOMMENDATIONS AND REQUIREMENTS IN MAJOR WORLD DEBRIS MITIGATION STANDARDS (PART 1) *(CONTINUED)*

	Measures	ESA Space Debris Mitigation for Agency projects	Space Sustainability (ECSS-U-AS-10C)	JAXA (JMR-003B)	NASA (NPR 8715.6 and NPR 8719.14)
Collision	With large objects	Risk assessment	–	Required (CAM, LOLA)	<0.001 (with >10 cm)
	With Small Objects	–	–	Required	Disposal success >0.01
Post mission disposal GEO	Reorbit at EOL	235 km+(1000$Cr(A/m)$)	235 km+(1000$Cr(A/m)$)	235 km+(1000$Cr(A/m)$)	235 km+(1000$Cr(A/m)$) $e < 0.003$
		$e < 0.005$	Success probability > 0.9 100 years guarantee	Success probability > 0.9	Success probability > 0.9 100 years guarantee
	GEO lower limit	–	<−500 km (within 25 years)	<−500 km (within 25 years)	GEO −500 km
	Protected Inclination	−15 deg < inclination < 15 deg	−15 deg < inclination < 15 deg	−15 deg < inclination < 15 deg	−15 deg < inclination < 15 deg
PMD LEO (MEO)	Reduction of orbital lifetime	EOL lifetime < 25 years	EOL lifetime < 25 years Success probability > 0.9	EOL lifetime < 25 years Success probability > 0.9	Total period < 30 years EOL lifetime < 25 years Success probability > 0.9
	Transfer to graveyard	Required (Excluding Galileo orbit)	Required	Required	2000 km ~ (GEO-500 km)
	On-orbit retrieval	–	–	Required	Retreive within 10 years
	Ground casualty	Ec < 10^{-4}	Required	Ec < 10^{-4}	Ec < 10^{-4}, (Count impact energy >15 J
Others	Tether	–	–	Required	Required

TABLE 2 RECOMMENDATIONS AND REQUIREMENTS IN MAJOR WORLD DEBRIS MITIGATION STANDARDS (PART 2)

	Measures	UN Guidelines	IADC Guidelines	ISO (DIS-24113)	European Code of Conduct for Space Debris Mitigation	U.S. Gov Standard Mitigation Practices
Mission-related objects	Operational debris	Addressed in Rec-1	Addressed in 5.1	Required	Required; not exceed, one for single payload launch, and two for multiple payloads	>5 mm (decay within 25 years)
	Slag from solid motor	–	–	Required	Slag < 0.01 mm (Changed to 1 mm)	–
	Pyrotechnics	–	–	Combustion products <1 mm	Objects <0.01 mm	–
	Secondary ejector	–	–	–	Required (SD-DE-07)	–
On-orbital breakups	Intentional destruction	Addressed in Rec-4	Addressed in 5.2.3	Required	Required (SD-DE-04)	–
	Accident during operation	Addressed in Rec-2	Addressed in 5.2.2	Probability of BU $< 10^{-3}$	Probability of BU $<$ 10^{-4} (SD-DE-05)	Required

TABLE 2 RECOMMENDATIONS AND REQUIREMENTS IN MAJOR WORLD DEBRIS MITIGATION STANDARDS (PART 2) *(CONTINUED)*

	Measures	UN Guidelines	IADC Guidelines	ISO (DIS-24113)	European Code of Conduct for Space Debris Mitigation	U.S. Gov Standard Mitigation Practices
	Postmission breakup (Passivation, etc.)	Addressed in Rec-5	Addressed in 5.2.1	Probability of BU $< 10^{-3}$	Required Inner press. $<$50% of critical press. Dispose within 1 year Success probability $>$0.9	Required
Collision	With large objects	Addressed in Rec-3 (CAM, COLA)	Addressed in 5.6	–	Required	–
	With small objects	–	Addressed in 5.6	–	(Recommended by other document)	–
PMD GEO	Reorbit at EOL	Addressed in Rec-7	235 km+ $(1000Cr(A/m))$ $e < 0.003$	235 km+ $(1000Cr(A/m))$ $e < 0.003$ Success probability > 0.9 100 years guarantee	235 km+ $(1000Cr(A/m))$ Success probability > 0.9	$>$36,100 km($>$300 km + GEO)

	GEO lower limit	–	<-200 km	–	-200 km	–
	Protected inclination	–	-15 deg $<$ inclination $<$ 15 deg	-15 deg $<$ inclination $<$ 15 deg	-15 deg $<$ inclination $<$ 15 deg	–
PMD LEO (MEO)	Reduction of Orbital Lifetime	Addressed in Rec-6	Addressed in 5.4 (Recommend 25 years)	EOL Lifetime $<$ 25 years Success Probability $>$ 0.9 100 years guarantee	EOL Lifetime $<$ 25 years Success Probability $>$ 0.9	EOL Lifetime $<$ 25 years
	Transfer to Graveyard	–	–	Required	Required	2000 km–19,700 km 20,700 km–35,300 km
	On-orbit retrieval	–	Addressed in 5.4	–	–	Retreive within 10 years
	Ground casualty	Addressed in Rec-6	Addressed in 5.4	Required	Ec $< 10^{-4}$ (Excluding France)	Ec $< 10^{-4}$
Others	Tether	–	–	–	–	–

are evaluated in a parameterized study to understand their effectiveness and influence.

A. POSSIBLE FUTURES

In order to assess space debris mitigation and remediation measures, the current space debris environment is propagated forward in time. For each propagation the virtual environment can be subdued to artificial influences that represent the counter debris measures, which are under assessment. Analysis of the results of these possible futures allows to draw conclusions about the measure's effectiveness. A high degree of randomness is inherent in SD simulations. Each individual simulation run, shows high fluctuation in the evolution of space debris population compared to the entire sample. The method of Monte Carlo simulations is employed to distill the general trend of debris population evolution under specific counter debris influences from the sample [3].

1. NO SPACE DEBRIS MITIGATION ACTIONS

The benchmark for all counter debris measures is the scenario where nothing is done to address the SD situation. This entails that spacecrafts are placed into orbit with a similar rate as has been done in the past 8 years. Figure 15 shows the

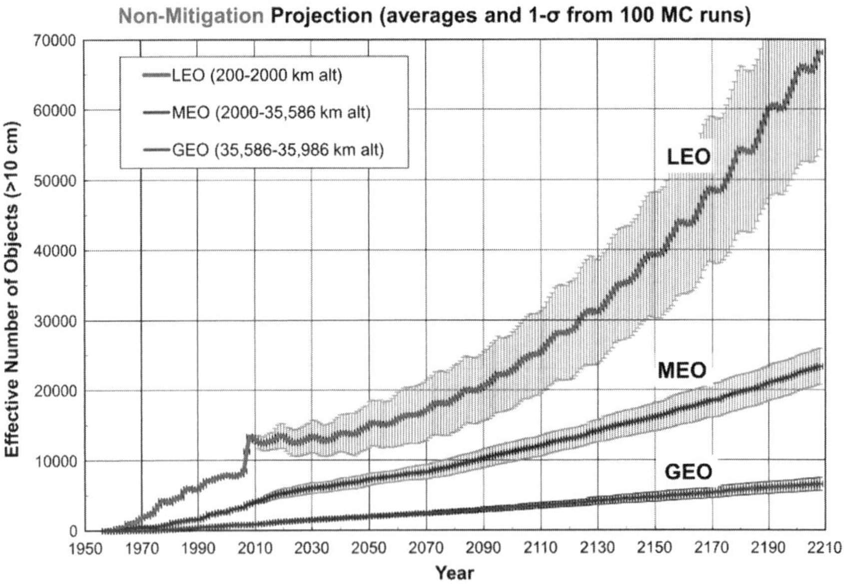

Fig. 15 Projection of debris population if no mitigation considered [4].

projected debris population grows as predicted by Liou et al. with their population simulation tool LEGEND for this scenario.

The figure distinguishes between the three orbital regions LEO, MEO, and GEO. The LEO region is dramatically affected by an effect called the Kessler Syndrome where the population enters a runaway situation in which debris multiplies itself through collisions. An exponential population growth is the result. The LEO population could grow by as much as 700% in the next 200 years compared to the population in 2013. MEO and GEO are much less densely populated relative to LEO and therefore experience a less severe, yet still significant, increase in debris population.

2. 90% POST MISSION DISPOSAL

The disposal of decommissioned satellites is a standard measure to address the debris problem. As has been shown in Sec. B, missions that are launched today are required to remove themselves from protected regions within 25 years of their end of life. For the following analysis, it is assumed that 90% of new missions successfully follow this practice.

Figure 16 shows the average population evolution as predicted by the six space agencies, which took part in the analysis. There is a high degree of agreement in the results. All simulations predict a debris population increase of about 30% in the next 200 years.

The chainsaw pattern in the plots corresponds to the solar cycle of approximately 11 years. When solar activity increases, earth atmosphere expands due to the elevated influx of solar power. In turn, drag on LEO objects increases and removes more objects faster. The population decreases. When solar activity declines, so does the drag effect. The sink function of atmospheric drag becomes less significant and thus, debris increases.

Figure 1 provides the results produced by JAXA's LEODEEM tool. In addition to the average population evolution, the 1-sigma uncertainty range is presented. The variation is determined through Monte Carlo analysis which in the case of JAXA is based on 60 runs [5]. Interestingly, the lower bound of the 1-sigma range shows a level population evolution. This means that there is an almost one in five chance that the debris population declines by applying 90% PMD alone.

The makeup of the average debris population is split into the three categories intacts, new fragments, and old fragments.

Their corresponding time histories is given. The intacts (intact satellites) remain constant with a slight increase due to new launches. The old fragments describe the debris population at the beginning of the simulation. This portion decreases overtime mainly due to atmospheric reentry. The dominating effect for debris evolution is the new fragments portion, which describes the generation of new debris particles due to collisions and break ups. This portion increases strongly overtime and drives the average debris population up.

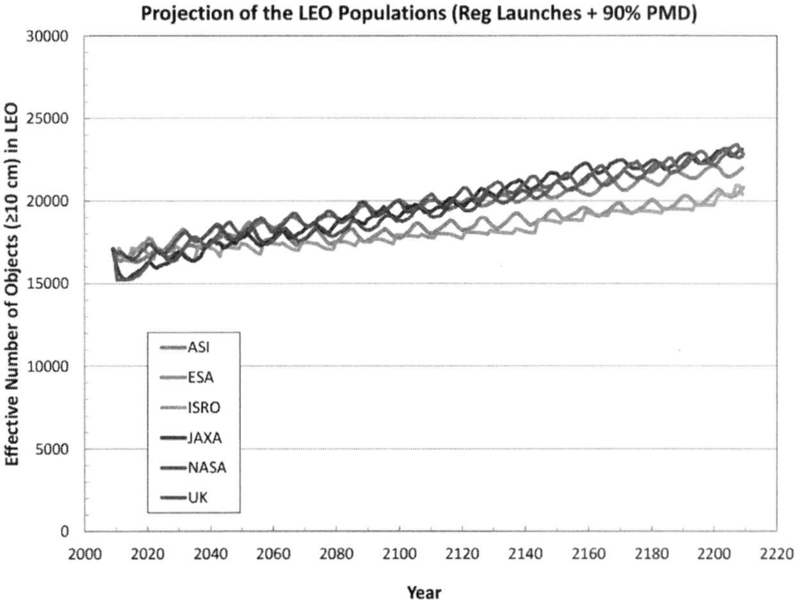

Fig. 16 Projection comparison of effective number of objects 10 cm and larger in LEO with 90% post mission disposal [4].

A closer look into the LEO regime is provided in Fig. 17. It shows the debris distribution over the LEO altitude regime for the year 2209. The solid colored lines are the results of the six simulation tools. The dashed gray line represents the situation at the beginning of the simulations in 2009. Up to an altitude of 750 km, the debris is removed effectively by atmospheric drag. Thus, no apparent change between the initial and final situation is discernible. However, debris density at an altitude of 900 km doubles during the simulation time frame. This is an important altitude because it corresponds to the sun synchronous orbit, which is paramount for various earth observation applications. Doubling spatial debris density in this region could compromise usage of this orbital regime with serious effects on a range of space-based services, which humans have grown reliant upon [7].[¶¶¶] Above a 900 km, a constant increase in spatial debris density is observable. The general distribution pattern remains unchanged.

The results presented thus far focused on trackable debris objects with a size of 10 cm or larger. Figure 18 shows the expected population evolution for objects smaller than 10 cm. It becomes clear that the number of smaller objects increases much faster than is expected for trackable pieces. Considering that 1 cm^3 aluminum at orbital velocities carries as much kinetic energy as is stored chemically

¶¶¶Weather forecast, environmental control, communication, earth science, ... [6].

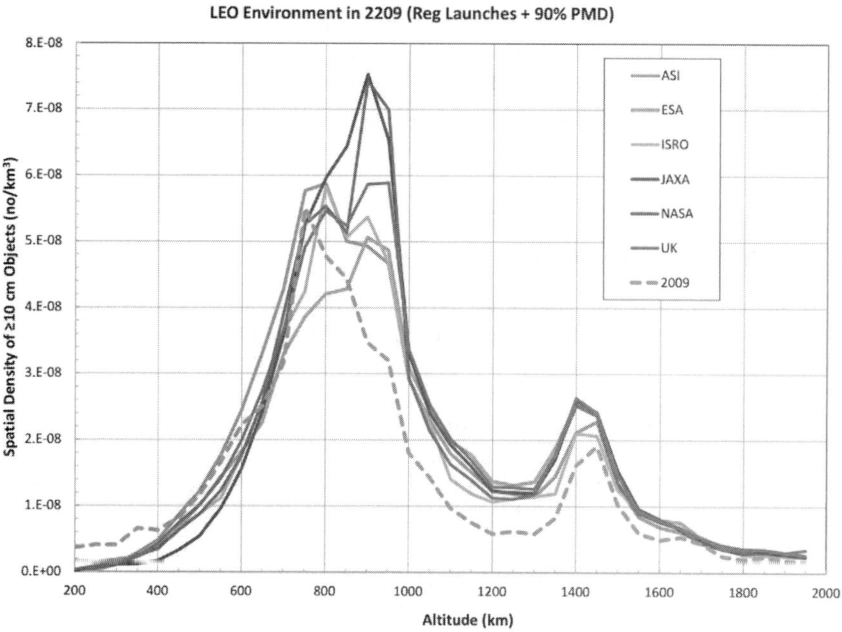

Fig. 17 Current and projected LEO environment in year 2209 [5].

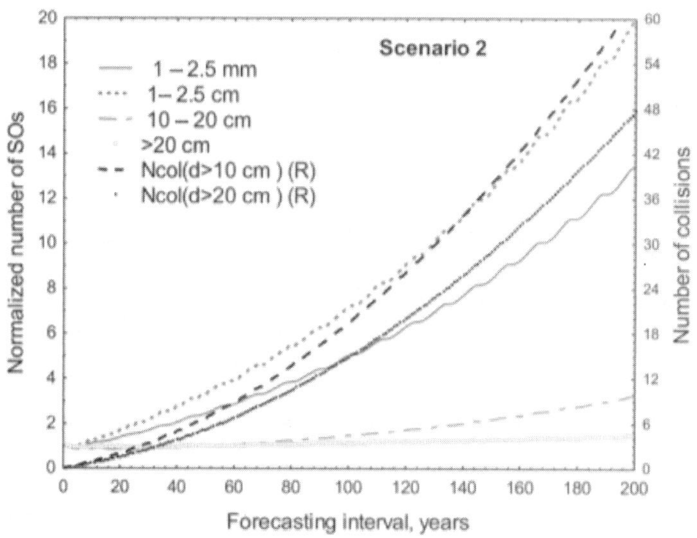

Fig. 18 Evolution of smaller-sized debris [9].

in a hand grenade,**** this finding is of great concern. It becomes clear that sub-centimeter particle can cause catastrophic damage to space assets. In addition to the small particles' potential to induce significant damage, they are hard to detect. Detectability of space debris is important because it allows to conduct evasive maneuvers before collisions take place with controlled objects [8]. This measure is unavailable when the particle location is unknown.

3. NO FUTURE SPACE MISSION LAUNCH

Even though a no future launch scenario is certainly an extreme measure, it is worth exploring the possible effect achieved by it. Liou and Johnson explored this scenario and came to the conclusion that orbital debris will increase even if no new objects are added to the environment artificially. Figure 19 illustrates the evolution of fragments generated by collisions (red), explosions (green), and by mission operations (blue). The gray line represents the superposition of the three components mentioned before. The dotted lines correspond to the evolution including the Chinese ASAT test conducted in 2009 that added significantly to the debris population. The solid lines disregard this event. It can be seen that collisions are the primary driver in increasing the space debris population.

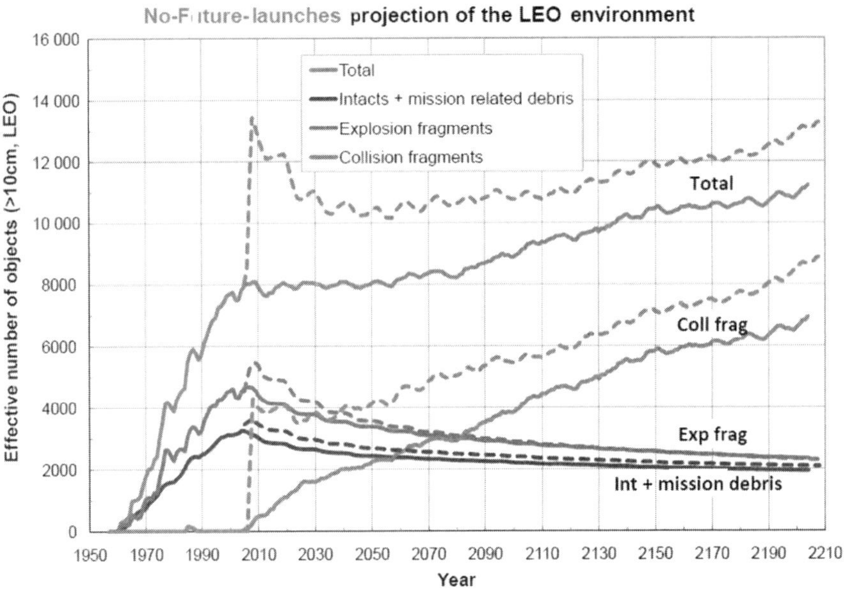

Fig. 19 Projection of the LEO environment if no future launches considered [4].

****1 g of TNT carries an equivalent chemical energy of 4184 J. There are about 380 g TNT in one grenade.

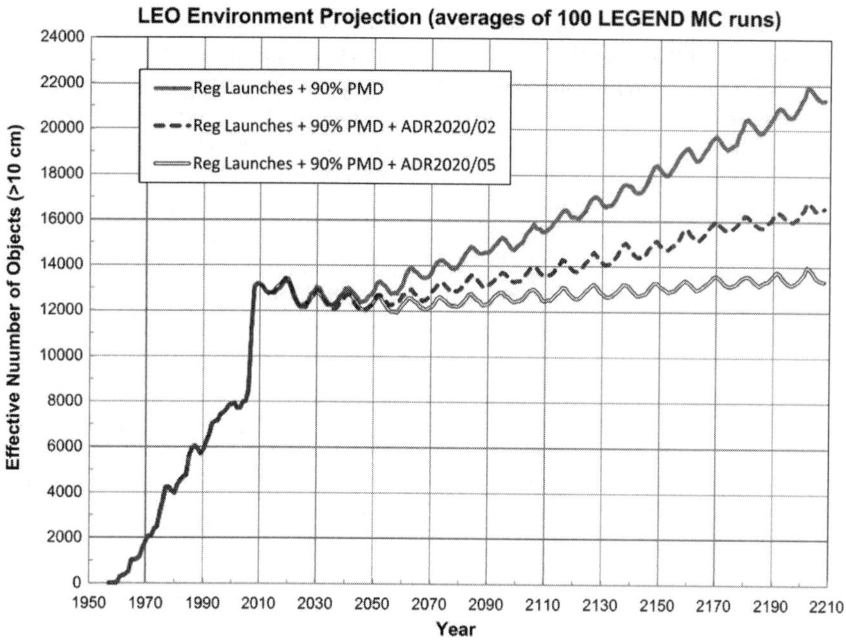

Fig. 20 Projection of the LEO environment with the consideration of two or five removed objects per year from 2020 [4].

4. ACTIVE DEBRIS REMOVAL

The measures discussed so far aimed at reducing the generation of new space debris. Active debris removal on the other hand describes an active measure to reduce the number of existing space debris objects. The effectiveness of this measure will be evaluated in the following paragraphs. In Fig. 20, the red line signifies the same assumptions made in Fig. 16; a normal launch schedule with 90% post mission disposal is assumed. The space debris population increases significantly. The dashed blue line represents a scenario where, in addition to PMD, active debris removal is employed. It is assumed that removals start in 2020 with two removals per year. A moderated debris population increase relative to the scenario without active debris removal is apparent.

Figure 20 shows that a constant average debris population may be achieved when active debris removal is employed starting in year 2020 with five removals per year. Considering the latter scenario, it is important to note that a constant average debris population translates into an equal chance of actual debris population decrease or increase. By applying active debris removal, the chances of a sustainable LEO environment can be increased, but certainty about this outcome cannot be gained at this point.

Figure 21 expands on Fig. 20 and adds an active debris removal scenario to the analysis where the removal of five objects per year is started in 2060. Because the measures are implemented at a later point in time, the debris population increases in parallel with the pure post mission disposal scenario up to 2060. In this year, the two scenarios diverge and the ADR simulation runs in parallel with the scenario in which ADR is employed starting in 2020, but remains at an elevated level.

The presented results seem to allow the conclusion that there is no pressing need for active debris remediation measures today, if we decide to accept more burden in the future. The debris population is likely to increase by only about 30% in total over the next 200 years which does not represent an imminent collapse of the usable space environment. However, there is still a high need for conclusive answers regarding which debris remediation measures are most effective and how many objects and which need to be removed. In addition, nontechnical aspects such as legal, political, and financial matters require sorting. The emerging debate among the academic communities is, therefore, none too soon and needs to be continued. The goal should be to have coherent answers to the open questions and to prepare enabling technologies in 5–20 years.

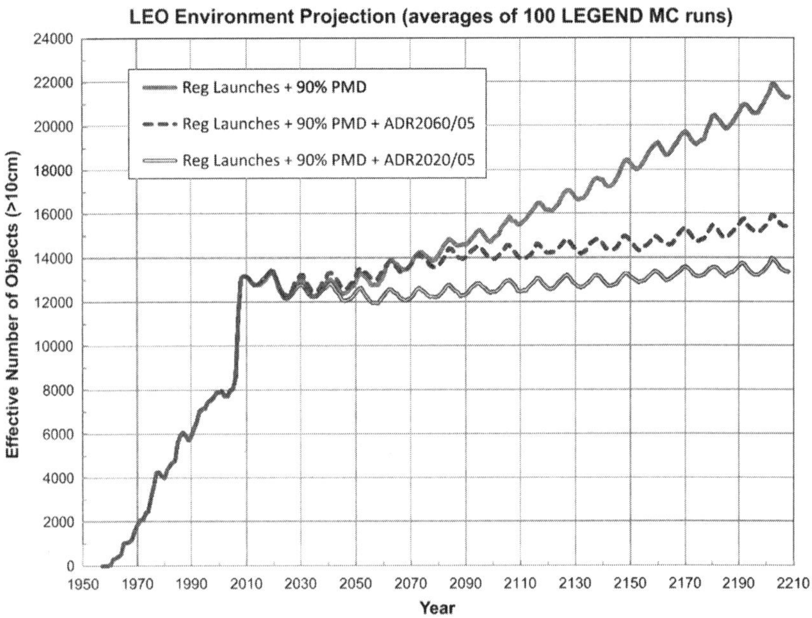

Fig. 21 Projection of the LEO environment with the consideration of five removed objects per year from 2060 [4]. To identify the lines in the grey colour plot consider that the lines originate from top to bottom in the following sequence: Reg launches + 90% PMD; Reg launches + 90% PMD + ADR2060/05; Reg launches + 90% PMD + ADR2020/05.

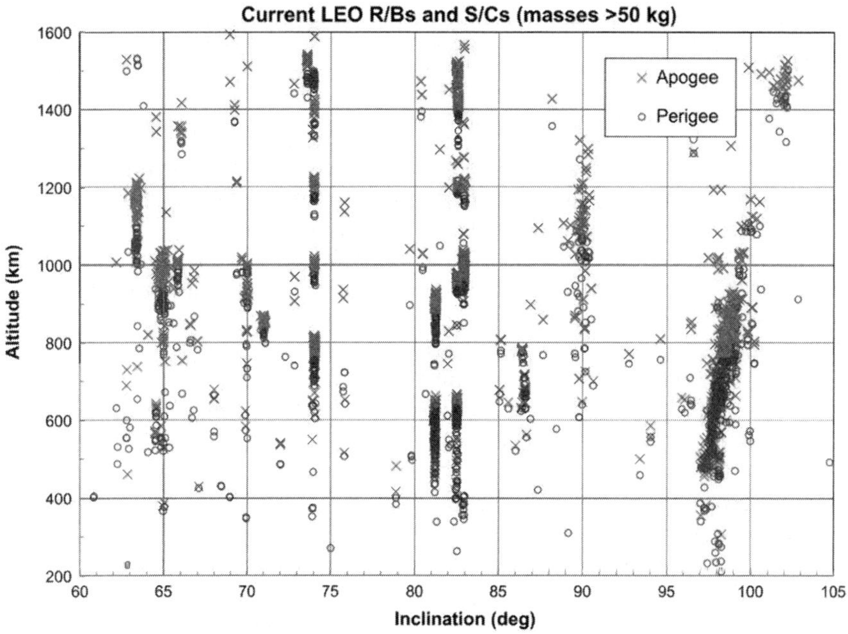

Fig. 22 Current LEO population more than 50 kg featured by their altitude, apogee, and perigee [4].

B. REMOVAL OBJECT PRIORITY

A considerable amount of research [4, 10] aims at establishing a priority list of objects, which should be removed first. It is impossible to predict which specific objects will collide or break up in the next couple years. If this would be known, it would be easy to pick the next removal targets based on the impact their collisions will have on the future debris situation. What is known today are the characteristics of the debris objects (mass, cross section area, and orbit) and also which orbital regions are more likely to see collisions. This knowledge is employed in variations to produce priority lists which hope to put the upcoming removal target selection on a rational basis.

In Fig. 22, the LEO orbital region is visualized using altitude and inclination. Perigee and apogee altitudes are plotted over their corresponding orbital inclinations. Ten dominant inclination bands are discernible (63, 65, 70, 71, 74, 81, 83, 87, 90 deg, SSO band starting at 97 deg). This observation gives away the densely populated regions in LEO. ADR missions should target these regions because they are the ones with highest collision probability. Additionally, because every region yields similar orbital parameters, it will be possible to conduct multi target missions because the maneuvers required to reach successive targets are small.

The specific target selection can be based on a variety of metrics. For of those are discussed in the following.

1. METRIC 1: TARGET MASS × P(COLL.) PER TARGET

One commonly used metric to rank objects for removal priority is multiplying the mass by the catastrophic collision probability associated with that object's orbital region P(Coll.). This metric is focusing on the effect of collision, that is, number of debris generated, to the debris environment. Sorting the objects in Fig. 22 according to this metric produces a list of 500 top priority objects. They are marked in Fig. 23. In each group, the highest priority objects are annotated next to the group. Many Cosmos series satellites as well as launcher upper stages and the ESA earth observation satellite Envisat are among them. Among the clustered groups, the orbit region in 900–1050 km altitude and 82.5–83.5 deg inclination is considered to be the most dangerous one.

In Fig. 24 a simulation has been conducted where ADR following prioritization according to Metric 1, Target mass × P(Coll.), was applied over all orbit regions. Removal operations start in 2020 with 5 objects per year and the results are represented by the green line. The dotted, blue line shows the case where only objects in the most dangerous region, inc = [82.5, 83.5 deg] and alt = [900, 1050 km], were removed. Prioritization follows once more Metric 1, with

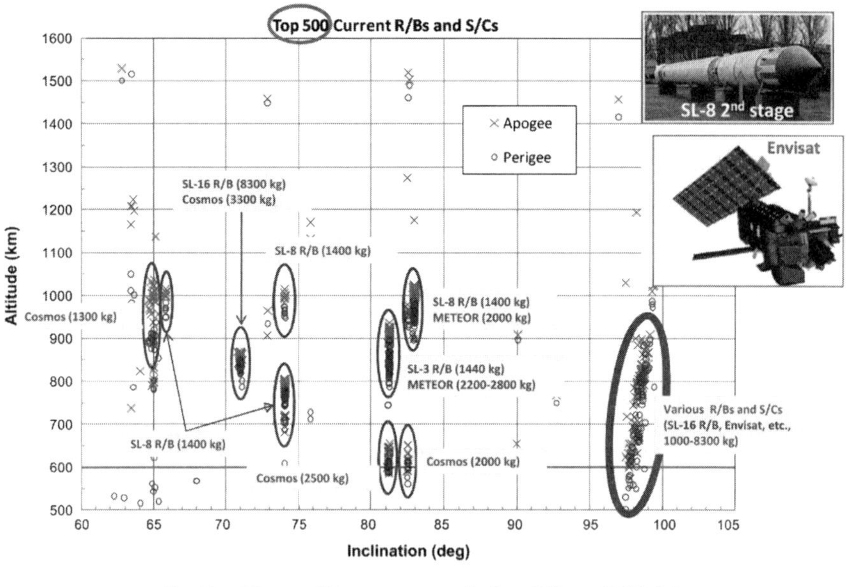

Fig. 23 The top 500 current populations R/Bs and S/Cs [4].

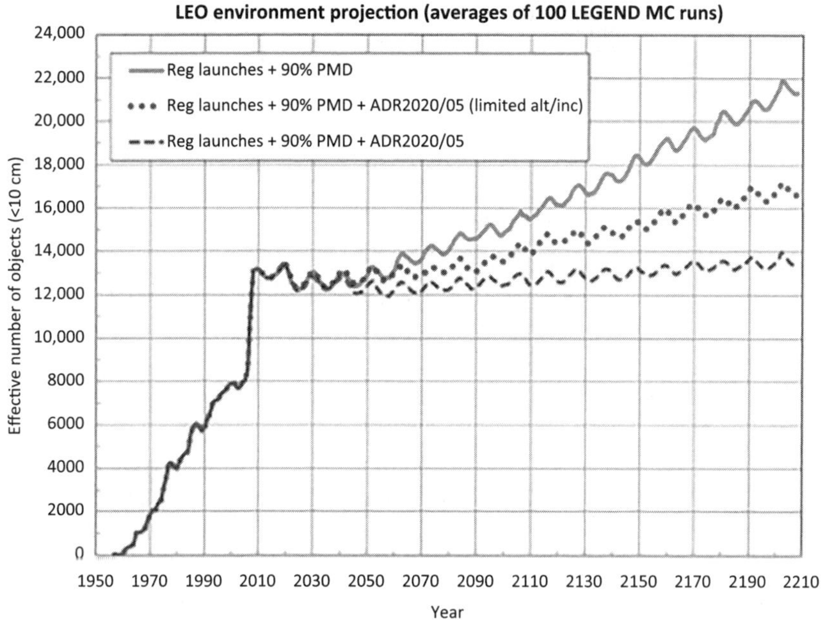

Fig. 24 Projection of LEO population with the consideration of five specific removed objects per year from 2020 [4].

removals starting in 2020 with five objects per year. The ADR scenarios show a significant improvement with respect to the no-ADR case. However, removing objects over all orbit regions still yields reliably better results. This finding suggests that a more elaborate approach for ADR employment is needed than just to focus on one high risk orbital region.

2. METRIC 2: TARGET CROSS SECTION × P(COLL.) PER ORBIT BIN

This metric produces the rate of catastrophic collisions. The highest priority regions are identified by applying this metric to the orbital region of 300–1300 km with inclinations 20–110 deg. When collecting the results into 2 deg inclination × 50 km altitude bins, two dominant peaks emerge. The first one is at 780 km altitude and 86.5 deg inclination. This region holds Tsyklon third stages, Vostok third stages, Delta II second stages and others [11]. The second peak is located at 920 km and 82 deg and is occupied by Cosmos satellites. Both peaks are enclosed by black rectangles in Fig. 25.

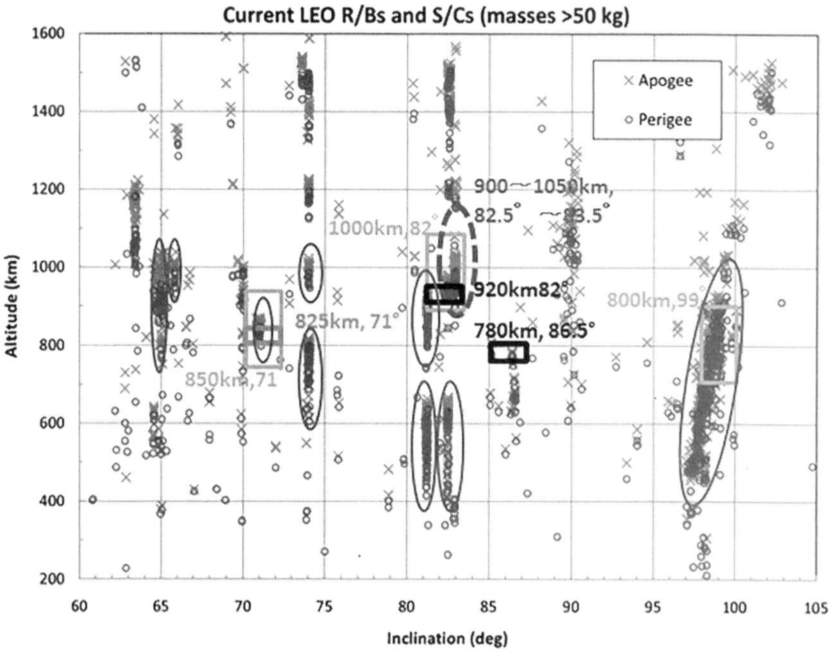

Fig. 25 Target selection in Fig. 23. (From Ref. [4] modified by Prof S. KIBE.)

3. METRIC 3: TARGET CROSS SECTION × TARGET MASS × P(COLL.) PER ORBIT BIN

The metric is focusing on the catastrophic collision risk and the number of debris particles generated. This is because target mass correlates with particle generation in collisions. Expressed in other words, the metric signifies the short-term risk to the environment due to collisions. Similarly to Sec. 2, objects are sorted according to this metric and the results are binned. The highest peak emerges at an altitude of 825 km and 71 deg inclination (Zenit second stages). This region is visualized in Fig. 25 by a green square.

4. METRIC 4: TARGET CROSS SECTION × TARGET MASS × P(COLL.) × ORBIT LIFE TIME PER ORBIT BIN

The results of Sec. IV.B.3 are multiplied by the orbital lifetimes associated with the targets and potential debris particles. It is a metric expressing the long-term risk to the environment due to SD. This metric considers that potential collision partners remain in orbit for long times and thus are more likely to be involved in collisions than other objects, which reenter the atmosphere quicker. Furthermore, the metric considers that the particle pollution generated in collisions would exist much longer because of the orbital lifetimes associated with the target's orbital

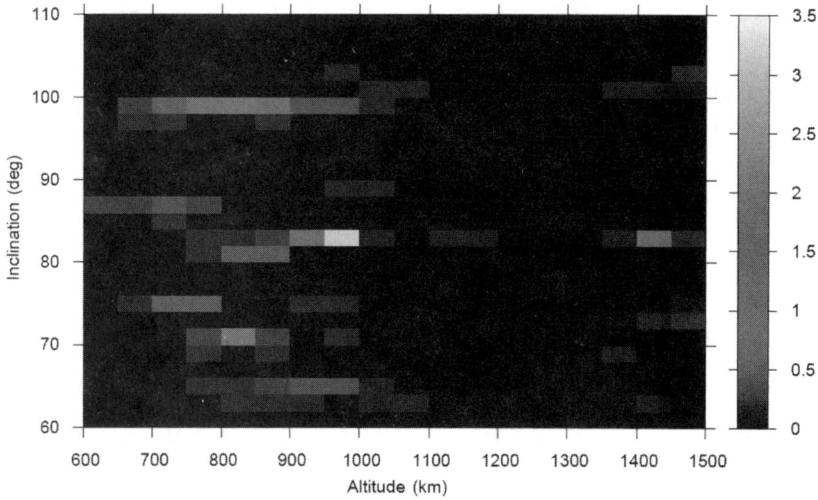

Fig. 26 Number of catastrophic collisions of targets after 200 years in the no-further-release scenario as a function of altitude and inclination [11].

region. With this metric the highest peak emerges at an altitude of 825 km and 71 deg inclination (Zenit second stages). A green square in Fig. 25 represents this result. This high priority region coincides with the prioritized region according to Metric 3.

5. METRIC 5: FUTURE COLLISION POSSIBILITY PER ORBIT BIN

Figure 26 presents the number of catastrophic collisions that are expected to occur within the next 200 years in the no future release scenario. The LEO regime from 600–1500 km altitude is shown with inclinations ranging from 60 to 110 deg. Each bin has a size of 200 km × 2 deg. Some regions are clearly more likely to yield collisions than others. The regions with very high collision expectations are the regions $H = 1000$ km at $i = 82$ deg, $H = 800$ km at $i = 99$ deg and $H = 850$ km at $i = 71$ deg and are also shown in Fig. 25 as orange squares. The results correspond to densely populated regimes recognized as prioritized orbit regions by the metrics mentioned in the previous sections.

6. APPLYING SPACE DEBRIS KNOWLEDGE TO ADR

It is insightful to compare a variety of ADR approaches to assess their relative effectiveness. A scenario in which no future launches are conducted serves as baseline for this analysis. In Fig. 27, the red line corresponds to a scenario where no future launches are conducted and no ADR is employed. The space debris population is likely to increase significantly. Below the red line are three graphs

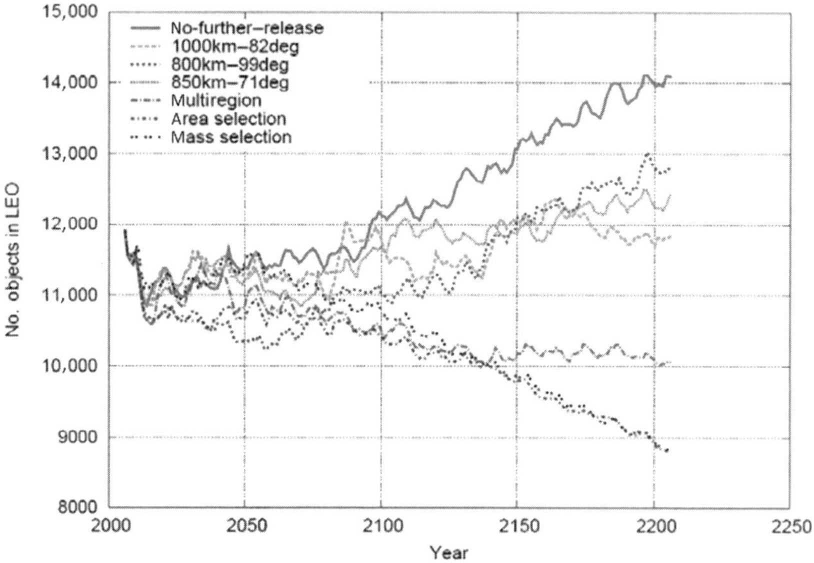

Fig. 27 Evolution of the LEO population larger than 10 cm for a no further release scenario, with active removal from individual orbital regions at 1000 km/82,800 km/71, for a multiregion removal strategy (switch to a different orbit region after every mission), and for a removal strategy ranked by mass or by area in any LEO region [11].

(dotted light green, solid purple, dotted blue) which exhibit a similar evolution. The triplet represents scenarios in which ADR is employed for one respective high risk region.[††††] The SD still increases on average over the next 200 years but at a moderated pace compared to no-ADR activities. Conversely, if each ADR mission is allowed to target a new high risk region, the overall space debris population evolution is declining (dashed green line). Two graphs which run in parallel in a strongly declining manner represent ADR mission scenarios where the targets are selected according to their area or mass, respectively. These two approaches are the most promising in this analysis. They are represented by the dotted purple and black lines in Fig. 27. A debris population decrease of 25% within the next 200 years is likely to occur on average. The conclusion to favor object removal from more than one region imposes itself. A single multitarget mission is unlikely to accomplish this because of more costly maneuvers involved to change between orbital regions. However, several multitarget missions can each focus on one orbital region where maneuvers to approach new targets are small.

[††††] Compare with results from Sec. 5.

C. ADDITIONAL CONSIDERATIONS FOR ADR TARGET SELECTION

The previous sections addressed the ADR target selection problem from a purely environmental perspective. However, a variety of additional aspects need to be considered before ADR missions can become routine. Targets show different geometries and attitude dynamics and will, therefore, pose individual technical challenges for grabbing. Some objects will be easier to grab than others. A layer of complexity is added when considering the possibility of multitarget missions. Here, target selection in one orbital region plays a crucial role for mission design. In Sec. IV.B it became clear that high risk objects are located in different orbital regions. It will, therefore, be difficult to rendezvous with the objects at the top of the priority list because of the maneuver cost involved to reach each object successively. Thus, more than one ADR mission needs to be conducted to remove a desirable amount of high-risk objects.

Additionally, nontechnical aspects are similarly challenging as the engineering hurdles we face. It is possible that target objects have nontechnical attributes associated with them that become the driving selection criteria. Envisat, for example, is a highly political target. The situation in which Envisat would be involved in a collision is perceived as disturbing by ESA. Furthermore, legal aspects will affect target selection. Already today, it is clear that it is illegal for a state to interact with a foreign satellite. These are some of the questions that will need to be answered in the future.

V. JAXA'S ADR SYSTEM CONCEPT AND R&D

For the following part, it describes the R&D work from JAXA dedicated to Active Debris Removal activities using the EDT[‡‡‡‡] concept as an example of ADR solutions.

A. EDT ADR OPERATION SCENARIO

JAXA has studied various concepts for removing debris. Figure 28 shows the one using the EDT.

After the remover is launched, a specific phase at the final operation consists in performing a rendezvous to a noncooperative debris object in order to obtain a sufficient closed distance to conduct some proximity operations such as motion estimation and attachment of the tether end. Thereafter, the debris will deorbit with the EDT until its reentry of the atmosphere and the remover will repeat its process with the next target.

[‡‡‡‡]ElectroDynamic tether.

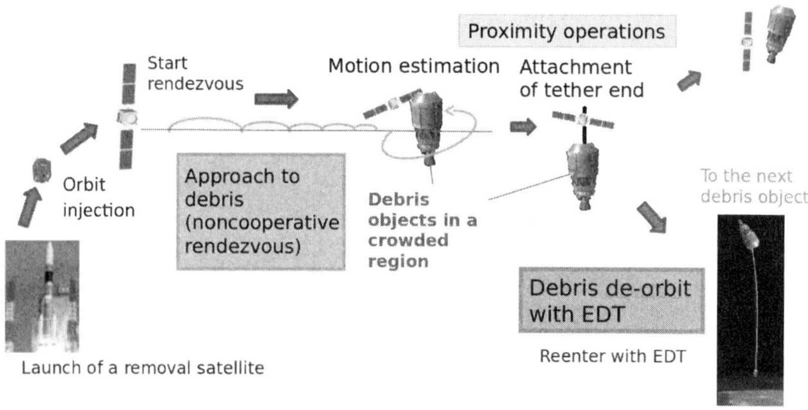

Fig. 28 An example of scenario for debris removal using EDT for ADR purpose.

B. RENDEZVOUS AND APPROACH TO THE NONCOOPERATIVE TARGET

The operation of the noncooperative rendezvous represents one of the most critical phases for an ADR mission, as it meets:

1. No communication link with the target
2. No image marker/laser reflector on the target
3. Uncertainty of the debris orbit
4. Change of optical property of the surface
5. Drastic change of lighting condition
6. No attitude and orbit control of the debris which is subjected to tumbling motion for instance

Therefore, this phase represents a big technological challenge during the design project of the remover. Indeed, Fig. 29 shows a baseline of this operation.

In the long distance over around 100 km, the remover approaches the target using the ground-based observation data such as TLE. And thereafter, it proceeds to approach by the DCR[§§§§] using only the angle data to the target from the on-board optical observation system in the meantime to achieve around 1.2 km of distance to the target. Then, it comes closer until 30 m by a V-bar hopping SOR[¶¶¶¶] using the object vision data and finishes by a forced motion circular orbit around the target.

[§§§§] Dual Co-elliptic rendezvous.

[¶¶¶¶] Stable Orbit rendezvous.

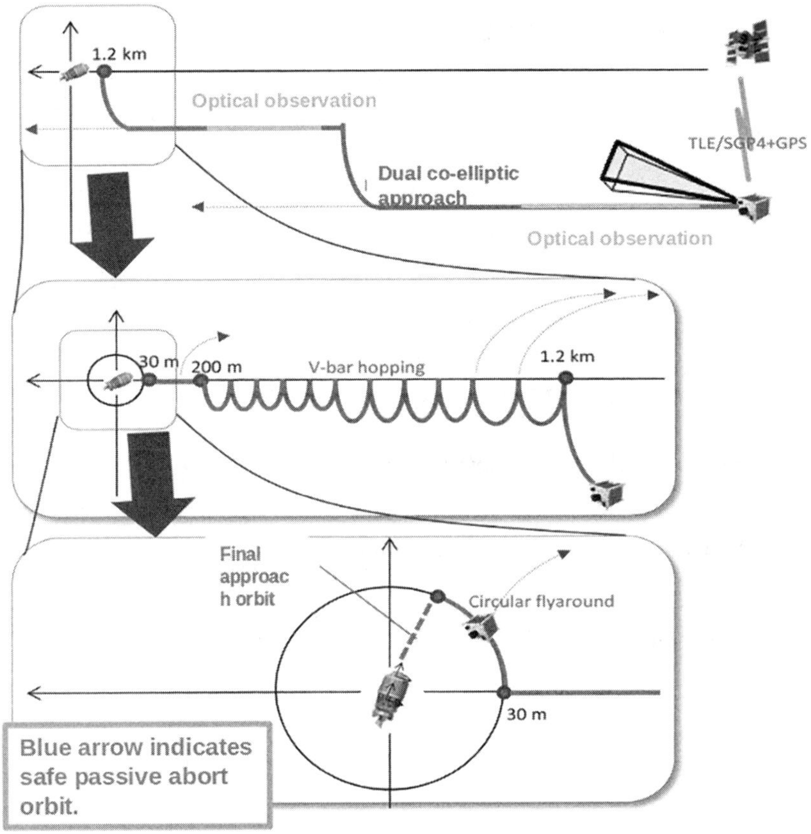

Fig. 29 Baseline of a rendezvous and approach scenario to a noncooperative object.

To perform such a complex procedure, various navigation sensors could be dedicated and featured according to the distance to the target (see Fig. 30). For instance, in Fig. 31, the visual camera,***** the infrared camera,††††† the ranging LIDAR and Flash/Scan LIDAR are listed with their applicable range, navigation technique and measured items.

A simulation was made with VISCAM at the beginning of the approach phase from 75 to 1 km. In Fig. 31, the blue and green lines are the true and estimated trajectories and the valid measurement was possible only during the red part due to the Sun light condition. The result indicates that both DCR and SOR rendezvous strategies are possible using only VISCAM in this approach phase.

*****VISCAM.
†††††IRCAM.

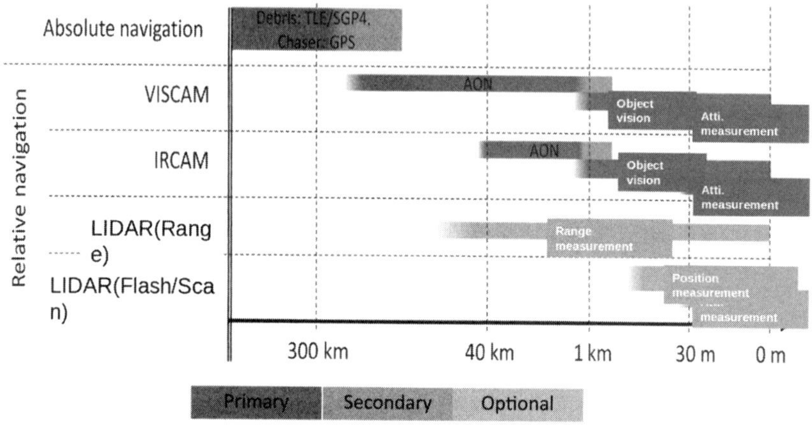

Fig. 30 Features of different navigation sensors as a function of the distance from a target.

To discuss performance and applicability of various sensors, simulations using VISCAM, IRCAM+VISCAM, and LIDAR+IRCAM+VISCAM were also conducted under the same situation previously mentioned. From Fig. 32, it can be easily understood that all cases provided nice range estimation (shown by the blue

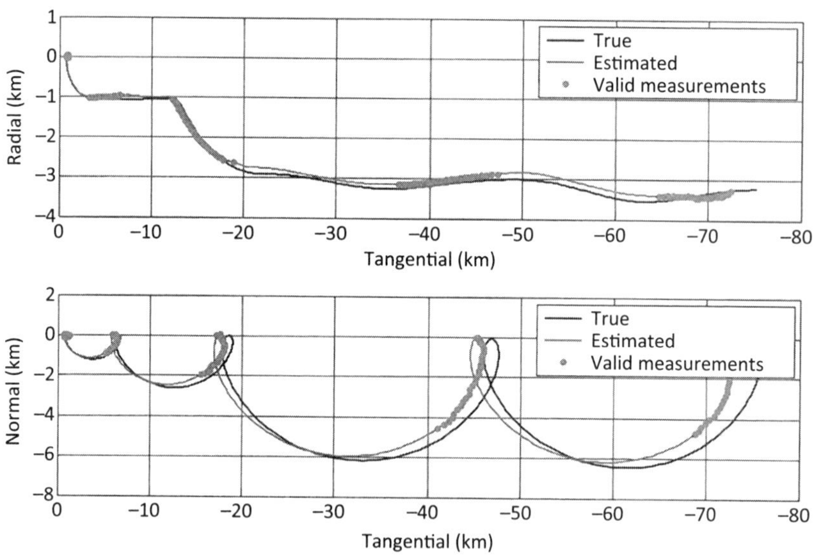

Fig. 31 Approach simulation using only VISCAM: Upper part = dual coelliptic rendezvous, Lower part = stable orbit rendezvous.

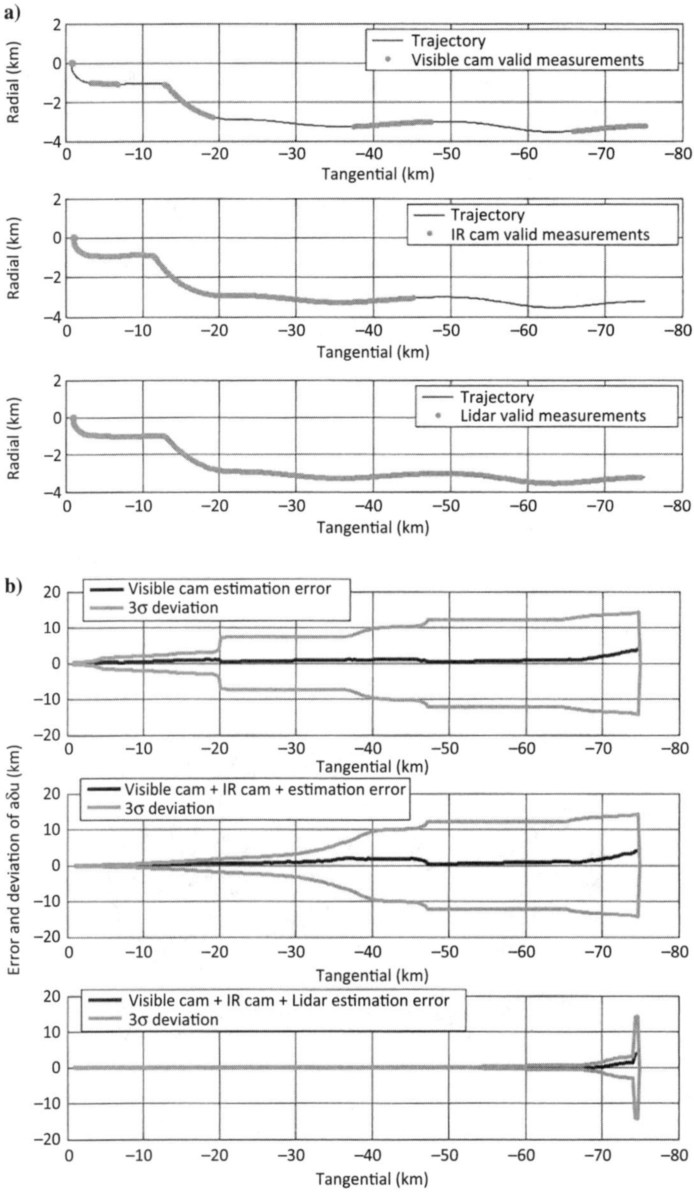

Fig. 32 Rendezvous and approach to the noncooperative target; comparison of sensors: a) range of valid measurement, Upper: VISCAM, Middle: IRCAM+VISCAM, Lower: LIDAR+IRCAM+VISCAM and b) 3σ deviation of range estimation, Upper: VISCAM, Middle: IRCAM+VISCAM, Lower: LIDAR+IRCAM+VISCAM.

lines on Fig. 33b). And the valid measurement (red lines on Fig. 33a) reduced the range estimation error variation with ramped changes of the red lines in Fig. 33b.

From these simulations above, considering the applicability of the object vision technique in a shorter distance than 1 km, it can be deduced that the visual camera can be used exclusively for every step of the approach procedure. But, obviously, adding IR camera and LIDAR would drastically improve the reliability for the range estimation. And the combined use of the three types of navigation sensors might be preferable but affordable only for missions such as multiremoval operation because of the sufficient resources onboard. In case of single-removal operation for instance, the visible camera system would be the most prospective candidate considering the size of the remover.

As the consequence, JAXA has investigated deeply on the image-based perception technologies for each phase at the final operation, whose sequence can be defined as the following.

1. APPROACH PHASE

This phase consists of finding the target and determining its orientation. From dozens of kilometers to the target, the target, which is projected as being one or a few pixels in size in a telescopic camera, is detected for estimating its orientation from the viewpoint of the removal satellite.

a)

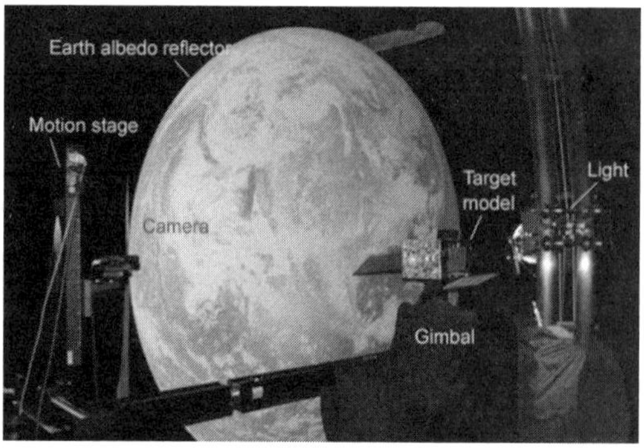

b)

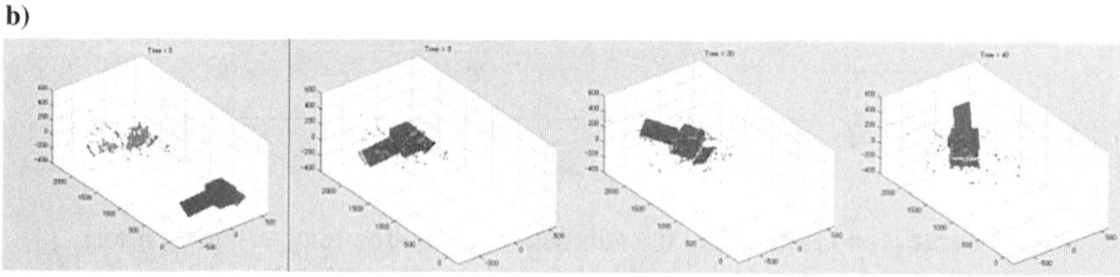

Fig. 33 Rendezvous and motion measurement sensor: a) a miniature target model on the optical simulator and b) model matching simulation with the data acquired.

Course range finding to the target and its shape. Within about 10 km, the target is projected to 10 pixels more on the image. Using a small projected target, the distance to the target and its shape are approximately estimated.

2. OBSERVATION PHASE

This phase consists of a precise 3-D reconstruction of the target and motion estimation. At around 50 m from the target, its shape and the distance from the viewpoint are precisely measured through image-based perception, that is, stereovision. The target motion, such as nutation/tumbling, is estimated using sequential images. For the final approach in the next phase, all perception information of the target should be estimated in this phase.

3. REMOVAL OPERATION PHASE

This phase is dedicated to image-based perception for robotic operations. Until contacting with the target, the visual perception or target tracking is continually executed for a robotic operation, that is, attaching a removal device on the body of the debris. After attaching the device, the performance of the device is monitored in the middle distance.

Considering a little availability of acquiring real images of a target debris, various kinds of mimic images, such as computer graphics, miniature scaled models, and the optical simulator (see Fig. 33a), the target pictures taken on ground and on board are used to investigate the applicability and the performance of various image-based perception algorithms.

In Fig. 33b, the results of the model matching simulation with the data acquired using the optical simulator are plotted: this shows that the attitude and the motion of a target debris can be estimated using simple and low-cost sensors such as stereo cameras, GPS receiver, and star tracker.

C. ANGULAR MOMENTUM REDUCTION

Because of various effects, for example, the residual angular momentum, the orbit motion and the solar pressure, the target debris is likely to have an angular motion such as simple rotation, nutation, and tumbling.

To facilitate the operations to the target, the angular motion should be reduced to some level or be stopped completely.

JAXA has proposed two different devices for this purpose: one is the "*Impact Thruster*," which reduces the angular momentum of the target by applying a series of impacts using small projectiles.

Another is the "*Brush Contactor*" mounted at the end of "*Extensible boom*" (see Fig. 34).

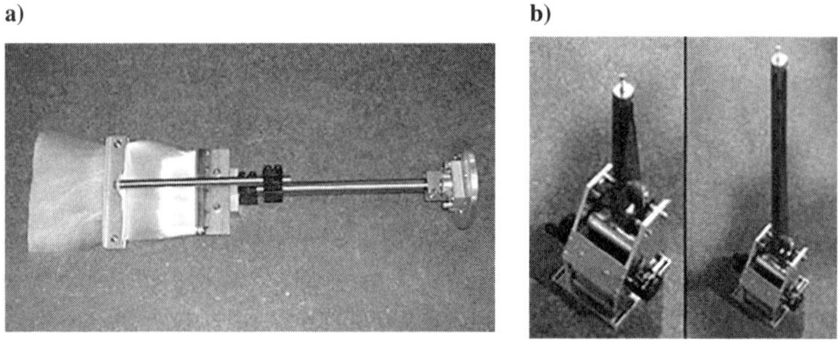

Fig. 34 **A device to control the angular momentum of the debris: a) Brush contactor with b) Extensible boom.**

D. PRINCIPLE OF THE THRUSTER

To deorbit a target debris, it should be decelerated and needs some thrusting device.

JAXA is developing the EDT for this purpose because it is a "*Propellant-free propulsion*" system, which is suitable for this kind of operation thanks to its high efficiency, simplicity, and low cost.

The concept of JAXA's EDT system is represented by a drawing in Fig. 35. Indeed, it consists of a long net type conductive tether with a bare part in one end and an electron emitting device of FEC[‡‡‡‡‡] on the other end.

The orbit motion of the conductive tether in the GEO-magnetic field generates an electro-motive voltage between the tether ends, and the electric current is induced by collecting electrons at the bare part forming the space plasma and emitting them at the another end using the FEC.

The interaction between the current and the GEO-magnetic field generates a "*Lorentz*" force in the retrospective direction for the deceleration of the target.

In fact, numerical simulations found that a large debris of 2 tons in the orbit with 1000 km of altitude and 83 deg of inclination can be reentered within one year using 10 km EDT with less than 1 A electric current.

A simple tether is a very vulnerable structure against debris impacts because it consists of a series of single failure points. To survive until the mission ends in the severe debris environment, JAXA accepted the net tether concept from the Japanese traditional fishing net. Hypervelocity impact test data verified its sufficient survivability until the end of mission.

Their net tether of 10 km consists of three cords, whose one of them is longer than the others, and the space between the cords is substantial for tasks such as larger current collection and survivability against small-sized debris impacts.

[‡‡‡‡‡]Field emission cathode.

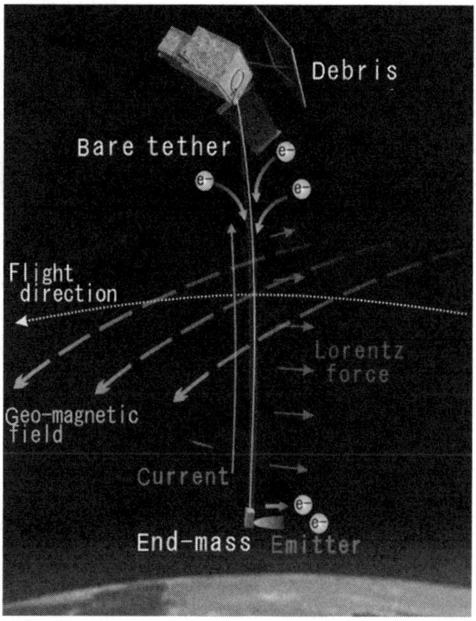

Fig. 35 Physical principle of the EDT concept. (From Ref. [11].)

The EDT needs simple and low-power electron emission device and so JAXA introduced Carbon Nanotube Cathode with 3-mA-emission capability.

Eight of this device are being integrated for KITE§§§§§ mission, which will be described in the following section.

E. JAXA'S ROADMAP FOR DEBRIS REMOVAL

Based on the R&D activity more than 10 years, JAXA recently drew up and updated its roadmap for debris removal (see Fig. 36). Assuming that ADR will be realized around 2030, a step-by-step approach is adopted:

1. The first step is the demonstration of the EDT system
2. The second step is the removal demonstration of one actual debris
3. The third step demonstrates the possibility of a multi-removal mission

The first step will be carried out in the so-called KITE§§§§§ project, whose the main specifications are detailed in Table 3. This mission is currently scheduled in

§§§§§ Konotori integrated tether experiment.

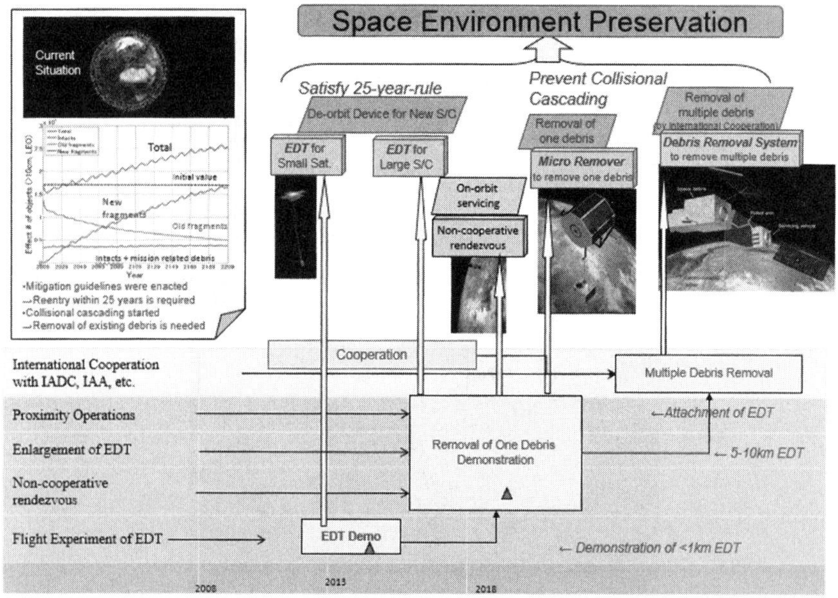

Fig. 36 Overview of the JAXAs roadmap for active debris removal.

late 2016 (planned, see Ref. [12]) and will experiment the EDT concept in HTV-6 platform (see Fig. 37).

VI. CONCLUSION

This chapter provides an understanding of why space debris constitutes a major concern for the space community. Professor Kibe provided the JAXA perspective

TABLE 3 KITE: EDT DEMONSTRATION EXPERIMENT

Launch Date	TBA Late in 2016
Orbit	$\geq$ 20 km below ISS orbit, 300–400 km altitude, 52 deg of inclination
Mission period	About 7 days
Tether length	700 m
Tether current	Max 10 mA
Mission equipment mass	45 kg (End mass 20 kg, On board mass 25 kg)
Required power	182.5 W

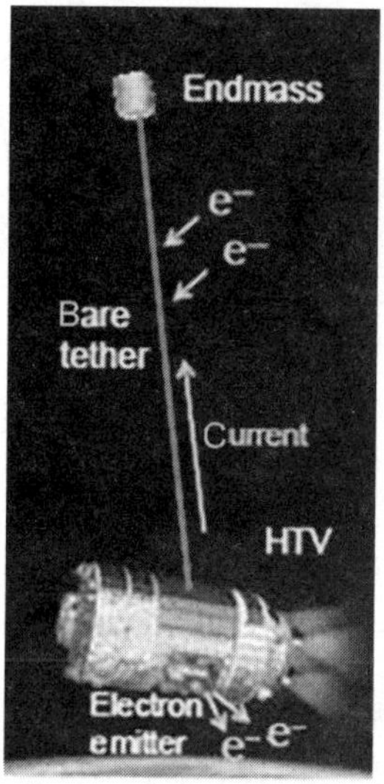

Fig. 37 KITE mission.

on the issue and how his agency is approaching the subject with guidelines and R&D efforts (Sec. II).

The current space debris situation and possible future scenarios are described in Sec. IV. It becomes clear that debris mitigation measures such as post mission disposal should consequently be implemented today to moderate future debris population growth. The more aggressive approach of active debris removal might not be urgent and pressing at this point but will be crucial in the future to address the space debris problem. To this end, aspects for target selection such as orbital region, object attributes, desired removal effect, multi- or single target mission, and political and legal considerations make this problem multidimensional. The research and discussion that enables a sensible and efficient future approach to the space debris problem needs to be promoted and intensified today. This is necessary because many questions, technical and nontechnical, need further elaboration to satisfy a holistic approach in the future.

In parallel with other IADC members, JAXA contributes actively to space debris research activities such as the space debris mitigation measures presented in Sec. III. JAXA has studied different concepts for removing space debris as shown in Sec. V. The example of the Electro Dynamic Tether concept is described in more detail and its operation is explained. JAXA plans to demonstrate the EDT concept in KITE[§§§§§] mission that is planned to be launched in late 2016 on the Japanese HTV-6.

ACKNOWLEDGMENTS

Discussions with H. Lewis from the University of Southampton, provided additional perspectives on the issue of space debris. The authors acknowledge this contribution to the quality of the chapter.

REFERENCES

[1] IADC, Liou, J.-C., Rossi, A., Krag, H., Raj, M. X. J., Anilkumar, A. K., Hanada, T., and Lewis, H. G., "Stability of the Future LEO Environment," Technical Report Jan., IADC, 2013.

[2] Nakajima, A., Yanagisawa, T., and Kurosaki, H., "The Stacking Method, the Technique to Detect Small Size of GEO Debris and Asteroids," Technical Report Feb., Japan Aerospace Exploration Agency, 2008.

[3] White, A. E., and Lewis, H. G., "The Many Futures of Active Debris Removal," *Acta Astronautica*, Vol. 95, Feb. 2014, pp. 189–197.

[4] Liou, J.-C., "An Active Debris Removal Parametric Study for LEO Environment Remediation," *Advances in Space Research*, Vol. 47, June 2011, No. 11, pp. 1865–1876.

[5] Liou, J.-C., Anilkumar, A. K., Bastida Virgili, B., Hanada, T., Krag, H., Lewis, H. G., Raj, M. X. J., Rao, M. M., Rossi, A., and Sharma, R. K., "Stability of the Future LEO Environment an IADC Comparison Study," *6th European Conference on Space Debris*, May, Darmstadt, Germany, 2013.

[6] Wertz, J. R., Everett, D. F., and Puschell, J. J., *Space Mission Engineering: The New SMAD*, Microcosm Press, Hawthorne, CA, 2011.

[7] Lewis, H. G., Swinerd, G., and Newland, R. J., "The Space Debris Environment: Future Evolution," *Aeronautical Journal*, Vol. 115, No. 1166, April 2011.

[8] Kelso, T. S., and Alfano, S., "Satellite Orbital Conjunction Reports Assessing Threatening Encounters in Space (Socrates)," Technical Report 1, Center for Space Standards & Innovation, Colorado Springs, CO, USA, May 2006.

[9] Yakovlev, M., Makarov, Y., Nazarenko, A., Raykunov, G., Trushlyakov, V., Gorobets, D., and Kudentsov, V., "Prediction of Near-Earth Space Debris Population and Future Space Object Disposal Measures," *International Astronautical Congress*, 2011.

[10] Braun, V., Lüpken, A., Flegel, S., Gelhaus, J., Möckel, M., Kebschull, C., Wiedemann, C., and Vörsmann, P., "Active Debris Removal of Multiple Priority Targets," *Advances in Space Research*, Vol. 51, No. 9, May 2013, pp. 1638–1648.

[11] Klinkrad, H., and Johnson, N., "Space Debris Environment Remediation," *Technical Report*, International Academy of Astronautics, Paris, France, 2011.

[12] Ohkawa, Y., Kawamoto, S., Okumura, T., Iki, K., Horikawa, Y., Kawashima, K., Miura, Y., Takai, M., Washiya, M., Kawasaki, O., Tsujita, D., Kasai, T., Uematsu, H., and Inoue, K., "Current Development Status of Electrodynamic Tether Experiment on HTV," *Proceedings of the 6th Space Debris Workshop*, Vol. JAXA-SP-14-013, 2015, pp. 237–245.

CHAPTER 13

Lasers for Asteroid and Debris Deflection

Marko Jankovic*
DFKI GmbH, Bremen, Germany

Nicolas Thiry†
University of Strathclyde, Glasgow, United Kingdom

Pierre Bourdon‡
ONERA—The French Aeropace Lab, Palaiseau, France

I. INTRODUCTION

Suddenly a light from hell appeared in the middle of the ruby. Then, from the end of a cylinder, a hundred thousand times brighter than the sun, burst forth a thin red light, a perfectly parallel monochromatic beam (...) Theodore Maiman and his assistants were silent for some time, enthralled by the beauty of this spectacle, the like of which no-one had ever seen before. "Einstein was right" he murmured, "light can be concentrated and coherent".

These were the words used by a journalist, upon the first demonstration of the working laser performed by T. Maiman in 1960 at the Hughes Laboratory in Malibu, California, U.S.A. They perfectly reflect the enthusiasm with which the general public received the materialization of a myth. Myth that has been in the collective imagination since ancient times, even before the concept of stimulated emission was predicted by Einstein in 1917. There was always something mysterious and magical that was related with the light. The science fiction has envisioned these "death rays" capable of grand and mysterious powers long before their first demonstration. H. G. Wells in his book "The War of the Worlds," in 1898, described a weapon used by the Martian fighting-machines (see Fig. 1) with properties remarkably similar to those of modern day lasers [1]:

...silent jet of light that knocked down everything it touched, and when this invisible line passed over them, even the pine trees and the bushes burst into flame with a deafening sound....

*Robotics Innovation Center.
†Department of Mechanical and Aerospace Engineering.
‡Theoretical and Applied Optics Department.

Fig. 1 Artwork for the H. G. Wells' book *The War of the Worlds* of a 1906 Belgian edition by the Brazilian artist Henrique Alvim Corréa. The image shows Martian fighting-machines destroying a town in England. (From Wikimedia Commons.)

And so it is of no surprise that the first working laser device has excited the public imagination to the point to make it a standard weapon, a tool in all future-based stories and movies. However, at the time the technology behind a laser device wasn't mature enough to make that science fiction dreams a reality.

In the 1930s, scientists already had all the necessary tools to build a laser. Theoretical knowledge was already there as well as the optics, but it was not until early 1950s that these two were used together to produce a coherent beam of microwave radiation, bringing to light a MASER, "Microwave Amplifier by Stimulated Emission of Radiation." This was the first step toward the laser. The first working prototype of a maser was developed in 1954 by H. Zeiger and J. P. Gordon based on the theoretical work of C. H. Townes. The emitted radiation from this device had a single frequency, however, its wavelengths had a size comparable to the dimensions of the device itself. And so, at the time, it wasn't easy to think of an extension of a maser to an optical-maser capable of emitting radiation in the optical region of the electromagnetic spectrum, which is five orders of magnitude smaller. Nevertheless, in 1958, with the publication of a paper written by C. H. Townes and A. Schawlow containing their ideas on how to extend the maser

Fig. 2 First ruby laser and its inventor Theodore Maiman. (From Wikimedia Commons.)

concept to optical frequencies, everything was set up for the development of the first LASER, expression coined in 1957 by G. Gould for the "yet-to-be-invented" device. Finally, on 16 May 1960 T. Maiman and his assistant I. d'Haenens, at Hughes Research Laboratories, observed pulses of red light coming out of a ruby crystal, used as an amplifier, fitted inside a powerful flashlamp, used as an energy source (see Fig. 2). After this discovery, other teams that were working on similar concepts, quickly developed their own prototypes and even new types of lasers. By the end of 1960, three different types of laser were developed and successfully demonstrated [2, 3].

Nevertheless, the laser technology at the time was just a "toy" for scientists and nobody could come up with a feasible idea on how to use this device. Of course, by 1965, people already envisioned useful purposes for it, for example in surgery instead of a scalpel, or in telecommunications for substituting the telephone network of the time, or even to store information. However, the technology wasn't ready yet. Moreover, the invention of the laser, unlike many other technologies developed in the twentieth century, didn't happen because of an actual need for it. To put it in the words of Pierre Aigran (secretary of state for research, researcher, and member of the Academy of Sciences): "We are used to having a problem and looking for a solution. In the case of the laser, we already have the solution, we just have to find the problem" [1]. Despite all this, its development continued inspired by the science fiction authors who envisioned the omnipresent usage of the laser technology in the upcoming future.

Today, more than 50 years after the first red pulses coming out of a ruby crystal, it seems almost impossible to envision modern life without lasers. In fact, although not every application envisioned in the late 1960s came to life, lasers widespread usage nowadays has earned them a title of one of the greatest inventions of the twentieth century. Today they are used in: communications, entertainment, commerce, industry, medicine, military, and science [1, 2]. Space sector, on the other hand, has yet to see a widespread usage of the laser technology although, its first application was imagined, onboard mighty spacecrafts such as Enterprise from the popular Sci-Fi franchise, Star Trek.

Nevertheless, in recent years, technological advancements and the renewed interest in lasers have enabled a wide range of space-based applications that were previously considered impracticable due to various factors such as costs, complexity, and efficiency of the system. Some of those applications include: communications, remote sensing, power transmission, propulsion, deflection of near-earth objects (NEOs), and de-orbiting of space debris. However, in order to materialize those concepts, further research and improvements of the laser beam quality, efficiency, and pointing accuracy are needed [4].

II. HIGH-POWER LASERS: AN OVERVIEW OF AVAILABLE TECHNOLOGIES

A. LASER FUNDAMENTALS

The word laser is an acronym for *Light Amplification by Stimulated Emission of Radiation* which describes, in short, the working principle of a laser device. Nevertheless, the acronym is widely used to indicate a device that generates a monochromatic, collimated, and coherent beam of light using the principle previously stated. Moreover, with the word laser, one usually means laser oscillator (more common type of device) but it can also intend laser amplifier. The distinction between the two types of lasers will be made in the following section.

1. LASER DESIGN

A laser, in its most simple form, consists of [1]:

1. *An amplifying medium*: necessary to amplify the light that travels through it by exploiting the process of *stimulated emission.* It can be solid, liquid, or gaseous.
2. *A power source*: necessary to provide ("pump") the energy to the amplifying medium in order to create right conditions for the light amplification. It can be optical (e. g., the Sun, flash lamps, diodes, or even other lasers), electrical (gas discharge tubes and electric current in semi-conductors), or even chemical (chemical reactions).
3. *An optical resonator*: required to amplify the light source making the process of stimulated emission predominant over the process of spontaneous emission within the amplifying medium. In its simplest form, it consists of two parallel mirrors between which the light can bounce back and forth augmenting the probability of stimulated emission to occur before it can leave the cavity through one of the mirrors that is not completely reflective.

Note, however, that in order to have a device capable of amplifying an existing light source, only the first two components are needed. This device is known as a laser amplifier. If we also include into the device an optical resonator, we obtain a laser oscillator. This kind of device is more common because it is the first component required to deliver highly coherent light beams [1].

Figure 3 illustrates the basic design of a laser, and in particular of a ruby laser. The amplifying medium (i. e., the ruby rod) is placed inside the optical resonator

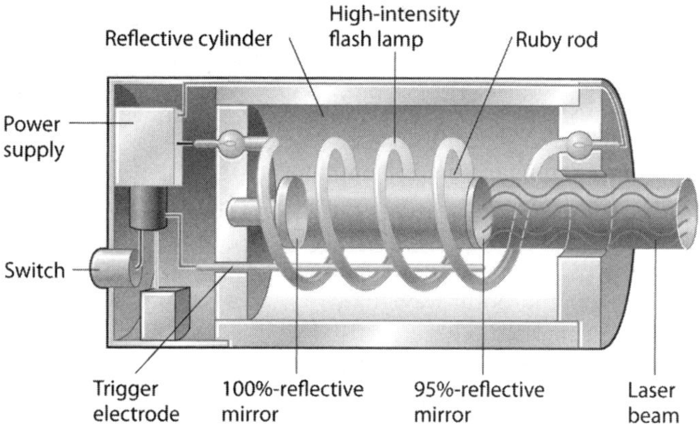

Fig. 3 Internal view of a ruby laser. (From Principles of General Chemistry v. 1.0 CC BY-NC-SA 3.0.)

(consisting of two parallel mirrors) and the energy source (i. e., the high-intensity flash lamp) is coiled around the ruby crystal. The latter is made of crystalline form of aluminum oxide in which some of the aluminum atoms have been replaced by chromium atoms. Upon the activation of the energy source, one part of those atoms will absorb the injected electromagnetic radiation (in the form of a flux of photons) and make the transition form their ground state to the excited state. However, after a short period of time those atoms will spontaneously return to their initial state emitting photons that will be bounced back and forth through the laser medium by the optical resonator. This will augment the probability of achieving the stimulated emission of photons and eventually, when the amplification of light has reached a certain level, a flux of photons will exit the laser cavity through the partially reflective mirror in a form of a characteristic red beam [5].

2. LASER PHYSICS

To comprehend better the operation of a laser, let us now consider the laser physics by analyzing the words composing the acronym laser (i. e., light and stimulated emission).

The term *light* is generally defined as an electromagnetic radiation having the range of wavelengths that goes from 1 nm to 1000 μm (i. e., 1 mm). Where $1\ \text{nm} = 10^{-9}$ m whereas $1\ \mu\text{m} = 10^{-6}$ m [6].

Within this range, several spectral domains can be defined. The visible spectrum goes from 400 nm to 700 nm and corresponds to the range of sensitivity of the human eye. The near infrared (NIR) spreads from 700 nm to 10 μm and anything beyond this is in the far-infrared (FIR) region. On the other hand, the

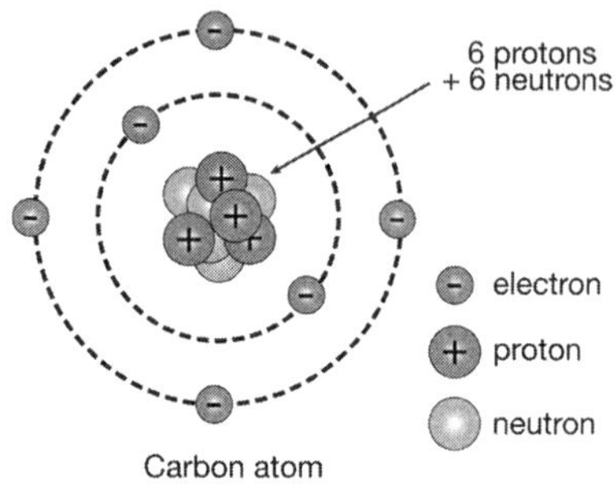

Fig. 4 Schematic representation of a carbon atom.

spectral domain between 200 nm and 400 nm is considered as the ultraviolet radiation whereas anything below 200 nm is in the deep ultraviolet region (DUV) [6].

Furthermore, light can be considered as a flux of discrete particles called photons, which exhibit both wave-like and particle-like properties. These particles have a constant speed in vacuum (the speed of light), $c = 2.998 \times 10^8$ m/s, and their energy is given by

$$E = h\nu$$

where ν is the wave frequency and h is Planck's constant ($h = 6.62 \times 10^{-34}$ J s).

To understand the process of *stimulated emission*, we first need to describe the structure of matter and its interactions with light.

The matter that surrounds us is made of atoms. Each atom is made of a nucleus and a cloud of orbiting electrons. A schematic representation of a carbon atom is visible in Fig. 4. This representation is the Bohr's model of an atom where each electron orbiting the nucleus is characterized by a certain energy level. Electrons are allowed to jump form one energy level ("orbit") to another only if a certain amount of energy is provided to the atom. They cannot stay between levels. The allowed energy states are defined as "quantum" states and are represented by "quantum numbers." It is worth noting, however, that this representation of an atom is not the most accurate one but is sufficient to describe in a simple manner the operating principle of a laser [5,6].

Interactions between matter and light can be split into three mechanisms: spontaneous absorption, spontaneous emission, and stimulated emission. To understand those mechanisms, consider an atom represented in Fig. 5, having just one electron orbiting a nucleus. The electron can, under certain circumstances,

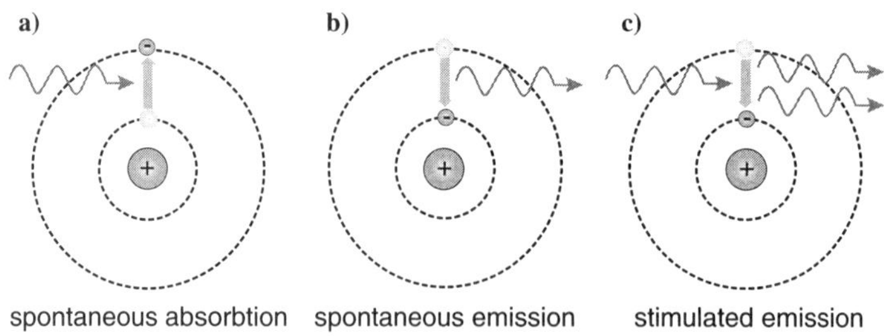

Fig. 5 Electron transitions in an atom.

transit only between two quantum states: the ground state (unexcited) E_1, and the higher (excited) state E_2. In Fig. 5, E_1 is the energy of the orbit closest to the nucleus whereas E_2 corresponds to a further away orbit.

For an electron to jump from its ground state to its higher energy state, it needs to absorb an incident photon that has an energy equal to the energetic gap between the two quantum states ($\Delta E = E_2 - E_1$). If this is the case, *spontaneous absorption* (Fig. 5a) of the incident light occurs and the electron jumps from E_1 to E_2. On the other hand, if the energy of the incident photon is too small or too big, the photon will not be absorbed by that particular atom. Macroscopically, the phenomenon of absorption is observable as an attenuation of the incident light after passing through the material [5–7].

Once absorbed, the electron will not stay in this excited state for ever. After a short period of time, 10^{-7} s, the electron will decay to a metastable level where it will stay for a longer period of time, around 10^{-3} s, before decaying permanently to its lowest quantum state, the ground state. In doing so, the electron will emit a photon with completely different properties from the absorbed one. This process is called *spontaneous emission* and Fig. 5b illustrates this phenomenon. Macroscopically this is observable in a material as an emission of luminescence [5,7].

On the other hand, the transition of an electron from its excited state to its ground state can be also achieved by interaction of the latter with a photon having energy equal to ΔE. This way the electron is stimulated to make the transition and in doing so, it will emit a photon having the exact same energy, direction, and phase as the incident photon. This process is called *stimulated emission.* Macroscopically, this phenomenon is observable as an exponential increase of the light intensity within the material [6,7].

Before proceeding, it is deemed necessary to define the meaning of the ground state atom and excited atom. An atom is defined at its ground state when all of its electrons are at their lowest possible energy level. If one or more electrons are at a higher energy level, than the atom is defined as in an excited state [5]. In a material, an atom at its ground state will be a part of the population, N_1, with the lower energy level whereas the excited ones will form the population, N_2, having higher energy level.

To obtain a laser medium, obviously, the stimulated emission has to be dominant over the other two mechanisms. However, this is not an easy task, given that in a material, for a radiative transition, the three mechanisms are always present at the same time. An incident photon has the same probability of being absorbed by a ground-state electron as of being amplified by an excited-state electron. Moreover, in a thermal equilibrium the Boltzmann's distribution indicates that a population of atoms with lower level of energy will always be larger than the one with the higher level of energy. Thus, in thermal equilibrium, any collection of atoms will constitute an absorber. To overcome this problem a condition of "population inversion" needs to be achieved in a medium so that the population N_2 is always larger than the population N_1. In other words, if this condition is met,

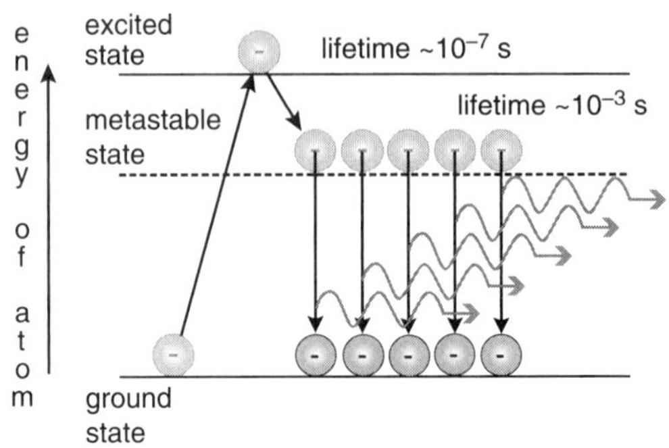

Fig. 6 Schematic representation of population transitions.

the number of atoms transiting form the metastable level to the ground state will always be bigger than those transiting the other way around (see Fig. 6). This condition can be met by pumping energy into the medium by means of an external energy source [1,5,6].

However, with the population inversion only the problem of absorption can be solved. The spontaneous emission will always tend to naturally lower the population N_2 by emitting photons in all directions. To solve this problem, the medium needs to be permeated with light and one way to achieve this is to put the medium in an optical cavity. By doing so, the laser cavity will at first act as a filter by retaining only the desired (on-axis) photons. Then, it will reflect those photons back and forth through the medium augmenting considerably the probabilities for the stimulated emission to occur. Thus, upon the activation of an energy source an amplifying medium will act at first as a "lamp." Photons will be emitted in all directions and only small amount of them will have the direction of propagation parallel to the axis of the laser cavity. Those, however, will be reflected by the mirrors back and forth through the amplifying medium, increasing the amount of light within the cavity. As a consequence, more and more atoms will be stimulated to release photons increasing the probability of the stimulated emission and making this process predominant along the axis of the cavity. Finally, one part of the photons will exit the cavity, creating a laser beam [1,6].

3. LASER BEAM PROPERTIES

Lasers can be seen as just another source of light, like candles or light bulbs. However, the light coming out of them has some unique and interesting properties that make it suitable for some specific applications described in more detail in the following pages. Unlike the light generated by generic light sources irradiating equally in all directions, the laser light can be concentrated in a really narrow beam making it possible to achieve extremely high levels of beam peak power (up to

10^{20} W). To put it into context, if one could make a laser out of a candle by concentrating its light into one single beam that has the dimension of a human eye pupil (around 3 mm in diameter) and direct it toward a target that is at one meter distance, the intensity of the light that would reach the target would be one million times higher than in the case of the normal irradiated light [3].

Main properties of a laser light (summarized in Fig. 7) are [5, 8]

1. *Narrow linewidth*: light that comes out of a laser device has narrow spectral bandwidth[§] which in case of a laser emitting in the visible spectrum means that the beam will be monochromatic (e.g. red, green, or blue). On the other hand, generic sources of light, such as a light bulb or a candle, emit light over a wide range of wavelengths of the electromagnetic spectrum, which is why we tend to perceive it as having a white color.
2. *Coherence*: laser light is composed of photons having all the same phase, which results in a good ability of a laser beam to produce interference patterns. An unwanted effect of this property, however, is a tendency of laser beam to form speckle patterns.
3. *Directivity*: laser light is generally emitted in the form of a beam that has a well-defined direction and a very low divergence, meaning that the radius of a laser beam does not change significantly during a medium propagation distance. This is due to the angular selectivity of the oscillator cavity and collinear photons produced during the stimulated emission.

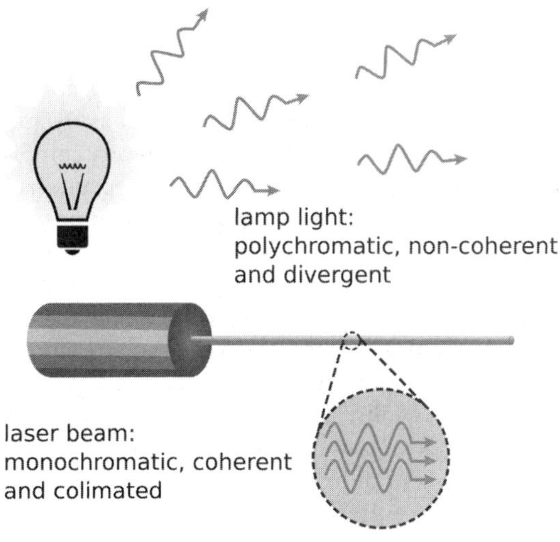

Fig. 7 Laser properties.

[§]This statement isn't true for lasers emitting trains of ultrashort pulses which can have large optical bandwidth.

4. LASER SPECTRAL DOMAIN

The laser spectral domain of currently existing lasers covers a range of wavelengths of almost six order of magnitude, from the FIR region to the soft-x-rays region (i. e., from 1000 μm to 3 nm) [3].

The choice of one wavelength over another is dictated by the required distance of propagation of a laser beam and the transmittance of the atmosphere in that particular wavelength band. These two factors will determine which laser to use in order to obtain minimum absorption of the beam during its propagation through the atmosphere. Ideally, one should choose the laser capable of emitting wavelengths within the atmospheric transmittance "windows" where little or no absorption of the propagating radiation occurs.

5. LASER TYPES

Since the first ruby laser, thousands of different lasers were discovered, using all possible means of pumping sources and materials for the lasing medium. To put it in the words of A. Schawlow: "Anything will lase if you hit it hard enough." Nevertheless, only a small amount of them has nowadays found practical applications [6].

Common types of lasers, classified according to the nature of the lasing medium, are semiconductor lasers, solid-state lasers, fiber lasers, dye lasers, and gas lasers [1, 8].

Semiconductor lasers (mostly diode lasers) have a semiconductor material as an amplifying medium and the stimulated emission is achieved by pumping electrical current inside the semiconductor. In the case of diode lasers, the electrical current is pumped in a region where the n-doped and p-doped materials are in contact. They exhibit very high efficiencies¶ (around 60%) and can reach very high output power (several kilowatts), although being very compact. Furthermore, a pulsed output can also be achieved with very high repetition rate. The covered spectral domain goes from near ultraviolet to NIR depending on the material used for the lasing medium. The only disadvantage is given by the poor beam quality** that, however, can be improved by lowering the output power of the laser [1, 8].

Solid-state lasers have ion-doped matrices as an amplifying medium and either diode lasers or discharge lamps as external energy sources. They are characterized by being extremely compact and robust. They can achieve very high output power or alternatively, high beam quality at the expense of the output power. Nevertheless, it is to be noted that at high power, heating, and dilatation of the laser medium can lead to strong impairment of beam quality. Ultra short pulse durations (i. e., of order of picoseconds or even femtoseconds) of the beam can be achieved.

¶ Defined as the ratio between the optical power emitted and the electrical power consumed.

** Defined as "a measure of how tightly a laser beam can be focused under certain conditions (e.g., with a limited beam divergence)" [8].

The covered wavelengths are in the red and NIR range of the electromagnetic spectrum [1,8].

Fiber lasers are a special type of solid-state lasers based on an ion-doped optical glass fiber. They have all the advantages of the solid-state lasers and less of their disadvantages, which means that they can achieve very high output power (up to several kilowatts) while retaining high-beam quality. However, thermal effects are also present in these lasers and can still limit their output power [8].

Dye lasers have organic molecules in a liquid solution[††] as an amplifying medium and a flash lamp or another laser as an energy source. They exhibit broad wavelength tunability and also the possibility of generating ultrashort pulses. The covered wavelengths include the whole visible spectrum, the UV and the near IR regions. Nevertheless, they have not seen a broad range of applications due to their low efficiency and difficulties to prevent the degradation of the amplifying medium during the operation of the laser.

Gas lasers have a gas as a lasing medium whereas the population inversion is usually achieved through an electrical discharge. The most used gases are carbon dioxide, argon, and krypton. *Excimer lasers* are a special type of gas lasers using an appropriate mixture of gases as the gain medium, which under the right conditions transforms to molecules called excimers responsible for ultraviolet and near infrared light output. Commonly used excimers are argon fluoride, krypton fluoride, xenon fluoride, and fluorine. Gas lasers are capable of emitting high-quality beams with very narrow spectral lines going from the UV to the FIR. They do not offer very high efficiencies (apart from carbon dioxide lasers) although they can achieve relatively high optical powers and offer wavelengths difficult to achieve otherwise [1,8].

Other less common types of lasers worth mentioning are chemical and nuclear pumped lasers, free electron lasers, and x-ray lasers [8].

6. LASER OPERATION MODES

Another distinction of the lasers described in Sec. II.A.5 can be made based on their operation mode. Two possible operation modes of lasers are the continuous-wave (cw) mode and the pulsed mode. The usage of one over the other depends on whether the output power of the laser beam needs to be continuously emitted or not. Differences in applying one mode or the other and examples of their applications can be found further on. Typical time scales used when talking about laser operation modes are the following [9]:

1. *Millisecond*: 0.001 s which is the time between two wing flaps of a housefly
2. *Microsecond*: 10^{-6} s, which is the duration of a high-speed, strobe light flash
3. *Nanosecond*: 10^{-9} s, which is the duration of molecules fluorescence

---

[††] Usually some kind of alcohol solution.

4. *Picosecond*: 10^{-12} s, which is the switching time of the world's fastest transistor
5. *Femtosecond*: 10^{-15} s, which is the pulse width of the world's fastest laser
6. *Attosecond*: 10^{-18} s, which is the shortest time measurable by a device

Continuous-wave (cw) operation mode generates, in time, a continuous emission of a laser beam. This also implies that the pumping source is constantly injecting energy into the lasing medium during the operation of the device. With this mode, very narrow linewidths of the output radiation can be achieved, typically of only a few kilohertz. The first cw laser was a helium–neon laser emitting a 1153 nm beam. Nowadays, different types of lasers can be operated in cw mode. The most common fall into the following categories: gas lasers, solid-state lasers, and dye lasers [8].

Pulsed mode, on the contrary, emits radiation in the form of short pulses repeating at a given rate. With this method very high pulse energies[‡‡] and peak powers can be obtained, whereas the optical bandwidth of the pulses tends to be wider compared to the cw mode.[§§] The generation of pulsed beams can be achieved by using a cw laser and a fast shutter that opens and closes at fixed intervals. However, this method, although intuitive, suffers from efficiency issues and the duration of the pulses would then be limited by the velocity of the shutter (around 1 ms for the fastest mechanical shutter). Thus, to obtain high energy pulses having very short duration, the following methods are commonly used [1,8]:

1. *Gain switching*: This method (usually employed for semiconductor lasers) uses the pump source to quickly change the gain of the lasing medium in order to obtain pulses that are from a few nanoseconds to a few picoseconds long and have energies around a few picojoules.
2. *Q-switching*: This method uses the modulation of the intracavity losses (i. e., the quality factor of the laser cavity) to obtain very short pulses (in the nanosecond range) with repetition rates of up to a few kilohertz and energies of a few millijoules or more. The peak power that can be achieved with Q-switched lasers can be a few orders of magnitudes higher than the power obtainable with cw lasers (in the megawatt range or even higher).
3. *Mode-locking*: This method (or better a series of methods) uses an active element (an optical modulator) or a nonlinear passive element (a saturable absorber), placed within the optical cavity in order to obtain ultrashort pulses (in the range of a few pico or even femtoseconds) having repetition rates of up to a few gigahertz and pulse energies that range from a few picojoules to a few nanojoules. The average power that can be achieved with mode-locked

---

[‡‡]The pulse energy is considered to be "equal to the integral of its optical power over time" [8].

[§§]The shorter the pulse the broader the spectrum. It is one instance of the Heisenberg principle.

laser oscillators is of the same order of magnitude as that obtainable with cw lasers. Femtosecond mode-locking laser amplifiers can reach the peak power in the terawatt range, with up to 100 mJ per 10 fs pulse.

B. APPLICATIONS OF HIGH-POWER LASERS

High-power lasers are yet another category of lasers having very high output powers. However, it must be noted that the true meaning of the attribute "high-power" depends greatly on the context in which the laser device is employed. This is due to the fact that there is not a commonly accepted definition for this property. For example, if a high-power laser is used in manufacturing, it means that the device should be able to generate a beam of up to a few kilowatts whereas the same attribute given to a device used for laser displays indicates that the beam power will be up to only tens of watts [8].

This particular attribute can also be used to differentiate a new laser technology (e.g., a "high-power" laser pointer) from a regular one (ordinary laser pointer). Moreover, if we take into account pulsed lasers, the "high-power" property could refer at the same time to several quantities. Nevertheless, whenever the attribute "high-power" is used for pulsed lasers, it is common to intend that devices should provide high average output power [8].

According to the encyclopedia of laser physics and technology [8], common applications of high-power lasers are

1. Manufacturing (e.g., welding, cutting, drilling, milling, engraving, etc.)
2. Military applications (e.g., anti-missile systems, laser weapons, etc.)
3. Large-scale laser displays
4. Remote sensing (e.g., LIDAR)
5. Medical applications (e.g., bloodless surgery, eye treatment)
6. Science (e.g., particle acceleration)
7. Laser-induced nuclear fusion

For the purposes of the following chapter, only the manufacturing and military applications will be described in more details. Information about the other applications can be found in the aforementioned encyclopedia (i.e., in Ref. [8]).

1. MANUFACTURING

Properties of laser beams, such as the high spatial coherence and potential to generate very short pulses, were and still are the drivers for a wide adoption of lasers in the manufacturing industry. Nowadays, almost every object that we daily come in contact with, has to some degree been processed by a laser device.

Laser welding (see Fig. 8), cutting, drilling (see Fig. 9), surface treatment, marking, engraving, micromachining (see Fig. 10), etc., are just some of the

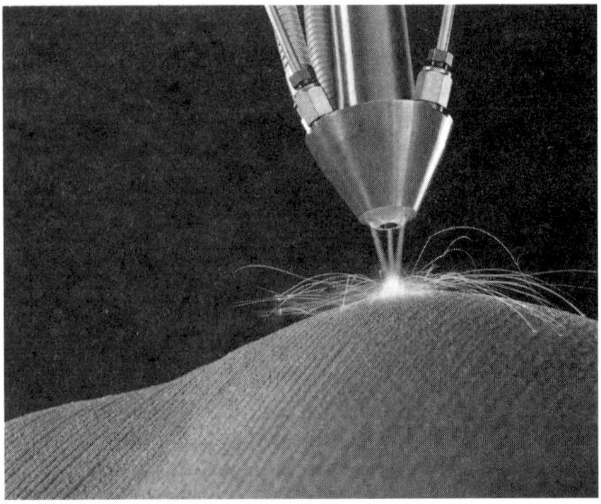

Fig. 8 Laser welding. (Credit: TRUMPF GmbH & Co. KG.)

main industrial applications dominated by laser devices. Compared to mechanical approaches, lasers offer unprecedented advantages being able to drill or cut through almost any type of material with ease, whereas not applying to it mechanical or thermal stresses. Moreover, taking the advantage of the possibility to concentrate high optical powers (up to 10 kW[¶¶]) to very small spots (50 μm–100 μm

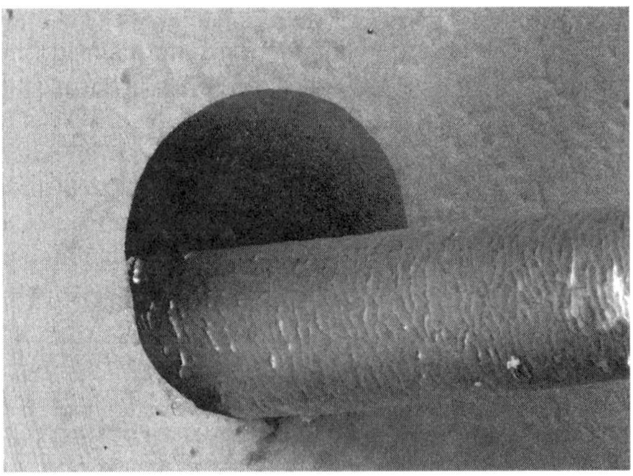

Fig. 9 Microhole drilled using ultrashort laser pulses. (Credit: Bosch GmbH.)

[¶¶] For continuous-wave lasers.

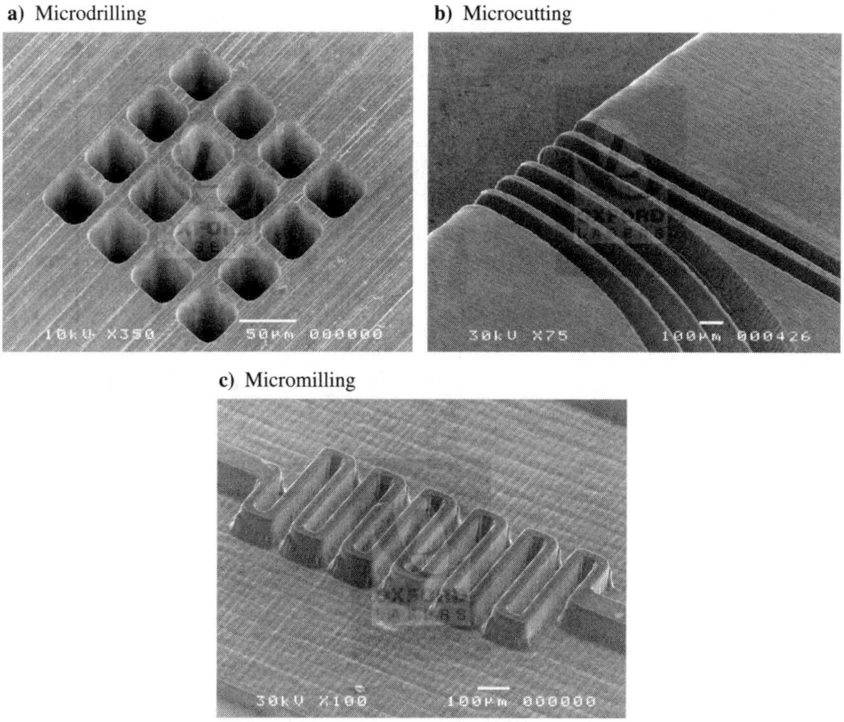

Fig. 10 Micromachining using lasers. (Credit: Oxford Lasers.)

in diameter), microscale structures can be obtained through athermal ablation. This process causes instantaneous removal of the desired portion of the material without affecting the surrounding substrate. Figure 9 illustrates the precision that can be obtained with this kind of method. For size comparison, a human hair, of a thickness of approximately 100 μm, can be seen inside the microhole. Figure 10 is also one of the examples of what can be done with high-power lasers without even touching the material [8].

2. MILITARY APPLICATIONS

Laser devices of all powers and sizes have been employed by the military for decades and their applications vary from target designation to communications. However, the usage of high-power lasers as directed energy weapons has only partially became a reality. To date, the only high-power laser device that has been employed in a battlefield is the ZEUS-HLONS, visible in Fig. 11. This 10 kW solid-state laser is mounted onto a Humvee (HMMWV) vehicle and its purpose is to neutralize roadside bombs from a distance of 300 m [10].

Other more futuristic devices, despite decades of development and appealing capabilities, have still to overcome the prototyping phase. Nevertheless, they can give us glimpses into the future of military applications of high-power lasers.

Much of these devices are developed to generate continuous-wave beams capable of burning through the skin and/or critical components of incoming missiles, drones, and artillery shells. Thus, the expected damage to targets is to be achieved mainly through the thermal effects: drilling, explosive deflagration, inflammation, etc. [11]. Figure 12 illustrates one of those devices currently being tested by the U.S. Navy.

The power threshold of 100 kW has been identified as necessary to consider a laser as "weapon grade" although useful military effects could be obtained also with powers of only 25 kW to 50 kW, if coupled with the high-quality of the beam [12]. Main difficulties of these systems are represented by atmospheric turbulence, beam quality, beam power, and size of the laser device itself.

C. COMPARISON OF LASER TECHNOLOGIES

In this subsection different laser technologies, shortly described in Sec. II.A.5, will be analyzed in more details in order to determine their advantages and disadvantages when used to generate high optical powers. However, before doing this, we

Fig. 11 ZEUS-HLONS engaging unexploded ordinance. (From Wikimedia Commons.)

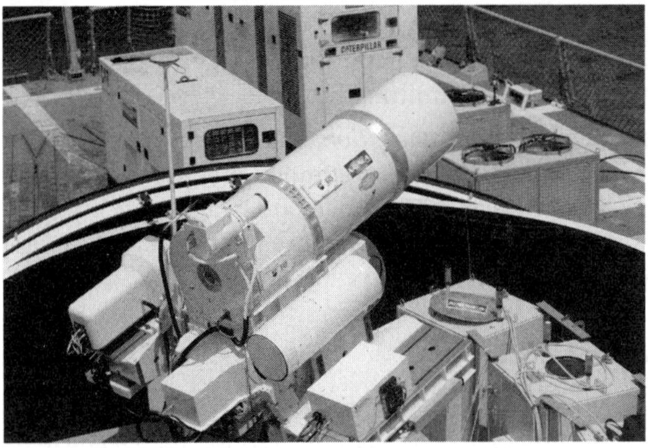

Fig. 12 Laser Weapon System (LaWS) installed aboard the guided-missile destroyer USS Dewey (DDG 105). (Credit: U.S. Navy.)

must first understand the technical challenges that arise whenever high output power is requested from a laser device. The most common ones are [8]:

1. *Powerful energy source(s)*: to generate very powerful laser beam, obviously, a powerful energy source is needed. This latter can consist of either just one source or several of them. Moreover, another laser can be used as an energy source of a high-power laser. In fact, to date, the highest output powers in cw mode are only achievable with diode-pumped lasers.
2. *High wall-plug efficiency*: this parameter is especially important if a laser is operated for extended periods of time in continuous-wave mode. Its high values are very difficult to reach due to variety of factors, thus a trade-off must be made.
3. *Limited thermal effects*: heat generation in a laser is inevitable. This is true, even if the laser medium is fairly efficient. Thus, to avoid degrading beam qualities, efficiency losses, or even damage to the lasing medium, thermal energy of the laser must be limited. This can be achieved through an efficient thermal management and cooling of the device during its operation.
4. *Limited nonlinear effects*: these effects are particularly relevant for fiber lasers and the ability to manage them is fairly important. They limit the level of power achievable by a laser, and have to be dealt with a proper choice of fiber structure and laser system architecture. Some of these effects are: stimulated Raman scattering, Brillouin scattering, and four-wave mixing.
5. *High optical damage threshold*: whenever high optical intensities are achieved, the risk of damaging the laser optics of the device increases. Thus, optics with very high optical damage threshold needs to be chosen whenever high-power

radiation is involved. Moreover, it is vital to keep the device sealed and free of dust particles, which can cause the damage to the optics even if optical intensities are well below limiting threshold.

6. *Limited misalignment and vibrations*: lasers are very sensitive to misalignment and vibration of their optical components. This is particularly true in the case of high-power lasers where huge optical intensities are produced. Thus, careful opto-mechanical design is required to preserve the laser performances once integrated on a carrier (aircraft or any vehicle).

1. UNCONVENTIONAL LASERS

Unconventional lasers are devices in which the external energy sources are unconventional. The most interesting lasers of this group, that will be described in more details, use solar light pumping and pumping with explosives.

A *solar-pumped laser* is a device that uses concentrated solar radiation as an external energy source. The lasing medium is usually a crystalline material, such as neodymium-doped yttrium aluminum garnet. The particularity of this laser is that it does not require any artificial energy source, thus being environmentally friendly. High optical intensities of the beam can be achieved although extremely large collection mirrors and very high quality lenses are required. Moreover, calculations prove that this method is not more efficient than the one that uses solar-cells to produce the electrical current, that is, than used to pump the gain medium [11, 13].

A *laser pumped with explosives* is a device, which uses the energy of an explosion to generate a laser beam. High efficiencies and high output powers are obtainable but the biggest disadvantage of this technology is that the laser can only be used once, making it unpractical for most of the applications [11].

2. NON SOLID-STATE LASERS

Non solid-state lasers are devices whose gain media are either in a gaseous or liquid form. Thus, within this group we consider: gas lasers, excimer lasers, chemical lasers, and dye lasers (see Sec. II.A.5 for a quick reference).

Gas lasers are devices based on a gaseous amplifying medium and, usually, electrical discharges as pumping mechanism. Most of them can achieve high-quality beams coupled with high optical intensities. Nevertheless, in order to be operated, they require high electrical voltage and often also high electrical power [8].

Different kinds of gas lasers exist depending on the type of gas used. Among all of them let us now consider the following three that stand out for their wall-plug efficiencies and achievable output powers:

1. *Carbon dioxide lasers* (CO_2 lasers) use a mixture of gasses (i. e., hydrogen dioxide (CO_2), helium (He), nitrogen (N_2), and possibly hydrogen (H_2)) as a

Fig. 13 Carbon dioxide electric discharge coaxial laser. (From Wikimedia Commons.)

gain medium and the energy pumping is achieved with electrical discharges. The output power can reach tens of kilowatts whereas the wall-plug efficiency of the device is around 20%. The emitted radiation is in the infrared part of the spectrum, centered mostly at 10.6 μm. They are widely used in manufacturing but due to their low pressure discharge property and necessity to be refilled after a while, these lasers are more and more replaced by fiber lasers [8, 11]. An example of this laser can be seen in Fig. 13 where the target on the right side of the image is set on fire by the operating laser on the left.

2. *Excimer lasers (also called rare gas halide lasers or exciplex lasers)* are devices using for the amplifying medium a gas mixture of a noble gas, a halogen and some other "buffer" gas whereas the pumping is achieved with electrical discharges, as in most of the gas lasers. However, unlike the other gas lasers, the emission of photons from these devices is obtained through the dissociation of unstable molecules created during the energy pumping. The emitted beam is a pulsed radiation in the ultraviolet domain (between 157 and 351 nm) and a repetition rate of up to few kilohertz. Average powers of a few watts are easily achieved through electric gas discharge pumping of excimer lasers with an efficiency varying between 0.2 and 2%. Through much more complex architectures and pumping configurations, some excimer lasers can even reach MW levels of power. Thus, they are considered the most powerful lasers in the ultraviolet spectrum, especially below 300 nm. Safety issues, however, are a concern when using these devices due to the toxicity of used gases and the usage of high voltages. Other shortcomings include the corrosive nature of the gases and the gas refueling. Nevertheless, the properties of their beams make them suitable for a number of applications such as material processing, micromachining, microelectronics, ophthalmology, and military [8, 11].

The Tactical High Energy Laser (THEL) illustrated in Fig. 14 is an example of deuterium fluoride laser (DFL) used for latter purpose.

3. *Chemical lasers* are devices exploiting chemical reactions of mixed gases to generate laser light. The latter is usually a continuous wave beam, typically in the mid- or near-infrared domain of the spectrum and can reach powers of up to 2–3 MW. This property coupled with high wall-plug efficiencies generally achievable has made them in the past perfect for military usage. The most famous chemical laser used by the military is the chemical oxygen iodine laser (COIL). The COIL was used on board the Boeing YAL-1 Airborne Laser Testbed. Disadvantages of these lasers are the same as for excimer lasers previously described [8, 11].

Dye lasers are devices using an organic dye, typically in a liquid solution, as lasing medium. Pumping of the gain medium can be achieved either using other lasers (e. g., an argon ion laser or an excimer laser) or flash lamps. The output radiation has a wide bandwidth that extends form the UV to the NIR region, which is particularly attractive for tunable lasers and pulsed lasers. Very high-power conversion can be achieved with these lasers, around 10 and 30% but only if another laser is used as a power source. On the other hand, dyes tend to chemically degrade during the operation of the laser, which requires a continuous circulation of the dye solution from and to a large reservoir. Moreover, after some period of usage, the solution needs to be exchanged. This operation isn't easy due to the toxicity and corrosivness of many dyes and some of their solvents [8, 11]. Despite all this, given their great tunability and high efficiency (i.e., laser gain), they have found many applications in astronomy, medicine, and manufacturing (just to name few). Figure 15 illustrates a dye laser.

Fig. 14 Tactical High Energy Laser (THEL). (From Wikimedia Commons.)

Fig. 15 Coherent model 899 ring dye laser. (Credit: Han-Kwang, AMOLF Institute, Amsterdam, Netherlands CC BY-SA 3.0.)

3. SOLID-STATE LASERS

Solid-state lasers are, in general, lasers having a solid lasing media based on doped crystals or glasses or even on semiconductor materials.*** The pumping mechanisms can be optical (flash lamps or laser diodes) or electrical (in case of semiconductor lasers). They are capable of very high output powers and variable beam qualities. Moreover, they can be operated either in continuous or pulsed mode [8].

Based on the type of lasing media and the type of pumping mechanisms, various solid-state lasers can be made. Describing all of them, however, goes beyond the scope of this chapter and so we will concentrate our attention, in particular, on two types of solid-state lasers: laser diodes and diode pumped solid-state lasers (DPSSL).

Diode lasers are a particular type of semiconductor lasers where an electrical current flows through a gain medium, made of a p-n junction. They are generally built as edge-emitting lasers, meaning that the semiconductor wafer emits the laser light from its edge after the laser beam has propagated along the direction of the wafer itself. This property makes them suitable to be stacked together in order to create one-dimensional bars or even two-dimensional arrays, known as "stacking." In this configuration several low-power sources can be arranged together to obtain a compact high-power device capable of emitting thousand of watts. The most common arrangement of these stacks is in vertical position as

***Note however that often semiconductor lasers are not included in the solid-state category. They are instead considered as a separate type of lasers [8].

Fig. 16 High-power laser diode array. (From Wikimedia Commons.)

illustrated in Fig. 16. The maximum obtainable wall-plug efficiency with these devices is around 60%[†††] although the beam quality and brightness are relatively low. Moreover, due to strongly asymmetric and high divergence of their beams, diode lasers require very complex beam forming optics. The beam itself cannot propagate very far, although it can be focused at a short distance. All this makes these devices perfect for being used as energy sources for other lasers. Alternatively, they can be used in all applications that do not require high beam quality such as material processing, welding, etc [8, 11].

Diode pumped solid-state lasers (DPSSL) are optically pumped devices that exploit the properties of diode lasers to obtain miniature devices capable of achieving excellent efficiencies and high output powers whereas at the same time maintaining good beam qualities. This is possible using diode lasers as a pumping source and some kind of crystal or piece of glass as the lasing medium. The advantages of this approach, when compared to the usage of flash lamps as energy sources, are [8]:

1. *Compactness of the laser device*: limited size of the energy source, power supply, and cooling mechanisms contribute to the overall compact size of DPSSL devices especially when compared to the output power that they are capable of emitting.
2. *Decreased electrical consumption*: electrical consumption of DPSSL devices is drastically lower if compared to solid-state lasers using lamps as pumping

[†††]This is true for devices emitting in the 700–1000 nm range of wavelengths. For higher wavelengths the efficiency quickly drops below 30%.

mechanisms. This is due to the high efficiency of the power source, which contributes to the overall very high wall-plug efficiency of these devices (i.e., around 30%).

3. *Decreased cooling demands*: these devices require a decreased thermal management due to their overall compact dimensions and high efficiency. Nevertheless, thermal effects in DPSSL remain one of the downsides of these systems, as it will be evidenced shortly.
4. *Extended lifetime*: diodes are very rugged energy source that can endure around 6000 h, which makes them very durable energy source. The downside is that this durability is easily influenced by a variety of factors and their eventual exchange in a device is very expensive.
5. *Wide variety of gain media*: lasing medium can be chosen from a wide range of materials based on the wavelengths that we would like to achieve. With discharge lamps this wouldn't be possible due to their lower brightness.

However, despite all the laudable properties, these devices also exhibit some disadvantages that need to be taken into account if the application requires high-power output. The main are the high thermal effects and higher cost per watt of pump power when compared to lamp pumping [8,11].

To date, the highest output power achievable with a DPSSL was demonstrated by Northrop Grumman. The obtained power was 105 kW and it was possible by coherently combining seven slab lasers (15 kW each). The demonstrated operation time was 300 s[‡‡‡] [11].

The applications of these devices vary greatly and encompass all possible areas of laser applications [8].

4. FIBER LASERS

Fiber lasers are essentially very long solid-state lasers in which the input laser light is guided through the fiber by total internal reflection (see Fig. 17). The laser medium is a doped fiber core (often ion-doped silica) whereas the internal clad (often undoped silica) acts as a pump guide. The laser dopants used in most cases are rare earth elements such as erbium (Er^{3+}), neodymium (Nd^{3+}), ytterbium (Yb^{3+}), thulium (Tm^{3+}), or praseodymium (Pr^{3+}). Typical length of a fiber is around 5 m [8,11]. The pumping source in most cases is a laser diode. Thus, this particular type of lasers can be seen as an evolution of previously described diode pumped solid-state lasers. This is the reason why they are considered separately in this chapter although they are effectively DPSSL devices.

[‡‡‡] Note that this operation time was repeatable so that it is not to be intended as the lifetime limit of the laser device.

Their advantages over the other solid-state lasers are essentially [8, 11]:

1. *High power capabilities*: diode pumped fiber lasers can achieve powers of several kilowatts from a single fiber. Physical limitations indicate that the maximum achievable power can be in the range from 20 kW to 40 kW (for cw operation mode).
2. *Improved thermal management*: high surface-to-volume ratio of a fiber provides extended area for the thermal exchange, which avoids excessive heating of one particular area and distributes the heat over the whole length of the fiber.
3. *Excellent beam quality*: beam quality of the fiber laser is intrinsically excellent due to the spatial filtering provided by the core of the fiber.
4. *Fiber splicing*: by using the principle of fusion splicing of fibers it is possible to obtain very long laser devices that have rugged fiber joints characterized by lower transition losses, weaker reflections, and are not affected by misalignment problems.
5. *Very high gain*: wall-plug efficiency of up to 50% can be expected whereas the optical to optical efficiency could be up to 85%, obtained with Yb^{3+} core-dopant.
6. *Compact and resistant design*: fiber glass diode pumped lasers consist of two very compact and rugged devices which in turn makes them very wear-resistant, sealed, and at the same time very compact devices. The fiber itself can be coiled and the laser beam can propagate over long distances without interfering with the surrounding environment.

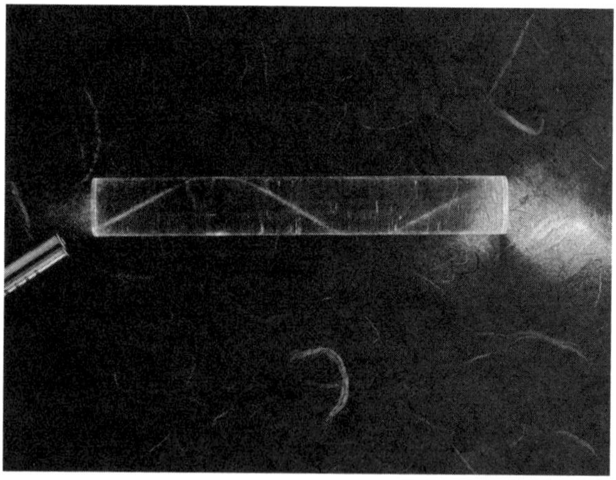

Fig. 17 Demonstration of the working principle of a fiber laser. (From Wikimedia Commons CC BY-SA 3.0.)

The limitations of this technology, on the other hand, are [8, 11]:

1. *Damage threshold*: the power intensities reached in recent years are very close to the damage threshold of the glass material itself indicating that in current state, fiber lasers have reached their maximum power limit.
2. *Cooling limits*: 100 W/m is the power of dissipation per unit length of a fiber reached with air-cooled devices. Thus, other methods of cooling (e. g., water-cooling) need to be considered in order to increase the output power. Otherwise lowering the concentration of laser dopants could ease the cooling process but it will at the same time increase the effects of nonlinearities.
3. *Nonlinearities*: nonlinear effects are present in fiber lasers and are one of the factors limiting their performance.
4. *Non trivial design*: in most cases, the design of a fiber laser is more difficult than other solid-state lasers due to a variety of factors.

However, despite those limitations, fiber lasers appear as a revolution of solid-state lasers and remain one of the candidates for any application that requires very high optical intensities, low consumption, high beam qualities, and wide range of wavelength tunability [8, 11].

III. USE OF LASERS FOR DEFLECTION PURPOSE

Since their first invention in the 1950s, the range of applications of lasers has not stopped to grow, reaching nowadays an impressive extent that includes, among others fields, surgery, computer sciences, disc/movie industry, drilling industry and telecommunications. In particular, the ability of lasers to focus energy in a very narrow beam over long distances in vacuum always made them very attractive tools for space applications. On 15 May 1987, the Polyus spacecraft was launched from Baikonur Cosmodrome Site 250 as part of the first flight of the Energia system, carrying a 1 MW carbon-dioxide laser as its primary payload. Fortunately, the launch failed due to a faulty inertial guidance system and the end of the cold war a few years later prevented (at least for the official version) further launches. As a more recent and peaceful example, the LLCD mission revealed last year the possibility of extremely fast broadband capability in space using laser communications demonstrating record-breaking data download and upload speeds to the moon at 622 megabits per second (Mbps) and 20 Mbps, respectively. In the early 1970s, fluid dynamicist Arthur Kantrowitz, more famous for his work on supersonic inlet velocities, introduced the idea of using lasers for space propulsion. As one can imagine, the physical principles underlying laser propulsion present many similarities with the ones required for the development of deflection technologies. As we are about to see, technology transfers are also possible between technologies used everyday on earth such as laser drilling, laser cutting, and researches that have been lead on the topic of laser weapons. The different analogies and the common challenges between these technologies and the ones that the development of a laser-based deflection method will require are reviewed in the following sections.

A. ANALOGY WITH LASER PROPULSION

Conceptually, the principles underlying laser propulsion are very similar to space object deflection. The general idea is to convert the energy carried by the laser beam into mechanical energy/momentum to the object (spacecraft or asteroid). Obviously, unlike a spacecraft, an asteroid was not designed to be used for laser propulsion and various constraints and uncontrolled parameters will add to the already many challenges faced by space propulsion. The different concepts can be regrouped in two different categories

1. Assisted techniques that will require the integration of specific components to the objects
2. Self-reliant techniques, which as their name indicate, do not require this integration

1. ASSISTED TECHNIQUES

For these techniques, the energy carried by the laser beam acts as a power source to enable the propulsion system onboard of the space object (see Fig. 18). In general, the laser source will not necessarily be part of the spacecraft and can be located on the ground. For laser electric propulsion, the light is converted into electricity by a power receiver to propel the vehicle. Alternatively, in the case of the heat exchange thruster, the energy source will be used to heat a solid exchanger and vaporize a liquid propellant to exhaust it through a nozzle. Similarly, for cw plasma propulsion, the laser beam is used to create and sustain a plasma in a gaseous flow, heat the gas, and expand it through a nozzle. A more exotic method is the space fountain, which consists in some kind of space elevator where a laser can be the pellet if placed in a Bose–Einstein condensate.

2. SELF-RELIANT TECHNIQUES

Following the same principle as a solar sail, the concept of photonic laser thruster uses the radiation pressure to propel an object:

$$F = \frac{P_{\text{laser}}}{c} \tag{1}$$

Another promising concept depicted in Fig. 19 is the approach followed by pulsed plasma propulsion, which induces plasma breakdown of a gas near the object. The transfer of momentum to the object is achieved by an expanding shock wave. Note that this approach was demonstrated by real flight experiments during the

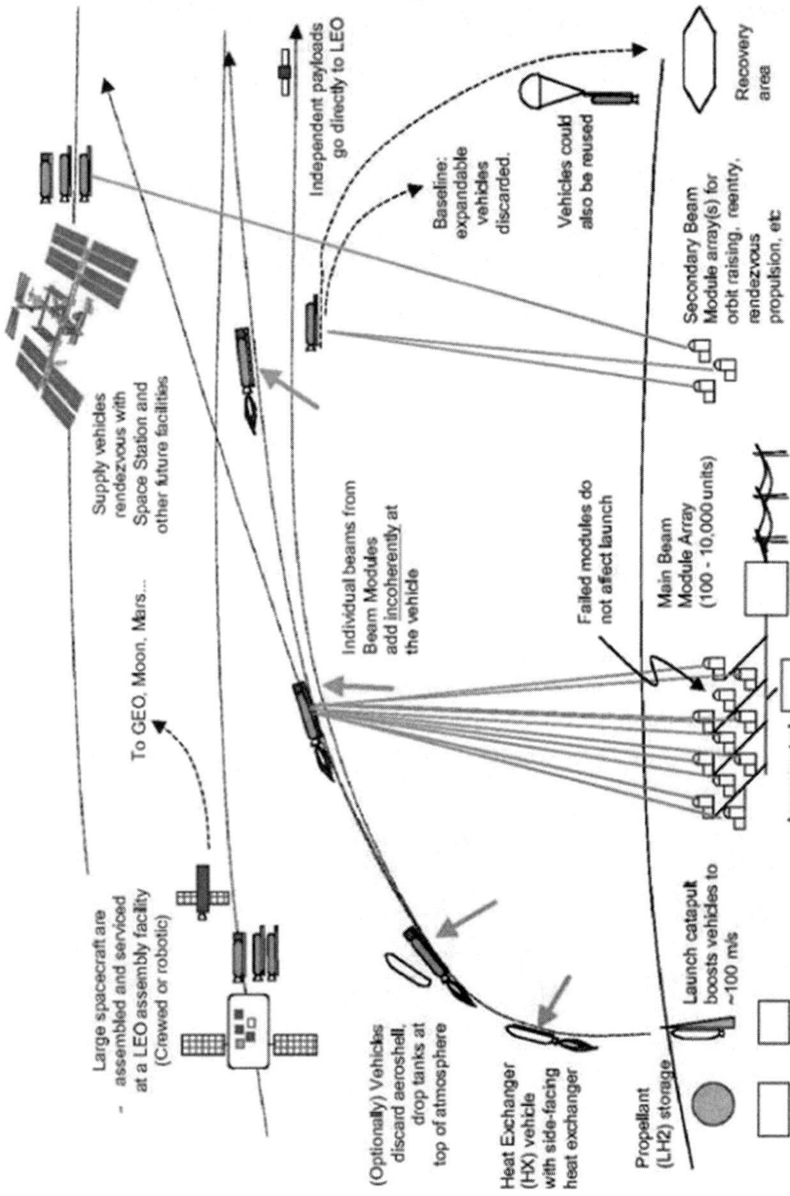

Fig. 18 Assisted techniques. (From Wikimedia Commons.)

a) Ikaros solar sail concept b) Light project (USA)

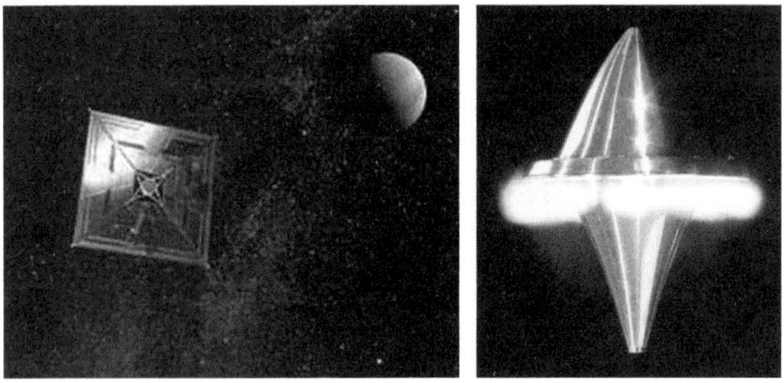

Fig. 19 Self-reliant techniques: a) Ikaros solar sail concept and b) Light project (USA).

lightcraft project. Unfortunately, this method requires the craft to operate in a gaseous environment and can thus not work in space. Another way of transferring momentum remotely to an object is to use the high-power density of the laser beam to ablate the surface material of the object. This propulsion method is called ablative laser propulsion and is illustrated on Fig. 20.

In the laser ablation process, ablated matter expands suddenly as it is removed from the object, thus resulting in high speed ejection of matter from the object. Not only this method can work in vacuum because the object material is used as its own propellant, but also laser ablation can in principle work with many different materials, even almost transparent ones. Both cw and pulsed lasers can be used, pulsed laser requiring less average power. Also, the thrust direction is not correlated with laser beam direction but only depends on the orientation of the

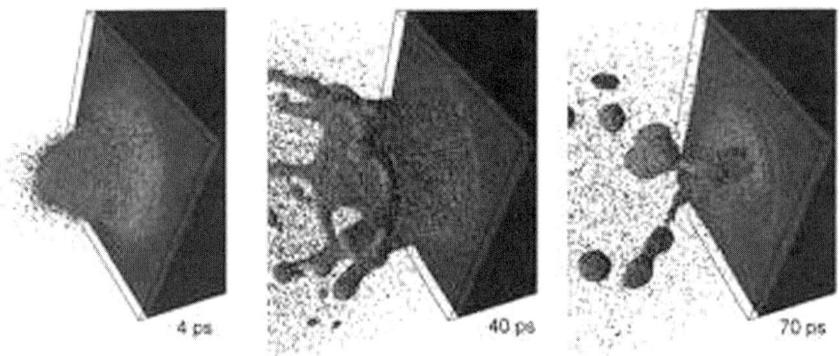

Fig. 20 Illustration of the laser ablation method taken from Dr. J. Roth Institute for Theoretical and Applied Physics, University of Stuttgart (Germany).

surface of the material. Eventually, it is worth mentioning that laser ablation is already used on earth for various terrestrial applications. Some relevant features of ablation-based techniques are reviewed in the next section.

B. SPIN-IN FROM TERRESTRIAL APPLICATIONS OF LASER ABLATION

Some useful knowledge can be gathered from selected terrestrial applications presenting indisputable similarities with the ablative laser propulsion concepts. Three of these techniques are

1. Laser cutting: ablation of metals for fabrication purpose
2. Laser weapons: ablation of material to heat and drill the target
3. Laser drilling: ablation of rocks for drilling purpose

1. LASER CUTTING

Numerous data on metal laser cutting are available. Historically, CO_2 lasers at 10.6 μm were first used for cutting purposes, but they tend nowadays to be replaced by fiber lasers at 1 μm, which are able to cut a larger collection of materials although using a similar or even lower amount of power.

Obviously the ability of a given laser to cut through an object is dependant of its material. An important parameter to gauge the complexity of laser cutting through a given material is the ratio between the energy required to reach the fusion temperature ($c_p \rho T_f$) and the absorption coefficient α:

$$\kappa = \frac{\rho c_p T_f}{\alpha} \tag{2}$$

Some materials can be compared from the data tabulated in Table 1 and from their cutting speeds in Fig. 21.

TABLE 1 FIGURE OF MERIT κ FOR DIFFERENT MATERIALS

Material	Titanium	Copper	Aluminum	Steel	Carbon Fiber	Fiber Glass
T_f, K	1940	1360	930	1800	3800	1800
ρ, kg/m^3	4540	8920	2700	7800	1700	2200
c_p, J/(kg K)	520	385	900	450	1200	1000
α	0.37	0.01	0.09	0.8	0.2–0.9	0.2–0.96
κ	12	467	25	8	39–9	20–4

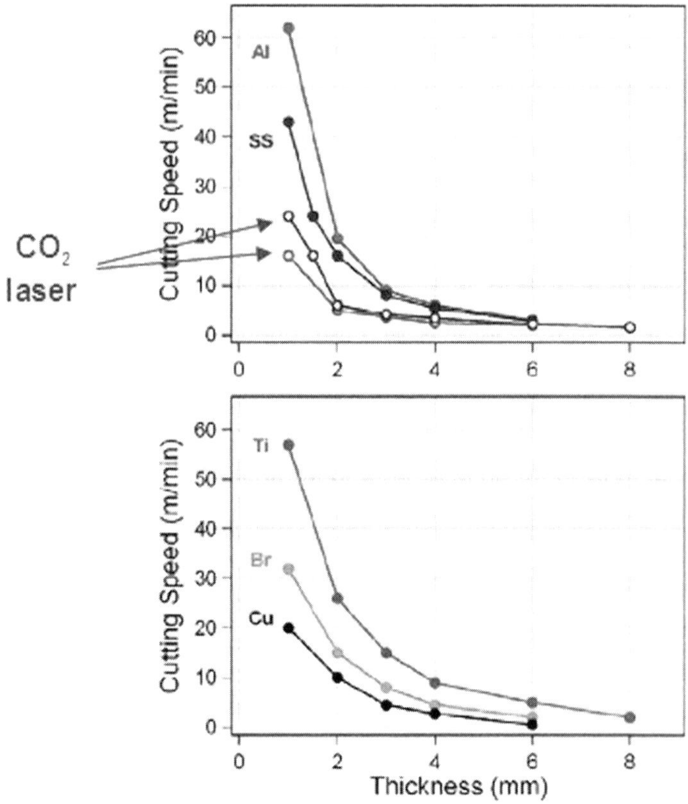

Fig. 21 Cutting speed in function of the thickness for different metals assuming a 100 μm diameter laser spot. (From Kliner et al. [14].)

2. LASER WEAPONS

The use of laser for military purposes started with the need from superpowers to defend themselves from a potential ballistic missile strike. In the United States, the Star Wars program was developed with the displayed intention to prevent a nuclear attack through an ICBM. Different methods were considered at the time:

1. Space-borne laser weapons were intended but their actual implementation did not succeed due to launch failures, the outer space treaty and the end of the cold war.
2. Ground-based laser weapons (THEL): MIRACL laser, which is a DF 2.2 MW cw laser at 4 μm (see Fig. 22).

Fig. 22 Ground-based laser weapons. (From Wikimedia Commons.)

Fig. 23 Airborne laser weapons.

3. Airborne laser weapons (see Fig. 23): COIL chemical laser at 1.3 μm, ABL (3 MW cw), ATL (few 100 kWs cw).

Following the cancellation of the program, the projects focus nowadays on the ground-based approach:

1. EXCALIBUR (see Fig. 24) is the most recent among many DARPA projects to develop a 100 kW solid-state laser by combining coherently the beams from high-power (kW class) fiber lasers. A first milestone aims to develop a 50 kW version.
2. Gimbal-mount: the U.S. Navy plans on developing a laser weapon system to equip its aircraft carriers (see Fig. 25). The platform would consist of a 6x8 kW fiber lasers from IPG with six beams overlapped on the same point of the

Fig. 24 EXCALIBUR. (From DARPA.)

Fig. 25 Laser Weapon System (U.S. Navy).

target. A trial campaign held in 2009–2010 demonstrated the feasibility of the technique with multiple drones destroyed by the system. The main effect visible on the videos was a gas tank drilling followed by fuel inflammation.

3. A recent demo showed "Qassam-type" rocket destroyed at 1.5 km with a 10 kW fiber laser.

3. LASER DRILLING

Already considered in the past, laser drilling (see Fig. 26) is gaining momentum rapidly these days. The principle underlying laser drilling is called spallation: the rock is broken into small particles and fragments that have to be removed for the laser head to go on drilling. As the lasers used are the same as for cutting, it is interesting to note that the material properties of a rock are already much closer

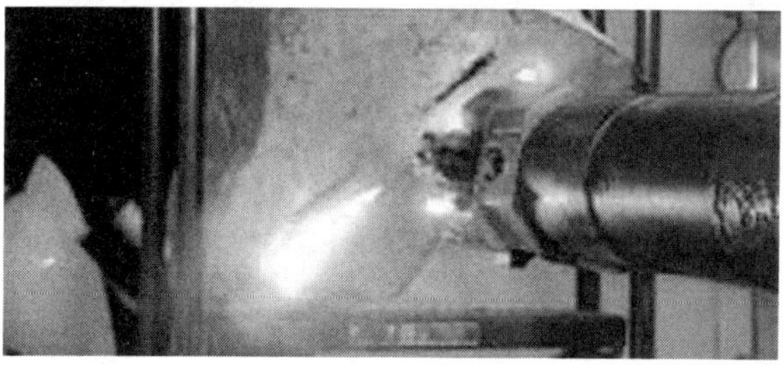

Fig. 26 Laser drilling. (From Zediker [15].)

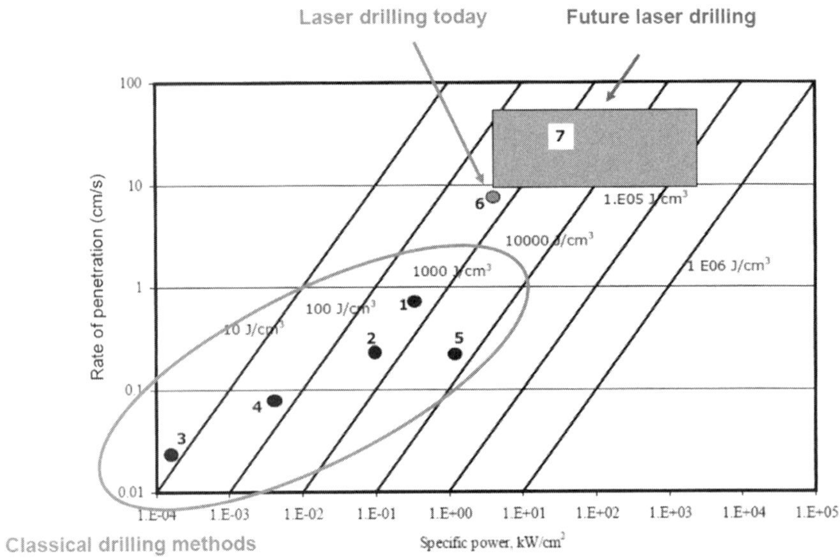

Fig. 27 Laser drilling in perspective. (From Xu et al. [16].)

to the type of materials encountered in asteroids. Therefore, important technological transfers between these techniques and ablative laser propulsion methods could be expected. As illustrated in Fig. 27, the future perspective of laser drilling are promising compare to the current techniques. Typical characteristics are summarized in Table 2. Note that levels of power lower than 0.6 kW will result in spallation without melting (rock breaking due to internal stresses).

C. PRELIMINARY TECHNOLOGICAL MILESTONES

As for the terrestrial applications, the laser technology will need to be optimized in function of the asteroid material parameters (see Fig. 28 for thermal properties of materials similar to those found in an asteroid) to yield the best ablation rate. A preliminary requirement will thus be to gain better knowledge of the physical properties of asteroid and how their material interact with the laser beam.

TABLE 2 TYPICAL CHARACTERISTIC OF LASER DRILLING TECHNIQUES

Type	Average Power	Melting Threshold	Pulse	Beam Diameter
Quasi-cw	0.3–3 kW	0.6 kW	0.001–0.5 s	1–2 cm

Rock type	Compositions	%	Bulk density (g/cm^3)	Thermal conductivity x 10^3 (cal/sec/cm^{0}C)	Heat capacity (cal/g^{0}C)	Diffusivity (cm^2/sec)
Berea Grey	SiO_2 Al_2O_3 Fe Rest	85 10 3 2	2.15	6.2	0.21	11.3
Limestone	$CaCo_3$ MgO Rock fragment	85 15 5	2.41	4.8	0.22	8.1

Fig. 28 Rock sample compositions and thermal properties. (From Xu et al. [17].)

In particular, the efficiency of the method will mainly be driven by the choice of an appropriate pulse sequence and beam size:

1. *Pulse sequence*: Continuous-wave lasers, when absorbed, deliver heat into the material and thus the ablation process depends purely on thermal effects. To reduce the losses by thermal conduction, it would be preferable to work with long-pulse (millisecond to nanosecond) duration laser that heat up the material but also quickly vaporize matter locally (this process is, for instance, used in laser-induced breakdown spectroscopy). Shorter pulses will deliver the same levels of intensity and generate the laser-induced breakdown for lower levels of energy per pulse. However, as the pulses shorten, the ionized fraction will become more and more important and the plasma will act as a screen shielding the laser beam.
2. *Beam size*: Because of the diffraction limit, there is a minimum divergence achievable with a laser (around 100 μrad at 1 μm wavelength). Thus, maintaining a small beam diameter on the space object maybe difficult for long distances of action. For the case of a space debris, with the laser system placed on the ground or on a plane, additional problems will arise linked with the small size of the object and the ability of the system to focus (point and track) the beam accurately through the atmospheric turbulence (at the best 30 cm accuracy with the current technologies).

When these preliminary issues are solved, the next difficulty will be to integrate the laser system on a spacecraft. Obviously, the weight, volume, cost, and power consumption are all growing functions of the output power as can be seen in Table 3. All technologies will need to be space-rated, especially their resistance to radiations. Note that many systems have been designed already for the space agencies (NASA, ESA) and technological solutions exist (e.g., ALADIN laser for ESA's Doppler wind Lidar).

TABLE 3 EXAMPLE: FIBER LASER SYSTEM INTEGRATION PRELIMINARY ASSESSMENT

Laser Power (kW)	Cost (M€)	Electric Power (kW)	Volume (m^3)	Weight (kg)
10	6.5	50	1.2	31
25	10	120	3	78
50	15	240	6	155
100	26	480	12	310
500	100	2.4	60	1600
1000	163	4.8	120	3100
5000	618	23.8	2100	20,600
20,000	1865	95.2	8400	82,000

Heat-capacity lasers could respond to these challenges. SSHCL are non actively cooled lasers that emit a short burst of laser light (generally for less than 1 s) and then rest for a few minutes before firing again. In principle, the technology would be less complex to integrate on a space vehicle, especially with applications like space object deflection where time is not of the essence.

IV. CONCLUSION

In conclusion, the basic principles and different types of laser technologies were exposed during the presentation. In link with asteroid deflection methods, parallels were drawn with current researches on laser propulsion and relevant terrestrial applications. From this, it appears that the most promising technology for deflection purpose is laser ablation. Although important work will be required to adapt laser ablation technologies for space applications, the risk associated with this type of technology is considerably low compared to the other mitigation strategies. Similarly to the terrestrial counterparts, the key drivers to improve the efficiency will be to understand better the material properties of an asteroid and adapt the pulse sequence in consequence as well as to maintain a small beam diameter over a long distance, allowing to concentrate the same amount of power in a narrower spot. Thus, the success or not of the laser ablation method will essentially rely on its ability to convince potential investors by proposing solutions that are both technically feasible and cost-effective.

ACKNOWLEDGMENTS

The authors of this chapter wish to express their gratitude to Dr. Bourdon for his help, corrections, and useful comments provided during the redaction of this chapter. Moreover, the authors would like to thank and Oxford Lasers, TRUMPF GmbH & Co. KG, Bosh GmbH, Oxford Lasers, for granting us the permission to use their images in this chapter.

REFERENCES

[1] Balembois, F., and Forget, S., "*Laser: Fundamentals*," Optique-ingenieur.org, 2007. Available: http://goo.gl/Qj3KQs [last accessed 25 Jan. 2016].

[2] Weart, S., "Bright Idea: The First Lasers—A History of Discoveries Leading to the 1960 Invention," Aip.org, 2010. Available: https://www.aip.org/history/exhibits/laser/sections/raydevices.html [last accessed 25 Jan. 2016].

[3] Silfvast, W., "Introduction," in: *Laser Fundamentals*, 2nd edition, Cambridge University Press, New York, U.S.A., 2004, Ch. 1, pp. 1–6. http://books.google.com/books?hl=en&lr=&id=x3VB2iwSaxsC&oi=fnd&pg=PR19&dq=LASER+FUNDAMENTALS&ots=20KGaDADb8&sig=6LC-n1da_6QFug2s6Q7RpA8xjPw

[4] Gibbings, A., Vasile, M., Hopkins, J.-M., Burns, D., and Watson, I., "Potential of laser-induced ablation for future space applications," *Space Policy*, 28 (3), 2012, pp. 149–153. http://www.sciencedirect.com/science/article/pii/S0265964612000641

[5] Kwok-san, L., Shiu-sing, T., and Leung, J., "Further Physics–What is Laser?," Hk-phy.org, 2000. Available: http://www.hk-phy.org/articles/laser/laser_e.html#note [last accessed 25 Jan. 2016].

[6] Melles Griot, "Laser Guide," Technical Guide, IDEX Corporation, 2016. Available: http://mellesgriot.com/Frontend/PDFs/TechGuide.pdf [last accessed 25 Jan. 2016].

[7] Foth, H.-J., "Principles of lasers," in: Lackner, M., (Ed.), *Lasers in Chemistry* Vol. 1: Probing Matter, Vol. 1, Wiley-VCH Verlag GmbH & Co. KGaA., Weinheim, Germany, 2008, Ch. 1, pp. 5–8.

[8] Paschotta, R., "Encyclopedia of Laser Physics and Technology—Lasers, Principle of Operation, Resonator, Cavity, Laser Beam, Stimulated Emission," RP Photonics Encyclopedia, 2008. Available: https://www.rp-photonics.com/lasers.html [last accessed 25 Jan. 2016].

[9] Lucas, L., and Zhang, J., "Femtosecond Laser Micromachining: A Back-to-Basics Primer," Industrial-lasers.com, 2012. Available: http://www.industrial-lasers.com/articles/2012/06/femtosecond-laser-micromachining-a-back-to-basics-primer.html [last accessed 25 Jan. 2016].

[10] Tierney, J., "Sci-Fi Ray Gun Debuts in Iraq," Tierney Lab, 2008. Available: http://tierneylab.blogs.nytimes.com/2008/11/21/sci-fi-ray-gun-debuts-in-iraq/?_r=0 [last accessed 25 Jan. 2016].

[11] Bourdon, P., High Power Lasers for Laser-Matter Interaction: An Overview of Available Technologies. Glasgow, Scotland, UK, 2013.

[12] Northrop Grumman, "Joint High Power Solid-State Laser," 2010. Available: http://www.northropgrumman.com/Capabilities/SolidStateHighEnergyLaserSystems/Pages/JointHighPowerSolidStateLaser.aspx [last accessed 25 Jan. 2016].

[13] Graham-Rowe, D., "Solar-Powered Laser," *MIT Technology Review*, 2007. Available: http://www.technologyreview.com/news/408698/solar-powered-laser/ [last accessed 25 Jan. 2016].

[14] Kliner, D. A., Chong, K., Franke, J., Gordon, T., Gregg, J., Gries, W. and Kliner, A., "4-kW Fiber Laser for Metal Cutting and Welding," *SPIE LASE*. International Society for Optics and Photonics, ISO 690, Feb. 2011, pp. 791418–791418.

[15] Zediker, M. S., "High Power Fiber Lasers in Geothermal, Oil and Gas," *SPIE LASE*. International Society for Optics and Photonics, Mar. 2014, pp. 89610D–89610D.

[16] Xu, Z., Reed, C. B., Parker, R., and Graves, R., "Laser Spallation of Rocks for Oil Well Drilling," *Proceedings of the 23rd International Congress on Applications of Lasers and Electro-Optics*, Oct. 2004, pp. 1–6.

[17] Xu, Z., Reed, C. B., Konercki, G., Parker, R. A., Gahan, B. C., Batarseh, S., and Skinner, N., "Specific Energy for Pulsed Laser Rock Drilling," *Journal of Laser Applications*, Vol. 15, No. 1, 2003, pp. 25–30.

CHAPTER 14

Robotic Active Debris Removal and On-Orbit Servicing

Marko Jankovic* **and Jan Paul***
DFKI GmbH, Bremen, Germany

I. INTRODUCTION

Before October 4, 1957, satellites and space technologies were just part of the science fiction narrative. Nowadays, even though we do not realize it, they became an integral part of our daily lives powering countless services that we greatly depend on. However, these modern conveniences are in constant danger due to the product of our human nature. Just like on Earth, after almost 60 years of space activities and 4900 launches, we have been able to create a staggering number (around 6300 t) of waste (visible in Fig. 1) constantly orbiting our planet, called the space debris [1, 2]. According to an internationally accepted definition developed by the International Academy of Astronautics (IAA), space debris is defined as: "Any man-made earth-orbiting object which is non-functional with no reasonable expectation of assuming or resuming its intended function or any other function for which it is or can be expected to be authorized, including fragments and parts thereof" [3].

Rapidly the space debris problem has evolved from theoretical studies and science fiction to a serious threat that needs to be considered from early stages of a mission development. In fact, today, every major mission has to consider:

1. The protection of sensitive equipment and/or manned modules, for particles that go up to 1 cm in size
2. Collision avoidance maneuvers, for particles equal or larger than 10 cm†

in order to increase the survivability of its spacecraft. Implementing these expedients, on the other hand, adds more cost to the mission than necessary by adding

*Robotics Innovation Center.
†Particles that have dimension from 1 to 10 cm are generally too small to be tracked and are too big to shield against. However, possible anticollision maneuvers are also performed for this category of objects.

Fig. 1 Distribution of cataloged objects in space based on actual density data. (Credit: ESA.)

more mass to the spacecraft and requiring a costly surveillance network that needs to track bigger particles regularly.

Considering a "business-as-usual" scenario, with 60–70 launches per year and an average of four to five future breakups a year, the number of objects in space will steadily increase augmenting the probability of future catastrophic collisions and thus taking us one step closer to the onsetting of the cascading process where fragments of one collision generate even more fragments, leading to a self-sustained process predicted in Ref. [4], in 1978, by Kessler and Cour-Palais and called *the Kessler syndrome*. This means that in the current situation, the creation of the space debris has overcome natural cleaning mechanisms, such as air drag and lunisolar perturbations and we can expect that this will lead to the situation where collisions will become the main source of space debris in less than 50 years, making the spaceflight in the lower Earth orbit (LEO) a dangerous task [2]. For example, in the time period between April 2011 and April 2012, the International Space Station (ISS), despite its big size, had to perform four collision avoidance maneuvers and it would have had two more but the warning came too late [5].

To mitigate this kind of situation and to eventually prevent the onsetting of the cascading process, the international space community has developed a set of nonbinding guidelines that essentially recommend the space-faring nations to [6]:

1. Limit the amount of debris during a nominal mission
2. Minimize the risk of potential in-orbit breakups and accidental collisions during the operational phases of a mission
3. To apply the postmission disposal concept to spacecrafts at the end of their mission

However, recent studies conducted by J.-C. Liou et al. [7] indicate that even with a 90% implementation of postmission disposal measures and a complete suppression of in-orbit explosions, the population of space debris bigger than 10 cm in the LEO is expected to increase by 75% in 200 years, visible in Fig. 2. The same study finds that the LEO population of debris appears to grow even in the "no-future-launches" scenario indicating that at certain altitudes, this population has already reached critical concentration levels. Moreover, these studies suggest that the proposed mitigation measures alone cannot be used to permanently stabilize the space debris environment. For that, active debris removal (ADR) and on-orbit servicing (OOS) of satellites needs to be considered. The benefits of the ADR application can be best seen in Fig. 2, which is the result of simulations made with the NASA long-term debris evolutionary model (LEGEND) [8]. The top curve of the figure represents a "business-as-usual" scenario with the applied

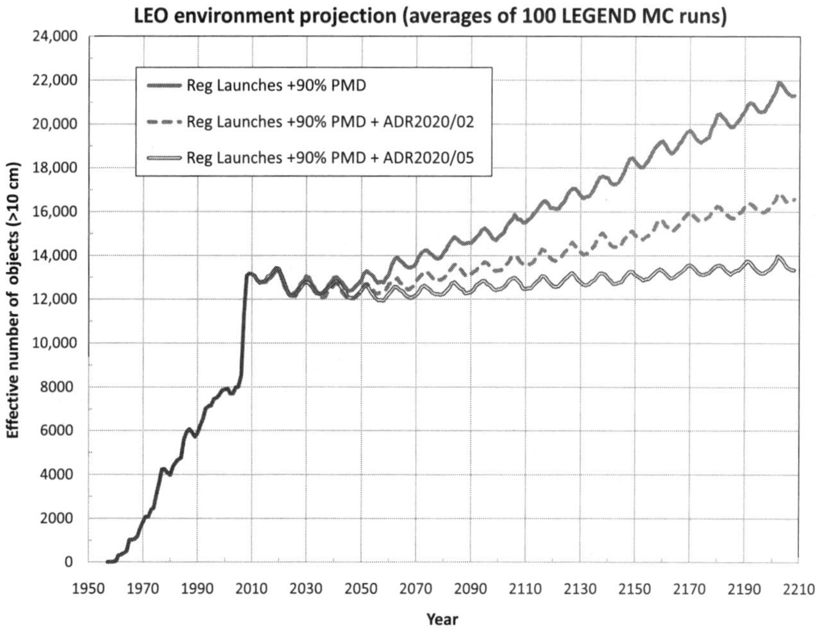

Fig. 2 Benefits of using ADR to limit the LEO population. (Credit: NASA Orbital Debris Program Office [8].)

90% of post-mission disposal measures whereas the bottom two curves represent the scenarios where ADR operations of two and five objects per year are added to the picture, starting from 2020. In case of removal of only two objects per year, the expected space debris population could be reduced by half in comparison with the previous scenario. Moreover, considering the case of five objects per year, the space debris environment could be completely stabilized in 200 years. To restore the space environment as it was before the Chinese antisatellite weapon test (ASAT), in 2007, the removal of more than five objects per year must be considered [8].

Some of the ADR methods proposed by the scientific community involve the use of:

1. High-powered ground lasers
2. Low-powered in-orbit lasers
3. Solar sails
4. Propulsive tugs
5. Foams and inflatable surfaces
6. Electrodynamic tethers
7. Ion beams
8. Robotic grappling devices

Among them, the ADR techniques involving orbital robotics appear to be the most mature ones since technologies and theories for in-orbit robotic capture and servicing of spacecrafts already exist and they have been successfully tested in-orbit in the past. In fact, robotic manipulators have been playing an important role in human spaceflight activities since the 1980s (see Sec. A) and the future missions indicate an even greater in-orbit presence of autonomous robotic systems. The future ADR and OOS missions such as the Phoenix and the *Deutsche Orbitale Servicing* Mission (DEOS) are just two examples of the rising trend that will pave the way to fully autonomous systems capable of safely approaching a target spacecraft and performing manipulation tasks.

This chapter is dedicated to the concise overview of the most relevant past as well as future ADR and OOS missions. Moreover, different hardware-in-the-loop (HIL) simulation systems will be briefly described in order to understand the difficulties of simulating those missions on ground. In particular, Sec. II is dedicated to the description of the historical space robotic missions and some of the most interesting future missions that should allow future broader usage of orbital robotics for the removal of space debris and OOS of spacecrafts. Moreover, within this section, some of the studies at DFKI-RIC concerning alternative capturing systems of free-floating satellites will be described. Section III, on the other hand, illustrates some of the simulation methods that can be used to simulate space operations on ground. Particular attention is dedicated to the description of the robotic simulation facilities European Proximity Operations Simulator (EPOS) of the German Aerospace Agency (DLR) and DFKI-RIC's INVERITAS.

II. ORBITAL ROBOTICS

Robotics as defined by B. Siciliano et al. in Ref. [9] is "The study of those machines that can replace human beings in the execution of a task, as regards both physical activity and decision making." In fact, the etymology of the word *robot* suggests that it comes from the Slavic word "rabota" meaning subordinate work and it was used for the first time in a 1920s play "Rossum's Universal Robots" (R.U.R.) of a Czech playwright, Karel Čapek. However, the realization of the centuries-long, human dream of creating an intelligent and skilled substitute similar to a human came into life only in the 1960s due to the convergence of two techniques: numerical control and teleoperation [10].

Space robotics represents one of the key technologies that enabled our species to safely explore and operate in harsh and extreme conditions of the space environment. Obviously, present-day robots appear to be clumsy in comparison to human beings. But if the comparison is made between skills of an astronaut during an extravehicular activity (EVA) and those of the best-available robot then the difference is almost negligible [11]. Furthermore, robotic systems can perform certain in-orbit tasks with less risk, work restlessly, and without the need for the costly transport and life support needed by the astronauts. It is needless to say that the human intuition and intelligence in certain, unpredictable situations make the latter irreplaceable, however, the mentioned positive attributes of the robotic systems, despite their many present limitations, make them perfect substitutes for risky tasks such as on-orbit debris removal and servicing [12].

A. HISTORY OF ORBITAL ROBOTICS

The concept of on-orbit servicing and assembly by a robotic free flyer dates back to the early 1980s when NASA published the Automation, Robotics, and Machine Intelligence System (ARAMIS) report. This idea was inspired by the successful use of a manipulator in an orbital environment in 1981 during an STS-2 mission where the Space Shuttle remote manipulator system was used for the first time. Later on, several manned missions followed in order to capture, repair, and deploy malfunctioning satellites (such as Anik-B, Intel-sat 6, and Hubble telescope). However, such unmanned and completely autonomous missions have yet to become a reality, despite several demonstration missions, such as the Engineering Test Satellite (ETS)-VII, Orbital Express, and Robotic Refueling Mission (RRM) (to be described later on in this section) [13].

1. SPACE SHUTTLE REMOTE MANIPULATOR SYSTEM

The *Shuttle Remote Manipulator System* (SRMS), or *Canadarm* was the first robotic manipulator arm used in Earth's orbital environment. It was a joint venture between the governments of Canada and United States and its development was largely carried out by a Canadian industrial team, led by Spar Aerospace Ltd., under the supervision of the National Research Council of Canada [13–15]. It made its space debut on the Space Shuttle Columbia (STS-2) on 13 November 1981 and was successfully used for 30 years retiring along with the Space Shuttle program after the mission STS-135, which marked the robotic arm's 90th flight. It was a mechanical arm that was 15 m long with 6 degrees of freedom (DOF) and, similarly to a human arm, it had shoulder yaw and pitch joints, an elbow pitch joint and wrist pitch, yaw, and roll joints. Its purpose was mainly to maneuver payloads from the payload bay of the space shuttle orbiter to its final position and then release it. However, Canadarm was also used to catch an in-orbit spacecraft and berth it to the payload bay of the orbiter as well as to assist human EVAs by means of a foothold that could be attached at the end of the manipulator. The standard end effector (SEE) of the SRMS was a special gripper called SEE, capable of grasping specifically designed pole-like fixture (GF) attached to payloads [13].

Servicing and maintenance missions of the Hubble Space Telescope (HST) and the construction of the ISS are just some of the most remarkable missions that used SRMS as a tool. An example of this can be seen in Fig. 3 showing Canadarm and Canadarm2 (SRMS' sibling on the ISS) work together to unload a cargo from the payload bay of Space Shuttle Endeavour on 15 August 2007.

After the accident of the Space Shuttle Columbia (STS-107), on 1 February 2003, SRMS played another important role being used, along with the Orbiter Boom Sensor System (OBSS), to inspect the external thermal shield of the orbiter for eventual damage that might have occurred during the launch [13].

2. ISS-MOUNTED MANIPULATOR SYSTEMS

As of June 2013, 135 launches have been made to the ISS since the launch of the first module, Zarya, on 20 November 1998, of which 89 were Russian launches, 37 were space shuttle launches, one test flight and two operational flights were made by SpaceX's Dragon, three were Japanese, and two were European launches [16]. It represents the largest human outpost in space as well as an international orbiting laboratory and as such, it has been equipped with robotic systems in order to facilitate various activities on the station whereas others are planned in the near future [13].

The *Space Station Remote Manipulator System* (SSRMS) or *Canadarm2*, launched in April 2001, is the next generation of the space shuttle's original SRMS, used on the ISS and developed by MacDonald, Dettwiler and Associates (MDA)

Fig. 3 Team work of the Canadarm and Canadarm2 to unload Space Shuttle Endevour. (Credit: NASA.)

Ltd. Since its installation, during the STS-100, it has played a major role alongside the SRMS in all the phases of construction and maintenance of the ISS both by assisting EVAs of astronauts and taking over the payload from the shuttle's SRMS (see Fig. 3). Canadarm2 is a 17.6-m-long manipulator equipped with seven joints and latching end effectors at both ends through which data and power can be transmitted allowing it to move itself end over end in an inchworm-like movement limited only by the number of power date grapple fixtures (PDGFs) on the station. To cover even more ground, it can use a work platform on rails, the mobile base system (MBS), installed outside the station aimed at providing lateral mobility to the arm [13, 17].

The *Special-Purpose Dexterous Manipulator* (SPDM), or *Dextre*, is a dual-arm manipulator system, much smaller than the previously mentioned SSRMS, and capable of performing delicate maintenance work and repairs done during the EVAs. Defined as "a robot handyman," it was developed by the same Canadian company that developed Canadarm2, MDA, and launched it in March 2008. On 4 February 2011, it successfully completed its first assignment that consisted of unpacking two critical pieces of equipment delivered by Japan's Kounotori2 spacecraft. As it can be seen in Fig. 4, the SPDM resembles a headless torso equipped with two extremely agile, 3.35-m arms, each of which has seven DOFs and special end effectors for handling standardized orbital replacement units (ORUs). The body has just one DOF and can be attached at the end

Fig. 4 SPDM or Dextre. (From Wikimedia Commons.)

of the SSRM or it can be positioned by the latter at the various ORUs work sites around the ISS. Dextre is aimed at being teleoperated from inside the space station's robotic work station (RWS) [13, 15].

The *Japanese Experiment Module Remote Manipulator System* (JEMRMS) is a robotic manipulator system developed by JAXA. It is intended for supporting experiments conducted on the exposed facility of the Japanese Experiment Module (JEM) or for supporting its maintenance tasks. JEMRMS is composed of two arms having both six DOFs: the main, 9.9-m arm, fixed to the Japanese module and the small fine, 1.9-m arm. The main arm was launched on May 2008 alongside the Kibo Pressurized Module whereas the small fine arm was launched a year after on September 2009 aboard the unmanned H-II Transfer Vehicle (HTV) [13].

The *Robotic Component Verification on the ISS* (ROKVISS), was a two DOF, 0.5-m-long manipulator arm with a dedicated test bench, installed on January 2005, on the outer wall of the Russian service module of the ISS, Zvezda. It was developed by the DLR in cooperation with the German companies EADS Space Transportation, Kaiser-Threde, and Hörner & Sulger and close collaboration of the Russian Federal Space Agency (ROSKOSMOS) and Rocket and Space Corporation (RSC) "Energia." The experiment was developed mainly as a precursor of the *European Robotic Arm* (ERA) currently under development by the European Space Agency (ESA). ROKVISS verified the feasibility of direct telemanipulation through a series of telepresence experiments and capabilities, under real space conditions, of the components of the future modular lightweight robots (LWRs) currently being developed at DLR's Institute of Robotics and Mechatronics. In particular, the telepresence experiments were conducted using stereo images and bilateral, high-fidelity force feedback that an operator on ground could feel via a force-reflecting joystick [13].

On September 2011, after six years, ROKVISS was brought back to Earth for analysis and after a series of tests, it confirmed that the current design of robots developed at DLR Institute of Robotics and Mechatronics is perfectly capable of dealing with the harsh space environment [18].

3. ROTEX

The *robot technology experiment* (ROTEX), developed by the DLR, represents one of the important milestones of space robotics. The first remotely controlled multisensory robotic arm was flown on the Space Shuttle Columbia (STS-55) in 1993 with the Spacelab-Mission D2. The ROTEX manipulator (see Fig. 5) was a small robot arm with six DOFs and able to reach in all directions and grasp objects within the enclosed work cell. Even so, the experiments proved several key technologies for orbital robotics in a microgravity environment, such as a multi-sensory gripper, teleoperation from the ground and the shuttle, shared autonomy, and time-delay compensation methods [13].

4. ETS-VII

The ETS-VII, visible in Fig. 6, was the first mission including a satellite equipped with a robotic manipulator, developed by the National Space Development Agency of Japan (NASDA, currently JAXA) and launched in November 1997.

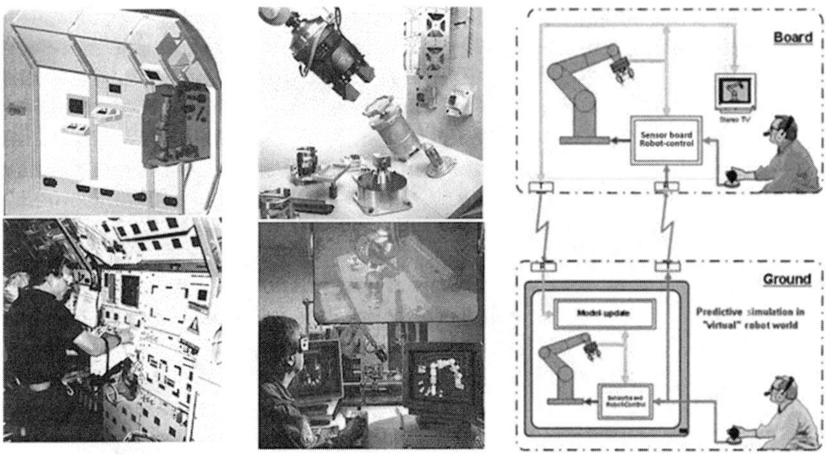

Fig. 5 ROTEX. (Credit: DLR, Institute of Robotics and Mechatronics.)

Together with ROTEX, it represents an important milestone in the development of space robotics and a precursor of later OOS missions that will be described hereafter. ETS-VII was an unmanned spacecraft, consisting of two main parts that could be separated for mission purposes (see Fig. 6). The main part, named Hikoboshi (after a prince in a Japanese classical tale), was equipped with a 2-m-long, six-DOF manipulator arm, whereas the smaller part, named Orihime (after a princess in a Japanese classical tale) acted as a target spacecraft [13, 19].

The mission main goal was to test free-flying robotics technology and demonstrate the feasibility of unmanned orbital operations. To reach those objectives, the mission needed to successfully complete the following two subtasks [13, 19]:

1. Autonomous rendezvous/docking (RvD) operations
2. A series of robot experiments (RBTs)

The RvD experiments were conducted by separating the spacecraft into two parts and using both R-bar and V-bar methods for the approach and docking phases. The onboard autonomy and a series of sensors, such as the global positioning system (GPS), a rendezvous laser radar, and vision-based proximity sensor were used to successfully accomplish the RvD experiments [19].

The RBT experiments were conducted using the onboard six-DOF manipulator allowing to obtain important flight data used subsequently to verify

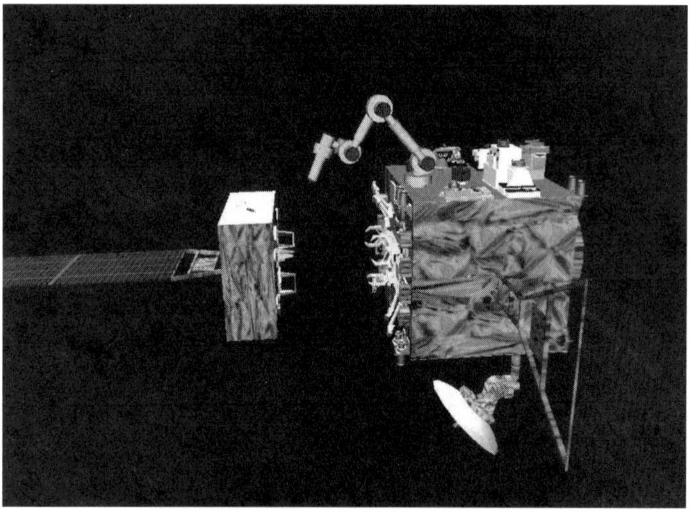

Fig. 6 Japanese ETS-VII. (Credit: Space Robotics Laboratory, Tohoku University.)

studies of different international research institutions regarding free-flying space robots [13]. These experiments as described in Refs. [13, 19] consisted of:

1. Remote observation of the spacecraft's surface using video cameras attached at different points of the robotic manipulator
2. Servicing and practical tasks, via teleoperation and/or autonomously, such as the exchange of an ORU, simulated fuel transfer, fastening a bolt, extracting a deployable truss mechanism, and so on
3. Dexterity testing of the manipulator using the dedicated task board
4. Autonomous chasing and capture of the free-floating target spacecraft[‡]
5. Berthing and releasing of the target spacecraft via the manipulator arm of the Hikoboshi
6. Control of the dynamic coupling between the manipulator and its base satellite using two different control approaches: the coordinated control and the reactionless manipulation

The communication between the spacecrafts and the ground control (Tsukuba Space Center, Japan) was made via Tracking and Data Relay Satellites (TDRSs) in geostationary Earth orbit (GEO). This choice extended the communication window length to up to 40 min during each orbit, in comparison to the direct communication window of 10 min, but at the same time, it introduced a time delay of 5–7 s to the signal, making the ground-based teleoperation very tricky. To overcome this difficulty, two methods were successfully tested. The first one consisted of a model-based assistance system able to predict and display the motion of the robotic manipulator in real time despite the signal delay. This method was verified using a surface-following and peg-in-hole tasks. Furthermore, the operator, thanks to this method, was not only able to see the correct kinematics of the manipulator delay, despite the signal, but he was also able to perceive the virtual reaction forces via a haptic device. The second method was the bilateral force feedback control and it proved to be more efficient than the conventional unilateral force control used in conventional teleoperation tasks [19].

5. RANGER

Ranger is a teleoperated dexterous robotic servicing system developed under the founding from NASA at the University of Maryland's Space Systems Laboratory. The design of Ranger is largely based on the requirements for robotic servicing of the HST that previously required human EVAs [13, 20].

Ranger, depicted in Fig. 7, consists of two dexterous manipulators, each having 10 actuators for eight DOFs, plus two actuation motions for the interchangeable end effectors.

[‡]It is important to note that for safety reasons, the free-floating of the Orihime spacecraft was limited within the space of the partially released docking mechanisms of the Hikoboshi [19].

Originally designed in 1993 for a free-flying experiment, the Ranger Telerobotic Flight Experiment (RTFX) evolved into the Ranger Telerobotic Shuttle Experiment (RTSX), in 1996, when a potential launch opportunity came up. Despite several tests and demonstrations for servicing missions in a neutral buoyancy facility of the University of Maryland (see Fig. 8), the effectiveness of the robot has still to be tested in space [13, 20].

6. ORBITAL EXPRESS

Orbital Express was a space mission developed by the U.S. Defense Advanced Research Projects Agency (DARPA). The objective of the program was to validate the readiness, cost-effectiveness, and utility of autonomous OOS technologies. The system was launched in March 2007 and consisted of two spacecrafts: a servicing satellite, an Autonomous Space Transport Robotic Operations (ASTRO) vehicle, developed by Boeing Integrated Defense Systems, and a satellite being serviced, a prototype of a modular next-generation serviceable satellite (NEXTSat), developed by Ball Aerospace & Technologies Corp. The ASTRO spacecraft was equipped with a 3-m-long robotic manipulator arm with six DOFs developed by MDA [19]. The purpose of the manipulator was to capture, video survey, correctly position, and service the target satellite, NEXTSat. The duration of the mission was 90 days.

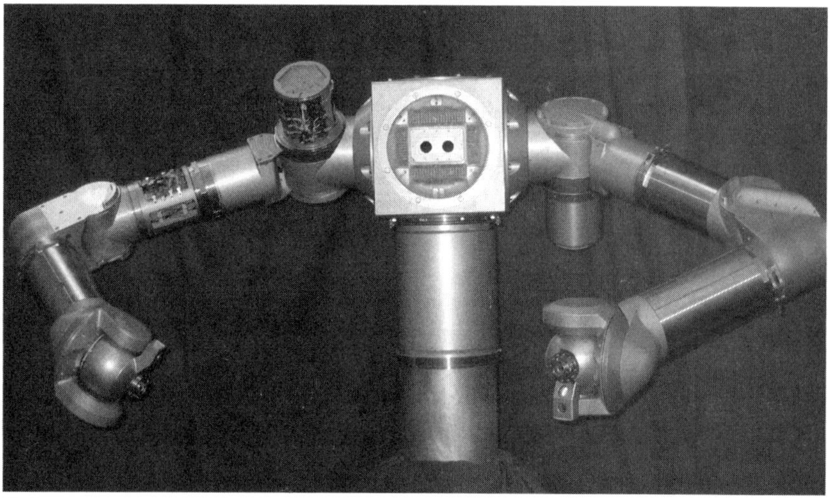

Fig. 7 Dexterous robotic servicing system Ranger. (Credit: Space Systems Laboratory, University of Maryland.)

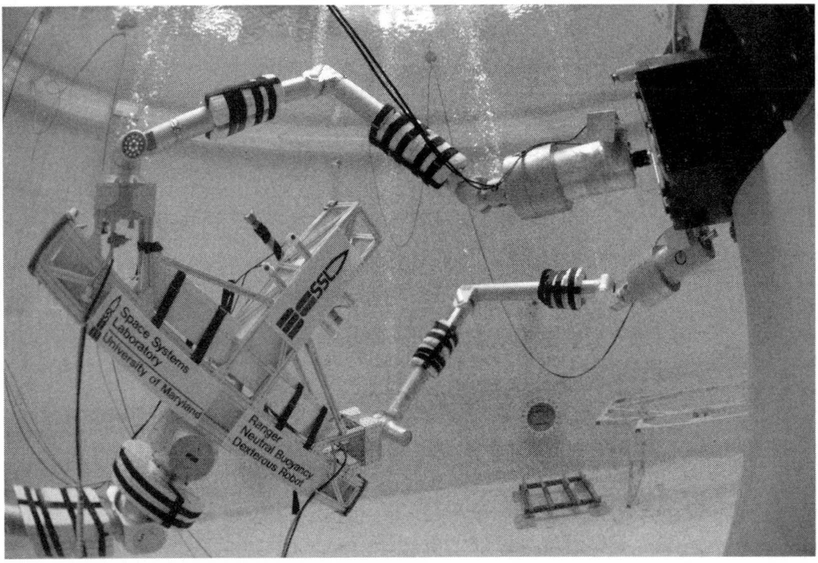

Fig. 8 Neutral buoyancy test of the Ranger system, reconfigured to simulate the servicing of the HST with the SPDM/Dextre. (Credit: Space Systems Laboratory, University of Maryland.)

The on-orbit demonstration involved the following experiments [19]:

1. Autonomous remote observation of the target satellite using video cameras attached to the manipulator
2. Autonomous ORU replacement
3. Autonomous transfer of a propellant
4. Autonomous rendezvous and capture (RvC) of the free-flying target satellite

Although similar scenarios were successfully explored 10 years before during the ETS-VII mission, Orbital Express demonstrated a higher level of autonomy performing a series of subtasks without any help from the ground station. For example, after ASTRO was separated from NextSat to a distance of 7 km, it returned to the target vehicle autonomously, performed a forced fly-around inspection, and finally approached, grappled and berthed the NextSat with the robotic arm. After that a servicing of the NextSat was performed by transferring the propellant to it and performing an ORU exchange (i.e., battery and computer replacement) [19].

It is important however to notice that although both ETS-VII and Orbital Express missions performed a successful autonomous capture/berthing and docking of the target spacecraft, the latter was always stabilized by an onboard Attitude and Orbit Control System (AOCS) and the spacecraft itself also presented dedicated grappling fixtures and optical markers [19].

However, in reality, this will not always be the case and if we consider a piece of space debris this will never be the case. And so, the past missions although representing important milestones for the orbital robotics are one solved piece of the puzzle that needs to be completed in order to have a fully functional autonomous system for robotic ADR and OOS. Thus, future missions and higher levels of technologies are needed to reach this future objective.

7. RRM

The RRM is an experiment mounted outside the ISS developed by the Satellite Servicing Capabilities Office (SSCO) at NASA's Goddard Space Flight Center in Greenbelt, Maryland, U.S.A. It was launched in July 2011, onboard the last space shuttle mission, the STS-135, and after its temporary installation on Dextre's Enhanced ORU Temporary Platform, it was transferred, in September 2011, to its permanent location on the ExPRESS (i.e., Expedite the Processing of Experiments to the Space Station) Logistics Carrier-4. Its purpose is to demonstrate and test, in a space environment, the robotic refueling and servicing of satellites. In particular of those that are not specifically designed for the on-orbit maintenance that are to date the majority of in-orbit satellites. The only spacecraft that was built with future maintenance tasks in mind was the HST [21].

As of August 2013, the mission is in its second phase after successful completion of the initial activities that involved extremely precise servicing tasks and the refueling of triple-sealed satellite fuel valves, common to current-day orbiting satellites [21].

The RRM, visible in Fig. 9, is a "washing-machine-sized" module that weighs approximately 250 kg and includes 1.7 l of liquid ethanol used as a fuel for the fuel transfer demonstration and a tank of nitrogen used to maintain certain system pressure. The module also includes activity boards, components, and four unique tools necessary for servicing tasks. The stowed tools are the Wire Cutter and Blanket Manipulation Tool, the Safety Cap Removal Tool, the Multifunction Tool, and the Nozzle Tool [21].

The robot used for delicate servicing tasks is the stations "robotic handyman," Dextre, remotely controlled by mission operators at NASA's Johnson Space Center in Houston, U.S.A. By using the mentioned tools, Dextre was able to successfully cut a peel-back protective thermal blankets, unscrew caps, gain access to valves, and finally perform the transfer of the ethanol [21].

The second phase of the mission will be carried out through 2015 with a new set of RRM tools and task boards that have been launched in summer 2013 and that will be launched to the ISS in early 2014 [21].

Past refueling missions, such as the ETS-VII or Orbital Express, have already demonstrated the feasibility and necessary technologies for this kind of task. However, they all included a "cooperative" spacecraft. The client spacecraft was specifically designed for the maintenance task. The innovation of RRM lies in

Fig. 9 RRM module on the ISS, temporarily installed on the Dextre robot's Enhanced ORU Temporary Platform. (Credit: NASA.)

the first on-orbit demonstration of technologies and techniques necessary to service a spacecraft not specifically designed for this purpose [22], making the whole mission more complicated but at the same time more realistic.

B. FUTURE OF ROBOTIC ON-ORBIT SERVICING AND ACTIVE DEBRIS REMOVAL

As mentioned at the beginning of Sec. A, robotic OOS has been a dream of the space community because of its conceptual design in the early 1980s. Nevertheless, this long-awaited dream has yet to become a routine in space activities [19]. One of the main reasons of this lies in the nonexistence of some of the required technologies and techniques as well as in-orbit tests of those that have been already developed.

However, this has not stopped us from dreaming and pursuing that aspiration. In fact, present and future missions of all major space actors have a common objective that is to encourage the development of the commercial satellite-servicing industry. This would eventually enable us, in the next few decades, to be able to extend the lifetime of spacecrafts or use decommissioned satellites as hardware stores for useful components to repair and/or upgrade the existing ones. Furthermore, given the similarities between ADR and OOS systems, this would also lead to the development of highly capable systems that could be used for ADR purposes.

However, it should be considered that the two systems although similar aren't the same, given that they present some differences that need to be thoroughly investigated. For example, during a typical OOS mission, the main objective is to capture, without damaging it, a cooperative spacecraft having known mass properties and eventual motion around its center of mass. The debris removal mission, on the other hand, involves capturing an unknown object, having an unknown motion around its center of mass, and the constraint of capturing it without damaging it is of less importance [12]. In addition, some of the objects could also contain leftover propellant and/or have significant tumbling rates indicating a nontrivial retrieval of those objects.

These problems, along with the nonholonomic behavior of a free-floating system and, most of all, problems of nontechnical nature, such as ownership, policy, and liability of abandoned spacecrafts have until today limited the widespread application of robotic OOS and ADR missions and need to be considered by the future missions if we want to unlock their full potential and pose the missing pieces to solve this long-awaited puzzle.

1. PHOENIX

Phoenix is a space program developed by the U.S. DARPA with the aim to demonstrate in-orbit technologies necessary for harvesting valuable components of decommissioned GEO satellites and reuse those components to create a new space system, a "satlet." This kind of satellites (currently under development) would be a new class of small satellites, similar to nanosatellites and thus with a wet mass between 1 and 10 kg. According to the project, satlets would be economically launched to the GEO region. They would be launched in a payload orbital delivery systems (PODSs) attached as a hosted payload on commercial satellites being launched to the same region. Once in-orbit satlets would be released and captured by a servicing satellite, a tender, in order to service retired satellites and create new systems (see Fig. 10). In case of a reused antenna, a satlet (or a series of satlets) will be attached to it acting as the power, communication, and control module, thus creating a new fully capable space system from reused old parts [23, 24].

The mission is planned for 2015 and it should be able to demonstrate the harvesting of an antenna aperture from a retired satellite, its reconfiguration, and reuse into the new free-flying space system. Currently, the mission is in phase II that involves the design, assembly, and validation of the flight components [24].

2. DEOS

DEOS is an OOS technology demonstration mission currently being developed by DLR and Astrium GmbH (as a prime contractor) with the aim to test in-orbit technologies and techniques necessary for a robotic capture of a noncooperative spacecraft, its servicing and subsequent disposal, and deorbiting [25, 26].

Fig. 10 Artist's concept of the Phoenix mission. (Credit: DARPA.)

The project is currently in the definition phase and according to the current planning it should be ready for launch in 2018 [26].

The DEOS' space segment consists of two spacecrafts, named the client and servicer (or target and chaser) designed having in mind different tasks that will be performed during the mission. The servicing spacecraft should have a 3-m robotic manipulator with seven DOFs, a docking, and berthing mechanism whereas the client spacecraft should exhibit a grappling fixture and also a docking and berthing mechanism.

The client should be designed to perform different attitude maneuvers in order to simulate a behavior of a noncooperative client satellite. At the beginning, it should have an Earth-oriented attitude mode, and then, increasing the complexity of the experimental program gradually, it should change to a spinning mode and finally it should change to a tumbling attitude behavior. Despite the fact that both the client and servicer will be equipped with GPS receivers, their use should be restricted for a case of emergency, such as for collision avoidance maneuvers, or as a reference. The vision-based system, using mono- and stereocameras, is instead planned to be used for navigational purposes [27].

The two satellites are expected to be launched rigidly connected to each other in an initial polar orbit with an altitude of 600 km, an inclination between 85 and 90 deg, and an eccentricity of 0 deg. The orbit should be then gradually decreased down to an altitude of approximately 400 km. This will reduce the contact time with the ground segment thus augmenting the complexity of the mission. The reentry (deorbit) within a predefined reentry corridor should be initiated from around 400 km. The near-polar inclination was chosen because it offers variable illumination conditions during each orbit, thus increasing even more the challenge of the mission. Depending on the nodal drift, identical illumination conditions will repeat every 4.3–6 months [28].

Similar to previous missions of this kind (e.g., ETS-VII and OE), the philosophy of DEOS will be to subsequently "crawl, walk, and run" [27] which means that with time the mission will gradually increase in complexity.

The set of individual tasks to be performed during the mission, as described in Ref. [25], should be:

1. *Far range formation flying:* During this initial mission phase, the servicer and client will fly in formation at a constant distance (of at least 2 km) from each other using just GPS sensors or angle measurements provided by different ground stations for navigation. During this phase, the dynamical behavior of each spacecraft will be accurately determined and in particular eventual residual spinning and tumbling rates of the client will be pointed out.
2. *Rendezvous:* During this task, the previous distance between the client and servicer will be reduced and the latter will be able to capture and berth the former. However, before this, a forced fly-around maneuver for inspection purposes will be preformed by the servicer.
3. *Capture and berthing of a noncooperative target:* During this task, at first several images of the client will be acquired and processed on ground in order to estimate its motion. Then once that the grappling fixture is inside the working space of the manipulator the actual grasping will be attempted. Finally, once the target spacecraft is successfully grasped and stabilized docking will take place.
4. *Docking with a cooperative target:* During this task, the two spacecrafts will be once again separated and this time, the client will be actively controlled in order to be approached slowly by the servicer so that the latter could perform the docking without the use of the manipulator.
5. *Flight in a coupled configuration and reentry:* Once that all previous mission phases are completed, different flight maneuvers of the coupled configuration will be performed in order to identify the dynamical parameters of system, OOS tasks will be demonstrated, and finally a controlled deorbiting of the rigidly coupled configuration is expected to be performed.

C. DFKI-RIC'S CONCEPTS FOR ON-ORBIT CAPTURING SYSTEMS

DFKI stands for German Research Center for Artificial Intelligence. It was founded in 1988 and with its research centers all over Germany, it is one of the largest nonprofit research institutes in the field of innovative software technology and applications based on artificial intelligence (AI).

The DFKI's Robotics Innovation Center (RIC) in Bremen, with more than 100 computer scientists, engineers, and specialists in mechatronics, is a renowned research institute in the area of robotics for space applications and planetary exploration, able to develop innovative solutions that combine the AI with customized robotic hardware.

In 2013, a study was performed at RIC about alternative capturing systems of free-floating satellites. The results are the concepts that will be briefly described in this section.

1. OCTOPUS ARMS

This concept, visible in Fig. 11, envisions an octopus-like robot able to grasp unknown objects by wrapping its tentacles around them.

The realization of this concept can be done using manipulators similar to those already proven for the spaceflight but with much more joints, that is, around 30 joints would be needed.

To contain the pressure applied to the grasped surfaces, an impedance control can be implemented and at the same time, it can be assumed that the contact points of the arms can be provided with some sort of springs and dampers.

The advantages of this concept are:

1. *Economical*: It could use proven and already-available components.
2. *Shape independent*: It could handle objects of any shape and it doesn't need particular grasping features on the surface of a satellite.

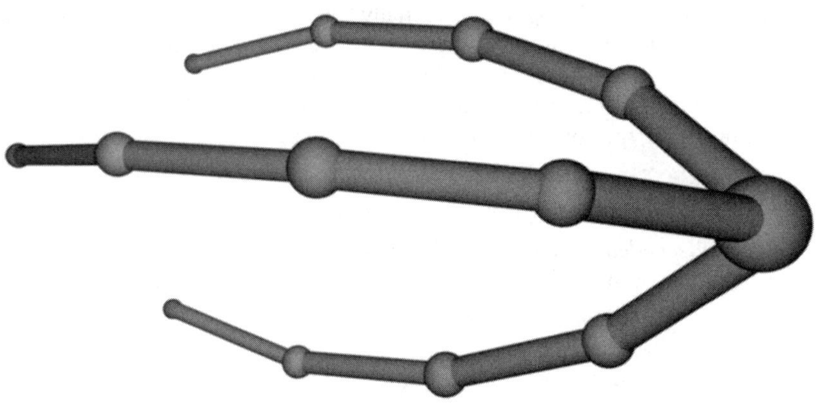

Fig. 11 Graphical representation of the "Octopus arms" concept. (Credit: DFKI GmbH.)

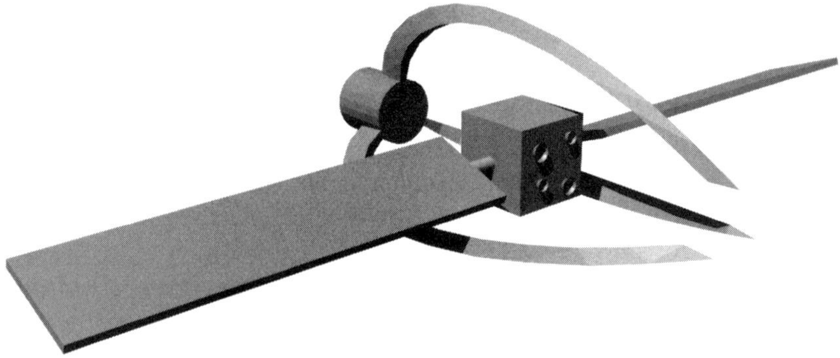

Fig. 12 Graphical representation of the "Elastic gripper" concept. (Credit: DFKI GmbH.)

The disadvantages, on the other hand, are:

1. *Complex*: A large number of actuated joints is needed and requires a big control effort.
2. *Energetically inefficient*: It requires high amounts of energy to actuate the manipulators.

2. ELASTIC GRIPPER

The following concept was imagined as one big gripper having an elastic internal structure of the fingers so that it can safely grasp a satellite without damaging it. The structure of the fingers could be similar to that of the wheels of planetary rovers.

The concept, visible in Fig. 12,[§] essentially takes some of the advantages of the previous one while reducing the complexity of the system by reducing the number of necessary joints to just one. As a consequence, the amount of the applied pressure can't be finely tuned according to different surfaces of a satellite.

To summarize, the advantages of this concept are:

1. *Economical*: It could use proven and already-available components.
2. *Shape independent*: It could handle objects of any shape and it doesn't need particular grasping features on the surface of a satellite.
3. *Simple design*: Just one actuator is required.

[§]For the sake of clarity, the figure does not include the elastic internal structure mentioned in the text.

Fig. 13 Graphical representation of the "Swarm of service vehicles" concept. (Credit: DFKI GmbH.)

The disadvantages are:

1. *Nonselective application of the pressure*: The amount of the applied pressure can't be finely tuned according to different surfaces of a satellite.
2. *Nontrivial manufacturing*: An appropriate internal structure could be difficult to manufacture given the required dimensions of the gripper.

3. SWARM OF SERVICE VEHICLES

In this concept, visible in Fig. 13, instead of using one big system to capture a satellite, a swarm of small-service vehicles is imagined. The vehicles could be a sort of nano- or picosatellites, such as CubeSats[¶] (see Fig. 14), capable of communicating between them and collectively applying thrust to a satellite in order to "capture it" or steer it in the desired direction. On Earth, this concept is used by tugboats that move vessels much bigger than themselves.

The advantages of this concept are:

1. *Economical*: It uses a relatively inexpensive approach given that CubeSats are nowadays widely used for research purposes by all major academic institutions in the world. Moreover, complete ready-for-launch CubeSat kits already exist.
2. *Modularity*: CubeSats can be combined to make a larger satellite.
3. *Resilient design*: The failure of one or few individuals of the swarm won't compromise the whole mission.
4. *Shape independent*: The swarm could handle objects of any shape although single vehicles of the swarm could require particular grasping features on the surface of a satellite that need to be captured.

[¶] CubeSats are picosatellites having standard dimensions: cubes of 10 cm per side and a maximum mass of 1 kg.

The disadvantages are:

1. *Technological immaturity*: Formation flying of a large number of spacecrafts and swarm robotics is still in its infancy.
2. *Difficult to capture*: Secure capture of the target satellite by individual vehicles of the swarm could prove to be difficult.

4. ADAPTIVE GRIPPER

The following concept represents a sort of mixture between the first and second concept described, respectively, in Secs. 1 and 2. It would consist of a large gripper capable of capturing the whole spacecraft without damaging it by adapting its fingers to the structure of the satellite. It would be based on the proprietary technology used by the adaptive gripper DHDG developed by Festo AG & Co. KG and called the *Fin Ray Effect®*. The gripper is made out of thin layers of polyamide powder that are applied onto a surface and fused using a selective laser-sintering process. This design allows its fingers to adapt easily to any form that they grasp applying minimum pressure on the gripping surfaces [29].

The basic structure of the fingers consist of two flexible bands, connected at one end to form a triangle, and intermediate struts connected to the bands via

Fig. 14 CubeSats launched from the ISS. (Credit: NASA.)

joints. The fingers are then controlled using a proportional valve that enables the correct amount of applied pressure [29].

This concept would eliminate the disadvantages of the second concept (see Sec. 2), although the manufacturing process of a gripper able to capture a satellite could be a problem.

The advantages of this concept are:

1. *Shape independent*: The gripper could handle objects of any shape and it doesn't need particular grasping features on the surface of a satellite.
2. *Simple design*: Just one actuator is required.
3. *Technological maturity*: Uses an already-existing technology.
4. *Reduced mass*: The adaptive gripper DHDG according to the producer [29] has 80% less weight than similar industrial grippers that could translate into reduced mass of this concept when compared to the second one.

The disadvantages are:

1. *Cost*: This kind of gripper would require the usage of the proprietary technology that in turn could be costly.
2. *Nontrivial manufacturing*: The manufacturing of the gripper having sizes necessary to grasp a satellite could be problematic.

5. SURFACE GRASPING WITH DIRECTIONAL ADHESIVES

In the following concept, the usage of gecko-inspired directional adhesives was envisioned without the discussion about the supporting structure to which these adhesives could be applied. However, recent studies on this topic [30, 31] suggest that for grasping purposes, directional adhesives could be applied to structures composed of either rigid tiles, loaded with central tendons, (see Ref. [30] for further information) or to foot-like flexible materials (see Fig. 15), loaded via push–pull control cables [31]. These systems can be then mounted either on a robotic manipulator or a service vehicle itself in order to perform a capture of a satellite.

Directional adhesives took the inspiration from nature and in particular from the dry adhesion used by geckos. This kind of adhesive is particularly indicated for the purposes of capturing a spacecraft because it works on many surfaces, requires minimal energy for attachment and detachment, and it could be used easily in space environment. In fact, contrary to all other methods of adhesion that are pressure sensitive and thus not suitable for space applications due to the radiation, temperature, and so on, the directional adhesion relies on the interactions, on molecular level, between tiny polymer hairs, and the surface, called Van-Der-Waals forces [30,32]. To put it in a way of the lead designer of the Stickybot, Mark Cutosky: "Other adhesives are sort of like walking around with chewing gum on

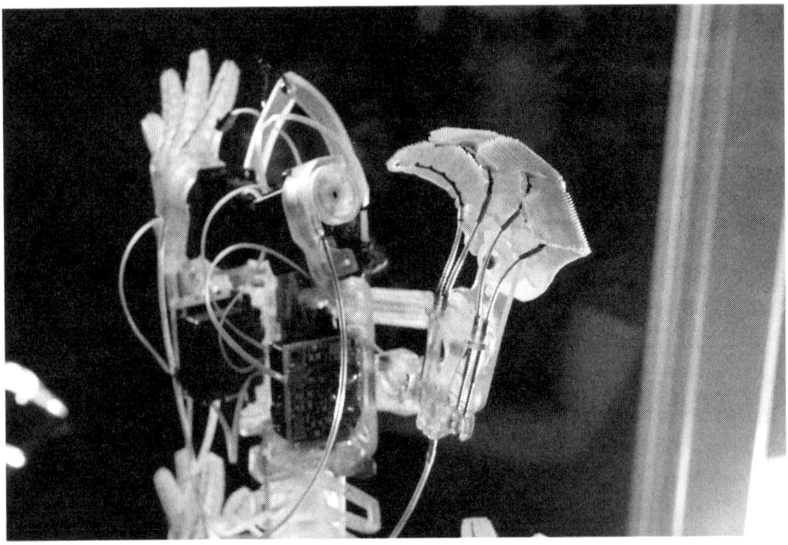

Fig. 15 Gecko-like Stickybot climbing a glass surface. (Credit: Cutkosky, M. R., Stanford University and Sangbae, K., MIT.)

your feet: You have to press it into the surface and then you have to work to pull it off. But with directional adhesion, it's almost like you can sort of hook and unhook yourself from the surface" [32].

The advantages of this concept are:

1. *Shape independent*: Devices using directional adhesives could attach themselves to a variety of flat or slightly curved surfaces without the need for particular grappling fixtures on the surface of the spacecraft.
2. *Relatively simple design*: As seen in Ref. [30], a device for this purpose can be relatively simple but very efficient.
3. *Energy efficient*: Dry adhesion requires a small amount of energy for attachment and detachment.
4. *Less reliant on precision*: Given that the device using this concept is almost surface independent, the precision of where the device could attach itself on the spacecraft is less stringent.
5. *Cost efficient*: A device using dry adhesion should be relatively light and the production costs shouldn't be prohibitive.

The only disadvantage of this concept is:

1. *Technological immaturity*: To date, the surface grasping with directional adhesion for space applications is still an object of academic studies although it is very promising.

6. EVOLUTIONARY ROBOTICS

Evolutionary robotics is a field of robotics that uses population-based, meta-heuristic optimization algorithms to develop novel hardware and software solutions for robotic systems. With this approach, it is possible to explore new and unexpected designs that wouldn't even be considered by a human. The expertise and knowledge of the latter, however, cannot be replaced by this method. Instead, for the best results, solutions of the evolutionary algorithms (EAs) should be used by a human expert as a starting point of the design process.

The added value of the evolutionary method, however, strongly depends on the formulation of the problem, performance evaluation, and generic description of solutions. Thus, a compromise must be made between the search space and the number of possible variations of solutions.

In the optimization study performed by RIC, the robotic system taken into account was the one visible in Fig. 16. This simple system, consisting of a robot having just one manipulator arm, was considered in order to obtain a solution as generic as possible. The objective of this study was to find the optimal solution of the robotic arm for firmly capturing a spacecraft, represented in Fig. 16 as a cube. The parameters used for the optimization process were 14 and some of them are number of elements, length of the first element, radius of the first element, and so on. The hypotheses were:

1. The target satellite is a free-floating object without an initial motion
2. The capture system is fixed with respect to the inertial frame of reference
3. The capture system is firmly attached to its base body

The requirements, on the other hand, were:

1. The capture system should be able to firmly catch the satellite
2. During the capture the satellite should remain intact

The optimization process was divided into two sessions of 480 s (or 8 min) each. In the first session, the evaluation was performed considering no applied external forces to the free-floating satellite and that the joints of the manipulator were free to move. In the second session, however, a force was applied to the satellite and an additional evaluation was performed based on:

1. The distance that the satellite traveled from the starting point
2. The sum of all collision forces between the arm and the spacecraft

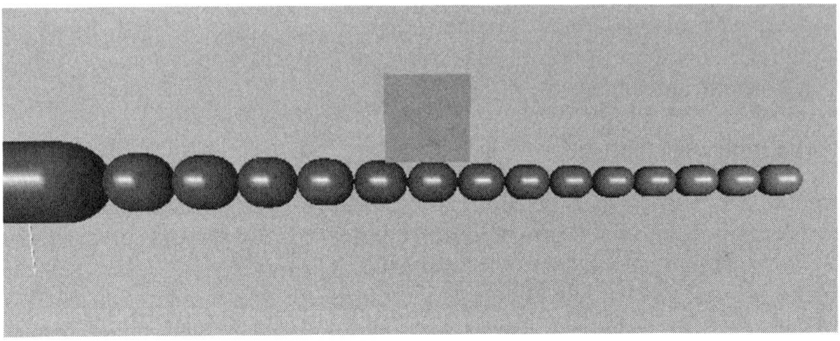

Fig. 16 Initial configuration of the arm optimization. (Credit: DFKI GmbH.)

Fig. 17 Best solution of the arm optimization. (Credit: DFKI GmbH.)

The whole process consisted of 70 independent optimizations and an evaluation of a total of 448,000 parameter configurations. The classification of the results was done using the growing-cell structures and 70 results were divided into 12 different classes. Among them, three classes presented a similar solution to the optimization process visible in Fig. 17. This solution was considered as the best possible given that it received the best average rating globally. Other solutions found can be seen in Fig. 18.

The conclusion of this study was that simulation-based optimizations of capture mechanisms can represent a valuable tool in the development process and the solutions could be used as a starting point for further steps in the development of new concepts.

III. MICROGRAVITY SIMULATION METHODS

The capture and berthing phase of any ADR/OOS mission represents one of the most critical and complex tasks of the whole mission. This is due to the difficulties in proper modeling of the dynamics of the system consisting of two connected spacecrafts of which one could be tumbling or have unknown mass properties. This is even more true in the case of a robotic mission, given the nonholonomic behavior of the servicer spacecraft due to the nonintegrability of the angular momentum equations. In other terms, due to the dynamical coupling between the spacecraft and its manipulator, the former experiences a motion due to the actuation of the latter. This effect can be compensated by using the servicer's AOCS. However, its extensive usage could severely limit the useful lifetime of the spacecraft. Therefore, other means should be used to either eliminate or deal with this disturbance whereas the AOCS should be used just to compensate the external torques acting in LEO environment.

To these circumstances, if we further add the difficulties in communications that are present in these kinds of missions (see Sec. 4), it becomes clear that the capture and berthing phase shouldn't be performed in space for the first time. All maneuvers should be thoroughly analyzed, simulated, and verified on ground in order to explore the capabilities and limitations of the planning and control algorithms. However, as simple as it may seem, reproducing a full, six-DOF, microgravity environment in a laboratory isn't an easy task and this section will cover some of the methods currently in use for such a purpose. Pure software simulations aren't considered in this section due to the difficulties to accurately model the real-time contact dynamics between the two spacecrafts and to achieve realistic sensor inputs. However, computer-based simulations can be particularly

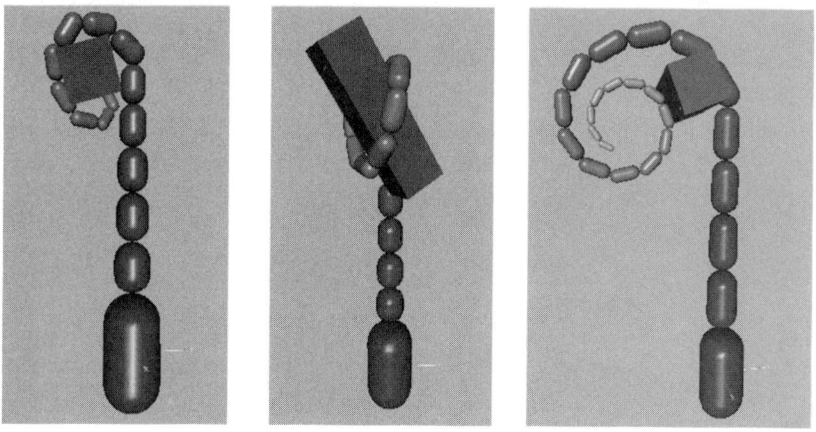

Fig. 18 Other solutions of the arm optimization. (Credit: DFKI GmbH.)

useful to estimate, in the first approximation, how the system will behave once that two spacecrafts are rigidly connected and thus what results to expect from the simulation methods that will be described hereafter.

A. NONROBOTIC SIMULATION METHODS

The following methods are standard nonrobotic methods for the simulation of microgravity conditions (i.e., weightlessness) on Earth and are widely used by aerospace agencies and companies to test their equipment.

1. AIR BEARINGS

Air bearings use a thin layer of pressurized air between moving objects of a mechanical system to transfer the load and eliminate the friction between the parts. Depending on the type of the bearing (i.e., linear, spherical, or cylindrical), certain force-free directions can be obtained. For example, using linear air bearings, it is possible to obtain three DOFs and simulate the microgravity environment in three directions: two translations and one rotation. Using spherical air bearings, visible in Fig. 19, it is possible to obtain nearly three rotational DOFs.

In the end, cylindrical air bearings provide the missing translation and a rotation around it, thus providing two DOFs.

To achieve all six DOFs, necessary for a full simulation of a microgravity environment, a combination of the three types of bearings must be taken into account. However, this means that the overall complexity of the system will increase considerably and with it the cost of the simulator [33].

Nevertheless, linear air bearings, despite their inherent characteristic of being a planar testbed, have been the most-used systems for testing on-orbit robotics in a laboratory. Notable space manipulators that have used an air-bearing table for simulation purposes are SRMS, SSRMS, and JEMRMS [34].

Fig. 19 Spherical air-bearing SRA250. (Credit: Specialty Components, Inc.)

2. NEUTRAL BUOYANCY

Neutral buoyancy refers to something that is neither floating nor sinking. It can be achieved by counteracting the force of gravity with the force of buoyancy. Practically, this can be accomplished in a water environment, for an unlimited amount of time, by specifically modifying the hardware that needs to be tested (e.g., spacesuits or spacecrafts) with a right amount of weights and floatation devices [35]. Thus, flight units of spacecrafts or space equipment cannot be directly tested. Instead, their custom-built versions need to be used for this purpose. Ranger Neutral Buoyancy Vehicle, similar to the robotic system illustrated in Fig. 8, is an example of such a kind of prototype. It was designed to be essentially identical to the Ranger TSX flight unit but it was also built to be used in the underwater environment. Moreover, the reaction forces of the surrounding fluid hinder the motion and this needs to be taken into account during the testing given that the characteristics of the dynamical motion of the system could be subjected to some changes [36].

3. PARABOLIC FLIGHTS, SOUNDING ROCKETS, AND DROP FACILITIES

Another way of creating a microgravity condition is to place an object in a state of free-fall. On Earth, this can be achieved by simply dropping it. However, this also holds true in orbit. In fact, the same object placed on the ISS or any other spacecraft in LEO will be subjected to the microgravity condition (commonly known as "0 g") given that the object and the spacecraft will be constantly falling around the Earth with the same rate [37]. Despite the fact that the module of the force of gravity in LEO is almost identical to that on the Earth's surface.

Ways of achieving the state of free-fall on Earth are by using specifically modified aircrafts (like the ESA's Airbus A300 Zero-G or NASA's "Vomit Comet"), sounding rockets, and drop towers (like NASA's Glenn Zero-Gravity Research Facility visible in Fig. 20 or Fallturm Bremen at the Center of Applied Space Technology and Microgravity at the University of Bremen).

In case of parabolic flights, the microgravity is achieved, as the name suggests, by flying an aircraft in parabolic arcs, each providing 20 s of microgravity ($\sim$0.02 g). A typical flight campaign consists of three individual flights, each consisting of 30 parabolas for a total of 90 flown parabolas. On each parabola, however, before and immediately after the weightlessness condition, two periods of increased force of gravity (1.8–2 g) are to be expected, each lasting for 20 s [38].

Sounding rockets are a kind of ballistic missiles having an almost vertical ascend and descend trajectories that can be used to provide microgravity conditions for a couple of minutes (up to 13 min). They were originally designed for sounding the physical properties of the atmosphere, hence the name of the rocket, but since 1982, ESA has used them also for microgravity research programs. The duration of the microgravity period depends on the apogee reached by the rocket. After the free-fall phase, the parachute is deployed to safely descend the experiment to the ground. The rocket configurations currently in use by ESA are MASER/TEXUS and MAXUS that could carry a payload of 0.64 m in diameter and 3.5 m in length to an altitude of 750 km and provide 13 min of weightlessness [39].

In case of drop towers, the weightlessness is achieved by dropping a capsule containing the experimental setup from a tower. This way, a microgravity condition can be obtained for a couple of seconds depending on the height of the tower itself. The NASA's Zero-Gravity Research Facility (see Fig. 20), located at the Glenn Research Center in Cleveland, Ohio, U.S.A., is a 142-m vertical steel vacuum chamber, largely below the ground, in which experiments can experience weightlessness for 5.18 s, during the free-fall of 132 m inside the drop vehicle. The latter is stopped in approximately 4.6 m using expanded polystyrene and the peak deceleration that it experiences is of 65 g. The mean deceleration is 35 g. The typical drop vehicle has a cylindrical shape, dimensions of 1 m in diameter, and the length of 4 m (see Fig. 20) so that to accommodate a payload that has 0.97 m in diameter and the height of 1.6 m. The maximum permitted payload weight is 455 kg [40].

4. SUSPENSION SYSTEMS

A suspension system is yet another method for creating a microgravity condition in a laboratory. It's working principle is to use a crane system to counteract the

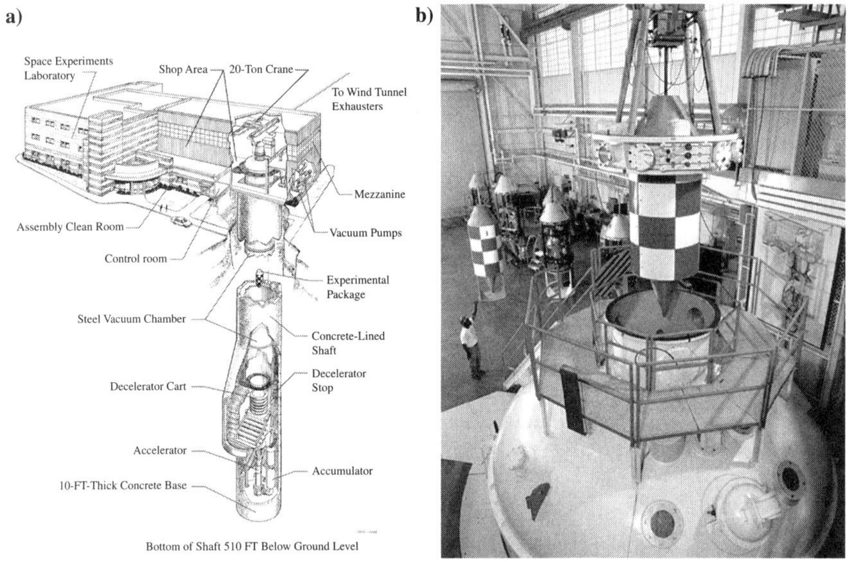

Fig. 20 Zero-Gravity Research Facility at NASA Glenn Research Center. (Credit: NASA.) a) Diagram of the Zero-Gravity Research Facility and b) positioning a drop vehicle on top of the vacuum chamber.

force of gravity and thus provide three translational DOFs whereas the remaining three rotational DOFs can be achieved with a gimbal system. Precise sensors are then used to guide the motor controllers of the suspension system to simulate the dynamics of a vehicle in a microgravity environment [41]. The system should, however, be able to apply the mentioned counteracting force constantly during the 3-D translational motion of every link of the spacecraft's robotic arm without applying additional forces to it. This is very important to acknowledge given that generally the joint actuators and control algorithms of the manipulators used in space are designed just for the microgravity environment [42].

An example of a such a kind of system is NASA's MicroGravity Test Facility [41]. Another one was developed by Xu et al. in 1994 for the testing purposes of Self-Mobile Space Manipulator (SM^2). The system consisted of a passive, vertical counterweight system and an actively controlled horizontal system with which it was possible to obtain a microgravity environment of 0.01 g [43].

B. ROBOTIC SIMULATION METHODS

Standard microgravity simulation methods, described in Subsec. A, while having been used widely since the beginning of the space era present some difficulties in simulating full six-DOF contact dynamics between spacecrafts involved in a generic robotic ADR/OOS mission.

For example, a testbed based on air bearings can simulate the microgravity condition just partially. Moreover, the system generally requires a quite large supporting structure that adds inertia to the system, which could create a problem during the contact phase due to the large accelerations that could arise from it. Neutral buoyancy method, while being able to recreate a full microgravity environment, needs specifically modified prototypes of systems being tested and the water drag could significantly alter the characteristics of the contact dynamics. The free-fall systems, on the other hand, are afflicted by the limitations regarding the dimensions of the testing systems. Moreover, if we exclude the sounding rocket method, the simulation time at disposition is at a maximum of around 20 s that is inappropriate for testing a full capture/docking phase of an ADR/OOS mission. In the end, suspended systems suffer from instability problems and effects of extra inertia just like the air bearings. The robotic HIL simulation systems, on the other hand, are not affected by mentioned problems and can effectively simulate the microgravity environment by an active gravity compensation control. They also do not present problems related to the dimensions and complexity of space systems being tested, as long as appropriate robotic systems are used [44].

Generic robotic HIL systems, in the context of microgravity simulation, consist of three fundamental elements [44]:

1. A computer-based dynamics model of the whole space system (i.e., a servicing and target spacecrafts)

2. A robotic system capable of delivering full six-DOF motions of the spacecrafts generated from the aforementioned model
3. Mock-ups of the spacecrafts, including the hardware necessary for the capture/docking task

This architecture makes this method a hybrid between the full software simulation and hardware simulation because it uses both in order to take the advantages from each of them. In particular, this method uses a software simulation for the satellite dynamics, which is relatively easy to model mathematically and simulate on a computer whereas it isn't to achieve on ground. Contact dynamics, on the other hand, are reproduced physically given that it has the opposite problem: it can be accurately reproduced by the real hardware whereas it is difficult to accurately model mathematically [44].

However, they are far from being the perfect method for on-ground simulation of robotic ADR/OOS missions. They also present some difficulties (described in more detail further on) that need to be addressed appropriately to avoid instability problems and be able to obtain reliable simulation results.

Nevertheless, in recent years, they have been intensively used in the aerospace environment and examples of such hybrid simulators can be found all over the world. The SPDM Task Verification Facility (STVF) is one of them. It was developed by Cloud Security Alliance (CSA) and is based on a giant sixDOF, custom-built manipulator for the verification of robotic tasks performed on ISS by Dextre (or SPDM) [44]. Lockheed Martin's facility, the Space Operations Simulation Center (SOSC), opened in 2011, is another one. It features a full-scale model of the international space station's docking area and a model of the Orion spacecraft. All this is supported by two gigantic six-DOF robotic manipulators capable of simulating full-scale spacecraft motion relative to another object or point in space while allowing the use of complete sensor suites and spacecraft systems. In fact, this facility is used for the testing of the new NASA's manned Orion space vehicle. In Germany, DLR has developed the EPOS. Its first version was used for the testing of rendezvous sensors of Automated Transfer Vehicles (ATVs) and Japanese H-II Transfer Vehicles (HTVs). Another system of this kind developed in Germany is the INVERITAS test facility, developed by DFKI-RIC. The system was developed in 2012 and was successfully used to test a variety of sensors and algorithms that could be used in a future ADR/OOS mission scenario. A more-detailed description of both EPOS and INVERITAS facilities can be found in the following sections.

1. EPOS

The EPOS facility (identified as EPOS 2.0) of the DLR in Oberpfaffenhofen, Germany, was a joint effort between DLR's two institutions: the German Space Operations Center (GSOC) and the Institute for Robotics and Mechatronics. It was finished in 2009, although DLR had a similar capability also two decades ago with EPOS 1.0. The latter was the previous version of the current HIL simulator, developed in the late 1980s as a joint effort between DLR and ESA. The facility was extensively used from 1991 for a series of sensor tests and in recent years notably for testing of ATV's sensors and systems used during the final RvD approach [45]. However, the old facility was lacking some features necessary for a successful testing of future robotic ADR/OOS missions (e.g., such as DEOS). Thus, starting from 2008, the new facility was developed using two industrial robots, capable of delivering full six DOFs each and performing a translation of 25 m on a rail system [44]. The complete architecture of the EPOS 2.0, visible in Fig. 21, consists of [46]:

1. A rail system providing translational capability to one of the robots (identified as a mobile robot) up to 25 m
2. Two KUKA industrial robots (KR100HA and KR240) capable of providing six DOFs each
3. A lightning system capable of providing illumination conditions similar to those experienced in the orbit
4. A control system used to monitor and control RvD simulations

The advantages over the old facility consist of 1) the usage of two industrial robots that are in comparison to the custom-built robots more reliable and cheaper, 2) the capability of simulating the last 25 m of the RvD operations thanks to the

Fig. 21 EPOS. (Credit: DLR CC-BY 3.0.)

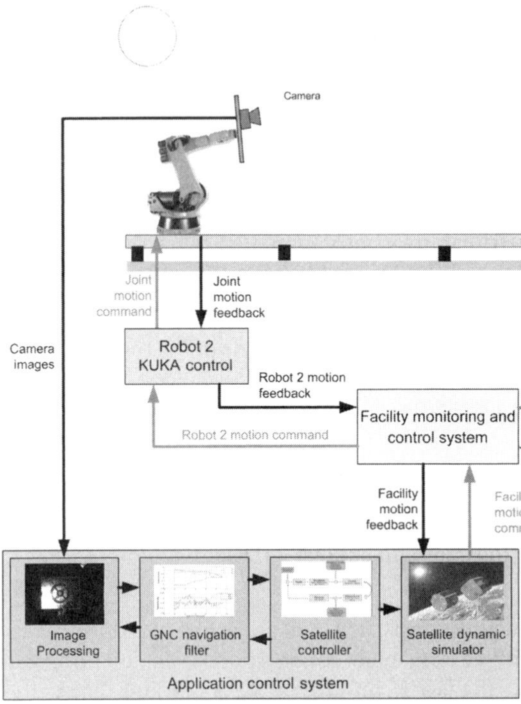

Fig. 22 EPOS simulation scenario. (Credit: DLR–EPOS [44].)

existence of the rail system, 3) the capability of simulating in-orbit illumination conditions, and 4) the capability of simulating the contact dynamics [44].

The achieved accuracy of the system is 2 mm for the position and 0.2 deg for the orientation [44].

In a typical simulation of RvD operations of an OOS, such as DEOS, the generic setup of the facility, visible in Fig. 22, consists of 1) a mock-up of the complete servicer spacecraft (manipulator and sensors included) mounted on the mobile robot, 2) a mock-up of the target satellite mounted on the fixed robot, and 3) a real-time dynamics simulator that calculates, at every sample update, the state vector (position and attitude) of the two spacecrafts based on the forces and torques (internal and external) applied to them. The forces and torques can arise either from the servicer's AOCS or from the physical contact. Either way, these quantities are fed to the simulator that in turn calculates the updated state vector of the two spacecrafts that is in turn fed to the control of the two robots [44].

To achieve a realistic simulation, the HIL system, however, must meet the following two fundamental conditions:

1. The responsiveness** of the robots must be very high to the control command
2. The dynamical behavior of the robots,†† during a physical contact, must be identical to the behavior of the satellites in space

The specifications of the two robots indicate that their responsiveness can be up to 32 ms. Moreover, their maximum sampling rate is 250 Hz. And while these numbers are quite high for industrial robots designed to be positioning machines for the automotive industry, they aren't enough to simulate real contact dynamics between two spacecrafts. For this application, the preferred sample rates should be at least 1000 Hz [44]. And so, special control must be considered to circumvent this problem [44].

The second requirement is obviously impossible to meet by the KUKA robots given their inherent high stiffness and the instability problem in case of a stiff contact.‡‡ Moreover, given that the low-level control software of the robots cannot be modified and that only the end-effector's position and velocity can be controlled, an additional external§§ control loop must be considered [44].

2. INVERITAS FACILITY

The INVERITAS facility, at DFKI-RIC in Bremen, Germany, was developed as a joint project between Astrium GmbH, a project coordinator, Jena-Optronik, and DFKI-RIC. The goal of the project was to develop a system capable of simulating a wide range of RvC operations. The developed system was successfully used in 2012 to test, during several different mission scenarios, various sensors and algorithms for autonomous guidance, navigation, and control (GNC).

The HIL system uses two robots (i.e., an industrial robot and a cable robot) to provide the motions of the target and servicer spacecraft, respectively. The choice of using this configuration, over the commonly used one for these kinds of facilities (see Sec. 1), was made in order to integrate the system inside the existing Space Exploration Hall (SEH) (visible in Fig. 23) and to maximize the approach distance between the servicer and target while not getting in the way if the system is unused. Furthermore, with this setup, the artificial Moon surface (visible in Fig. 23) could also be used to simulate landing maneuvers to the surface of the Moon or other celestial bodies.

The overall architecture of the INVERITAS HIL system consists of:

1. An industrial robot KUKA KR60-3 (visible in Fig. 24a)
2. A spidercam® cable robot (visible in Fig. 24b)

**Intended here as the time period between the control command and the actual movement of the robot.

††At their tips.

‡‡Note that the stiffness of the contact operation entirely depends on the satellites and their docking hardware [44].

§§Outside the inner, low-level control loop of the robots.

Fig. 23 Overview of the Space Exploration Hall at DFKI-RIC. (Credit: DFKI GmbH.)

3. A motion-tracking system (MTS)
4. A lighting system
5. Protection mechanisms
6. A control system

The robotic arm used in the INVERITAS facility is an industrial robot, KUKA KR60-3 (see Fig. 24a), capable of providing full six DOFs to a mounted mock-up of a spacecraft. The maximum admitted mass of the mock-up is around 40 kg. The control frequency of the manipulator is of 83.33 Hz and it can be controlled either in Cartesian or joint space. This robot was used to simulate the tumbling conditions of the client spacecraft (see Fig. 24a) and the latter could be either passive (i.e., without electronic parts) or active [47].

The spidercam cable robot (see Fig. 24b) used in the facility is a modified, proprietary system developed by SPIDERCAM GmbH in order to meet the precision and stability required by the HIL system. It uses a central mount supported by eight cables driven by four winches attached at the corners of the SEH. The robot can perform any translational motion and a rotation around its vertical axis. The workspace of the robot is a rectangular cuboid that is 15 m long, 5 m wide, and 6 m high. The maximum mass that the robot can withstand is 150 kg. The system is used to mount the servicer spacecraft and it is also capable of providing, through its cables, the power and data connection to the latter. The robot is controlled in Cartesian space and the control frequency is of 250 Hz. The motor used for the rotation of the servicer was developed at the DFKI-RIC and it also has the control frequency of 250 Hz [47].

The MTS deployed in the SEH uses seven Vicon motion capture cameras and retroreflective markers for off-line calibration and online measurements of the state vector (position and rotation) of the cable robot or better of the servicing spacecraft during its motion [47].

The lighting system of the INVERITAS simulator consists of six mobile spotlights, each having an 800-W halogen lamp. The spotlights are capable of providing a light of 3200 K with a maximal intensity of 3900 Lux when at 5-m distance from the illuminated object. Moreover, to recreate the lighting conditions in the orbit where large differences exist between the illuminated and dark parts of a spacecraft, all visible components of the HIL system have been covered with special light-absorbing paint. The result of this system can be seen in Fig. 25, representing a real image of the target spacecraft illuminated by only one spotlight as seen from the servicer's cameras.

The protection mechanisms integrated into the system were necessary in order to guaranty the safety of operators as well as of robots used for the simulation purposes. They include the light curtain that immediately stops the simulation (and hence the robots) once somebody enters the workspace of the robots; laser protection foil applied to the windows of the control room to prevent potential damage

a) b)

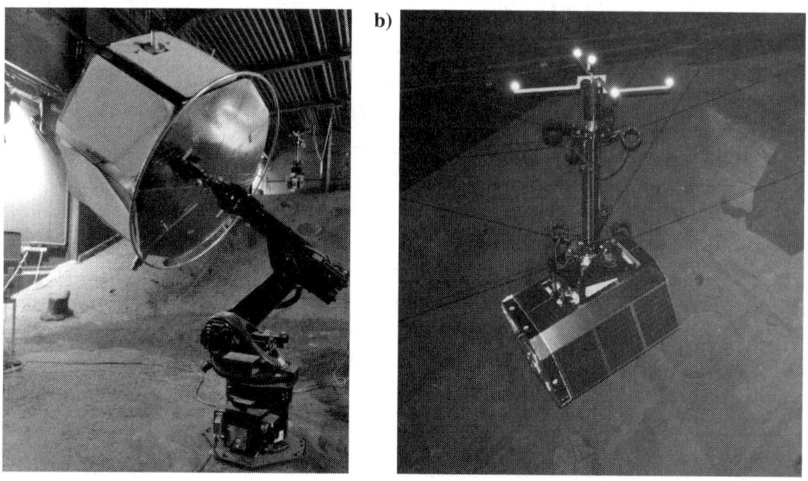

Fig. 24 Robotic systems of the INVERITAS HIL system. (Credit: DFKI GmbH.) a) KUKA KR60-3 used to move the target satellite and b) spidercam cable robot used to move the servicer satellite.

Fig. 25 Real image of the target satellite acquired from the servicer. (Credit: DFKI GmbH.)

to the vision of an operator due to the usage of a LIDAR; and collision avoidance system that runs in real time to ensure that the requirement of the minimal imposed distance between the parts of the system's hardware is met.

The overall accuracy of the INVERITAS HIL simulator resulted to be around 45 mm for a distance of 9 m between the spacecrafts. The major sources of error in the system were identified to be the cable robot and the MTS. Their initial absolute accuracies were 267 and 8.9 mm, respectively. However, after improvements applied to their measuring systems, the reached accuracies were of 6.6 and 5.3 mm, respectively. These proved to be sufficient to evaluate the efficiency of the pose estimation algorithms of the servicer's GNC system although there weren't adequate to measure its precision [47]. Thus, future optimization of the system is needed in order to reach higher accuracies. This can be done, for example, by improving or replacing the MTS or even by using different mock-ups of the client spacecraft to avoid deformations that occurred to its lightweight structure made of polystyrene during the testing phases [47]. Moreover, the control frequency of the used hardware needs to be appropriately addressed, just like in EPOS facility, in order to be able to simulate the contact phase of an ADR/OOS mission.

The test facility was used in cooperation with Astrium GmbH and Jena-Optronik to simulate autonomous rendezvous scenarios (see Fig. 26) between the servicer and client spacecrafts. For this purpose, a full-size mock-up of the client was used based on the specifications of the DEOS mission (see Sec. 2) and mounted on the KUKA robot. The servicer was a scaled version of the DEOS's servicing spacecraft[¶¶] and it was mounted under the cable robot. The mock-up although scaled had embedded sensors (i.e., a LIDAR, two stereocameras, and an observation camera) and processing units. The relative position of the spacecrafts during several mission scenarios was always determined by a real-time simulator so that the servicer's sensors could observe the target spacecraft as they would in the orbit. This enabled the testing of various sensors, data-processing algorithms, as well as of the GNC system of the servicer that was able to adapt the trajectory of the latter in a closed loop according to real-time measurements. Obviously, the whole approach phase of the servicer spacecraft, from 100 m, couldn't be realized given the limited length of the diagonal of the SEH. To overcome this limitation, computer simulation was used to create artificial sensor data up to a distance of approximately 12 m. Then the simulation was passed to the INVERITAS system [47].

The simulated mission scenarios included a station keeping at a distance of approximately 9 m, straight approach, and a fly around for inspection of the

Fig. 26 Rendezvous scenario simulation with INVERITAS facility. (Credit: DFKI GmbH.)

[¶¶]Without its robotic manipulator.

servicing spacecraft. In case of the target satellite, a stable free-floating and several tumbling motions were successfully simulated.

IV. CONCLUSION

The space surrounding our planet seemed almost infinite at the beginning of the space exploration. Nowadays, that space is densely populated by the leftovers of past missions, spent upper stages and fragments of in-orbit explosions, and collisions, called space debris. To prevent the onsetting of the Kessler syndrome and preserve the space environment for future generations, active removal of the in-orbit mass must be preformed.

Within this context, this chapter at first examined the readiness of current robotic technologies to tackle the space debris issue, by describing some of the most-relevant past and near-future missions. The conclusion that can be drawn is that despite the first milestones posed by the pioneering missions such as ETS-VII and Orbital Express, more needs to be done in order to be able to successfully address the challenging tasks of ADR. Phoenix and DEOS are just some of the planned missions pointing in that direction and that will take us one step closer to the realization of a fully capable robotic system able to tackle the problem of space debris.

The second part of this chapter explained the necessity of a HIL simulation and evaluated some of the most-common methods, focusing in particular on those involving robotic manipulators. The systems illustrated in more detail were DLR's EPOS and DFKI's INVERITAS. These systems were used to evidence the advantages and main limitations of existing, state-of-the-art robotic systems. The conclusion that can be drawn is that despite the difficulties in properly setting up such robotic systems, their potential and benefits are significant when compared to pure software simulation or even to other HIL simulation methods.

ACKNOWLEDGMENTS

This chapter is based on the lecture titled "Overview of DFKI-RIC research topics, focused on space robotics," held by Dr. Jan Paul at the Stardust Open Training School. Furthermore, we would like to thank the researchers S. Kim, M. Cutkosky, T. Boge, K. Yoshida, and D. Akin, as well as the companies DFKI, ESA, DLR, NASA, DARPA, and Specialty Components, for granting us the permission to use their images in this chapter.

REFERENCES

[1] Gaziyev, J., "Space Debris: Orbiting Debris Threatens Sustainable Use of Outer Space," July 2009, http://www.un.org/en/events/tenstories/08/spacedebris.shtml.

[2] E. Space Debris Office, Feb. 2013, http://www.esa.int/Our_Activities/Operations/Space_Debris/About_space_debris About Space Debris.

[3] Rex, D., "Will Space Run Out of Space? The Orbital Debris Problem and Its Mitigation," *Space Policy*, Vol. 14, No. 2, 1998, pp. 95–105, doi: http://dx.doi.org/10.1016/S0265-9646(98)00004-6, http://www.sciencedirect.com/science/article/pii/S0265964698000046.

[4] Kessler, D. J., and Cour-Palais, B. G., "Collision Frequency of Artificial Satellites: The Creation of a Debris Belt," *Journal of Geophysical Research*, Vol. 83, No. A6, 1978, pp. 2637–2646.

[5] Liou, J.-C., "Increase in ISS Debris Avoidance Maneuvers," *NASA Orbital Debris Quarterly News*, Vol. 16, No. 2, 2012, pp. 1–2, http://orbitaldebris.jsc.nasa.gov/newsletter/pdfs/ODQNv16i2.pdf.

[6] Parliamentary Office of Science & Technology, Space Debris, POST Note 355, The Parliamentary Office of Science and Technology, 7 Millbank, London SW1P 3JA, U.K. (Mar. 2010), www.parliament.uk/parliamentary_offices/ post/pubs2010.cfm.

[7] Liou, J.-C., Johnson, N. L., and Hill, N. M., "Controlling the Growth of Future LEO Debris Populations with Active Debris Removal," *Acta Astronautica*, Vol. 66, No. 5–6, 2010, pp. 648–653, doi: http://dx.doi.org/10.1016/j.actaastro.2009.08.005, http://www.sciencedirect.com/science/article/pii/S0094576509003981.

[8] Liou, J.-C., "Project Review: An Update on LEO Environment Remediation with Active Debris Removal," *NASA Orbital Debris Quarterly News*, Vol. 15, No. 2, 2011, pp. 4–6.

[9] Siciliano, B., Sciavicco, L., Villani, L., and Oriolo, G., "Robotics: Modeling, Planning and Control," *Advanced Textbooks in Control and Signal Processing*, Springer, London, 2009, Chap. 1, p. 1.

[10] Siciliano, B., and Khatib, O., "Introduction," *Springer Handbook of Robotics*, edited by B. Siciliano, and O. Khatib, 1st ed., Springer-Verlag, Berlin/Heidelberg, 2008, pp. 1031–1063.

[11] Reintsema, D., Landzettel, K., and Hirzinger, G., "DLR's Advanced Telerobotic Concepts and Experiments for On-Orbit Servicing," *Advances in Telerobotics*, edited by M. Ferre, M. Buss, R. Aracil, C. Melchiorri, and C. Balaguer, Vol. 31, Springer Tracts in Advanced Robotics, Springer, Berlin/Heidelberg, 2007, pp. 323–345, http://dx.doi.org/10.1007/978-3-540-71364-7_20.

[12] Ueno, H., Dubowsky, S., Lee, C., Zhu, C., Ohkami, Y., Matsumoto, S., and Oda, M., "Space Robotic Mission Concepts for Capturing Stray Objects," *Proceedings of the 23rd International Symposium on Space Technology and Science*, Matsue, Japan, 2002, pp. 1–8.

[13] Yoshida, K., and Wilcox, B., "Historical Developments and Advances of Orbital Robotic Systems," *Springer Handbook of Robotics*, B. Siciliano, and O. Khatib, 1st ed., Springer-Verlag, Berlin/Heidelberg, 2008, pp. 1031–1063.

[14] Canada, I., Canadarm—The SRMS Technical Details, http://www.ieee.ca/millennium/canadarm/canadarm_technical.html.

[15] Doetsch, K., and Lindberg, G., The Canadian Encyclopedia—Canadarm, 2013, http://www.thecanadianencyclopedia.com/en/article/canadarm/.

[16] Dunbar, B., Facts and figures, http://www.nasa.gov/mission_pages/station/main/onthestation/facts_and_figures.html.

[17] Canadian Space Agency, Canadarm and Canadarm2—Comparative Table (Dec. 2002), http://www.asc-csa.gc.ca/eng/iss/canadarm2/c1-c2.asp.

[18] German Aerospace Agency (DLR), DLR ROKVISS Robotic Arm Returns from Space, 2011, http://www.dlr.de/dlr/presse/en/desktopdefault.aspx/tabid-10312/475_read-1564/year-all/#gallery/3297.

[19] Yoshida, K., "Achievements in Space Robotics," *Robotics Automation Magazine*, IEEE, Vol. 16, No. 4, 2009, pp. 20–28, http://dx.doi.org/http://dx.doi.org/10.1109/MRA.2009.934818 doi:http://dx.doi.org/ 10.1109/MRA.2009.934818.

[20] U. o. M. Space Systems Laboratory, Dexterous Robotics at the Space Systems Laboratory (Jan. 2005), http://robotics.ssl.umd.edu/ranger/.

[21] Alessandro, A., Brill, J., and Reed, B., NASA Robotic Refueling Mission, 2013, http://ssco.gsfc.nasa.gov/robotic_refueling_mission.html.

[22] Alessandro, A., Brill, J., and Reed, B., Satellite Servicing Capabilities Office—Quick facts, http://ssco.gsfc.nasa.gov/quickfacts.html.

[23] Barnhart, D., Phoenix Mission, http://www.darpa.mil/Our_Work/TTO/Programs/Phoenix.aspx.

[24] SpaceRef, Honeybee Robotics Begins Developing Flight Systems for DARPA Phoenix On-Orbit Satellite Assembly and Servicing Program, 2013, http://www.spaceref.com/news/viewpr.html?pid=41976.

[25] Rupp, T., Boge, T., Kiehling, R., and Sellmaier, F., "Flight Dynamics Challenges of the German On-Orbit Servicing Mission DEOS," *21st International Symposium on Space Flight Dynamics*, Toulouse, France, 2009, pp. 1–13.

[26] Astium, Astrium wins DEOS Contract to Demonstrate In-Orbit Servicing, 2012, http://www.astrium.eads.net/en/press_centre/-kx9.html.

[27] Sellmaier, F., Boge, T., Spurmann, J., Gully, S., Rupp, T., and Huber, F., "On-Orbit Servicing Missions: Challenges and Solutions for Spacecraft Operations," *SpaceOps 2010 Conference*, AIAA, Huntsville, AL, 2010, pp. 1–11.

[28] Wolf, T., "The In-Flight Technology Demonstration of German's Robotics Approach to Dispose Malfunctioned Satellites," *11th Symposium on Advanced Space Technologies in Robotics and Automation*, ESA, ESA/ESTEC, Noordwijk, Netherlands, 2011, pp. 1–8.

[29] Festo AG & Co. KG., "Festo Adaptive Gripper DHDG," Tech. Rept., Festo AG & Co. KG., 2011, http://www.festo.com/net/SupportPortal/Files/53886/AdaptiveGripper_DHDG_en.pdf.

[30] Hawkes, E. W., Christensen, D. L., Eason, E. V., Estrada, M. A., Heverly, M., Hilgemann, E., Jiang, H., Pope, M. T., Parness, A., and Cutkosky, M. R., "Dynamic Surface Grasping with Directional Adhesion," *2013 IEEE/RSJ International Conference on Intelligent Robots and Systems*, Vol. 2013, IEEE, Tokyo, Japan, 2013, pp. 5487–5493, doi:10.1109/IROS.2013.6697151. http://ieeexplore.ieee.org/xpls/abs_all.jsp?arnumber=6697151 http://ieeexplore.ieee.org/lpdocs/epic03/wrapper.htm?arnumber=6697151

[31] Kim, S., Stickybot, http://www-cdr.stanford.edu/ sangbae/Stickybot.htm.

[32] Blackman, C., Secrets of the Gecko Foot Help Robot Climb—Stanford University, http://news.stanford.edu/news/2010/august/gecko-082410.html.

[33] Schwartz, J., Peck, M., and Hall, C., "Historical Review of Air-Bearing Spacecraft Simulators," *Journal of Guidance, Control, and Dynamics*, Vol. 26, No. 4, 2003, pp. 513–522.

[34] Xu, W., Liang, B., and Xu, Y., "Survey of Modeling, Planning, and Ground Verification of Space Robotic Systems," *Acta Astronautica*, Vol. 68, No. 11–12, 2011, pp. 1629–1649, doi: http://dx.doi.org/10.1016/j.actaastro.2010.12.004. http://www.sciencedirect.com/science/article/pii/S0094576510004352.

[35] Collins, J. W., and Elick, L. D., NASA—Neutral Buoyancy Laboratory, 2012, http://dx12.jsc.nasa.gov/about/whatIsNB.shtml?link=2.

[36] Xu, W., Liang, B., Xu, Y., Li, C., and Qiang, W., "A Ground Experiment System of Free-Floating Robot for Capturing Space Target," *Journal of Intelligent and Robotic Systems*, Vol. 48, No. 2, 2007, pp. 187–208, http://dx.doi.org/10.1007/s10846-006-9087-8.

[37] Zona, K., NASA—What Is Microgravity?, 2009, http://www.nasa-usa.de/centers/glenn/shuttlestation/station/microgex.html.

[38] E. Human Spaceflight Research Office, Parabolic flights, 2013, http://www.esa.int/Our_Activities/Human_Spaceflight/Human_Spaceflight_Research/Parabolic_flights2.

[39] E. Human Spaceflight Research Office, 2010, http://www.esa.int/Our_Activities/Human_Spaceflight/Human_Spaceflight_Research/ Sounding_rockets.

[40] Maier, S., Zero Gravity Research Facility (Jan. 2008), http://facilities.grc.nasa.gov/zerog/index.html.

[41] Dino, J., NASA—MicroGravity Test Facility, 2008, http://www.nasa-usa.de/centers/ames/research/technology-onepagers/microgravity_test_facility.html.

[42] Menon, C., Busolo, S., Cocuzza, S., Aboudan, A., Bulgarelli, A., Bettanini, C., Marchesi, M., and Angrilli, F., "Issues and Solutions for Testing Free-Flying Robots," *Acta Astronautica*, Vol. 60, No. 12, 2007, pp. 957–965, doi: http://dx.doi.org/10.1016/j.actaastro.2006.11.014, http://www.sciencedirect.com/science/article/pii/S0094576506004279.

[43] Xu, Y., Brown, J. H. B., Friedman, M., and Kanade, T., "Control System of the Self-Mobile Space Manipulator," *Control Systems Technology, IEEE Transactions on*, Vol. 2, No. 3, 1994, pp. 207–219, http://dx.doi.org/ http://dx.doi.org/10.1109/87.317978, doi:http://dx.doi.org/10.1109/87.317978.

[44] Boge, T., and Ma, O., "Using Advanced Industrial Robotics for Spacecraft Rendezvous and Docking Simulation," *Robotics and Automation (ICRA), 2011 IEEE International Conference on*, 2011, Shanghai, PRC, pp. 1–4, http://dx.doi.org/http://dx.doi.org/10.1109/ICRA.2011.5980583, doi:http://dx.doi.org/10.1109/ICRA.2011.5980583.

[45] Wimmer, T., European Proximity Operations Simulator (EPOS), 2009, http://www.weblab.dlr.de/rbrt/OOS/EPOS/EPOS.html.

[46] Boge, T., Wimmer, T., Ma, O., and Zebenay, M., "EPOS—A Robotics-Based Hardware-in-the-Loop Simulator for Simulating Satellite RvD Operations," *10th International Symposium on Artificial Intelligence, Robotics and Automation in Space*, Sapporo, Japan, 2010, pp. 267–274, http://elib.dlr.de/74391/1/Kopie_ISAIRAS_2010_050 _paper.pdf.

[47] Girault, B., Bartsch, S., and Kirchner, F., "Multifunctional Robot Test Facility for On-Orbit and Extraterrestrial Surface Exploration," *Proceedings of Ground-Based Space Facilities Symposium (GBSF-2013)*, Paris, France, O.A., 2013, pp. 1–6.

CHAPTER 15

Recent Developments in Asteroid and Space Debris Mitigation Techniques

Fabien Gachet* and Ioannis Gkolias*
University of Rome "Tor Vergata," Rome, Italy
Hiroshi Yamakawa†
Kyoto University, Kyoto, Japan

I. INTRODUCTION

The awareness concerning the space debris population on one hand and the near-Earth objects threat on the other hand has increased in the recent years. Following this, different ways of tackling these problems have risen. In this chapter, we first present the means of dealing with asteroids that could be a danger for the Earth. The concepts of a kinetic impactor are presented through the study of the uncertainty in the velocity at impact. Solar sailing and electric-sailing impactors are then introduced using retrograde orbits improving the deflection. A discussion on how to use solar sails to not only deflect asteroids but discover them fast is also presented. Finally, Coulomb force attractors using artificial halo orbits are presented. The second part focuses on the mitigation of uncooperative space debris. The use of a charged chaser satellite is proposed to rendezvous with target debris in the Earth's magnetic field. Then a periodic charge control is proposed, in order to deorbit it. The charged satellite, also compared with other propellantless methods, seems efficient enough for this kind of missions.

II. MOVING ASTEROIDS

In this section, various ideas of orbital deflection methods of potentially hazardous asteroids are discussed. First, we present the concept of a kinetic impactor. Then new concepts of such kinetic impactors using a solar sail or electric sail are introduced. We discuss how these new concepts can also be used in a novel near-Earth asteroid (NEA) flyby survey mission to increase the observability. The discussion focuses not on the solar sail spacecraft itself but on its orbital dynamics to realize the missions. In the last section, some detailed discussions on the combined use of gravity and electrostatic forces applicable for small asteroids are provided.

---

*Department of Mathematics.
†Research Institute for Sustainable Humanosphere.

A. MOTIVATION

The recent 2013 Chelyabinsk incident in which a NEA entered Earth's atmosphere and airburst over Chelyabinsk in Russia, or the older Tunguska (Fig. 1) event that was caused by the impact of a small asteroid over a forest in Siberia in 1908, show that small bodies do impact Earth and could cause severe damages if they happened over densely populated areas. Efficient asteroid deflection methods are therefore needed.

B. KINETIC IMPACTOR CONCEPT

There are many different types of planned missions to deflect an asteroid. One of the ways to do it is using a *kinetic impactor*, which will change the momentum of the asteroid by impacting it with a high-velocity mass. This will result in a small deviation of its trajectory, translating later at the point where an impact with Earth is assumed, in a miss distance large enough to avoid the said impact. This is illustrated in Fig. 2.

One important thing to take into account is the diversity in the imparted momentum at the asteroid impact. The shape, structure, and surface of the asteroid result in an uncertainty in the ΔV that gives a distribution of its magnitude and direction as illustrated in Fig. 3.

To design a kinetic impactor mission, one needs to optimize the impactor trajectory. *In the following sections, we assume that the kinetic impactor is actually the whole satellite itself.* The optimal spacecraft trajectory is then obtained by maximizing the worst value of the closest-approach distance, in terms of the Earth departure date and the asteroid arrival date of the spacecraft, under the constraints on the square of the escape velocity at Earth departure C3, due to launcher performance.

Fig. 1 Results of different asteroid encounters with Earth: a) result of the Tungunska event. http://www.sott.net/article/148819-Tunguska-the-Horns-of-the-Moon-and-Evolution and b) meteorite airburst above Chelyabinsk. http://trend-hunter.blog.so-net.ne.jp/2013-02-15-4.

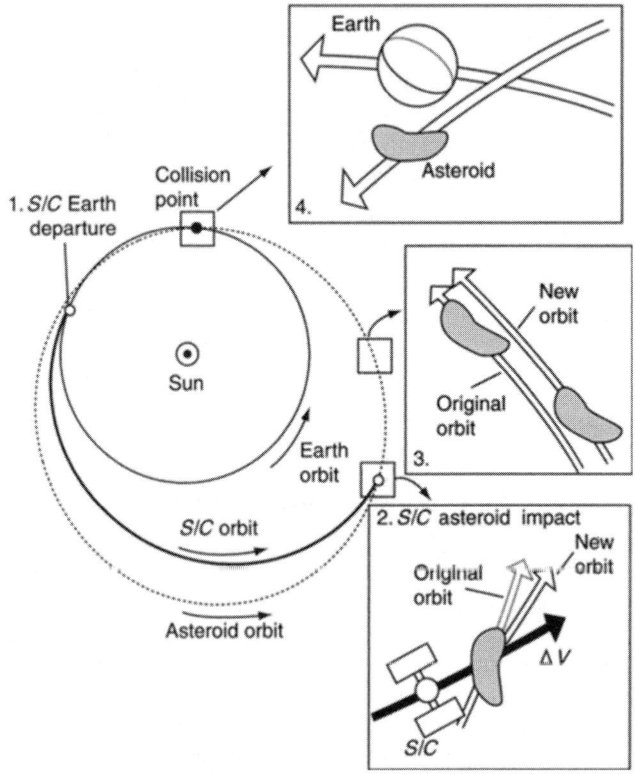

Fig. 2 The different phases of a mission involving a kinetic impactor [1].

The optimal problem is then defined by

1. Its objective function: The minimum Earth distance, which is to be maximized
2. The design variables being the spacecraft's Earth departure date, and asteroid arrival dates
3. Under the constraint that the Earth escape velocity—C3—is smaller than a given value depends on the launcher

Example 1

As an example, we can consider, as demonstrated in Ref. [1], a fictitious asteroid supposed to impact the Earth in 2060 with the orbital elements shown in Table 1.

We consider two cases, first without including any uncertainties in the transfer of momentum at impact between the spacecraft and the asteroid, and then considering uncertainties in terms of direction of the

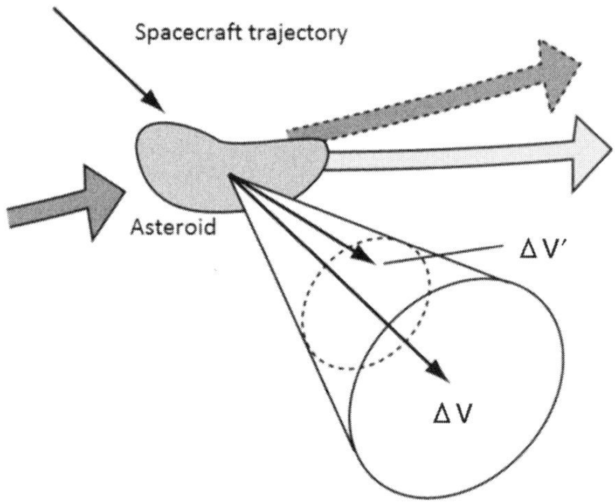

Fig. 3 ΔV uncertainty after impact [1].

ΔV (the optimal trajectory has been proven to be insensitive to uncertainties in the magnitude of the ΔV represented by α, the ratio between the effective momentum, and the momentum for a perfectly inelastic collision, in Ref. [1]). The constraints, identical for both cases, and uncertainties are shown in Table 2.

The results of the search for the optimal Earth departure and asteroid arrival dates in terms of the maximum closest-approach distance can be seen in Figs. 4a and 4b. The optimal result for the nominal case is the largest closest-approach distance of about 22,000 km, obtained at an Earth departure date of around 7 January 2011 for an asteroid arrival date of around 1 December 2011. This is a deterministic optimal design as no uncertainties were considered.

For the case including uncertainties, we notice that the largest closest- approach distance is significantly smaller at only 7000 km, but this is a result of the α parameter being set at 0.3. Otherwise, the distribution of the closest-approach distance with uncertainty is different from the one where uncertainty is not considered, with an optimal Earth departure date a little later, and an optimal asteroid arrival date a little later too.

TABLE 1 ORBITAL ELEMENTS OF ASTEROIDS [1]

a, AU	1.283	1.209
e	0.3226	0.3226
i, deg	2.683	2.683
Ω, deg	100.4	189.7
ω, deg	301.6	301.6
$\nu 0$, deg	58.42	346.0
Epoch	1/1/2060	9/13/2000

TABLE 2 ASSUMPTION ON FICTITIOUS ASTEROID, SPACECRAFT, AND ΔV UNCERTAINTIES [1]

Earth Collision Data of the Asteroid	1/1/2060
Asteroid diameter, m	150
Asteroid density, g/cm^3	3.00
Asteroid mass, tonne	5.30E+06
Spacecraft mass, tonne	1.00
C3 limit at Earth, km^2/s^2	50.0
Ratio of momentum transferred α	0.3
Average of ΔV magnitude μ_a	0.60
Standard deviation of ΔV magnitude σ_a	0.10
Average of ΔV direction μ_r, deg	0.0
Standard deviation of ΔV direction σ_r, deg	10.0
Time of departure range	01 Dec. 2010–28 Feb. 2011
Time of arrival range	01 Nov. 2011–29 Jan. 2012

Remark: Given their shape, these figures are called "pork chop plots" in the literature.

The different trajectories generated by the optimal solutions of the two different cases can be illustrated in Fig. 5 where the delay in the Earth departure and asteroid arrival is shown.

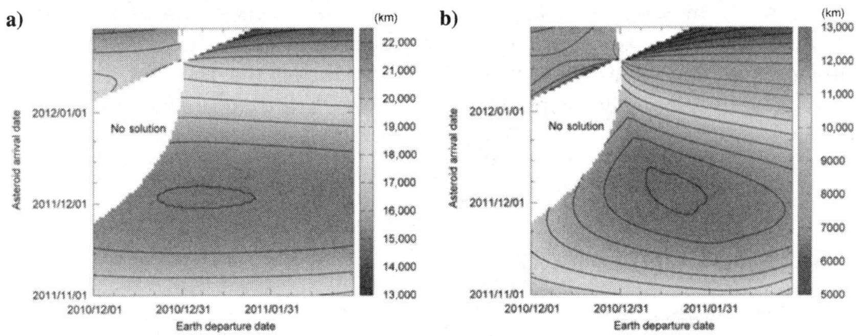

Fig. 4 Closest-approach distance distribution for a nominal case and including uncertainty [1]: a) closest-approach distance distribution for a nominal condition and b) closest-approach distance distribution including uncertainty in the ΔV direction.

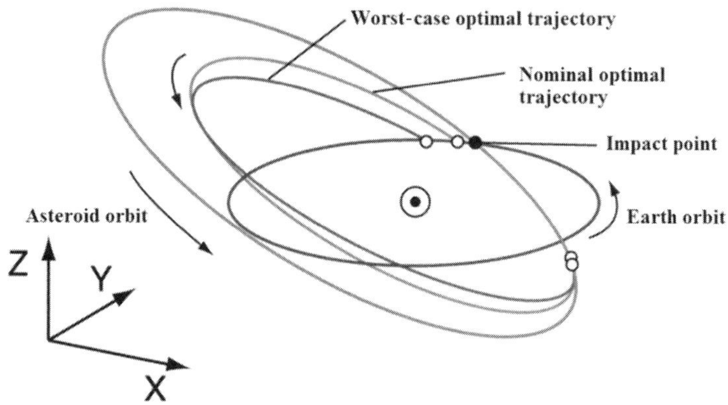

Fig. 5 Optimal trajectories for kinetic impactor missions, with and without considering uncertainties [1].

C. SOLAR AND ELECTRIC SAILS AS KINETIC IMPACTORS

A regular spacecraft using chemical propulsion designed to be a kinetic impactor spends a nonnegligible part of its mass to get to the asteroid. The mass of the satellite being a critical point to be able to launch, new propulsion designs have been presented including solar and electric sails. Compared with chemical and electric propulsion schemes, they increase the relative velocity with respect to the asteroid when the impact occurs, in addition to reducing the time to a kinetic impact [1].

The spacecraft trajectories of conventional asteroid missions move on nearly ballistic paths even when electric propulsion is used. With solar or electric sails, investigations have been led on heliocentric transfer trajectories departing from the Earth with a circular heliocentric velocity to a certain point at the same heliocentric distance. During the transfer, the angular velocity is gradually decreased whereas the heliocentric distance is kept constant.

1. SOLAR SAILS AS KINETIC IMPACTORS

a. Description of a Solar Sail

Solar sails (illustrated in Fig. 6) use the transfer of momentum transported by solar photons that encounter the sail; therefore, they do not need to take any kind of reaction mass onboard drastically reducing the total mass compared to other kinds of chemicals or even electric propulsion spacecrafts.

b. Description of the Retrograde Orbit Strategy

A good strategy to use solar sails as kinetic impactors is using a retrograde orbit that greatly improves the speed of the impactor: from $10\,\mathrm{km}\cdot\mathrm{s}^{-1}$ for a spacecraft using regular propulsion, to $60\,\mathrm{km}\cdot\mathrm{s}^{-1}$ for one using a retrograde orbit [2]. The idea being to first spiral inward from the Earth orbit to an orbit closer to the Sun at 0.25 AU, and then change the inclination there. Inclination change is indeed more efficient at radii closer to the Sun. From there, spiral outward to the targeted asteroid's orbit. This is illustrated in Fig. 7. During all this time, appropriate attitude control through maneuvers is applied.

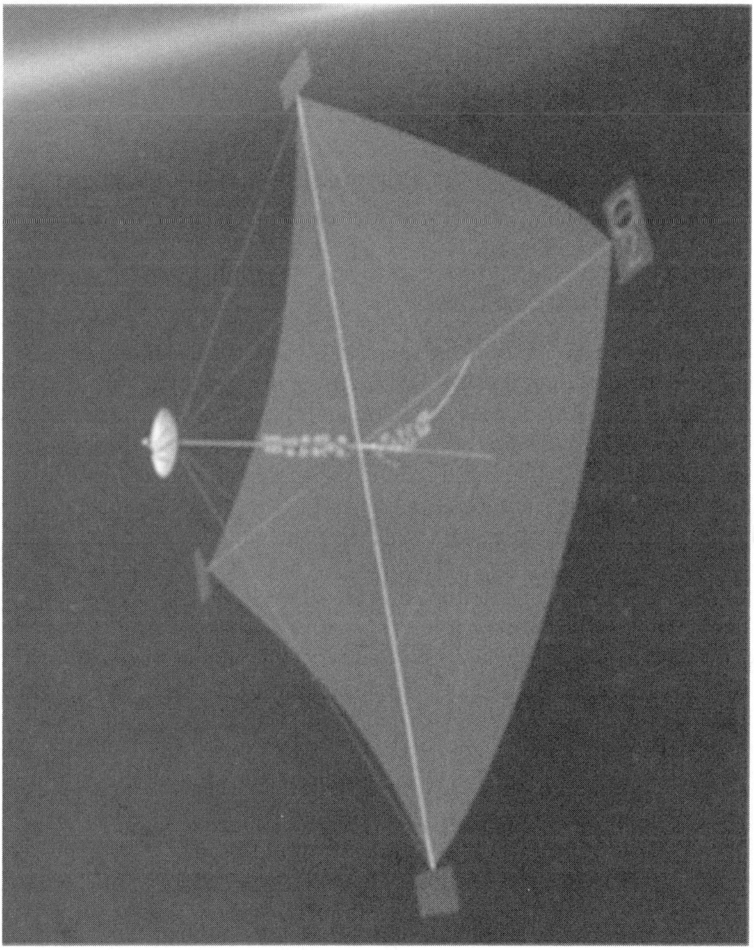

Fig. 6 Proposed model for a solar sail. http://antwrp.gsfc.nasa.gov/apod/ap030308.html.

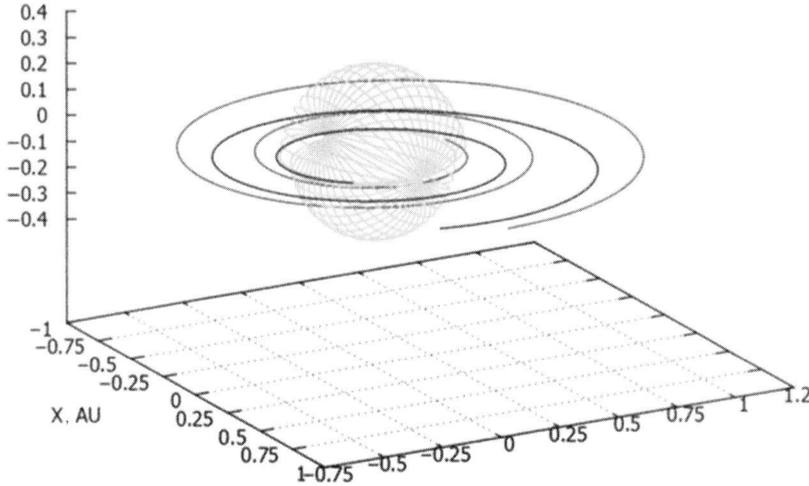

Fig. 7 Scheme of a retrograde orbit: spiral inward to the Sun, then change the inclination over the course of several orbits, and then spiral outward in a retrograde orbit [3].

2. ELECTRIC SAILS AS KINETIC IMPACTORS

The principle of retrograde orbits to increase impact speed can also be used with another type of propulsion that is called *electric sailing*.

Electric sailing, first proposed by Janhunen [4] consists of using the solar wind dynamic pressure to generate thrust without the need of reaction mass. To this purpose, long tethers are deployed around the spacecraft and kept at a high positive potential by an onboard electron gun and interact with protons of the solar wind stream. To give an order of magnitude, if the potential to be kept is 20 kV, the power requirement for the electron gun is 300 W. The total thrust force of an electric sail with a hundred of tethers that are 20 km long is 1 N at 1 AU.

Using electric sails as kinetic impactors with a retrograde orbit has been studied by Yamaguchi and Yamakawa [3]. Assuming, on the contrary to the solar sails that the spacecraft attitude cannot be changed, and is therefore fixed in an inertial frame, an ON/OFF control has been suggested, activating an electron gun to charge the tethers only at given times. The results of this ON/OFF control maneuvers during the change of the inclination phase, close to the Sun is shown in Fig. 8.

D. NEAR-EARTH ASTEROID FLYBY SURVEY MISSION USING SOLAR-SAILING TECHNOLOGY

Another proposed way to use the trajectory strategies using solar sailing developed in Sec. 1 is a flyby survey mission of NEAs. The heliocentric, non-Keplerian, circular transfer problem starting from the Earth and moving to

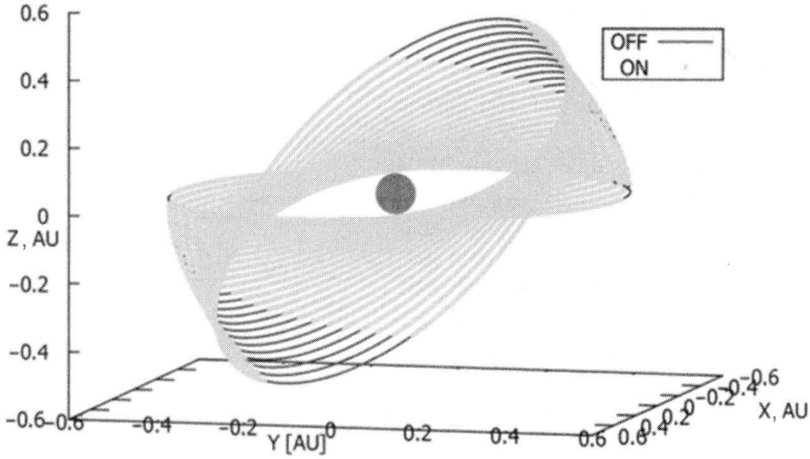

Fig. 8 Change of the inclination phase close to the Sun during a retrograde orbit for an electric sail. An ON/OFF strategy is used [3].

a certain point at the same heliocentric distance, in such a way that the solar sail remains stationary with respect to the Sun, has been studied by McInnes [5] and extended by Bando and Yamakawa [6] for cases where the solar sail is not stationary but appears to move slower than the Earth in the heliocentric frame. Figure 9 depicts the geometry of the mission.

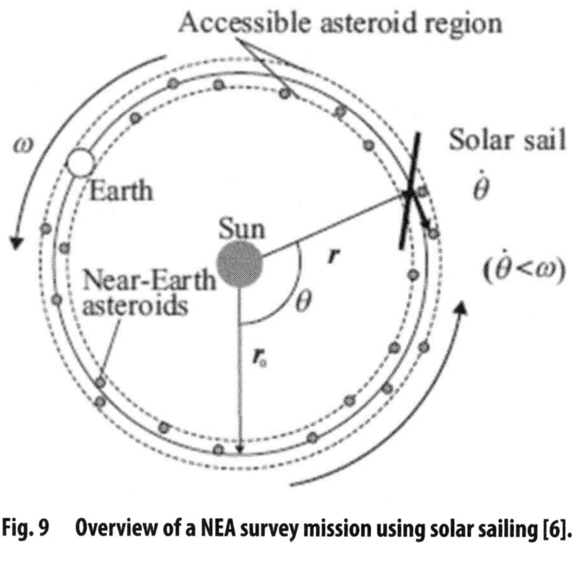

Fig. 9 Overview of a NEA survey mission using solar sailing [6].

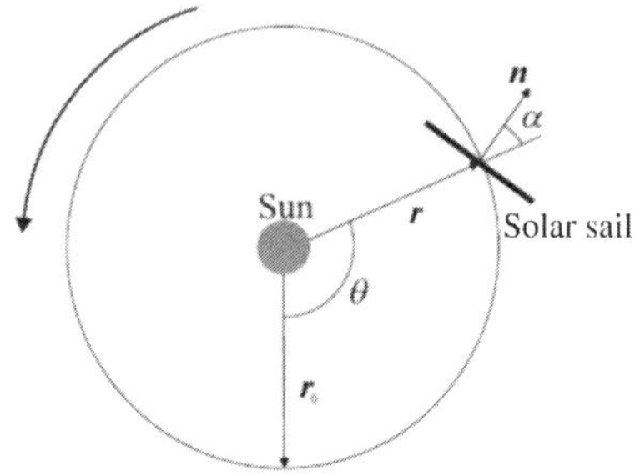

Fig. 10 Definition of the angles and distances of the problem [5, 6].

As said before, the spacecraft trajectories of conventional asteroid missions move on nearly ballistic paths; actually, this means that the spacecraft angular velocity is close to that of the given asteroid, which is the reason why the period between successive encounters of the spacecraft and the asteroid is several years. Using a solar sail and changing the orbital angular velocity allows instead to flyby asteroids in the vicinity of Earth's orbit in approximately one terrestrial year from spacecraft launch.

The equations governing the movement of the solar sail in polar coordinates (r, θ) in the heliocentric reference frame as shown in Fig. 10 are the following:

$$\ddot{r} - r\dot{\theta}^2 = -\frac{\mu}{r^2}(1 - \beta\cos^3\alpha) \tag{1}$$

$$r\ddot{\theta} + 2\dot{r}\dot{\theta} = \beta\frac{\mu}{r^2}\cos^2\alpha\,\sin\alpha \tag{2}$$

with β the sail lightness number, defined as the ratio of the solar radiation pressure force to the solar gravitational force acting on the solar sail. It is assumed to be constant. α is the pitch angle between the normal to the solar sail and the anti-Sun direction parallel to the Sun-sail line as depicted in Fig. 10. μ is the gravitational constant of the Sun.

From Eq. (2), a steering law is derived by taking $\ddot{r} = \dot{r} = 0$

$$\dot{\theta}^2 = \omega^2(1 - \beta\cos^3\alpha) \tag{3}$$

$$\ddot{\theta} = \omega^2\beta\cos^2\alpha\,\sin\alpha \tag{4}$$

with $\omega = \sqrt{\mu/r^3}$ the circular angular velocity.

Such a trajectory allows to encounter from 0.14% to 0.69% of the total NEA that are within 0.02 AU of the Earth's orbit in one year depending on what the sail lightness number is.

E. COULOMB FORCE ATTRACTOR

1. INTRODUCTION OF THE CONCEPT

Another possible concept to deflect asteroids recently proposed by Yamaguchi and Yamakawa [3] and previously by Bombardelli et al. [7] is the Coulomb force attractor concept. A Coulomb force attractor can deflect an asteroid using the Coulomb force produced by artificially charging both a spacecraft and the asteroid thanks to a conductive charging body and an electron gun. The concept is close to

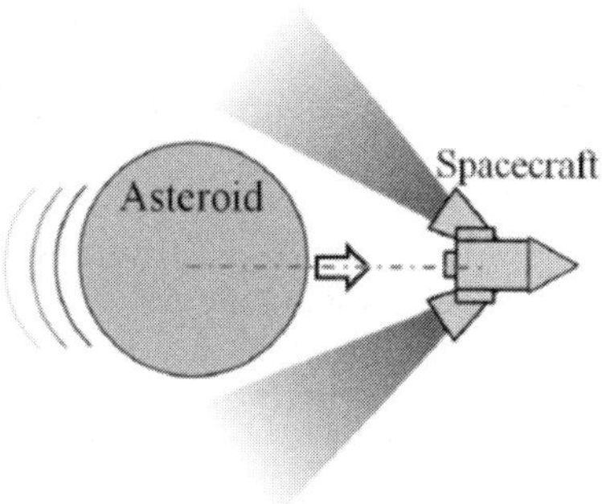

Fig. 11 Depiction of a spacecraft with two canted thrusters to avoid plume impingement [3].

that of a gravity tractor, with the expectation that an effective asteroid deflection using a spacecraft with a smaller mass than necessary for the latter is possible.

This method has been proven to be efficient for small asteroids. For instance, to maintain a 20-kV potential, which is the value usually considered, on an asteroid with a radius of 50 m, 5.21 kW are required; and for an asteroid with a 100-m radius, 16.61 kW are necessary [7]. It is to be added that 75% of known asteroids are C-type mainly composed of carbon and 8% are M-type, mainly composed of metal, which makes 83% of the known asteroids electrifiable and thus towable thanks to the electrostatic force. An example of a spacecraft equipped to be a gravity tractor can be seen in Fig. 11. One of the problems faced by Coulomb force attractor concepts it using classical propulsion with a reaction mass (chemical or electric), is the plume impingement due to the thrusters. Indeed, if the asteroid and the spacecraft are charged with opposite polarities, which is proven to be more efficient at high voltages [7], the thrusters need to face the asteroid to tract it. If all the mass ejected from the thrusters ends up on the surface of the asteroid, then the center of mass of the system would not move, which is not the goal as we want to tow the asteroid. This led to canted thrusters studies that solve the problem at the cost of a little inefficiency due to the cant angle.

2. ARTIFICIAL HALO ORBITS

Yamaguchi and Yamakawa [3] proposed a new concept, more fuel efficient, in considering an artificial halo orbit around an asteroid. The concept is depicted in Fig. 12. Even though being more fuel efficient, this method needs a greater towing force than with the classical hovering spacecraft approach, because as the asteroid is orbiting on a halo orbit to avoid plume impingement, the distance to the asteroid is greater. This is why a concept involving many spacecrafts on artificial halo orbits is later introduced.

3. INTRODUCTION OF THE CONCEPT

First, let us look at the case of one Coulomb force attractor. The equations describing the motion of the asteroid under the influence of a spacecraft orbiting such an artificial halo orbit are derived from the Clohessy–Wiltshire equations of motion. In the rotating reference frame defined by its center O on the original asteroid

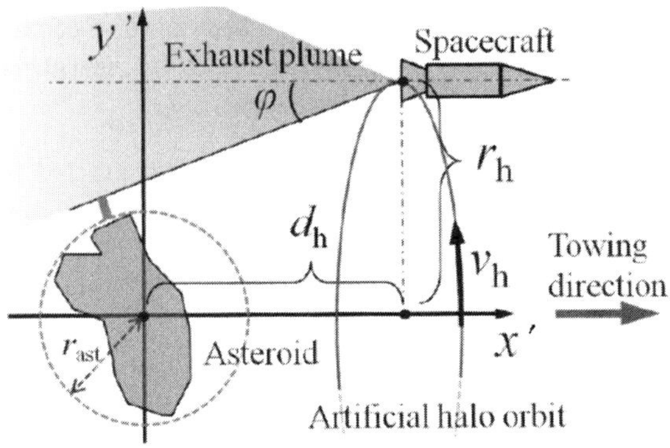

Fig. 12 Geometry of the artificial halo orbit concept, with the spacecraft positioned as to avoid plume impingement on the asteroid [3].

orbit, and rotating with the orbital motion of the asteroid, the equations of motion are

$$\begin{aligned}\ddot{x} &= 2\omega\dot{y} + \sum_{i=1}^{n} \frac{Gm_{SCi}(\mathbf{x_i} - \mathbf{x})}{x_i{}^3} + \sum_{i=1}^{n} \frac{F_d(\mathbf{x_i} - \mathbf{x})}{M_{ast}} \\ \ddot{y} &= 2\omega\dot{y} + \sum_{i=1}^{n} \frac{Gm_{SCi}(\mathbf{y_i} - \mathbf{y})}{y_i{}^3} + \sum_{i=1}^{n} \frac{F_d(\mathbf{y_i} - \mathbf{y})}{M_{ast}} \\ \ddot{z} &= 2\omega\dot{y} + \sum_{i=1}^{n} \frac{Gm_{SCi}(\mathbf{z_i} - \mathbf{z})}{z_i{}^3} + \sum_{i=1}^{n} \frac{F_d(\mathbf{z_i} - \mathbf{z})}{M_{ast}}\end{aligned} \tag{5}$$

where (x, y, z) are the coordinates of the asteroid with respect to the rotating reference frame. ω is the orbital rate that can be simplified by $\Omega = \sqrt{GM_{sun}/r_0{}^3}$ if a circular heliocentric motion at distance r_0 from the Sun is assumed, with G the gravitational constant and M_{sun} the mass of the Sun.

The first sum represents the contribution of the gravitational force from the satellites with (x, y, z) the asteroid position vector in the reference frame and (x_i, y_i, z_i) the ith spacecraft position vector in the reference frame, and $r = \sqrt{(x - x_i)^2 + (y - y_i)^2 + (z - z_i)^2}$.

The second sum is the Coulomb force contribution with F_d the Coulomb force in a plasma environment derived in Ref. [7].

The following assumptions made for the numerical simulations are summed up in Table 3.

Two cases are considered: in case A, the spacecraft orbits on an orbit with radius $r_h = 300$ m and a displacement distance $d_h = 300$ m. In case B, $r_h = d_h = 150$ m, so that the spacecraft is closer and the Coulomb force should therefore be

TABLE 3 ASSUMPTION ON A FICTITIOUS ASTEROID AND SPACECRAFT [3]

Asteroid diameter, m	50
Asteroid density, g/cmA^3	2.00
Spacecraft mass, kg	500
Surface electric potential, kV	20
Duration of the mission, days	1 year

stronger and also the deflection. The results of numerical simulations including the Coulomb force action and compared to the gravity force action is presented in Fig. 13. The deflection and change in velocity of the asteroid are clearly greater using a Coulomb attractor than using a gravity tractor, except on the z-axis where it is similar. This fact can be explained by looking at Eq. (5) and noticing that the z-axis motion is decoupled from motion along other axes. We can also note that the deflection and change of velocity are even greater if the spacecraft is closer as expected.

4. ARTIFICIAL HALO ORBITS USING MANY SPACECRAFTS

As said before, to increase the towing force, a concept involving many spacecrafts has also been studied. The spacecrafts are put either in the same orbit but not on the same position (diametrically in the example here) or on a different halo orbit at a different displacement distance. This is illustrated in Figs. 14a and 14b.

In principle, in numerical simulations, one should study the impact of the mutual attraction of the spacecrafts, but it has been proved [3] that this mutual attraction amounts to 0.5% of the towing force applied to the asteroid, so that it is neglected in the numerical simulations.

Two cases were considered again, and compared to the single-spacecraft strategy. The same assumptions are made for the asteroid and the spacecrafts as before. In case 1, two spacecrafts are put diametrically opposite in an artificial halo orbit with $r_h = d_h = 150$ m. In case 2, two spacecrafts are put diametrically opposite on an artificial halo orbit with $r_h = d_h = 150$ m, and two other spacecrafts are put diametrically opposite in an artificial halo orbit with $r_h = d_h = 200$ m.

The results can be seen in Fig. 15. As expected, introducing more Coulomb force attractors increases the towing force. For example, using four spacecrafts, the deflections in Δx, Δy, Δv_x, and Δv_y are 3 times bigger. It is to be noted that the results in Δz and Δv_z are small as we saw before due to the uncoupled equations along the z-axis, but this time, they even cancel out, due to the geometry of the spacecrafts put in diametrically opposite positions.

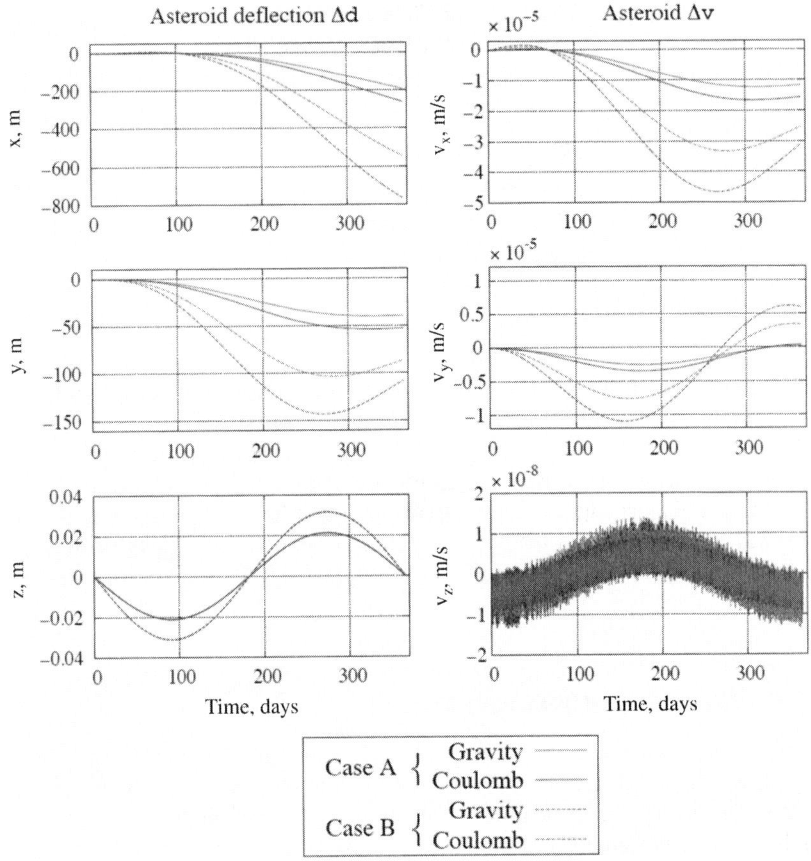

Fig. 13 Asteroid deflection Δd and change in velocity Δv thanks to a single Coulomb force attractor compared to the gravity force action [3].

III. TACKLING SPACE DEBRIS

In the previous section, various techniques to deflect potentially hazardous asteroids were presented. However, asteroids are not the only threat to human assets in Earth's neighborhood. Because of the dawn of the space era, the near-Earth space environment has been populated by a great number of man-made objects and unfortunately only a tiny fraction of them are still in operation. The vast majority of objects in low-Earth orbit (LEO) are nonfunctional, also known as space or orbital debris (Fig. 16). The continuous growth of the space debris population threatens operational satellites as well as upcoming missions, manned or unmanned.

The space debris population consists of rocket bodies or other pieces of equipment, that manage to get into the orbit around the Earth during the launch procedure. In addition, the orbital lifetime of satellites in LEO is several times longer than their operational lifetime. When a postmission reentry maneuver is not performed, a defunct satellite could orbit Earth for several years as a space debris, before it is disposed naturally by the atmospheric drag.

However, the deposit of new objects in LEO is not the only reason for the constant increase of space debris. Moreover, the debris population is proliferated by explosion and collision events. Explosions may occur to on-orbit fuel tanks that have a remaining quantity of high-pressure fluids. That was the main reason for the increase in the fragmentation debris up to 2007. Then two catastrophic collisions between satellites totally changed the evolution of the debris population (Fig. 17).

In the last few years, earlier scientists had already predicted that, if the density of objects in LEO becomes high enough, there is the possibility for a cascade of collisions to trigger. This kind of event would render the LEO environment totally unusable. The scenario, proposed in 1978 by Kessler and Cour-Palais [8], is also known as the "*Kessler Syndrome*" and nowadays seems more threatening than ever. High-fidelity population models, even in the best-case scenario that postmission disposal is assumed for almost all the future missions, predict an increasing trend for the debris population. It is clear that an active debris removal strategy should be considered.

The design process of a possible active debris mitigation mission is presented in Fig. 18. At first, the chaser satellite is launched and injected into an orbit. In the next step, a rendezvous maneuver is performed in order to approach the

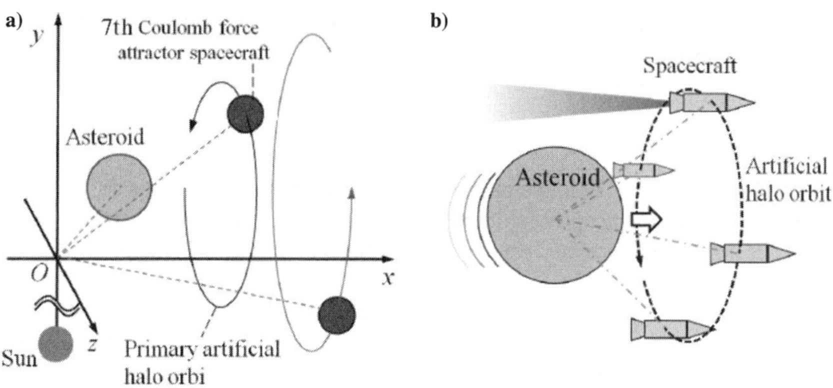

Fig. 14 Coulomb force attractors using many spacecrafts, either on different halo orbits or on the same orbit at different positions [3]: a) spacecrafts on different artificial halo orbits and b) spacecrafts set in a different position on the same artificial halo orbit.

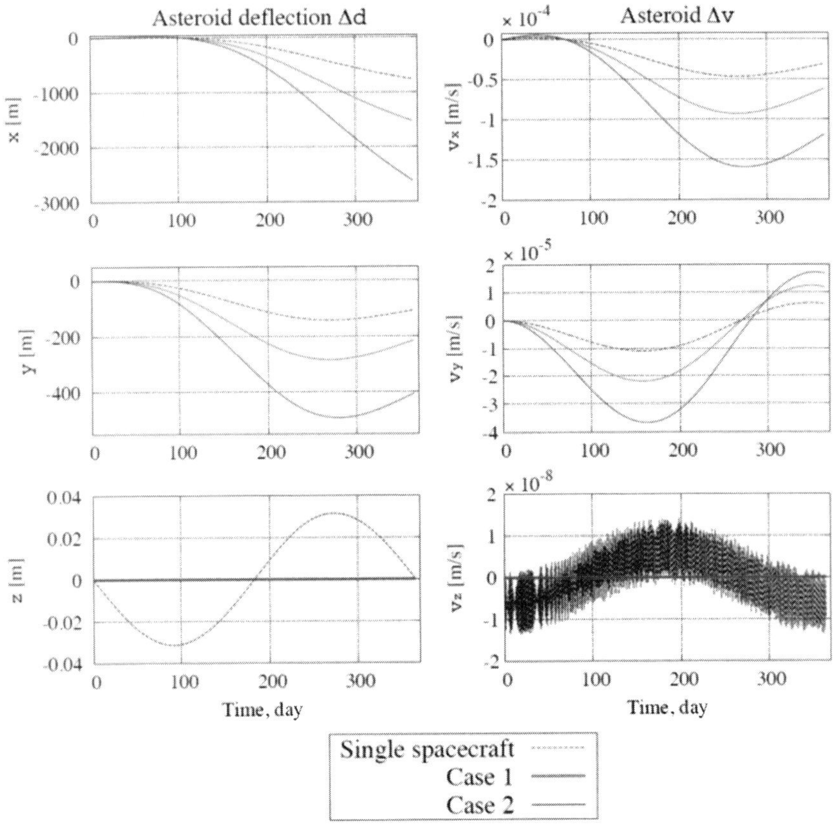

Fig. 15 Asteroid deflection Δd and change in velocity Δv thanks to different numbers of Coulomb force attractors [3].

uncooperative target. Then, the orbital and attitude motion of target debris is estimated and the chaser captures the debris or attaches a device to it. In the final stage, the debris reenters the atmosphere due to influence of the chaser satellite or the attached device.

In this section, the possibility to use a charged satellite as a chaser, in a debris removal mission, will be explored. Two parts of the mission will be investigated in detail, the approach to the uncooperative target and the deorbit phase. To design the rendezvous trajectory between the charged satellite and target debris, the orbital dynamics of a charged body in Earth's magnetic field should be explored. Then, assuming that target debris is captured, a periodic charge control to the chaser will be used, in order to dispose it.

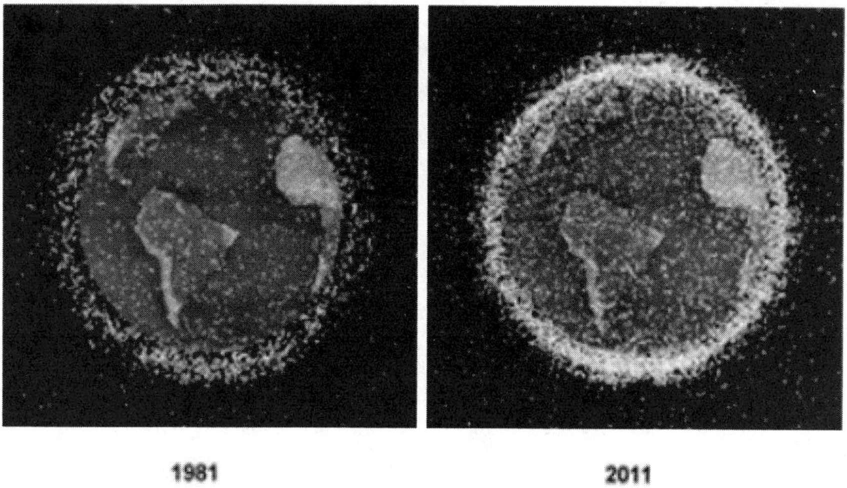

Fig. 16 View of the population of space debris in LEO in 1981 (*left panel*) and in 2011 (*right panel*). (Image courtesy of NASA.)

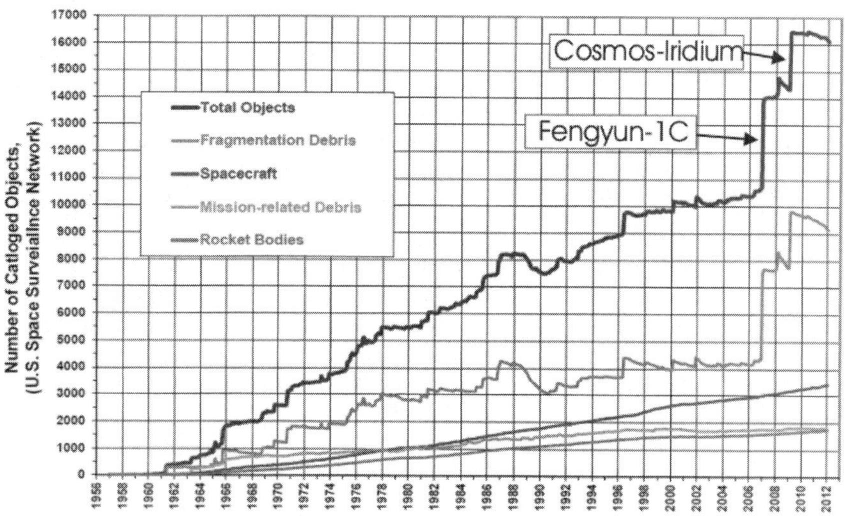

Fig. 17 The growth of the cataloged space debris population in the last 56 years. The intentional breakup of Fengyun-1C and the catastrophic collision of Cosmos-2251 and Iridium-33 dramatically changed the number of space debris in the last 5 years. (Image courtesy of IADC.)

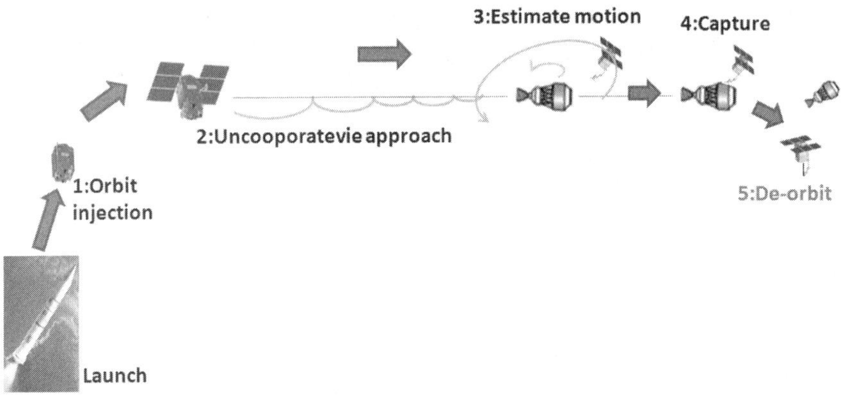

Fig. 18 An active debris mitigation scheme. (Image courtesy of JAXA.)

A. REACHING SPACE DEBRIS WITH A CHARGED SATELLITE

To apply most of the deorbit techniques for uncooperative debris, a chaser satellite should be maneuvered to the vicinity of target debris. Of course, traditional propulsion systems could be used to control its orbit. However, a new method will be discussed here, proposed in Ref. [9], namely the possibility to use a charged chaser satellite to approach target debris.

The feasibility of a satellite's charge control is rather challenging. Fortunately, various instruments have been proposed [10] in order to store and control a satellite's charge amount (Fig. 19). The dynamics of a charged satellite in a LEO are strongly influenced from the geomagnetic Lorentz force, which describes the interaction between the satellite's charge amount and the Earth's magnetic field. The use of the Lorentz force provides a new concept of propellantless electromagnetic propulsion.

The Lorentz force experienced by a charged satellite moving through the Earth's magnetic field is a function of its velocity vector and its position within the field. Modeling the Earth's magnetic field with a perfect dipole, the Lorentz force for a circular or elliptic orbit around Earth can be calculated, although the formulation might be complicated. To acquire a better insight of the problem,

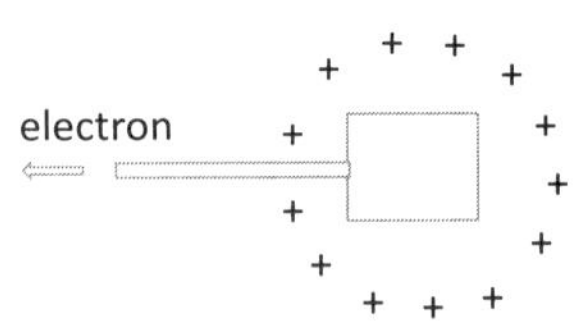

Fig. 19 Illustration of a device used to control a satellite's charge amount.

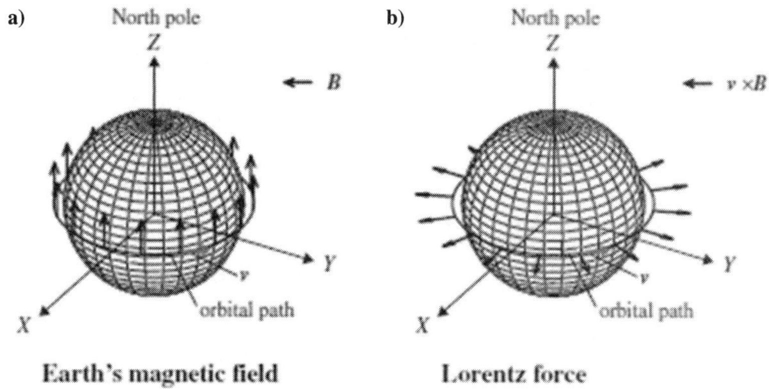

Fig. 20 Vector plot of Earth's magnetic field and resulting Lorentz force for an equatorial orbit [9].

the Earth's magnetic field and the resulting Lorentz force is calculated for three different cases.

In Fig. 20, an equatorial orbit, that is, an orbit with zero inclination, is assumed. In this case, the Earth's magnetic field is always toward the north and consequently the Lorentz force acts outward in the radial direction.

In Fig. 21, an orbital plane of 30-deg inclination is assumed. The Earth's magnetic field can be estimated under our assumptions and consequently the vector field for the Lorentz force can be calculated.

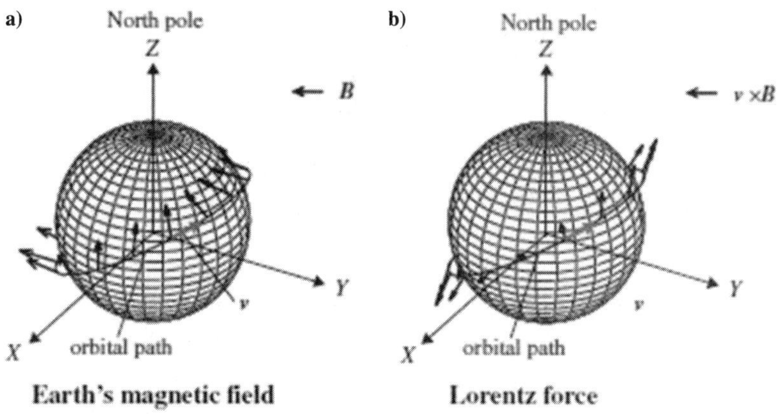

Fig. 21 Vector plot of Earth's magnetic field and resulting Lorentz force for an orbit with an inclination of 30 deg [9].

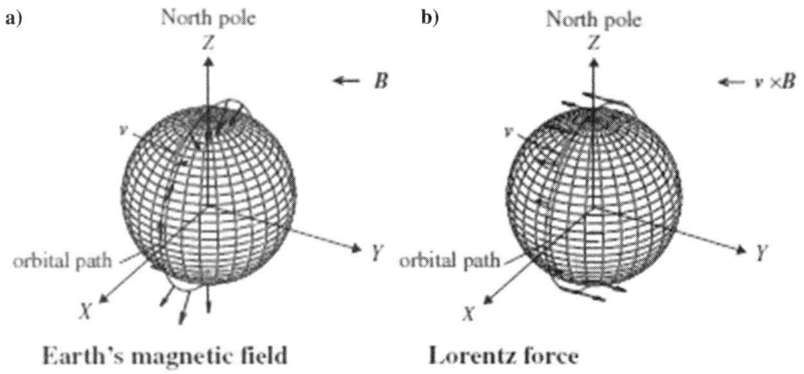

Fig. 22 Vector plot of Earth's magnetic field and resulting Lorentz force for a polar orbit [9].

Finally, a polar orbit, that is, an orbit with 90-deg inclination, will be assumed. In this case, as shown in Fig. 22, the Lorentz force is always perpendicular to the orbital plane.

From the previous section, it is obvious that, although the Lorentz force acting on a charged satellite is very complicated, its effect can be calculated. Now, let us derive a dynamical model of a charged satellite in the vicinity of the target in a circular or elliptic orbit. The Hill–Clohessy–Wiltshire (HCW) equation will be used and the Lorentz force will also be introduced in addition to the gravitational potential. In Fig. 23, the coordinated system of the HCW formulation is presented.

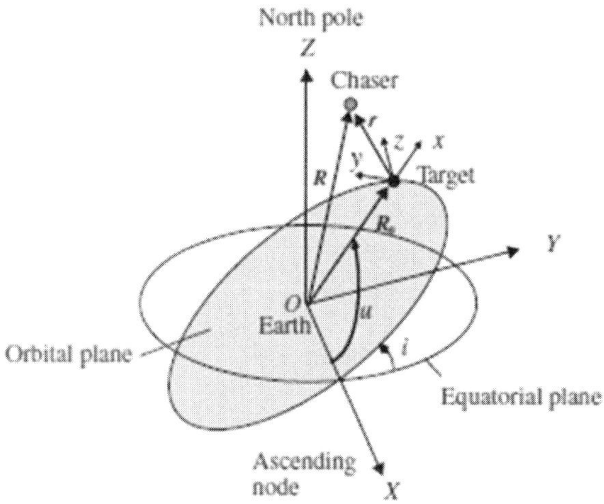

Fig. 23 Coordinate systems for a charged satellite in the Earth's magnetic field [9].

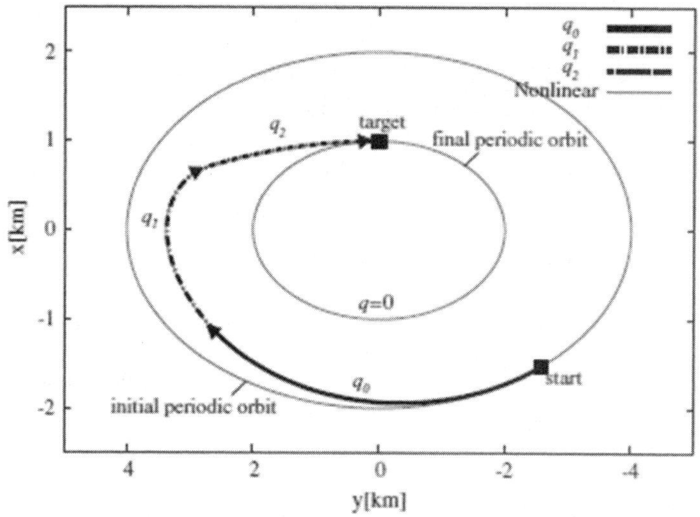

Fig. 24 Controlled trajectory for the transfer from a periodic orbit to another [9].

Notice that the origin of the coordinate system corresponds to the position of the target satellite.

In this context, the equation of motion in the rotating frame are

$$\frac{d^2x}{du^2} - 2\frac{dy}{du} - 3x = \frac{q}{m}\left(A_c \cos^2 u + B_c \sin^2 u + C_c + a_1\right) \tag{6}$$

$$\frac{d^2y}{du^2} + 2\frac{dx}{du} = \frac{q}{m}\left(D_c \sin u \cos u + a_2\right) \tag{7}$$

$$\frac{d^2z}{du^2} + z = \frac{q}{m}\left(E_c \sin u + a_3\right) \tag{8}$$

where the coefficients are given from the formulas

$$A_c = \frac{B_0}{\mu} R_0 \omega_E, \quad B_c = \frac{B_0}{\mu} R_0 \omega_E \cos^2 i$$

$$C_c = -\frac{B_0}{\mu} R_0 \omega_s \cos i, \quad D_c = 2\frac{B_0}{\mu} R_0 \omega_E \sin^2 i$$

$$E_C = -2R_0 \frac{B_0}{\mu}\left(\omega_s \sin i - \omega_E \sin i \cos i\right)$$

In the previous equations, q is the charge amount of the satellite and m is its mass. The ratio q/m is a parameter that could play a significant role in terms of orbit control. Changing the value of this parameter, enables us to control the orbital behavior of the satellite in Earth's magnetic field.

Using the linearized version of the HCW equation, known as Tschauner–Hempel (TH) equations, an optimal control problem can be formulated for the orbital motion of the charged satellite. The trajectory can be optimized by a direct method and solved using a nonlinear-programming solver. Of course, suitable boundary conditions should be provided, that is, the minimum and maximum values for the satellite charge amount.

There are numerous periodic orbits for the charged chaser around the target satellite. Using the ratio q/m as a control parameter, an optimal solution can be found for the transfer from one periodic orbit to another. In Fig. 24, an example of an optimized transfer trajectory is presented.

The same control scheme can be used in order to navigate between drift orbits. In Fig. 25, an example of an optimized transfer trajectory is presented. Controlling the amount of charge allows us to move from a drifting orbit to another one with a different drifting size.

Finally, an optimized three-dimensional rendezvous trajectory between the chaser and the target is presented. In Fig. 26, the three-dimensional trajectory, as well as the in-plane and out-plane motion are presented. Although the orbit looks

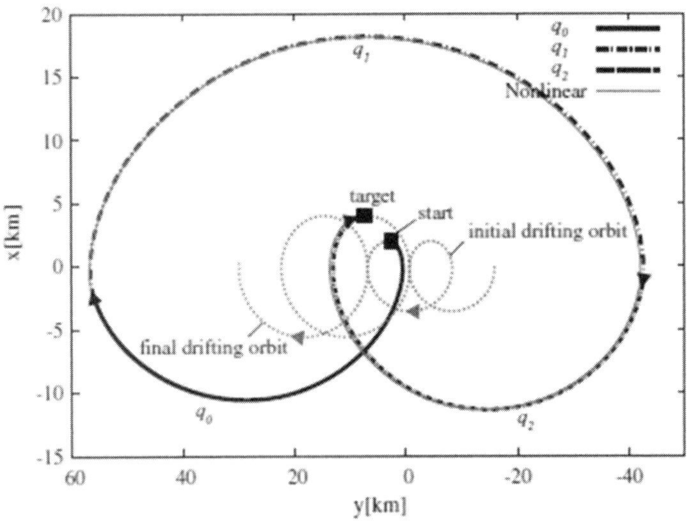

Fig. 25 Controlled trajectory for the transfer from a drifting orbit to another [9].

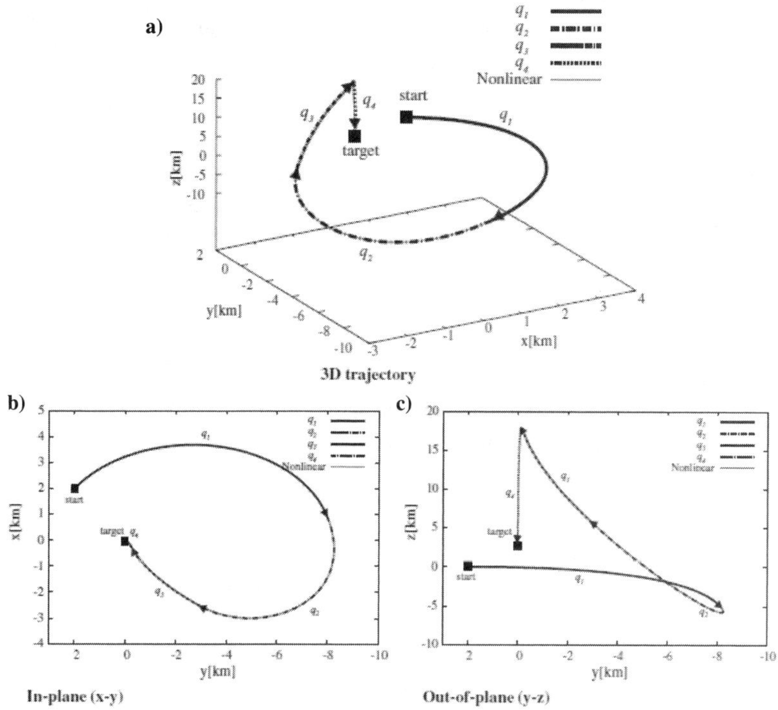

Fig. 26 A three-dimensional rendezvous trajectory controlling the chaser satellite's charge amount [9].

unconventional, due to the complicated dynamics induced by the Lorentz force, an optimal solution is calculated.

B. MOVING SPACE DEBRIS WITH A CHARGED SATELLITE

In the previous section, the idea to use a satellite's charge amount as a parameter to control its orbital motion was introduced. As it was discussed, this idea can be utilized to design rendezvous missions between a charged chaser satellite and a target debris. Assuming that there is also the technological and scientific capability to capture the space debris, the final phase in this kind of mission would be to deorbit it. A method will be presented here, proposed in Ref. [11], which enables a captured debris to reenter the atmosphere by periodically changing the charge amount of the chaser satellite.

To simplify the analysis, a perfect-dipole approximation will be used for Earth's magnetic field (Fig. 27, *left panel*), with the magnetic north pole coinciding with the geographic north pole. However, this is not true, because in reality, the

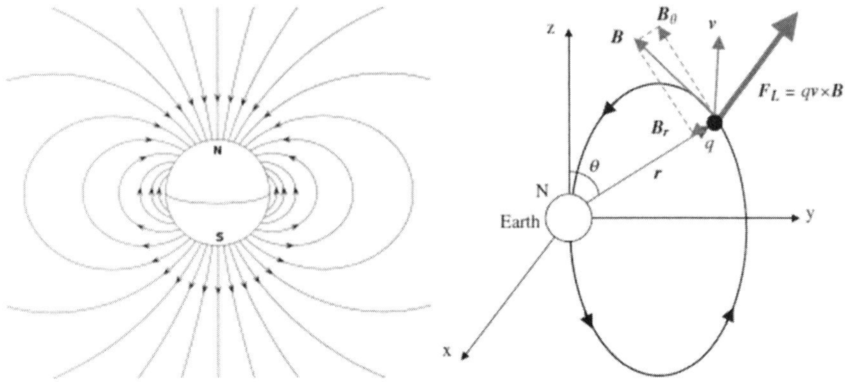

Fig. 27 Illustration of the perfect-dipole approximation used for Earth's magnetic field (*left panel*) [11]. The coordinate system for a charged satellite in the orbit around Earth is also shown (*right panel*) [11].

magnetic axis is tilted with respect to the Earth's rotational axis. It is also assumed that the magnetic field is rotating with Earth. Under this prism, the magnetic field can be represented as

$$\boldsymbol{B} = \frac{B_0}{r^3}(2\cos\theta\hat{r} + \sin\theta\hat{\theta}) \tag{9}$$

The Lorentz force can then be calculated from the equation

$$\boldsymbol{F_L} = q(\mathbf{v} - \omega_E \times \boldsymbol{r}) \times \boldsymbol{B} \tag{10}$$

where $\boldsymbol{v}$ is the velocity vector of the satellite, q is the charge amount, and ω_E is the Earth's angular velocity vector. If the gravitational force F_G and the electrostatic force F_D are also taken into account, then the equation of motion in vectorial form yields

$$\boldsymbol{a} = \frac{1}{m}(\mathbf{F_G} + \boldsymbol{F_L} + \boldsymbol{F_D}) \tag{11}$$

However, the solution of Eq. (11) for the case of a charged satellite is rather unexciting. Let us assume a captured debris and a chaser satellite with a charge of $-10C$ and total mass equal to 1000 kg. Also, suppose that the two bodies revolve in a circular orbit around Earth's equator in an altitude of 800 km above the surface of Earth. In Fig. 28, the time evolution of their altitude is presented. Although it varies in time, the variation is sinusoidal with a very small amplitude. It is impossible in this case to achieve a reentry of the captured debris.

A different approach should be considered in order to have a dramatic change in the debris altitude. An interesting idea is to try to control the charge amount periodically. If the frequency of the control period is chosen carefully, then remarkable results can be achieved due to resonance. Assuming the preservation of angular momentum, the equation of motion for the radial component of Eq. (11) is

$$\ddot{r} = r\dot{\theta}^2 - \frac{\mu}{r^2} + \frac{B_0}{m}\left(\frac{R_S}{r}\right)^3 r\left(\dot{\theta} - \omega_E\right) q_0 \sin \omega t \tag{12}$$

The linearized equation around the initial altitude, r_0, is representing a forced oscillation

$$\ddot{\Delta r} + \omega_0^2 \Delta r = q_1 \sin \omega t \tag{13}$$

If the control period ω is chosen to be equal to the orbital period ω_0, then a resonance occurs. The solution for the altitude variation then yields

$$\Delta r = \frac{q_1}{2\omega_0}\left(-t \cos \omega_0 t + \frac{1}{\omega_0} \sin \omega_0 t\right) \tag{14}$$

In the *left panel* of Fig. 29, both the analytic result of Eq. (14) and a numerical integration of the equation of motion Eq. (12) are presented. In this case, a resonant periodic charge control is used. The satellite is assumed to orbit Earth in a circular, equatorial orbit with an altitude of 800 km. Its mass is 1000 kg and the amplitude of the charge oscillation is 10*C*.

First, we should notice that the analytic solution is in perfect agreement with the numerical solution, proving that our analysis is valid. Also, the amplitude of the altitude oscillation increases due to the resonance, providing us a possible

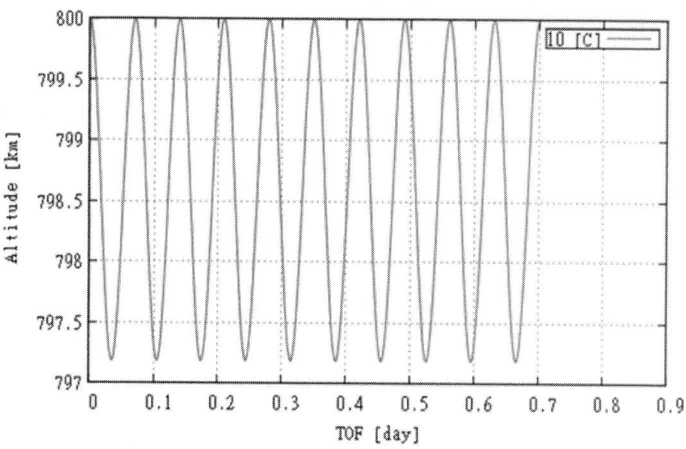

Fig. 28 The altitude evolution for a satellite with constant charge amount [11].

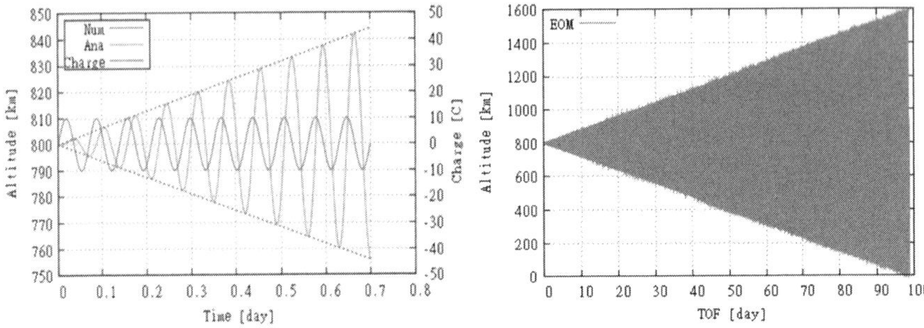

Fig. 29 The altitude evolution for a satellite with a periodic charge control [11].

debris removal technique. Indeed, in the right panel of Fig. 29, a longer numerical simulation is presented, showing that within three-months time span, the debris will hit the surface of Earth. It should also be noticed that because of the oscillation in the altitude, as the minimum altitude is decreased, the maximum altitude is increased at the same time.

A more-accurate modeling should also incorporate the effect of the Earth's atmosphere. In Fig. 30, the evolution of the altitude for the same initial conditions as in Fig. 29 are used, but this time, the atmosphere is also simulated with a modified exponential model. Appropriate values for the area of the satellite and the drag coefficient are chosen. In this case, the evolution is slightly different. Because of the atmospheric drag effect, the final phase of the deorbit is altered. Specifically, in the *right panel* of Fig. 30, a sudden drop to the Earth's surface is observed, if we focus our attention only in the last two days of the evolution.

Finally, the method of a charged satellite will be compared with another propellantless method, which is of the attachment of an electrodynamic tether to target debris. A debris in a circular orbit of altitude 800 km around the equator of

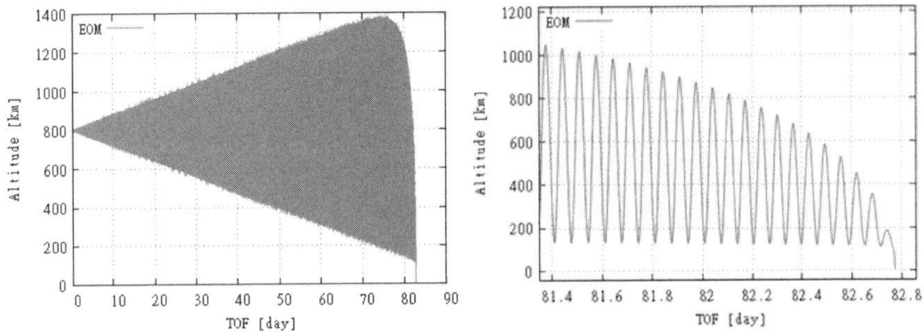

Fig. 30 The altitude evolution for a satellite with a periodic charge control in the atmosphere [11].

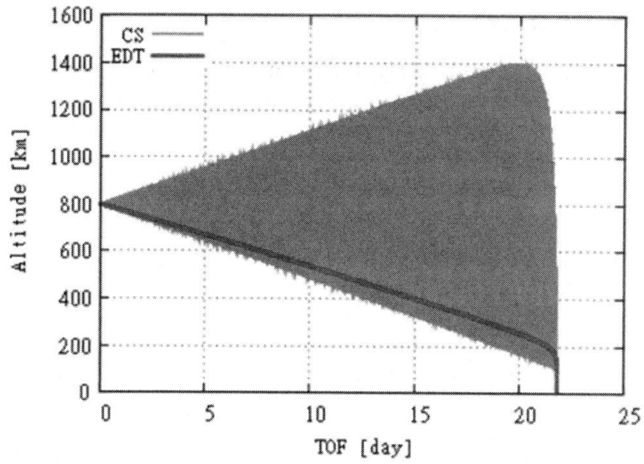

Fig. 31 A comparison between the charged satellite (CS) and the electrodynamic tether (EDT) techniques for debris disposal [11].

Earth will be considered. Then the parameters for the two methods are adjusted in order to achieve a reentry in almost 20 days. Using a charged satellite, a charge amount of $-3.91C$ is enough to achieve this task. On the other hand, as shown in Fig. 31, if an electrodynamic tether is attached to the debris, then in order to have the same result, the tether should have a length of 10 km and the value of the current should be 0.7 A.

C. SUMMARY

In this section, the problem of active debris removal was explored. The concept of a charged satellite was used to both rendezvous with and dispose of the target debris. The effect of the Lorentz force on a charged satellite in Earth's magnetic field gives an alternative way of propellantless navigation. Also, it was demonstrated that using a periodic charge control, it is feasible to mitigate uncooperative debris.

IV. CONCLUSION

NEAs and space debris represent a significant hazard for human assets, both in space and on Earth. A constant awareness is required from the scientific community, in order to be prepared in the case of an uncooperative object. In this chapter, a selection of advanced deflection techniques was presented. Using alternative propulsion systems, like solar sails and the Lorentz force, innovative missions were designed, able to mitigate any potential hazard.

ACKNOWLEDGMENTS

Ioannis Gkolias is supported by the European Grant MC-ITN Stardust. Fabien Gachet is supported by the European Grant MC-ITN Stardust.

REFERENCES

[1] Yamaguchi, T., Kogiso, N., and Yamakawa, H., "Optimal Interplanetary Trajectories for Impulsive Deflection of Potentially Hazardous Asteroids Under Velocity Increment Uncertainties," *Japan Society of Aeronautical Space Sciences Transactions*, Vol. 51, 2008, pp. 176–183.

[2] McInnes, C. R., "Deflection of Near-Earth Asteroids by Kinetic Energy Impacts from Retrograde Orbits," *Planetary and Space Science*, Vol. 52, No. 7, 2004, pp. 587–590.

[3] Yamaguchi, K., and Yamakawa, H., "Orbital Deflection of Potentially Hazardous Asteroids Using a Coulomb Force Attractor," *Proceedings of the ISTS*, Nagoya, Japan, 2013, pp. Tr_27–Tr_33.

[4] Janhunen, P., "Electric Sail for Spacecraft Propulsion," *Journal of Propulsion and Power*, Vol. 20, No. 4, 2004, pp. 763–764.

[5] McInnes, C. R., "Inverse Solar Sail Trajectory Problem," *Journal of Guidance, Control, and Dynamics*, Vol. 26, No. 2, 2003, pp. 369–371.

[6] Bando, M., and Yamakawa, H., "Near-Earth Asteroid Flyby Survey Mission Using Solar Sailing Technology," *The Journal of the Astronautical Sciences*, Vol. 58, No. 4, 2011, pp. 569–581.

[7] Murdoch, N., Izzo, D., Bombardelli, C., Carnelli, I., Hilgers, A., and Rodgers, D., "Electrostatic Tractor for Near Earth Object Deflection," *In 59th International Astronautical Congress*, Vol. 29, October 2008.

[8] Kessler, D. J., and Cour-Palais, B. G., "Collision Frequency of Artificial Satellites: The Creation of a Debris Belt," *Journal of Geophysical Research*, Vol. 83, 1978, pp. 2637–2646.

[9] Tsujii, S., Bando, M., and Yamakawa, H., "Spacecraft Formation Flying Dynamics and Control Using the Geomagnetic Lorentz Force," *Journal of Guidance, Control, and Dynamics*, Vol. 36, Jan. 2013, pp. 136–148.

[10] Streetman, B., and Peck, M. A., "New Synchronous Orbits Using the Geomagnetic Lorentz Force," *Journal of Guidance, Control, and Dynamics*, Vol. 30, Nov. 2007, pp. 1677–1690.

[11] Nakamiya, M., Akashi, Y., and Yamakawa, H., "Active Debris Removal Using Charged Satellite," IAC-13,A6,S.4x17706, Beijing, China, 2013, pp. 137–141.

CHAPTER 16

Rotational Dynamics and Attitude Control for Space Debris Rendezvous and Capture

Scott J. Walker* **and Natalia Ortiz Gómez***
Southampton University, Southampton, United Kingdom

I. INTRODUCTION

The attitude control subsystem (ACS) of a spacecraft is an important component of any mission and is directly linked to the success or failure of the mission objectives. Each mission has to comply with payload requirements, which will drive the design of the ACS of the spacecraft. For instance, an Earth observation mission that will acquire optical images of the Earth surface needs to point within a certain accuracy toward specific targets on the surface. In addition to this, every satellite has to achieve pointing requirements that supports the operation of other subsystems, such as pointing the antennas accurately for Earth communications or pointing the solar panels toward the Sun to acquire power. Attitude control is, therefore, necessary to meet both the payload and "housekeeping" requirements of a satellite. The ACS is in charge of changing the attitude of the satellite and achieving certain accuracy requirements by overcoming the external and internal torque disturbances. This is fundamentally achieved by managing the overall angular momentum of the satellite. Attitude control for active debris removal missions present additional challenges due to the uncertainties regarding the physical properties of the uncooperative target. This chapter discusses and reinforces the concepts of rotational dynamics and angular momentum in general as well as discuss these concepts when applied to an active debris removal mission.

II. ROTATIONAL DYNAMICS IN SPACE

A. SPACECRAFT ATTITUDE

A satellite's attitude refers to its angular orientation with respect to a certain reference frame. The most usual angular magnitudes used to describe the satellite attitude are Euler angles and Quaternions. This chapter shows a quick overview of these angular magnitudes. For a full description, please refer to Ref. [1].

---

*Department of Astronautics.

1. *Euler angles*: These consist of three angular magnitudes that describe the rotation of an initial reference frame (X_1 Y_1 Z_1) to a final reference frame (X_2 Y_2 Z_2). Euler angles represent a sequence of three elemental rotations such as rotations about the axis of a coordinate system. There exist 12 possible sequences of rotation axis:
   a. where the first and third rotation axis are the same: *z-x-z, x-y-x, y-z-y, z-y-z, x-z-x, y-x-y*
   b. where all the rotation axis are different: *x-y-z, y-z-x, z-x-y, x-z-y, z-y-x, y-x-z*

   Throughout this chapter, we consider the following 3-1-3 or *z-x-z* Euler sequence of rotations defined by angles ψ, ϕ, and θ.
   a. ψ: Angle between X_1-axis and the line of nodes axis (Intersection between X_1 Y_1 plane and X_2 Y_2 plane). This angle defines the *precession rate.*
   b. ϕ: Angle between the line of nodes and the X_2-axis in the body local frame. This angle defines the *intrinsic* or *spin rotation* of the body.
   c. θ: Inclination angle or angle between the Z_1-axis and Z_2-axis. This magnitude defines the *nutation rate* of the body.

   A scheme of the Euler angles is shown in Fig. 1.
2. *Quaternions*: A quaternion is a [4 × 1] matrix that consists of a scalar magnitude and a vector. Quaternions describe the rotation of a body as a rotation angle θ around an axis of rotation $\boldsymbol{e}$ (as described in Euler's rotation theorem), depicted in Fig. 2, and the quaternion acquires the following form:

$$\boldsymbol{q} = \begin{bmatrix} \boldsymbol{e}\sin\left(\frac{\theta}{2}\right) \\ \cos\left(\frac{\theta}{2}\right) \end{bmatrix} = \begin{bmatrix} q_1 \\ q_2 \\ q_3 \\ q_4 \end{bmatrix} \tag{1}$$

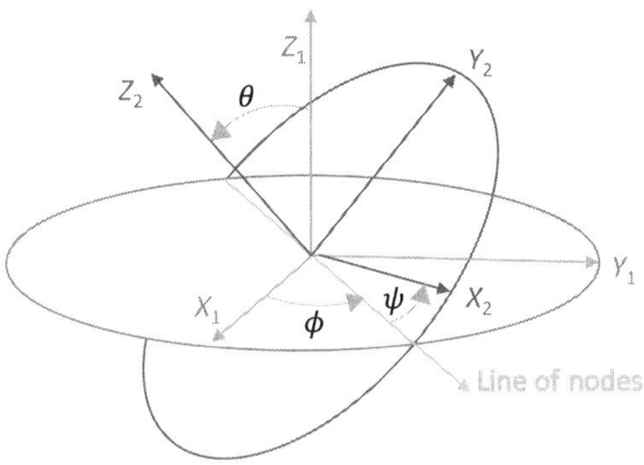

Fig. 1 Euler angles.

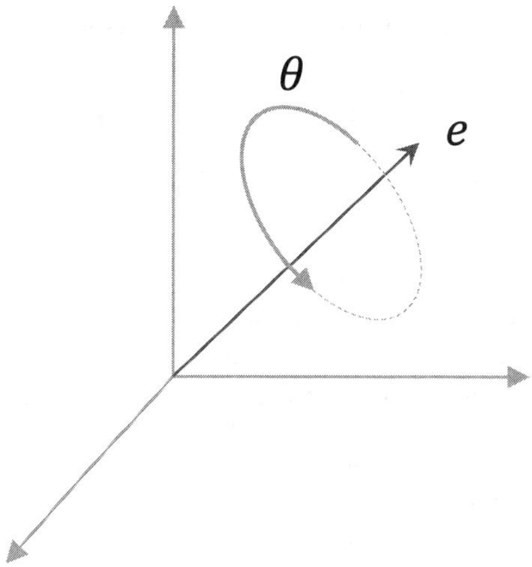

Fig. 2 Quaternions parameters: rotation θ around Euler axis.

Clearly quaternions norm is equal to 1.

$$q_1^2 + q_2^2 + q_3^2 + q_4^2 = 1 \tag{2}$$

The conversion from a transformation quaternion $\boldsymbol{q}$ to a direction cosine matrix R is shown below:

$$\mathbf{R} = \begin{bmatrix} q_1^2 - q_2^2 - q_3^2 + q_4^2 & 2(q_1q_2 + q_3q_4) & 2(q_1q_3 + q_2q_4) \\ 2(q_1q_2 - q_3q_4) & -q_1^2 + q_2^2 - q_3^2 + q_4^2 & 2(q_2q_3 + q_1q_4) \\ 2(q_1q_3 - q_2q_4) & 2(q_2q_3 - q_1q_4) & -q_1^2 - q_2^2 + q_3^2 + q_4^2 \end{bmatrix} \tag{3}$$

The inverse transformation from matrix R to quaternions is

$$\begin{cases} q_1 = \dfrac{1}{4q_4}(R_{23} - R_{32}) \\ q_2 = \dfrac{1}{4q_4}(R_{31} - R_{13}) \\ q_3 = \dfrac{1}{4q_4}(R_{12} - R_{21}) \\ q_4 = \pm\dfrac{1}{2}(1 + R_{11} + R_{22} + R_{33})^{1/2} \end{cases} \tag{4}$$

The combination of a rotation q followed by a rotation q' can be expressed as:

$$q'' = \begin{bmatrix} q_4' & q_3' & -q_2' & q_1' \\ -q_3' & q_4' & q_1' & q_2' \\ q_2' & -q_1' & q_4' & q_3' \\ -q_1' & -q_2' & -q_3' & q_4' \end{bmatrix} q \tag{5}$$

If we compare both attitude representation methods, we will find several pros and cons. The main advantages of Euler angles are that they have a clear physical interpretation and they have no redundant parameters. However, the Euler angles attitude representation method has several drawbacks. It uses trigonometric functions, it has singularities, and there is no convenient product rule for successive rotations. Although quaternions do not have an obvious physical interpretation and possess one redundant parameter, they do not have singularities, do not use trigonometric functions and have a convenient product rule for successive rotations. It is for these reasons why they are the most used representation for spacecraft attitude control.

B. SPACE ENVIRONMENT ROTATIONAL PERTURBATIONS

A spacecraft rotating in free motion will be affected by the space environment causing changes in its attitude. The main sources of environmental perturbation are described as:

1. *Atmospheric drag*: This force has a bigger impact on spacecraft in low Earth orbit (at attitudes $<\sim 500\,\text{km}$) and depends on the atmospheric density, on the area perpendicular to the air drag flow, the drag coefficient C_D and the satellite's velocity relative to the atmosphere. The drag coefficient is a dimensionless magnitude that describes the interaction of the gas flow with the body surface and a typical value for spacecraft is $\sim$2.2. The aerodynamic force acting on the elementary area dA defined by its perpendicular direction $\boldsymbol{u}_N$ is

$$\mathrm{d}\boldsymbol{F}_i = -\frac{1}{2}\rho C_D {V_a}^2(\boldsymbol{u}_N \bullet \boldsymbol{v}_a)\boldsymbol{v}_a \,\mathrm{d}A \tag{6}$$

where $\boldsymbol{v}_a$ is the velocity relative to the atmosphere of element dA and it is defined as the sum of the relative velocity of the center of mass and the cross product of the angular velocity $\boldsymbol{\omega}$ and the distance to the center of mass:

$$\boldsymbol{v}_a = \boldsymbol{v}_G + \boldsymbol{\omega} \times \boldsymbol{r} \tag{7}$$

The torque induced by the air drag has the expression:

$$\boldsymbol{M}_{\text{aero}} = \int_S \boldsymbol{r}_i \times \mathrm{d}\boldsymbol{F}_i \,\mathrm{d}A \tag{8}$$

The integral is evaluated for all the surfaces of the satellite for which the product $\boldsymbol{u}_N \bullet \boldsymbol{v}_a$ is positive. This means that surfaces not exposed to the aerodynamic flow do not intervene in the torque computation.

2. *Solar radiation pressure*: This force appears due to the incidence of solar radiation on an illuminated surface and has a bigger impact on the spacecraft, the closer it gets to the Sun (at attitudes $>\sim$ 600–700 km). The radiation flux can be computed as the radiation flux divided by the speed of light multiplied by the reflective constant s, which will vary between 0 and 2 [Eq. (9)]. The radiation flux mean value in the vicinity of the Earth is approximately 1400 W/m^2 and follows the equation

$$P = s\frac{W}{c} \tag{9}$$

Radiation forces can be modeled assuming that part of the incident radiation is absorbed, part is reflected specularly and part is reflected with diffusion (see Fig. 3). Here, it is assumed that the incident radiation has no transmitted component. A specular reflection is a mirror-like reflection of light from a surface, whereas in a diffuse reflection the incoming light is reflected at many angles.

The three components of the force acting on a surface element dA having orthogonal direction $\boldsymbol{u}_N$ are

$$\begin{cases} \mathrm{d}\boldsymbol{F}_{\text{absorbed}} = -P\,c_a\,\cos\vartheta\,\boldsymbol{u}_{\text{sat}-\text{Sun}}\,\mathrm{d}A \\ \mathrm{d}\boldsymbol{F}_{\text{diffuse}} = -P\,c_d\,\left(-\frac{2}{3}\cos\vartheta\,\boldsymbol{u}_N - \cos\vartheta\,\boldsymbol{u}_{\text{sat}-\text{Sun}}\right)\,\mathrm{d}A \\ \mathrm{d}\boldsymbol{F}_{\text{specular}} = -2P\,c_s\,(\cos\vartheta)^2\,\boldsymbol{u}_N\,\mathrm{d}A \end{cases} \tag{10}$$

where $\boldsymbol{u}_{\text{sat}-\text{Sun}}$ is the unit direction of the satellite–Sun vector and ϑ is the angle between the surface orthogonal direction $\boldsymbol{u}_N$ and the direction $\boldsymbol{u}_{\text{sat}-\text{Sun}}$. When ϑ is greater than 90 deg, the surface is not illuminated and not subject to radiation forces. The coefficients c_a, c_d, and c_s are respectively the coefficient of absorption, diffuse reflection, and specular reflection, and are constrained by the following equality:

$$c_a + c_d + c_s = 1 \tag{11}$$

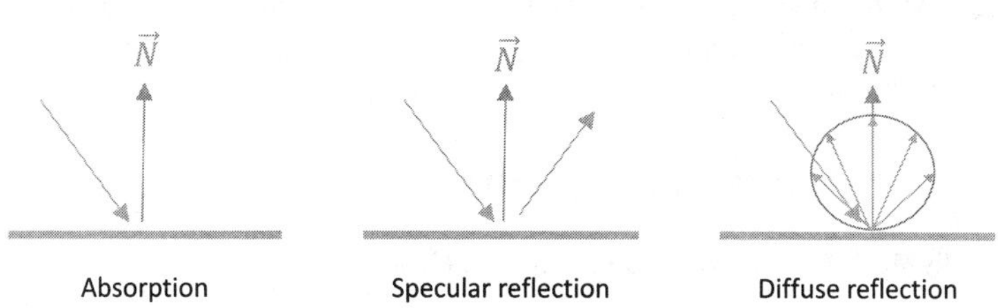

Fig. 3 Absorption, specular reflection, and diffuse reflection.

Merging the three components due to the three possible effects, the total force due to radiation is evaluated as:

$$\mathrm{d}\boldsymbol{F}_i = -P \int_S \left[(1 - c_s)\boldsymbol{u}_{\text{sat-Sun}} + 2\left(c_s \cos\vartheta + \frac{1}{3} c_d \right) \boldsymbol{u}_N \right] \cos\vartheta \, \mathrm{d}A \quad (12)$$

The torque due to radiation will then be:

$$\boldsymbol{M}_{\text{SRP}} = \int_S \boldsymbol{r}_i \times \mathrm{d}\boldsymbol{F}_i \, \mathrm{d}A \quad (13)$$

3. *Magnetic force*: This force appears due to the interaction of the Earth magnetic field with the satellite at attitudes between $\sim$ 500–35,000 km and generates a torque, which can be computed as the cross product of the magnetic moment $\boldsymbol{m}$ (Amp m^2) and the local flux density vector $\boldsymbol{B}$ (Teslas). The Earth magnetic field varies from 26 to 65 μT and the residual magnetic moment of a satellite can be 0.1 Amp m^2.

$$\boldsymbol{M}_{\text{MAG}} = \boldsymbol{m} \times \boldsymbol{B} \quad (14)$$

4. *Earth gravity gradient*: The gravitational field changes as the inverse square law of the distance between the Earth and the satellite and therefore reduces its effect with respect to the distance. This makes the body parts of the spacecraft closer to the Earth feel a higher attraction force than those that are further away from the Earth. This effect results in the following torque around the center of mass of the spacecraft where $\boldsymbol{u}_r$ is the unit vector from Earth to the satellite, $n = \sqrt{\mu/a^3}$ is the mean circumferential translational velocity of the spacecraft and I is the tensor of inertia of the body. The range of attitudes where its effect is noticeable is $\sim$ 500–10,000 km.

$$\boldsymbol{M}_{\text{GRAV}} = -3n^2 \boldsymbol{u}_r \times I \boldsymbol{u}_r \quad (15)$$

A body subjected to this torque will be in a stable position if the axis of minimum inertia is aligned with the local vertical. For instance, the gravity gradient makes the Moon face the Earth with the same orientation. If the body in equilibrium is slightly disturbed, an oscillatory mode will occur called a "libration mode." The motion that takes place is similar to a conical pendulum.

III. SPACECRAFT ROTATIONAL DYNAMICS

A. LINEAR MOMENTUM AND ANGULAR MOMENTUM

To analyze the rotational dynamics of a spacecraft, it is first necessary to introduce parameters such as linear momentum, angular momentum, and the inertia matrix of a body that are explained in this section. A particle with a mass m moving with a velocity $\boldsymbol{V}$ is said to have a linear momentum:

$$\boldsymbol{p} = m\boldsymbol{V} \tag{16}$$

The momentum is an additive value and it can be represented for a spacecraft as the sum of multiple particles:

$$\boldsymbol{p} = \sum m\boldsymbol{V}_i = m\boldsymbol{V}_G \tag{17}$$

According to Newton's second law, the momentum $\boldsymbol{p}$ will change if an external force is applied:

$$\frac{\mathrm{d}\boldsymbol{p}}{\mathrm{d}t} = m\frac{\mathrm{d}\boldsymbol{V}}{\mathrm{d}t} = \sum \boldsymbol{F}_{\text{ext}} \tag{18}$$

Therefore, free motion is achieved and the linear momentum is conserved if $\sum \boldsymbol{F}_{\text{ext}} = 0$.

Analogously, the angular momentum of a particle is defined by its moment of momentum. The next formula shows the angular momentum with respect to a point A and the vectors $\boldsymbol{r}$ and $\boldsymbol{V}$ must be relative to A:

$$\boldsymbol{H}_A = \boldsymbol{r} \times m\boldsymbol{V} \tag{19}$$

This parameter is equivalent to the linear case so that it is also an additive value. By representing the spacecraft as a sum of particles, the total angular momentum can be expressed as the sum of the angular momentum of all the particles. The angular momentum describes the rotational inertia based on the mass distribution and angular rotation rates of the body following the equation:

$$\boldsymbol{H} = \sum \boldsymbol{r}_i \times m_i\boldsymbol{V}_i \tag{20}$$

It may be useful to express the angular momentum about a point other than the center of mass. The expression that relates the angular momentum expressed about the center of mass and any other point has a second term, which expresses that there are particles that move relative to the center of mass.

$$\boldsymbol{H}_A = \boldsymbol{H}_G + \boldsymbol{h}_A \tag{21}$$

$$\boldsymbol{H}_A = \boldsymbol{r}_G \times m\boldsymbol{V}_G + \sum \boldsymbol{R}_i \times m_i\boldsymbol{v}_i \tag{22}$$

From Newton's second law of motion, we can derive the variation of the angular momentum when an external torque is applied.

$$\frac{\mathrm{d}\boldsymbol{H}_G}{\mathrm{d}t} = \sum \boldsymbol{T}_{\text{ext}} \tag{23}$$

Therefore, in free motion, no external torques are applied and the angular momentum is conserved. In the case of a rigid body (assembly of particles that do not move relative to each other), the second term in Eq. (22) is zero and the angular momentum can be expressed as the product of the inertia matrix referred to its center of mass I_G and its angular velocity $\boldsymbol{\omega}$ relative to an inertial reference frame:

$$\boldsymbol{H}_G = \boldsymbol{r}_G \times m\boldsymbol{V}_G = \mathbf{I}_G\boldsymbol{\omega} \tag{24}$$

$$\boldsymbol{\omega} = (\omega_x, \omega_y, \omega_z) \tag{25}$$

$$\mathbf{I}_G = \begin{bmatrix} I_{xx} & -I_{xy} & -I_{xz} \\ -I_{xy} & I_{yy} & -I_{yz} \\ -I_{xz} & -I_{yz} & I_{zz} \end{bmatrix} \tag{26}$$

The inertia matrix shows the distribution of mass of the body about a reference point, being typically its center of mass G. The moments of inertia are defined as:

$$I_{xx} = \int (y^2 + z^2)\mathrm{d}m, \quad I_{yy} = \int (x^2 + z^2)\mathrm{d}m, \quad I_{zz} = \int (x^2 + y^2)\mathrm{d}m \tag{27}$$

The moment of inertia of a body around an axis directly relates to the distribution of the mass around that axis. For a body of mass M, the smaller the inertia, the more the mass is concentrated close to the axis. The products of inertia measure the symmetry of the body about the defined reference axes and they can be expressed as:

$$I_{xy} = \int (xy)\mathrm{d}m, \quad I_{yz} = \int (yz)\mathrm{d}m, \quad I_{xz} = \int (xz)\mathrm{d}m \tag{28}$$

For instance, if xz plane is a symmetry plane, the products of inertia I_{xy} and I_{yz} are zero. Therefore, if two of the coordinate planes are planes of symmetry, the three products of inertia are zero. If products of inertia are present, they will cause cross-coupling, which means a torque about one axis will produce response about another axis. On certain occasions, it may be convenient to manipulate the inertia matrix for different translational or rotational positions of reference. The parallel axis theorem enables the

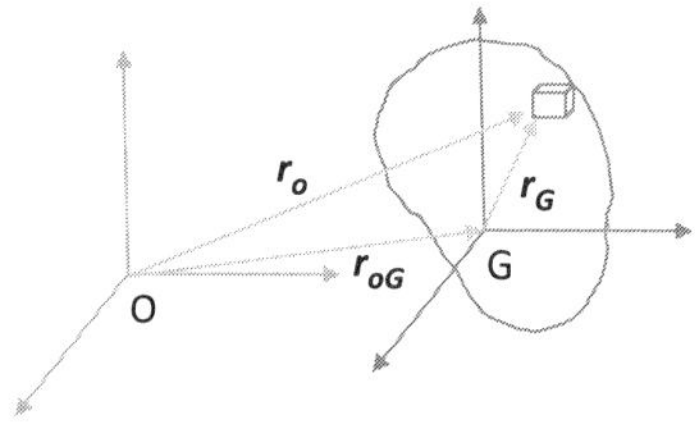

Fig. 4 Reference frame change for inertia matrix computation.

moment of inertia to be referred to a parallel axis as it is stated as follows:

$$I_{xx} = I_{Gxx} + m(y^2 + z^2) \tag{29}$$

Having two parallel reference frames, one with its origin at the center of mass of the body and the other one with its origin in an arbitrary point O (see Fig. 4), the relationship between the two inertia matrices referenced at each frame can be written as:

$$\mathbf{I}_O = \mathbf{I}_G + \mathbf{I}_t \tag{30}$$

where $\mathbf{I}_t$ has the form:

$$\mathbf{I}_t = \begin{bmatrix} \int (y_{OG}^2 + z_{OG}^2)dm & -\int (x_{OG}y_{OG})dm & -\int (x_{OG}z_{OG})dm \\ -\int (x_{OG}y_{OG})dm & \int (x_{OG}^2 + z_{OG}^2)dm & -\int (y_{OG}z_{OG})dm \\ -\int (x_{OG}z_{OG})dm & -\int (y_{OG}z_{OG})dm & \int (x_{OG}^2 + y_{OG}^2)dm \end{bmatrix} \tag{31}$$

The body frame of the inertia matrix referred to the center of mass can also be rotated using the rotation matrix R as follows:

$$\mathbf{I}_{G2} = \mathbf{R}\mathbf{I}_{G1}\mathbf{R}^T \tag{32}$$

The principal axes are defined as those in which all three products of inertia are zero and the corresponding moments of inertia are called principal moments of inertia. Thanks to the properties of the inertia matrix, there always exits a rotational matrix R, which can achieve a diagonal tensor. If an inertia tensor is specified for the axes x, y, and z, the moment of inertia of the body about an inclined axis can be computed using:

$$\mathbf{I}_{Oa} = \mathbf{I}_{xx}\boldsymbol{u}_x{}^2 + \mathbf{I}_{yy}\boldsymbol{u}_y{}^2 + \mathbf{I}_{zz}\boldsymbol{u}_z{}^2 - 2\mathbf{I}_{xy}\boldsymbol{u}_x\boldsymbol{u}_y - 2\mathbf{I}_{yz}\boldsymbol{u}_y\boldsymbol{u}_z - 2\mathbf{I}_{xz}\boldsymbol{u}_x\boldsymbol{u}_z \tag{33}$$

For this calculation the direction cosines u_x, u_y, and u_z of the axes must be determined. These numbers specify the cosines of the coordinate direction angles made between the inclined axis and the x, y, z axes, respectively.

B. TORQUE-FREE PRECESSION

Precession can be defined as a change in the orientation of the rotational axis. If a body of revolution rotates about its axis of symmetry, the angular momentum vector and angular velocity will be parallel. If the angular velocity vector is deviated slightly from the principal axis, precession takes place. In torque-free precession, the total angular momentum of the body remains constant.

Let us consider a body with two of the three principal moments of inertia are equal, such as a cylinder. I_s is the inertia of the axis of symmetry and $I_\perp$ is the inertia of the two axis perpendicular to the axis of symmetry. We use three frames of reference with their origins in the center of mass of the body:

1. An inertial reference frame XYZ. The total angular momentum remains constant along the $Z-$ axis. $\boldsymbol{L} = L\boldsymbol{k}$
2. An intermediate reference frame $X'Y'Z'$. To obtain this frame, first we carry out a rotation around the $Z-$ axis of an angle ϕ and then we carry out a rotation around the $X'-$ axis of an angle θ. It can be depicted from Fig. 5 that the angle between X' and X is ϕ and the angle between Z and Z' is θ.
3. A body reference frame $X_bY_bZ_b$. This frame has rotated an angle ψ around Z', which is equal to Z_b.

The angular velocity of the body and the total angular momentum vectors are deviated from the axis of symmetry of the body by an angle α and θ, respectively. The angular rotation vector in the intermediate reference frame $X'Y'Z'$ is

$$\omega = \omega_x \boldsymbol{i}' + \omega_y \boldsymbol{j}' + \omega_z \boldsymbol{k}' = \dot{\theta}\boldsymbol{i}' + \dot{\phi}\sin(\theta)\boldsymbol{j}' + (\dot{\phi}\cos(\theta) + \dot{\psi})\boldsymbol{k}' \tag{34}$$

The angular momentum $\boldsymbol{H} = H\boldsymbol{k} = H\sin\theta\boldsymbol{j}' + H\cos\theta\boldsymbol{k}'$ is invariable. Using Eq. (24) that relates the angular momentum and the angular velocity, we reach the conclusion that the nutation angle θ remains constant:

$$H_{x'} = 0 = I_{\perp}\dot{\theta} \longrightarrow \boldsymbol{\theta = \text{const}} \tag{35}$$

$$H_{y'} = H\sin(\theta) = I_{\perp}\theta\ \omega_{y'} \tag{36}$$

$$H_{z'} = H\cos(\theta) = I_s\theta\ \omega_{z'} \tag{37}$$

Merging the expressions Eqs. (34–37), the following expression for $\boldsymbol{\omega}$ is reached:

$$\boldsymbol{\omega} = \frac{H\sin\theta}{I_{\perp}}\boldsymbol{j}' + \frac{H\cos\theta}{I_s}\boldsymbol{k}' \tag{38}$$

We have chosen the intermediate reference frame such that the rotational velocity lies in the plane $Y'Z'$. In addition, we can express $\boldsymbol{\omega}$ as a function of the angle α, which is the angle between $\boldsymbol{\omega}$ and the axis of symmetry of the body Z_b

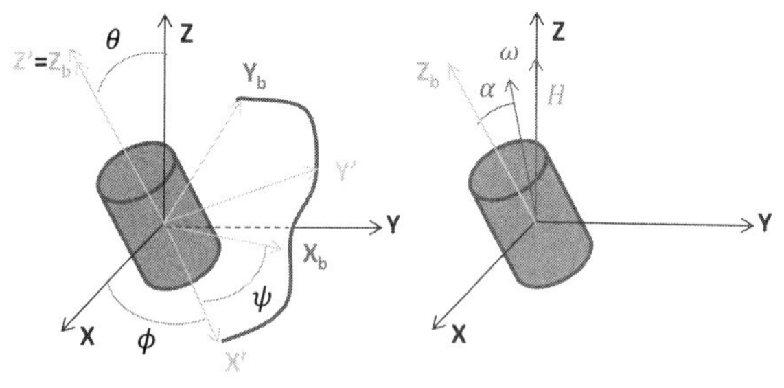

Fig. 5 Reference frames for precession motion.

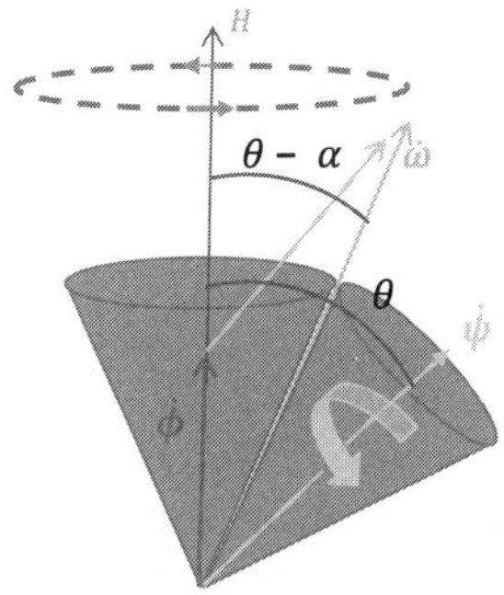

Fig. 6 Torque-free precession for a prolate object.

(see Fig. 5). Equation (40) gives us the relationship between the angles θ and α:

$$\boldsymbol{\omega} = \omega \sin \alpha \boldsymbol{j}' + \omega \cos \alpha \boldsymbol{k}' = \frac{H \sin \theta}{I_\perp} \boldsymbol{j}' + \frac{H \cos \theta}{I_s} \boldsymbol{k}' \tag{39}$$

$$\tan \theta = \frac{I_\perp}{I_s} \tan \alpha \tag{40}$$

For prolate objects (e.g., rod) the relationship between the inertias is $I_\perp > I_s$, and therefore $\theta > \alpha$ and for oblate objects (e.g., coin), $I_\perp < I_s$ and therefore $\theta < \alpha$. In the absence of external torques, the axis of symmetry of the body and the angular velocity will describe two cones around the fixed direction of the angular momentum with the same apex equal to the center of mass, as shown in Fig. 6. The moving cone is the one associated with the body and it rolls without slipping over the surface of the immovable cone, associated to the angular momentum vector. The points of the rotating body that are located, at a given time, in the intersection between the two cones, have zero linear velocities.

The precession $\dot{\phi}$ and the rotational spin $\dot{\psi}$ of the body remain constant:

$$\omega_{y'} = \dot{\phi} \sin(\theta) = \frac{H \sin \theta}{I_\perp} \longrightarrow \dot{\phi} = \frac{H}{I_\perp} = \text{const} \tag{41}$$

$$\begin{aligned} \omega_{z'} &= \dot{\phi} \cos(\theta) + \dot{\psi} = \frac{H \cos \theta}{I_s} \longrightarrow \dot{\psi} = \frac{H \cos \theta}{I_s} \frac{I_\perp - I_s}{I_\perp} \\ &= \omega_{zb} \frac{I_\perp - I_s}{I_\perp} = \text{const} \end{aligned} \tag{42}$$

The total angular velocity of the body can be divided into two components. One component goes along the axis of symmetry of the body $\dot{\psi}\boldsymbol{k}'$ and it is referred as the axial rotation (spinning) of the body. The other component is parallel to the angular momentum $\dot{\phi}\boldsymbol{k}$ and it is referred as precession rate. The total angular velocity acquires the form:

$$\boldsymbol{\omega} = \dot{\phi} \boldsymbol{k} + \dot{\psi} \boldsymbol{k}' \tag{43}$$

So the vectors $\boldsymbol{k}_b$ (axis of symmetry of the body) and $\boldsymbol{\omega}$ precess with constant angular velocity $\dot{\phi}$ about the fixed angular momentum vector $\boldsymbol{H}$. If we analyze the system in the rotating reference frame, we reach similar conclusions. Starting from the angular velocity expressed in the body frame, we apply Newton's second law:

$$\boldsymbol{\omega} = \omega_{xb} \boldsymbol{i}_b + \omega_{yb} \boldsymbol{j}_b + \omega_{zb} \boldsymbol{k}_b \tag{44}$$

$$\frac{\mathrm{d}\boldsymbol{H}}{\mathrm{d}t} = \frac{\partial \boldsymbol{H}}{\partial t} + \boldsymbol{\omega} \times \boldsymbol{H} = \frac{\partial I \boldsymbol{\omega}}{\partial t} + \omega \times (I \boldsymbol{\omega}) = \boldsymbol{T}_{\text{ext}} \tag{45}$$

And we reach Euler equations

$$I_{xb}\dot{\omega}_{xb} + (I_{zb} - I_{yb})\omega_{yb}\omega_{zb} = T_{xb} \tag{46}$$

$$I_{yb}\dot{\omega}_{yb} + (I_{xb} - I_{zb})\omega_{xb}\omega_{zb} = T_{yb} \tag{47}$$

$$I_{zb}\dot{\omega}_{zb} + (I_{yb} - I_{xb})\omega_{yb}\omega_{xb} = T_{zb} \tag{48}$$

As we are dealing with a free motion system, there are no external torques $\boldsymbol{T}_{\text{ext}} = 0$. Moreover, we consider two of the inertias to have an equal value $I_{xb} = I_{yb} = I_{\perp}$ and $I_{zb} = I_s$. Taking into account these considerations, Euler equations can be written as

$$I_{\perp}\dot{\omega}_{xb} + (I_s - I_{\perp})\omega_{yb}\omega_{zb} = 0 \tag{49}$$

$$I_{\perp}\dot{\omega}_{yb} + (I_{\perp} - I_s)\omega_{xb}\omega_{zb} = 0 \tag{50}$$

$$I_s\dot{\omega}_{zb} = 0 \longrightarrow \boldsymbol{\omega_{zb} = \text{const}} \tag{51}$$

Introducing the constant $\Omega = \omega_{zb}\,(I_s - I_{\perp})/(I_{\perp})$ in the equations we get the following expressions:

$$\dot{\omega}_{xb} + \frac{(I_s - I_{\perp})\omega_{zb}}{I_{\perp}}\,\omega_{yb} = \dot{\omega}_{xb} + \Omega\omega_{yb} = 0 \tag{52}$$

$$\dot{\omega}_{yb} + \frac{(I_{\perp} - I_s)\omega_{zb}}{I_{\perp}}\,\omega_{xb} = \dot{\omega}_{yb} - \Omega\omega_{xb} = 0 \tag{53}$$

$$\omega_{xb} = \omega_0\,\cos(\Omega t) \tag{54}$$

$$\omega_{yb} = \omega_0\,\sin(\Omega t)II \tag{55}$$

where t is the time and ω_0 corresponds to the value of ω at $t = 0$. If $\boldsymbol{\omega}$ is projected in the X_bY_b plane of the body, it describes a circle with angular frequency Ω. Therefore, in the rotating reference frame, $\boldsymbol{\omega}$ rotates with constant angular frequency Ω. Comparing the equations of motion in the rotating and inertial reference frame, it can be derived that

$$\Omega = -\dot{\psi} = \omega_{zb}\frac{I_s - I_{\perp}}{I_{\perp}} \tag{56}$$

A precession motion has been observed in some rocket bodies. For instance, Yanagisawa et al. [2] describe in detail the motion observed via lightcurves of Cosmos 2082 (the second stage of a Zenit rocket). This rocket body has a spin rate of 8.6 deg/s and precesses with a nutation angle θ of 29.5 deg and a precession rate of 0.2 deg/s.

C. STABILITY OF A SPINNING BODY

Let's consider a body that is spinning around the Z-axis at a constant rate ω_z and no external torques are present. Two of the Euler equations have the following form:

$$I_{xx}\dot{\omega}_x - (I_{yy} - I_{zz})\omega_y\omega_z = 0 \tag{57}$$

$$I_{yy}\dot{\omega}_y - (I_{zz} - I_{xx})\omega_x\omega_z = 0 \tag{58}$$

The characteristic polynomial of these equations is

$$s^2 + \left(1 - \frac{I_{zz}}{I_{xx}}\right)\left(1 - \frac{I_{zz}}{I_{yy}}\right)\omega_z = 0 \tag{59}$$

The rotational motion will be stable if I_{zz} coincides with the maximum or the minimum inertia axis. However, it must be highlighted that in the long-term, a rotation about the maximum axis of inertia is stable whereas a rotation about the minimum axis of inertia is not. This occurs due to the dissipation of energy that takes place and diminishes the kinetic energy of the body. To keep the angular momentum constant $H_G = I_G\omega$, the body will tend to move toward a rotation about its maximum axis of inertia with a lower ω.

D. TORQUE-INDUCED PRECESSION

In a torque-induced precession (gyroscopic precession), an external torque changes the total momentum of the system. To better understand the torque-induced precession, we consider a symmetrical top, which rotates about a fixed location (see Fig. 7). The symmetrical axis of the body is tilted by an angle θ with respect to the local vertical and is under the influence of the gravitational acceleration g. If l is the distance between the center of mass of the body and the fixed point at the base and m is the mass of the top, the torque induced with respect to the fixed point is

$$\boldsymbol{M} = mgl \sin\theta \, \boldsymbol{i}' \tag{60}$$

The angular rotation vector in the auxiliary frame $X'Y'Z'$ is

$$\boldsymbol{\omega} = \dot{\theta}\boldsymbol{i}' + \dot{\phi} \sin\theta\boldsymbol{j}' + (\dot{\phi} \cos\theta + \dot{\psi})\boldsymbol{k}' \tag{61}$$

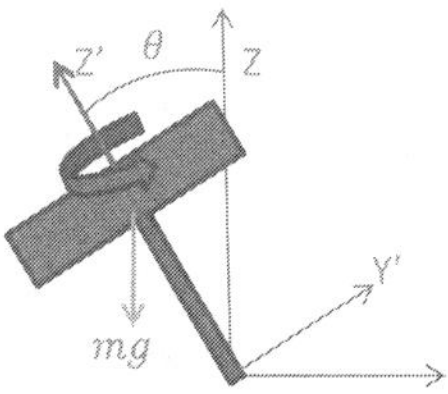

Fig. 7 Symmetrical top submitted to torque-induced precession.

The Lagrangian, expressed in the intermediate reference frame, is computed from the kinetic energy and potential energy.

$$T = \frac{1}{2} I_\perp \left(\omega_{x'}^2 + \omega_{y'}^2 \right) + \frac{1}{2} I_s \omega_{z'}^2 = \frac{1}{2} I_\perp \left(\dot{\theta}^2 + (\dot{\phi} \sin\theta)^2 \right) + \frac{1}{2} I_s (\dot{\phi} \cos\theta + \dot{\psi})^2 \tag{62}$$

$$V = mgl \cos\theta \tag{63}$$

$$L = T - V = \frac{1}{2} I_\perp (\dot{\theta}^2 + (\dot{\phi} \sin\theta)^2) + \frac{1}{2} I_s (\dot{\phi} \cos\theta + \dot{\psi})^2 - mgl \cos\theta \tag{64}$$

It can be derived from the Lagrangian that ψ and ϕ are cyclic variables and therefore their conjugate angular momenta are constants:

$$p_\phi = \frac{\partial L}{\partial \dot{\phi}} = I_\perp \dot{\phi} (\sin\theta)^2 + I_s (\dot{\phi} \cos\theta + \dot{\psi}) \cos\theta = \text{const} \tag{65}$$

$$p_\psi = \frac{\partial L}{\partial \dot{\psi}} = I_s (\dot{\phi} \cos\theta + \dot{\psi}) = \text{const} \longrightarrow \boldsymbol{\omega_{zb} = \omega_{z'} = \text{const} \longrightarrow L_{z'} = I_s \omega_{z'} = \text{const}} \tag{66}$$

Thus, the spin in the reference body frame remains constant and $\dot{\phi}$ and $\dot{\psi}$ can be expressed as a function of θ.

$$\dot{\phi} = \frac{p_\phi - p_\psi \cos\theta}{I_\perp (\sin\theta)^2} \tag{67}$$

$$\dot{\psi} = \frac{p_\psi}{I_s} - \frac{p_\phi - p_\psi \cos\theta}{I_\perp (\sin\theta)^2} \cos\theta \tag{68}$$

Unlike the torque-free precession, the nutation θ is not constant. From the Lagrange equations, we can obtain a second-order equation in θ.

$$\frac{\partial}{\partial t} \left(\frac{\partial L}{\partial \dot{\theta}} \right) - \frac{\partial L}{\partial \theta} = 0 \tag{69}$$

$$I_\perp \ddot{\theta} - I_\perp (\dot{\phi})^2 \sin\theta \cos\theta + I_s (\dot{\phi} \cos\theta + \dot{\psi}) \dot{\phi} \sin\theta - mgl \sin\theta = 0 \tag{70}$$

If we consider a steady precession top ($\theta = \text{const}$), the symmetrical top will have a constant precession rate $\dot{\phi}$ and spin rate $\dot{\psi}$. This means that the angular momentum and the rotational axis of the body will undergo a regular precession, rotating around the vertical line.

There exist two possible solutions for the precession $\dot{\phi}$:

$$I_\perp (\dot{\phi})^2 \sin\theta \cos\theta - I_s (\dot{\phi} \cos\theta + \dot{\psi}) \dot{\phi} \sin\theta + mgl \sin\theta = 0 \tag{71}$$

$$\dot{\phi} = \frac{I_s \omega_{z'} \cos\theta \pm \sqrt{(I_s \omega_{z'} \cos\theta)^2 - 4mgl \cos\theta (\sin\theta)^2 I_\perp}}{2 I_\perp \sin\theta \cos\theta} \tag{72}$$

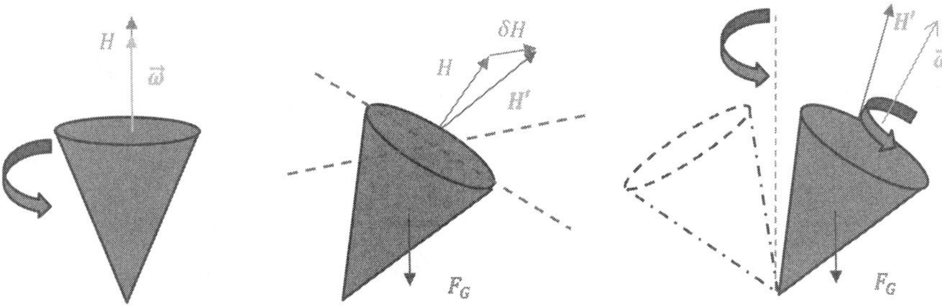

Fig. 8 Torque-induced precession.

The radical must be positive:

$$(I_s\omega_{z'})^2 > 4mgl\, I_{\perp}\frac{(\sin\theta)^2}{\cos\theta} \tag{73}$$

The body will keep precessing if the spin of the body $\omega_{z'}$ is sufficiently large. Otherwise, if the spin of the symmetric top slows down, it will reach a point where the gravity effect dominates and makes the body fall (see Fig. 8). In reality, a symmetric top rotating on the ground, will be subjected to friction that will cause dissipation of kinetic energy and therefore a decrease in the spin rotation. This explains why the symmetric top finally falls on to the ground. In general, the nutation θ won't be constant. As well as the precession mode, there may exist a nutation mode that makes θ change periodically. From the conservation of the equation of energy, it can be derived that θ has a periodic motion. The nutation mode is depicted in Fig. 9.

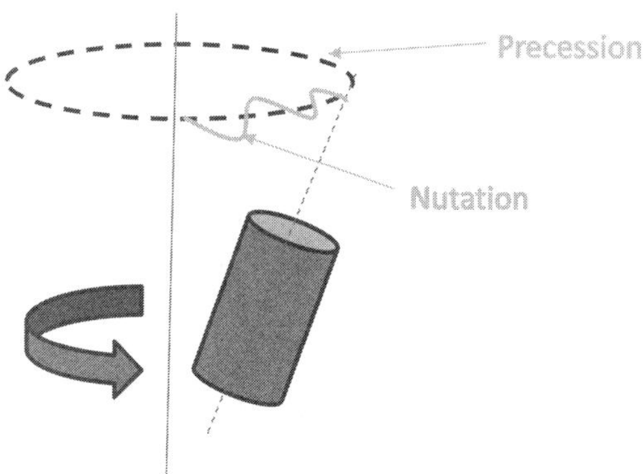

Fig. 9 Nutation mode.

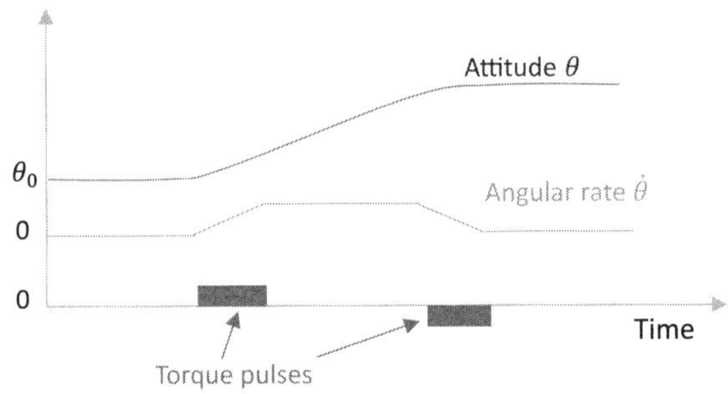

Fig. 10 Torque response without bias.

E. MOMENTUM BIAS DEFINITION

As explained previously in Sec. III.A, an external torque will modify the angular momentum of the system. However, the larger the angular momentum of the system, the less perturbed it will be by an external torque. This is called gyroscopic stiffness or gyroscopic rigidity. This "tool" has been exploited by ACS subsystem engineers in the design of satellite missions and it is called "momentum bias." It consists of having an axis of rotation with a very large angular momentum so that the precession rate caused by external torques will be very small. This methodology is applied to satellites that are invariantly pointing toward a certain direction. For many applications this direction tends to be the axis normal to the orbital plane. A system with bias will have a different response to torques compared with a system with no bias as it can be depicted from Figs. 10 and 11. The system with bias needs a continuous torque to achieve the same attitude rate as the system with no bias.

It can be observed in Fig. 11 that an oscillatory mode takes place called nutation mode. Satellites with active stabilization methods that use momentum bias

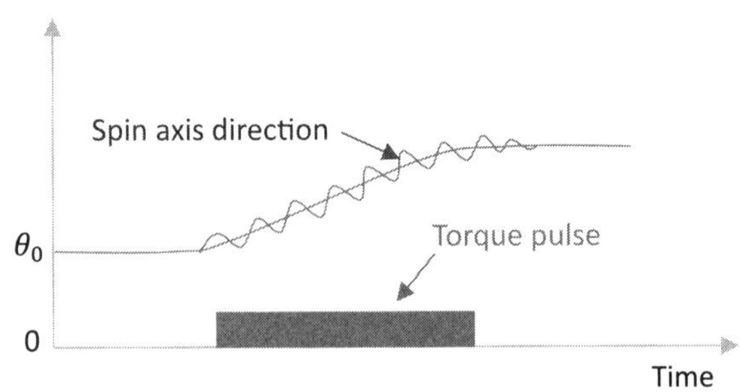

Fig. 11 Torque response with bias.

need to have passive or active dampers that will counteract it. This mode is activated when a disturbance causes a misalignment between the spinning axis and the angular momentum vector. The spin axis adopts a coning type movement around the angular momentum vector (precession). The nutation mode is excited on top of the precession causing an oscillatory movement.

F. ACTIVE STABILIZATION METHODS

As explained in Ref. [3], active stabilization methods can be classified into four groups:

1. Hybrid
2. Dual spinner
3. Pure spinner
4. Three-axis stabilization

The first three uses some form of momentum bias whereas the fourth method has no bias. Each method has certain advantages and disadvantages and they are chosen depending on the mission and system requirements.

1. DESCRIPTION OF ACTIVE STABILIZATION METHODS

Spacecraft with momentum bias can be classified into three groups as shown previously:

1. *Pure spinner*: The whole spacecraft rotates to acquire momentum bias.
2. *Dual spinner*: A significant part of the spacecraft is spinning to acquire momentum bias.
3. *Hybrid*: A small part of the spacecraft is spinning to acquire momentum bias.

For the first two groups the momentum bias is achieved with large inertias moving at slow angular velocities whereas the last group uses small inertias spinning at high angular velocities. A pure spin spacecraft means that the whole spacecraft is spinning to achieve momentum bias. The spin axis must be an axis that does not need to be changed in terms of orientation throughout the mission and it must be a principal axis. Therefore, there will be constraints in the design to avoid having products of inertia (asymmetries in the mass distribution). The spin axis must coincide with the maximum moment of inertia to have long-term stability and the other two must be equal and lower. It must be noted that the spin axis may coincide with the axis of least inertia for spin motion of limited duration. Examples of pure spin satellites are Intelsat-1 or Meteosat SG. Hybrid and dual-spin stabilized satellites consist of bodies that rotate at different rates. They have a part of the spacecraft that rotates achieving momentum bias, whereas the other part can be nonrotating. In the case of a dual spinner a significant part of the structural body of the spacecraft moves at a low rate providing momentum bias. To achieve stability, the spin axis may be aligned with the axis of least inertia, provided that the major source of energy dissipation is in the non-spinning part. The main advantage of a dual spinner compared to a pure spin is that part of the spacecraft is de-spun. This means that a de-spun payload can be used, for instance, to point constantly toward a certain direction (e.g., antenna pointing toward the Earth). The spinning part needs to have equal inertias about axes orthogonal to the spin axis. Examples of dual spinners are Giotto, Intelsat 2-4,6 or Galileo (Jupiter). In the case of hybrid spacecraft a small part of the spacecraft is spinning at a high rate. For example, momentum wheels can be used. Examples of hybrid satellites are the Navstar GPS (2R) or some of the Comsat satellites. The fourth method of actively stabilizing spacecraft is the 3-axis stabilization method. They achieve attitude stability without the use of momentum bias. For instance, they may use thrusters or reaction wheels (see Sec. III.G) to control the spacecraft orientation. This method imposes less stringent constraints on the inertial properties of the satellite and therefore the final configuration varies greatly depending on the mission criteria. Examples of 3-axis stabilized satellites are Magellan, Spot 5, or the Hubble Space Telescope.

2. HISTORICAL SELECTION OF STABILIZATION METHODS

In the past, the spacecraft stabilization choice was limited by the accuracy of the components and the processing speed of the computers available at the time. Therefore, spacecraft tended to exploit the properties of momentum bias to improve the inherent stability of the vehicle. Nowadays, technology has progressed to the extent that these limitations no longer exist. Satellites are more widely exploiting the increased freedom offered in the configuration design by the use of the 3-axis stabilization method. Nevertheless, momentum bias methods are still widely used for missions that may have an invariantly pointing direction vector or require high ΔVs, which lead to significant torque disturbances as a result of thrust misalignment. In addition, the long heritage of previous missions that have used momentum bias still makes this tool attractive. A trade-off between pointing design requirements, accuracy needed, and heritage must be carried out in order to choose the optimum spacecraft stabilization type for most missions. However, if there is no invariantly pointing direction vector then momentum bias cannot be used and the satellite must employ the 3-axis stabilization method.

G. MOMENTUM MANAGEMENT

A spacecraft can be subjected to both external and internal torque and each can be subcategorized into controllable and uncontrollable (or disturbance) torques [4].

External torques follow Netwon's second law:

$$\frac{\mathrm{d}\boldsymbol{H}_G}{\mathrm{d}t} = \sum \boldsymbol{T}_{\mathrm{ext}} \tag{74}$$

whereas internal torques comply with the conservation of momentum:

$$\frac{\mathrm{d}\boldsymbol{H}_G}{\mathrm{d}t} = 0 \tag{75}$$

As explained previously, the total angular momentum of a body will only be changed if external torques are applied. If a spacecraft is subjected to internal torques then the total angular momentum will be conserved but the distribution of angular momentum between the parts will be changed. ACS design engineers use controllable external and internal torques for the spacecraft momentum management. However, uncontrollable torques (external and internal) must be counteracted.

The main sources of external torques are

1. *Uncontrollable (disturbance) torques*: air drag, magnetic torque, gravity gradient torque, and SRP torque
2. *Controllable torques*: Thrusters such as gas jets or magnetorquers

Typical sources of internal torques are

1. *Uncontrollable (disturbance) torques*: fuel sloshing, deployment of mechanism (e.g., solar array), crew movement, etc
2. *Controllable torques*: reaction wheels and momentum wheels

Controllable internal torques are used to redistribute the angular momentum of the spacecraft and therefore control its attitude. It must be noted that devices that create controllable internal torques are not enough for the attitude control of a spacecraft. They cannot modify the total angular momentum of the spacecraft and therefore cannot counteract external disturbances, which create external torques. Furthermore, their main disadvantage is that they may reach a speed saturation, which appears if there are significant secular torques. Additional systems that provide external torques, such as thrusters, are necessary for desaturation of controllable internal torques devices. The two main systems to control internal torques are reaction wheels and momentum wheels. Reaction wheels have operational rotational speeds which can go in the positive or negative direction and this allows for changing their spin orientation. The main advantages are that no propellant is needed and they provide good pointing accuracy. Their main disadvantage is that they produce micro vibrations. Momentum wheels always

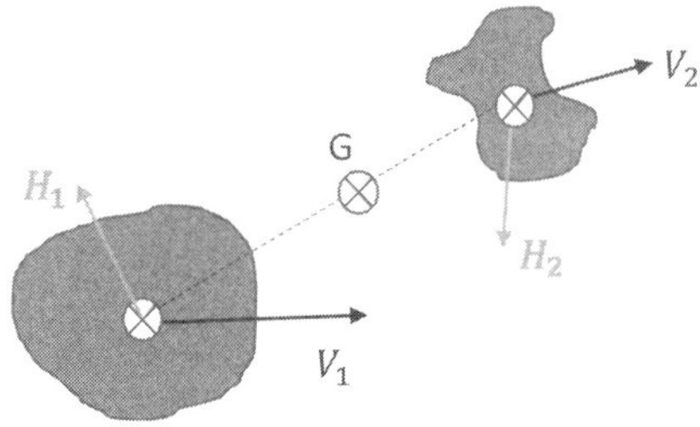

Fig. 12 Chaser and target system.

work at their nominal rotational velocity providing gyroscopic rigidity. They usually work at high rotational speeds. Their main advantage is that they don't need propellant. However, they present dynamical nutational modes, which must be damped.

H. ATTITUDE CONTROL FOR RENDEZVOUS CAPTURE

In previous sections, we have studied the physical principles that apply to the rotational dynamics of one body. Attitude control for active debris removal requires a chaser spacecraft that will rendezvous, capture, and control a secondary object. Therefore, this particular problem incorporates linear and angular momentum applied to two bodies and places unique demands on the requirements of the ACS system. The conservation of linear and angular momentum applies to the combination of several bodies in the absence of external torques [5]. Therefore, the total angular momentum about their combined center of mass won't change even if the bodies interact with each other (collision, separation, and tethering). Given the following two-body system, if one body is moved with respect to the other, the total absolute linear momentum and total angular momentum referred to the center of mass G remains constant.

The chaser-target system is depicted in Fig. 12 and the total linear and angular momentum with respect to the center of mass are

$$MV_G = m_1 V_1 + m_2 V_2 \tag{76}$$

$$\boldsymbol{H}_G = \boldsymbol{H}_1 + \boldsymbol{H}_2 + \frac{m_1 m_2}{M}(\boldsymbol{r}_{12} \times \boldsymbol{v}_{12}) \tag{77}$$

Regarding the design of the chaser spacecraft for active debris removal missions, two critical changes must be taken into account for ACS:

1. The design must minimize thrust misalignment of the combined system (chaser-target),
2. ACS authority over both objects is crucial during the capture and de-orbiting phase.

In the case of active debris removal missions where a chaser spacecraft will grab and de-orbit existing space debris, several challenges arise. First of all, the chaser must grab a noncooperative rotating object. A proposed solution is to achieve a state with zero relative rotation between the two bodies and grab the target. The new combined system will have a new center of gravity and a new inertia matrix. The chaser spacecraft must be able to control and de-tumble the target and once a stable configuration has been achieved, apply the final maneuvres for the de-orbiting phase. Any thrust vector should go through the center of mass of the system (which is not accurately known) in order to avoid unnecessary additional disturbance torques. Another difficulty that appears is finding a suitable grabbing spot as the target is noncooperative. In addition, there is a lack of knowledge of the precise physical properties of the target and therefore of the behavior of the overall system chaser-target (i.e., uncertainties in the center of gravity and moments of inertia of the combined system). The next necessary step is to look into the technologies that are currently being researched for active debris removal.

I. ACTIVE DEBRIS REMOVAL TECHNOLOGIES

Space debris can be either re-orbited into a graveyard orbit or de-orbited to burn up in the atmosphere. The first option is preferred for geosynchronous satellites ($\sim$35, 000 km), whereas the latter is preferred for low Earth orbit satellites ($<$2000 km). Current active debris removal technologies can be divided in contact and contactless methods for complete removal. Noncontact methods have some advantages as they avoid all the problems related to grabbing a noncooperative target that may be tumbling. However, these methods lead to an uncontrolled reentry, which is only adequate for small targets that will disintegrate in the atmosphere (mass $<\sim$ 500–700 kg). These type of methods may also have applications for putting large space debris into graveyard orbits. The controlled reentry of large objects requires the use of chemical propulsion for the de-orbiting phase. Therefore, existing noncontact methods are not suitable for a controlled reentry. It must be noted that contact and contactless methods can be combined. For instance, a contactless method can be used for de-tumbling the spacecraft, followed by a grabbing phase and the de-orbiting phase using a contact method. Contactless methods can be based on momentum exchange such as the ejection of chemical propellant toward the target, the use of lasers or an ion-beam. Moreover, contactless methods can be based on increasing the area to mass ratio of the target. By ways of illustration, Pergola et al. [6] describes how foam could be ejected from a chaser spacecraft to increase the volume of the target without requiring any docking mechanism. Potentially, this method could be used to visit and de-orbit several targets within the same mission. The main drawback of this method is the limited knowledge of foam behavior in vacuum. An example of a contactless method for active debris removal is the ion-beam shepherd. This is a concept being analyzed by the Polytechnical University of Madrid. The shepherd spacecraft will point an ion-beam of an ion-thruster toward the space debris and thanks to the momentum exchange derived from the impingement, the orbit of the target is modified. A secondary propulsion system is needed to compensate the reaction force on the shepherd [7]. Contact methods will grab the space debris by means of rigid or flexible capture and attach a device that will help the combined system to decay at a quicker pace. Examples of rigid capture systems are tentacles or a robotic arm. Examples of flexible capture systems are nets, a harpoon, or a clamp. The attached device can be a propulsive or non-propulsive device. Propulsive devices may be chemical or electrical. Non-propulsive devices may be, for instance, drag augmentation systems or an electrodynamic tether. The electromagnetic ether (EDT) is based on the Lorentz force principle, shown in Fig. 13. A conductive tether is deployed in space that has plasma contactors at opposite sides of the tether and this will make a close circuit of electric current thanks to the ambient plasma. In addition, the interaction of the electric current and the geomagnetic field will generate a Lorentz force that will help the system to de-orbit. The system has to be attached to the space debris prior its de-orbiting. In Ref. [8] different attachment methodologies are studied stating that the EDT does not need to be rigidly attached thanks to its low thrust system. Regarding the existing risks for this method, there is a threat of debris impact on the tether, which could damage or even cut the tether.

Drag augmentation devices such as solar sails, deployable structures, balloons or foam can be used to achieve large area to mass ratios that increase the atmospheric drag effect and therefore make the system decay quicker. The main advantage of these methods is that they require no propellant for the de-orbiting phase. The University of Surrey in the United Kingdom presented a project in 2011 [9] called Hybrid Solar Sails for active debris removal that consists of a satellite with

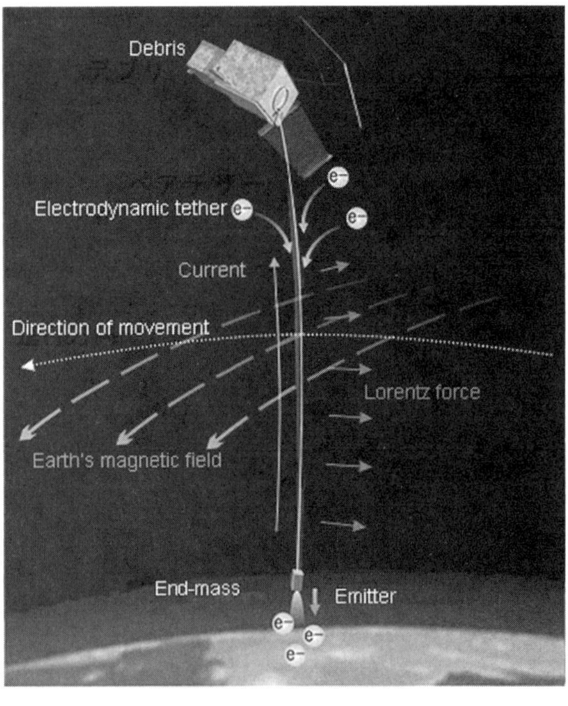

Fig. 13 Electrodynamic Tether Principle. (Jaxa, [8]).

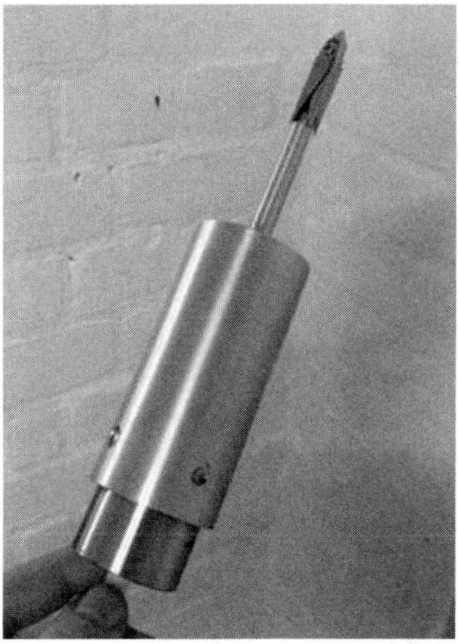

Fig. 14 Harpoon system (Courtesy of Airbus Defence & Space).

docking capability that attaches itself to the space debris object and deploys a reflective sail with tethers. The system decays thanks to the drag augmentation as well as the Lorentz force exerted on the charged tethers. After docking, the chaser de-tumbles the space debris with a magnetic attitude control system and later deploys the sail and tethers for the de-orbiting phase.

Additional systems such as the use of a net, harpoon, or grabbing devices (i.e., a robotic arm, tentacles, and clamps) are being envisaged as adequate systems to de-orbit large space debris that require a controlled reentry. Once the spacecraft is grabbed using one of these methods, the chaser uses chemical thrusters to de-orbit the system. Airbus Defence & Space in UK has already developed a prototype of a harpoon system (see Fig. 14) and it is being tested to analyze the effect of penetration for different materials as well as its proposed shape and length to avoid any additional debris generation or tank puncture.

The net capture process, depicted in Fig. 15, starts with the deployment of the net that reaches its full size just before impact with the target and then wraps itself around either passively thanks to the motion of corner masses or actively with an additional closing mechanism. Afterward, the tether is tensioned by the chaser and the de-orbit phase initiates. This method encounters difficulties during the pulling phase where the system chaser-target must avoid potential collisions and reach stability [10].

Grabbing systems with a hard mechanical interface such as robotic arms are being studied by DLR in Germany and MDA in Canada among others. The DEOS mission (Deutsche Orbitale Servicing Mission) developed by the German Aerospace Center DLR plans to launch a client and a servicer where the client will act as a noncooperative target that needs maintenance and disposal (see Fig. 16). This mission will test capturing techniques with a robotic arm using on-board autonomous control. Moreover, it will demonstrate servicing applications and the final de-orbit of the captured satellite. The ETS-VII (Engineering Test Satellite VII) developed by JAXA already demonstrated in 1997 unmanned autonomous rendezvous and docking operations between two cooperative spacecraft using a robotic arm. In 2007, STP-1 (Space Test Program-1) sponsored mostly by DARPA (Defense Advanced Research Projects Agency) was launched and carried out soft capture and mating operations using a robotic arm developed by MDA.

Fig. 15 Deployed net to capture space debris. (Courtesy of European Space Agency).

J. DE-TUMBLING TECHNOLOGIES

One of the main problems of capturing systems for ADR applications is being able to grab and control a noncooperative spacecraft that may be rotating around all three axes at rotation rates as potentially as high as 50 deg/s. Ground observations of space debris rotational motion are scarce at present. There are ground measurements that show some space debris objects have a non-stable attitude,

Fig. 16 DEOS mission (Courtesy of Airbus Defence & Space).

not even spinning around their principal axes. Most of the technologies previously described that need a docking phase have poor robustness with respect to the rotational dynamics of the debris object. When a satellite or launcher upper stage reaches its end of life, the fuel tanks are usually emptied to reduce the risk of explosions due to overpressure. The outgassing of propellant may lead to high rotational rates (tens of degrees per second), whereas current methods are limited to capturing objects that do not rotate faster than a few degrees per second. At present, there are some studies on how eddy currents induced by the Earth's magnetic field may damp the spinning of conducting bodies. At attitudes of around 500 km, the eddy current torques can outweigh atmospheric drag torques by several orders of magnitude and they dominate at attitudes down to 160 km. N. Praly et al. [11] carried out an analysis on the rotational dynamics of Ariane 4 upper stages (H10), which were initially in a sun-synchronous orbit between 500 and 800 km attitude. Their simulations take into account the gravity gradient and the eddy current torques and show an exponential decrease of the rotation due to these perturbations (from 20 up to 200 days). Currently, there exists no database regarding the rotational state of space debris but a big effort is being made by the scientific community to carry out new observations and acquire data regarding the rotational motion of space debris objects. If observations validate these results, it would be possible to find conducting debris with low angular rotation rates. From the equation of motion of a rigid body (78), it can be derived that in order to change the rotational state, we can either perform changes in the inertia matrix of the system, redistribute the total angular momentum of the system with internal torques or apply external torques. In practice, spacecraft can only use devices to apply internal and external torques for momentum management described in Sec. III.G:

$$\frac{\mathrm{d}\boldsymbol{H}_G}{\mathrm{d}t} = \frac{\mathrm{d}\boldsymbol{I}_G\boldsymbol{\omega}}{\mathrm{d}t} = \sum \boldsymbol{T}_{\mathrm{ext}} \tag{78}$$

De-tumbling options can be divided into contactless methods such as the use of exhaust products onto the tumbling object and contact methods that will apply external torques to control the tumbling motion. Nishida S. et al. [12] proposed the use of a brush type contactor, shown in Fig. 17, to reduce the rotational motion of the space debris object to a rate that enables the object to be captured by a robotic arm. The brush at the tip of the robotic arm makes contact with the rotating target and decelerates it using contact friction braking.

Dealing with tumbling objects using flexible contact methods such as a harpoon, net or tether-like system is still an open point. Once the target is captured, the tension achieved on the tether with impulsive thrusts during the de-orbiting phase may dominate and control the rotational motion of the target [10, 13] McKinight D. et al. [14] present different methods to de-tumble a body, taking advantage of the magnetic field, using propulsive nanosatellites adhered to the space debris object and making use of mass extension or foam in order to increase the inertia and therefore slow the rotational rate.

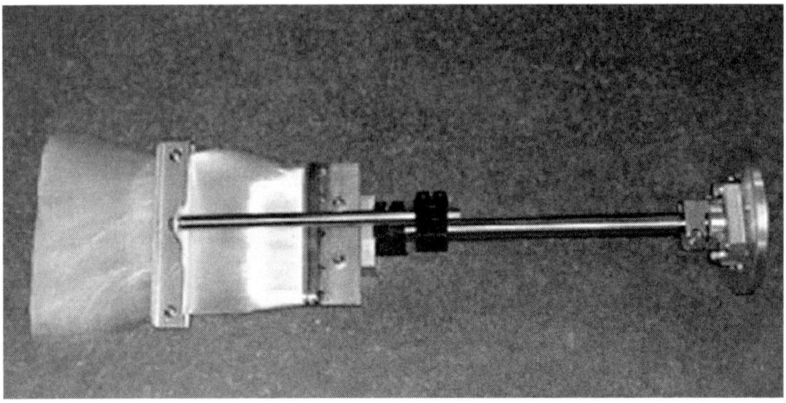

Fig. 17 Prototype of brush contactor [12].

IV. CONCLUSION

Space missions for active debris removal pose great challenges with regard to the ACS. At present, there is still great uncertainty on the rotational dynamics of current space debris objects. Additional ground observations should be carried out in order to increase our knowledge of the rotational state of the debris and be able to validate the existing simulations of dynamical motions of these bodies. In addition, there is a wide variety of technologies being researched for active debris removal. Although each one shows different advantages and weaknesses, there is no single method that stands out with respect to the rest. Each approach could be suitable depending on the type of target, its orbit, and rotational motion. In addition, a de-tumbling process before capture may be needed for bodies submitted to high rotational rates. Despite the differences of each method, the common goal is to minimize the risk when capturing a noncooperative object. All in all, there are still a lot of questions to be answered. Further research needs to be carried out and in-orbit demonstration missions are required to better understand the challenges that active debris removal presents.

REFERENCES

[1] Wertz, J. R., *Spacecraft Attitude Determination and Control*, Kluwer Academic Publishers, Dordrecht, 2002.

[2] Kurosaki, H., and Yanagisawa, T., "Shape and Motion Estimate of LEO Debris using Light Curves," *Advances in Space Research*, Vol. 50, 2012, pp. 136–145.

[3] Stark, J., Fortescue, P., and Swinerd, G., *Spacecraft Systems Engineering*, John Wiley & Sons, Ltd., Chichester, West Sussex, England, 2011.

[4] Hallmann, W., Ley, W., and Wittmann, K., *Handbook of Space Technology*, John Wiley & Sons, Ltd., Munich, 2009.

[5] Lifshitz, E. M., and Landau, L. D., *Mechanics. Course of Theoretical Physics*, 3rd ed., Vol. 1, Pergamon, Oxford, England, 1976.

[6] Andrenucci, M., Pergola, P., Ruggiero, A., and Summerer, L., "Low-Thrust Missions for Expanding Foam Space Debris Removal," *32nd International Electric Propulsion Conference*, Wiesbaden, Germany, Vol. IEPC-2011-126, Sept. 2011, pp. 1–13.

[7] Bombardelli, C., and Peláez, J., "Ion Beam Shepherd for Contactless Space Debris Removal," *Journal of Guidance, Control, and Dynamics*, Vol. 34, 2011, pp. 916–920.

[8] Nakanishi, H., Kawamoto, S., Ohkawa, Y., "Active Debris Removal by a Small Satellite," *63rd International Astronautical Congress*, Naples, Italy, IAC-12-A6.7.8, 2012, pp. 1–8.

[9] Theodorou, T., Visagie, L., and Summerer, L., "Hybrid Solar Sails for Active Debris Removal Final Report," Technical Report, Surrey Space Centre, University of Surrey and ACT, European Space Agency, *32nd International Electric Propulsion Conference*, Wiesbaden, Germany, Vol. IEPC-2011-126, 2011, pp. 1–59.

[10] Wormnes, K., and De Jong, J.H., "Throw- Nets and Tethers for Robust Space Debris Capture," *64th International Astronautical Congress*, IAC-13-A6.5.2, Beijing, China, 2013.

[11] Bonnal, C., Praly, N., Petit, N., and Laurent-Varin, J., "Study on the Eddy Current Damping of the Spin Dynamics of Spatial Debris from the Ariane Launcher," *Acta Astronautica*, Vol. 76, 2012, pp. 145–153.

[12] Kawamoto, S., and Nishida, S., "Strategy for Capturing of a Tumbling Space Debris," *Acta Astronautica*, Vol. 68, 2011, pp. 113–120.

[13] Schaubb,H., and Jaspera, L., "Input Shaped Large Thrust Maneuver with a Tethered Debris Object," *Acta Astronautica*, Vol. 96, 2013, pp. 128–137.

[14] Kaczmarek, A., McKnight1, D., Pentino, F., and Knowles, S., "Detumbling Rocket Bodies in Preparation for Active Debris Removal," *6th European Conference on Space Debris, European Space Operations Centre*, Darmstadt, Germany, 2013, pp. 1–13.

CHAPTER 17

Methods and Techniques for Asteroid Deflection

Massimiliano Vasile* and Nicolas Thiry*
University of Strathclyde, Glasgow, United Kingdom

I. INTRODUCTION

Among all minor celestial objects in the solar system, near-Earth objects (NEO), the majority of which are asteroids, are those with perihelion lower than 1.3 Astronomical Units (AU) and aphelion greater than 0.983 AU. A subclass of these minor objects is deemed to be potentially hazardous and are designated as potentially hazardous asteroids (PHA). PHAs have a minimum orbit intersection distance (MOID) with Earth's orbit that is less than or equal to 0.05 AU and their diameter is larger than 150 m. As of the 15 January 2014, 10521 NEOs have been detected; of those, more than 3100 have a diameter between 0.3 and 1 km, and 1447 are listed as PHA. Impacts from asteroids of about 1 km or more in diameter are considered to be capable of causing global climate change and the destruction of ozone, with a land destruction area equivalent to a large state or a country [1]. Those with an average diameter of 100 m are expected to release over 10^2 Mt of energy and can cause significant tsunamis and land destruction of a large city [2]. It is estimated that there are between 30,000 and 300,000 NEOs with diameters around 100 m, meaning that a large number of NEOs are still undetected. The reader has to consider, however, that these numbers are rapidly changing.

In the past few decades, several, different methods have been proposed and studied to deflect asteroids [3]. One can group deflection methods in broad classes depending on the type of interaction between the deflecting spacecraft and the minor body. One possible classification distinguishes between impulsive and slow-push methods, depending on the duration of the deflection action. Among the impulsive methods one can find the kinetic impactor [4–6] (in all its variants using single or multiple spacecraft, chemical or low-thrust engines [7], retrograde solar [8] or electric sails), the nuclear blast and the smart cloud [9], whereas slow-push methods are the low-thrust tug [10], the gravity tug [11], the laser ablation [12–15], the solar ablation [16] or photon beaming [17], the solar

*Department of Mechanical & Aerospace Engineering.

sail Tug, the Ion Beaming [18], the Electrostatic or Electromagnetic Tug [19], the mass driver [20], the electromechanic tether [21]. Some methods envisage flying a formation (or swarm) of spacecraft in the proximity of the asteroid. Vasile and Maddock first proposed a formation of solar concentrators [17, 22, 23] and then extended the concept to a swarm of solar-pumped laser systems [24]. Other concepts propose the use of multi-gravity tractor formations [25]. Finally some methods consider fragmenting the asteroid using a rotating tether or swarms of robots. Other methods are generally classified as passive and aim at changing the reflectivity of the asteroid or the Yarkovsky effect [26] using paints, blankets, and flying shields.

This chapter covers the basic principles of asteroid deflection and for some deflection methods presents simple analytical models that can be used to evaluate the performance of the method and to design a potential deflection mission.

The chapter is divided in two parts. The first part explains how a given acceleration or a given Δv can be used to divert the trajectory of an asteroid and avoid a collision with the Earth. Analytical formula are provided for the case of an impulsive and low-thrust deflection action in the case in which the variation of the orbit of the asteroid remains small and the altered trajectory remains proximal to the original one. The second part tackles the technological aspect of the different deflection methods. Physical models are derived and formula are presented to compute the Δv or the acceleration resulting from a given deflection strategy. Some performance indicators are also introduced, which provide an objective framework to compare different deflection strategies.

II. BASIC DEFLECTION PRINCIPLES AND COMPUTATIONAL TOOLS

The orbit of an asteroid can be modified in different ways depending on the direction in which the deflection action is applied. With reference to Fig. 1a one can define a local reference frame centered in the center of mass of the asteroid, with the y-axis aligned with the velocity vector $\boldsymbol{v}$, the z-axis aligned with the angular momentum $\mathbf{h}$ and the x-axis aligned with the local normal $\mathbf{n}$ contained in the orbit plane. The deflection action can be projected on these axes to give the three deflection components a_t, a_n, and a_h. By aligning the deflection action with one of these three coordinate directions one can produce the following modifications of the orbit:

1. Change the energy of the orbit by aligning the deflection action along the velocity direction
2. Rotate the orbit in its plane by aligning the deflection action in the normal direction
3. Modify the inclination of the orbit or rotate the line of the nodes by aligning the deflection action along the z-axis

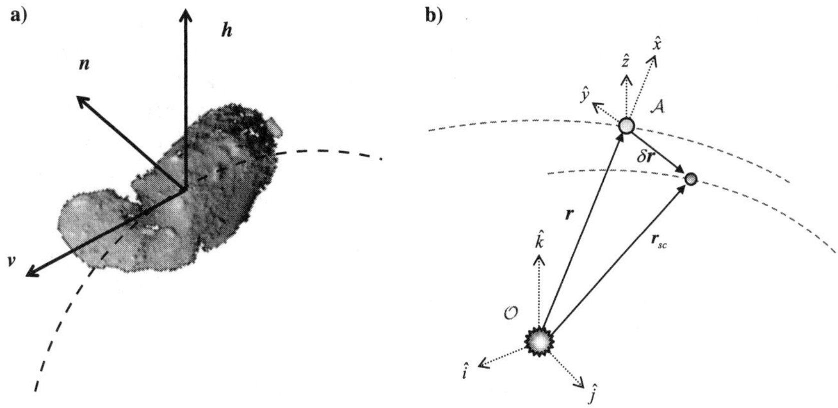

Fig. 1 a) Tangential *v*, normal *n*, and bi-normal *h* reference frame and b) definition of the *r-t-h* Hill's reference frame $\mathcal{A}$ centered on the undeflected asteroid.

The result of a deflection action can be deduced by looking at Gauss' planetary equations:

$$
\begin{aligned}
\frac{\mathrm{d}a}{\mathrm{d}t} &= \frac{2a^2 v}{\mu} a_t \\
\frac{\mathrm{d}e}{\mathrm{d}t} &= \frac{1}{v}\left[2(e+\cos\theta)a_t - \frac{r}{a}\sin\theta a_n\right] \\
\frac{\mathrm{d}i}{\mathrm{d}t} &= \frac{r\cos\theta^*}{h\sin i} a_h \\
\frac{\mathrm{d}\Omega}{\mathrm{d}t} &= \frac{r\sin\theta^*}{h\sin i} a_h \\
\frac{\mathrm{d}\omega}{\mathrm{d}t} &= \frac{1}{ve}\left[2\sin\theta a_t + \left(2e + \frac{r}{a}\cos\theta\right) a_n\right] - \frac{r\sin\theta^*\cos i}{h\sin i} a_h \\
\frac{\mathrm{d}M}{\mathrm{d}t} &= -\frac{\sqrt{1-e^2}}{ve}\left[2\left(1+\frac{e^2 r}{p}\right)\sin\theta a_t + \frac{r}{a}\cos\theta a_n\right] + n
\end{aligned}
\tag{1}
$$

where a is the semimajor axis, e the eccentricity, i the inclination, Ω the right ascension of the ascending node, ω the argument of the pericenter, θ the true anomaly and M the mean anomaly of the orbit of the asteroid. The quantity θ^* is the sum of the true anomaly and the argument of the pericenter, μ is the gravity parameter of the Sun, r is the distance of the asteroid from the Sun, $v = \|\boldsymbol{v}\|$ the modulus of the velocity of the asteroid on its orbit, h the modulus of the orbit angular momentum and $n = \sqrt{\mu/a^3}$.

A deflection action along the y-axis would produce a variation of a and M and a variation of e and ω if not applied at $\theta = k\pi$ with $k = 0, 1, \ldots$. A deflection action along the z-axis would produce a variation of Ω and i and a variation of ω if not applied at $\theta^* = k\pi$ with $k = 0, 1, \ldots$. Finally, a deflection action along the x-axis would produce a variation of ω and a variation of e and M if not applied at $\theta = k\pi$ or $\theta = \pi/2 + k\pi$ with $k = 0, 1, \ldots$, respectively.

Looking at the last equation in Eqs. (1), one can see that a deflection action can have two effects on M: one does not modify n and is what we will call a *geometric variation* of M, the other is a variation of n induced by a variation of a. This latter effect produces a change in the time of arrival of the asteroid at the point of minimum orbit interception distance (MOID), or point where the geometric distance between the orbit of the Earth and the un-deviated orbit of the asteroid is at its minimum. A deflection action a_t along the velocity of the asteroid would yield both a change in n and a geometric variation of M.

By inspecting Gauss planetary equations one can derive a qualitative indication of how to act on an asteroid to modify its orbit. The next step is to quantitatively compute the deflection produced by a given deflection action. Here and in the following the word deflection will refer to the variation of the position of the asteroid with respect to its nominal (un-deflected) orbit at a given time.

A. FIRST-ORDER IMPULSIVE DEFLECTION ANALYTICAL FORMULAS

In the case of an impulsive variation of the velocity of the asteroid, the deflection action generates an instantaneous modification of the orbital parameters. Gauss' planetary equations (1) can be rewritten as follows:

$$
\begin{aligned}
\delta a &= \frac{2a^2 v}{\mu}\delta v_t \\
\delta e &= \frac{1}{v}\left[2(e+\cos\theta)\delta v_t - \frac{r}{a}\sin\theta\,\delta v_n\right] \\
\delta i &= \frac{r\cos\theta^*}{h\sin i}\delta v_h \\
\delta\Omega &= \frac{r\sin\theta^*}{h\sin i}\delta v_h \\
\delta\omega &= \frac{1}{ve}\left[2\sin\theta\,\delta v_t + \left(2e + \frac{r}{a}\cos\theta\right)\delta v_n\right] - \frac{r\sin\theta^*\cos i}{h\sin i}\delta v_h \\
\delta M &= -\frac{\sqrt{1-e^2}}{ve}\left[2\left(1+\frac{e^2 r}{p}\right)\sin\theta\,\delta v_t + \frac{r}{a}\cos\theta\,\delta v_n\right] + \delta n(t_f - t_d)
\end{aligned}
\tag{2}
$$

where:

$$\delta n = \sqrt{\frac{\mu}{(a + \delta a)^3}} - \sqrt{\frac{\mu}{a^3}}$$

t_d is the time of application of the impulsive deflection action, t_f is the time at which the variation of the orbital elements is measured and $\delta v_t = a_t \delta t$, $\delta v_n = a_n \delta t$, $\delta v_h = a_h \delta t$. The evolution of the deviated orbit can be described in the Hill's reference frame centered in the undeviated asteroid (see Fig. 1b) where now the x-axis is aligned with the radial direction. In this reference frame the position vector of the deviated asteroid is $\delta \mathbf{r} = [x\ y\ z]^T$. If δr is small compared to r, the components of $\delta \mathbf{r}$, at a given θ along the undeviated orbit, can be related to the variation of the orbital parameters (2), through the following set of linearized proximity equations [27]:

$$\begin{aligned} x &\approx \frac{r}{a}\delta a + \frac{ae\sin\theta}{\nu}\delta M - a\cos\theta\delta e \\ y &\approx \frac{r}{\nu^3}(1 + e\cos\theta)^2\,\delta M + r\delta\omega + \frac{r\sin\theta}{\nu^2}(2 + e\cos\theta)\,\delta e + r\cos i\delta\Omega \\ z &\approx r\left(\sin\theta^*\delta i - \cos\theta^*\sin i\delta\Omega\right) \end{aligned} \tag{3}$$

with $\nu = \sqrt{1 + e^2}$. When $t_f = t_{\mathrm{MOID}}$ and the true anomaly in Eqs. (3) is the one of the MOID then $\delta \mathbf{r}$ is the deviated position of the asteroid at the point of the MOID.

By combining Eqs. (2) with Eqs. (3), one can compute the deflection at the time of the MOID as a function of the deflection impulse at time t_d in the following matrix form:

$$\delta \mathbf{r}(t_{\mathrm{MOID}}) = \mathbf{A}_{\mathrm{MOID}} \mathbf{G}_d \delta \boldsymbol{v} = \mathbf{T} \delta \boldsymbol{v} \tag{4}$$

where the matrix $\mathbf{A}_{\mathrm{MOID}}$ and $\mathbf{G}_d$ are

$$\mathbf{A}_{\mathrm{MOID}} = \begin{bmatrix} \frac{r_{\mathrm{MOID}}}{a} - \frac{3}{2}\frac{e\sin\theta_{\mathrm{MOID}}}{\nu}\frac{\sqrt{\mu}}{a^{3/2}}\Delta t & -\frac{3}{2}\frac{r_{\mathrm{MOID}}}{\nu^3}(1 + e\cos\theta_{\mathrm{MOID}})^2\frac{\sqrt{\mu}}{a^{3/2}}\Delta t & 0 \\ -a\cos\theta_{\mathrm{MOID}} & \frac{r_{\mathrm{MOID}}\sin\theta_{\mathrm{MOID}}}{\nu^2}(2 + e\cos\theta_{\mathrm{MOID}}) & 0 \\ 0 & 0 & r_{\mathrm{MOID}}\sin\theta^*_{\mathrm{MOID}} \\ 0 & r_{\mathrm{MOID}}\cos i & -r_{\mathrm{MOID}}\cos\theta^*_{\mathrm{MOID}}\sin i \\ 0 & r_{\mathrm{MOID}} & 0 \\ \frac{ae\sin\theta_{\mathrm{MOID}}}{\nu} & \frac{r_{\mathrm{MOID}}}{\nu^3}(1 + e\cos\theta_{\mathrm{MOID}})^2 & 0 \end{bmatrix} \tag{5}$$

$$\mathbf{G}_d = \begin{bmatrix} \dfrac{2a^2 v_d}{\mu} & 0 & 0 \\ \dfrac{2(e+\cos\theta_d)}{v_d} & -\dfrac{r_d}{a v_d}\sin\theta_d & 0 \\ 0 & 0 & \dfrac{r_d\cos\theta_d^*}{h} \\ 0 & 0 & \dfrac{r_d\sin\theta_d^*}{h\sin i} \\ \dfrac{2\sin\theta_d}{e v_d} & \dfrac{2e+(r_d/a)\cos\theta_d}{e v_d} & -\dfrac{r_d\sin\theta_d^*\cos i}{h\sin i} \\ -\dfrac{\sqrt{1-e^2}}{e v_d}2\left(1+\dfrac{e^2 r_d}{p}\right)\sin\theta_d & -\dfrac{\sqrt{1-e^2}}{e v_d}\dfrac{r_d}{a}\cos\theta_d & 0 \end{bmatrix} \tag{6}$$

and

$$\delta n(t_f - t_d) = \delta n \Delta t \cong -\frac{3}{2}\frac{\sqrt{\mu}}{a^{5/2}}\Delta t\, \delta a \tag{7}$$

The subscript d indicates that quantities are taken when the deflection is applied. The norm of the deflection at the time of the MOID is

$$\|\mathbf{r}(\mathbf{t}_{\mathrm{MOID}})\| = \delta\mathbf{r}^T\delta\mathbf{r} = \delta\boldsymbol{v}^T\mathbf{T}^T\mathbf{T}\delta\boldsymbol{v} \tag{8}$$

which is a quadratic form in $\delta\boldsymbol{v} = \delta v\hat{\boldsymbol{v}}$. As such, if $\mathbf{V}$ is the matrix of the eigenvectors of the matrix $\mathbf{T}^T\mathbf{T}$ one can write:

$$\|\mathbf{r}\| = \delta v^2\hat{\boldsymbol{v}}^T\mathbf{V}\mathbf{T}^T\mathbf{T}\mathbf{V}^T\hat{\boldsymbol{v}} = \delta v^2\sum_{i=1}^{N}\lambda_i(\mathbf{q}_i^T\hat{\boldsymbol{v}})^2 \leqslant \delta v^2\lambda_{\max}\sum_{i=1}^{N}(\mathbf{q}_i^T\hat{\boldsymbol{v}})^2 \tag{9}$$

and by taking the unit vector $\hat{\boldsymbol{v}}$ aligned with the eigenvector $\mathbf{q}_{\max}$ corresponding to the maximum eigenvalue $\lambda_{\max}$ one gets:

$$\|\delta\mathbf{r}_{\mathbf{t}_{\mathrm{MOID}}}\|_{\max} = \delta v^2\lambda_{\max} \tag{10}$$

Figure 2 compares the achievable deflection for two different asteroids with low (2000SG344) and high eccentricity (1979XB) Table 1 (see Fig. 1, and for an impulse applied along the optimal direction (solid line), and along the tangent (dotted line), normal (dash-dotted line), and out-of-plane direction (dashed line). The modulus of the deflection impulse is $\delta v = 0.07$ m/s in all cases. The δr associated to δv_{opt} is the maximum one and overlaps to the deviation achieved with a normal impulse, for low Δt, and the one with a tangent maneuver, for longer Δt. An impulsive action at the pericenter is found to be the most effective one, whereas a δv at the apocenter gives a deviation close to the minimum one.

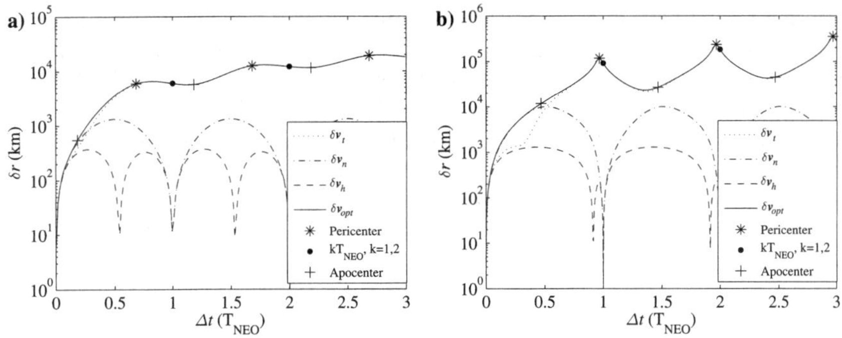

Fig. 2 Deflection achieved with $\delta v = 0.07$ m/s for: a) asteroid 2000SG344 and b) asteroid 1979XB [28].

The choice of an optimum timing along the orbital period is more significant for highly eccentric orbits than for nearly circular ones.

As a result of this analysis we can derive that, for a Δt smaller than a specific $\Delta t_{\text{NEO}} < T_{\text{NEO}}$, different for every asteroid, the component perpendicular to the motion dominates the other two whereas for larger Δt, the tangential component becomes dominant. These conclusions are consistent with the results in Ref. [4]. It can be noted that the value of the normal component of the optimal deviation impulse goes periodically to zero with a period equal to the one of the asteroid. Therefore, a deviation impulse given in the normal direction yields no deviation after an exact number of revolutions.

We can now consider a deflection applied with a relatively large Δt and only in the tangential direction (i.e., $\delta v_n = \delta v_h = 0$). In this case the components of the deflection in the Hill's reference frame simplify and reduce to

$$
\begin{aligned}
x &\simeq r\frac{2av}{\mu}\delta v_t + \frac{ae\sin\theta}{v}\left(-\frac{3\Delta t v}{\sqrt{a\mu}}\right)\delta v_t \\
y &\simeq \frac{r}{v^3}(1+e\cos\theta)^2\left(-\frac{3\Delta t v}{\sqrt{a\mu}}\right)\delta v_t
\end{aligned}
\tag{11}
$$

TABLE 1 ORBITAL PARAMETERS AND MASS OF ASTEROIDS 2000SG344 AND 1979XB

Name	*a*	*e*	*i*	*ω*	Ω	*M*	Epoch (MJD)	Mass (kg)
1979XB	2.35	0.72	25.1	75.7	85.5	62.0	53700	4.4e11
2000SG344	0.97	0.06	0.11	274.9	192.3	132.3	53800.5	7.1e7

where for small deflections and large warning times Δt, the variation of eccentricity δe can be considered negligible compared to the variation in semimajor axis and in mean anomaly. From Eqs. (11) one can see that both the radial and the transversal components of the deflection have a secular term due to the variation of M. The term in the radial direction, however, is multiplied times $\sin\theta$ and therefore is null for a MOID at $\theta = k\pi$ with $k = [0, 1, \ldots]$. On the contrary the transversal component can be written as

$$y \simeq -\frac{a^{3/2}}{r}\frac{\sqrt{1-e^2}}{\sqrt{\mu}}(3\Delta t v)\,\delta v_t \tag{12}$$

For almost circular orbits the radial components is almost aligned with the normal component and the transversal almost coincide with the tangential direction. Eqs. (11) then become

$$\begin{aligned} x &\simeq \frac{2a^{3/2}}{\sqrt{\mu}}\,\delta v_t \\ y &\simeq -3\Delta t\,\delta v_t \end{aligned} \tag{13}$$

From which one can derive that for a warning time $\Delta t > (T_{\mathrm{NEO}}/3\pi)$ the secular component exceeds the geometric component.

1. REPRESENTATION ON THE IMPACT PLANE

The deflection $\delta\mathbf{r}$ does not fully describe what happens at the point of the MOID. A better understanding can be gathered from the projection of the deflection on the impact plane or the plane perpendicular to the incoming relative velocity of the small body at the Earth at the time of the expected impact (see Fig. 3).

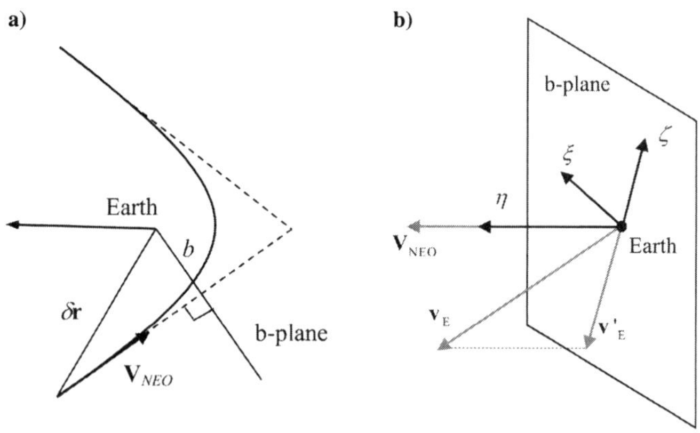

Fig. 3 The impact plane and *b* parameter.

We can define a local reference system centered in the Earth with the axis perpendicular to the b-plane and aligned along the unperturbed velocity of the asteroid relative to the Earth, the axis ξ along the direction opposite to the projection of the heliocentric velocity of the Earth $\boldsymbol{v}_{\mathrm{E}}$ onto the b-plane, and the axis η that completes the reference system (see Fig. 3a). The general transformation from the radial, transversal, out-of-plane Hill's reference frame to the b-plane reference frame is

$$\mathbf{x}_b = [\hat{\xi} \;\; \hat{\eta} \;\; \hat{\zeta}]\delta\mathbf{r} \tag{14}$$

where

$$\hat{\eta} = \frac{\mathbf{V}_{\mathrm{NEO}}}{\|\mathbf{V}_{\mathrm{NEO}}\|} \qquad \hat{\xi} = \frac{\boldsymbol{v}_E \times \hat{\eta}}{\|\boldsymbol{v}_E \times \hat{\eta}\|} \qquad \hat{\zeta} = \hat{\xi} \times \hat{\eta}$$

The assumption is that the velocity along the deviated trajectory at the time of the MOID remains parallel to $\mathbf{V}_{\mathrm{NEO}}$. The proper representation would be on the instantaneous b-plane, perpendicular to the deviated relative velocity of the asteroid, however, the maximum relative error between the plane perpendicular to the nominal relative velocity and the perturbed one is around 0.01, thus in the following we are using the former one, which avoids the additional calculation of the velocity of the deflected asteroid. Moreover, for this analysis, we set to zero the distance at the MOID in order to have the Earth at the origin of the reference system on the b-plane. On the b-plane one can represent the distance b (called impact parameter) from the Earth to the intercept of the asymptote of the hyperbola of the deviated orbit of the asteroid:

$$b = \sqrt{\xi^2 + \zeta^2} \tag{15}$$

As for the maximum deflection problem the maximum impact parameter can be derived from the quadratic form

$$b^2 = \delta\boldsymbol{v}^T\mathbf{T}^T\mathbf{B}^T\mathbf{Q}^T\mathbf{Q}\mathbf{B}\mathbf{T}\delta\boldsymbol{v} \tag{16}$$

where $\mathbf{Q}$ is a simple projection matrix defined as

$$\mathbf{Q} = \begin{bmatrix} 1 & 0 & 0 \\ 0 & 0 & 0 \\ 0 & 0 & 1 \end{bmatrix} \tag{17}$$

that is used to select only the ξ and ζ components and $\mathbf{B} = [\hat{\xi} \; \hat{\eta} \; \hat{\zeta}]$.

From the impact parameter b one can compute the point of minimum distance r_p from the the center of the Earth post deflection:

$$r_p = \sqrt{\mu_E^2/V_{\mathrm{NEO}}^4 + b^2} - \mu_E/V_{\mathrm{NEO}}^2 = \left[\sqrt{A^2 + b^2} - A\right] \tag{18}$$

where here $A = \mu_E / V_{\mathrm{NEO}}^2$. The minimum distance is asymptotically equal to the b-parameter for high V_{NEO}, although for small ratios of b/A it is proportional to the square of b whereas for high values of b/A, tends to grow linearly with b. To avoid a collision the following must hold

$$b^2 > (r_E + h_{\mathrm{atm}} + A)^2 - A^2 \tag{19}$$

with h_{atm} the altitude of the atmosphere.

Given the monotonic relationship between minimum distance and impact parameter one can derive an optimal strategy using the impact parameter and then calculate the exact minimum distance using Eq. (19). Note, however, that for shallow crossing asteroids, low V_{NEO}, values of the impact parameters that are above the Earth limit do not guarantee a successful deflection. Furthermore, for very low V_{NEO}, a different approach taking into account the full n-body dynamics is required.

Now, given the transformation (16) one might wonder whether using the deflection or the b parameter, as optimality criterion, can lead to different conclusions on the evaluation of a successful deflection strategy. Figure 4 shows the impact parameter (bold lines) for a highly elliptical orbit (1979XB) together with the norm of the deviation $\|\delta\mathbf{r}\|$ (thin lines). The figure suggests that the maximization of the b-parameter and the deflection lead to the same conclusion on the optimal deflection strategy, although for a time-to-impact Δt above a specific value.

For smaller Δt the b-plane suggests a different strategy. This can be appreciated in Fig. 5 where the result for the δr-maximization strategy (thin line) is compared to the ones for the b maximization (bold line), for 2000SG344 in the

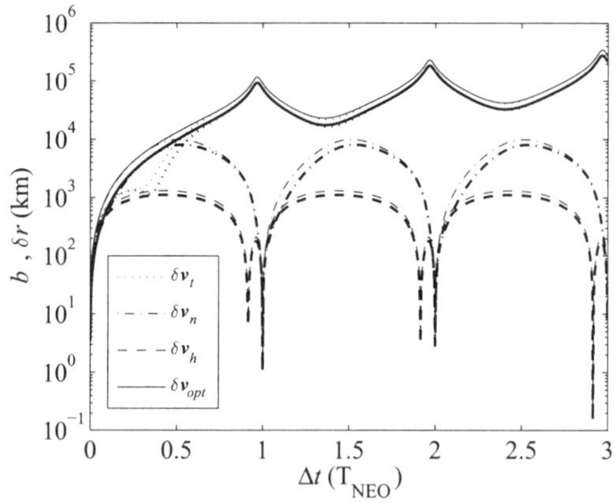

Fig. 4 Impact parameter and magnitude of the deviation for 1979XB with $\delta v = 0.07$ m/s. The bold lines show the b-parameter, the thin lines the deviation.

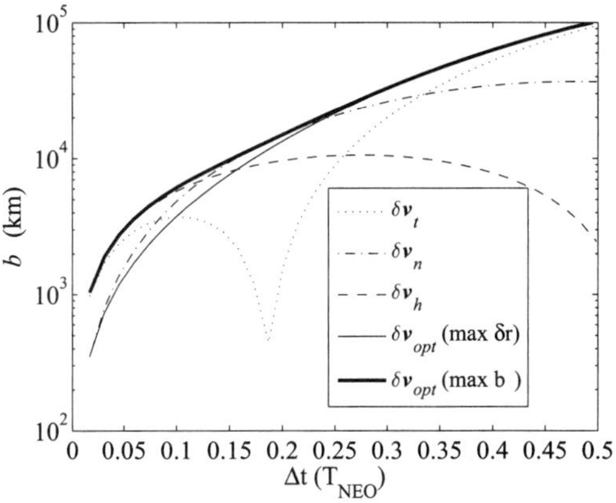

Fig. 5 Impact parameter for 2000SG344 $\delta t < 0.5T_{neo}$. Strategy of maximum deviation (solid line) and maximum b-parameter (bold line) [28].

range $\Delta t > 0.5T_{\rm NEO}$. The maximization of the impact parameter would lead to choose the h-direction strategy for very small Δt, the n-direction for a range of $0.15T_{\rm NEO} < \Delta t < 0.25T_{\rm NEO}$ and the tangential direction for higher Δt. Note that for small Δt the angle of the optimal impulse changes depending on the orbital parameters of the asteroid, but for higher Δt the optimal strategy approaches the tangential direction.

2. FULL THREE-BODY ANALYSIS

A full three-body dynamic model was used (i.e., considering the Sun and the Earth as gravitational bodies) and the trajectory was propagated, after a 2 m/s impulse, over the interval $I = [t_d\, t_{\rm MOID} + 0.1T_{\rm NEO}]$. Then, the closest point to the Earth was computed as follows:

$$\delta r_{\min 3b} = \min_{t \in I} \|\mathbf{r}_{\rm E} - \mathbf{r}_{\rm NEO}\| \tag{20}$$

Figure 6 shows the comparison of the strategy maximizing δr or b when a full three-body representation is used. The dashed line and the bold line represent, respectively, the deviation and its projection on the b-plane, calculated through the two-body problem. The continuous thin line represents the minimum deviation computed through the three-body problem. The maximization of b provides better results, closer to the full numerical simulation.

3. NON-LINEAR PROXIMITY EQUATIONS

The error in the evaluation of the deflection depends on the accuracy of the proximity motion equation. A more accurate solution can be obtained by the following

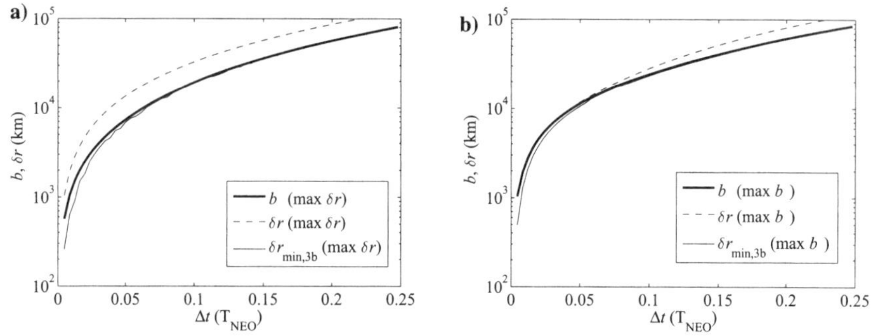

Fig. 6 a) Maximum deviation strategy for 1979XB and b) Maximum *b*-parameter strategy for 1979XB.

set of nonlinear proximity equations [29]:

$$\Delta \mathbf{r} = \mathbf{r}_{A_{\text{dev}}} \Psi - \begin{bmatrix} \mathbf{r}_{A0} \\ 0 \\ 0 \end{bmatrix} \tag{21}$$

with

$$\Psi = \begin{bmatrix} \xi \cos(\theta_{A0} + \Delta\theta) + \sin(\theta_{A0} + \Delta\theta) \left(\rho \sin\theta_{A0} - \cos(i_{A0} + \Delta i) \sin\Delta\Omega \cos\theta_{A0}\right) \\ -\xi \cos(\theta_{A0} + \Delta\theta) + \sin(\theta_{A0} + \Delta\theta) \left(\rho \sin\theta_{A0} - \cos(i_{A0} + \Delta i) \sin\Delta\Omega \cos\theta_{A0}\right) \\ \cos(\theta_{A0} + \Delta\theta) \sin\Delta\Omega \cos\theta_{A0} + \bar{\omega} \sin(\theta_{A0} + \Delta\theta) \end{bmatrix} \tag{22}$$

and the additional terms

$$\bar{\omega} = \cos i_{A0} \sin(i_{A0} + \Delta i) - \cos\Delta\Omega \cos(i_{A0} + \Delta i) \sin i_{A0} \tag{23}$$

$$\rho = \sin i_{A0} \sin(i_{A0} + \Delta i) + \cos\Delta\Omega \cos(i_{A0} + \Delta i) \cos i_{A0} \tag{24}$$

$$\xi = \cos\Delta\Omega \cos i_{A0} + \cos i_{A0} \sin\Delta\Omega \sin\theta_{A0} \tag{25}$$

The Δ parameters appear nonlinearly in these expressions, hence by combining Eqs. (1) with Eqs. (21), one obtains a nonlinear relationship between the deflection $\delta\mathbf{v}$ and $\Delta\mathbf{r}$. The nonlinear equations tends to the linear ones when the variation of the elements become small.

B. FIRST-ORDER SLOW-PUSH DEFLECTION FORMULAS

For a slow-push action the variation of the orbital elements can be obtained by integrating Gauss' planetary equations in (1) over the time span $[t_i, t_e]$ for which the deflection action is applied. If the deflection action stops at time $t_e < t_{\mathrm{MOID}}$, the total variation of the orbital elements a, e, i, ω, Ω, at the MOID, is given by the integration of Gauss' equations up to time t_e. The total variation of M at the MOID is given by

$$\delta M = \delta M_0 + (n_e - n)t_{\mathrm{MOID}} + n_i t_i - n_e t_e \tag{26}$$

where δM_0 is the variation of M over the interval $[t_i, t_e]$ and n_e is given by

$$n_e = \sqrt{\frac{\mu}{(a + \delta a)^3}} \tag{27}$$

Once the variation of the elements at the MOID is available, proximity Eqs. (3) or (21) can be used to compute the deflection at the time of the MOID and Eqs. (14) can be used to project the deflection on the impact plane and compute the b-parameter.

The numerical integration of Gauss' equations can be computationally expensive and make a parametric or Monte Carlo analysis of the deflection time consuming. Most slow-push actions, however, produce a very low level of acceleration compared to the local gravity field. For this case, efficient analytical solutions can be used instead of a full numerical integration. An example can be found in the work of Colombo and Vasile [30] where a semi-analytical formulation was developed for a tangential acceleration with a dependency on the distance from the Sun. Later on, a more general solution based on perturbation theory was derived by Zuiani et al. and applied to the deflection of asteroids [31] and the removal of space debris [32]. In the following section this analytical theory will be briefly introduced.

1. ANALYTICAL PROPAGATION OF LOW-ACCELERATION ARCS

In Ref. [33] a first-order analytical theory was proposed to derive the variation of the orbital parameters under a generic constant acceleration. In this section, the approach proposed in Refs. [32,33] and further extended in Ref. [34] is presented.

Gauss' planetary equations (1) are rewritten in terms of equinoctial non-singular elements to accurately treat also zero inclination and zero eccentricity orbits:

$$\frac{\mathrm{d}a}{\mathrm{d}t} = \frac{2}{B}\frac{a^3}{\mu}\left[(P_2 \sin L - P_1 \cos L)\, a_r + \Phi(L) a_\theta\right]$$

$$\frac{\mathrm{d}P_1}{\mathrm{d}t} = B\sqrt{\frac{a}{\mu}}\left[-a_r \cos L + \left(\frac{P_1 + \sin L}{\Phi(L)}\right) a_\theta - P_2 \frac{Q_1 \cos L - Q_2 \sin L}{\Phi(L)} a_h\right]$$

$$\frac{\mathrm{d}P_2}{\mathrm{d}t} = B\sqrt{\frac{a}{\mu}}\left[-a_r \sin L + \left(\frac{P_1 + \cos L}{\Phi(L)}\right) a_\theta - P_1 \frac{Q_1 \cos L - Q_2 \sin L}{\Phi(L)} a_h\right]$$

$$\frac{dQ_1}{dt} = \frac{B}{2}\sqrt{\frac{a}{\mu}}(1 + Q_1 + Q_2^2)\frac{\sin L}{\Phi(L)}a_h$$
$$\frac{dQ_2}{dt} = \frac{B}{2}\sqrt{\frac{a}{\mu}}(1 + Q_1 + Q_2^2)\frac{\cos L}{\Phi(L)}a_h \tag{28}$$

where the non-singular elements are defined as

$$\begin{aligned}
P_1 &= e\sin(\Omega + \omega) \\
P_2 &= e\cos(\Omega + \omega) \\
Q_1 &= \tan\left(\frac{i}{2}\right)\sin\Omega \\
Q_2 &= \tan\left(\frac{i}{2}\right)\cos\Omega \\
L &= (\Omega + \omega) + \theta \\
B &= \sqrt{1 - P_1^2 - P_2^2} = \sqrt{1 - e^2} \\
\Phi(L) &= 1 + P_1\sin(L) + P_2\cos(L)
\end{aligned} \tag{29}$$

and the control acceleration is expressed as

$$\begin{aligned}
a_r &= \epsilon\cos\alpha\cos\beta \\
a_t &= \epsilon\sin\alpha\cos\beta \\
a_h &= \epsilon\sin\beta
\end{aligned} \tag{30}$$

In this expression, ϵ is assumed to be small compared to the local gravity field, typically 10^{-12}–10^{-8} m/s^2 for asteroid deflection, and the two angles α and β are expressed with respect to a local r-t-h reference frame. Further it is assumed that along an arc, both the magnitude and the angles are constant or vary with a known law of the true longitude L. Note, however, that not all possible functions of L can lead to a solution in the form presented here next.

A solution will be found by expressing the non-singular elements as power series expansions of the small parameter ϵ:

$$\begin{aligned}
a &= a_0 + \epsilon a_1 + \mathcal{O}(\epsilon^2) \\
P_1 &= P_{10} + \epsilon P_{11} + \mathcal{O}(\epsilon^2) \\
P_2 &= P_{20} + \epsilon P_{21} + \mathcal{O}(\epsilon^2) \\
Q_1 &= Q_{10} + \epsilon Q_{11} + \mathcal{O}(\epsilon^2) \\
Q_2 &= Q_{20} + \epsilon Q_{21} + \mathcal{O}(\epsilon^2)
\end{aligned} \tag{31}$$

For a first-order approximation, the series can be truncated to retain only first-order terms in ϵ, and the elements (31) can be inserted in Eqs. (28). After expanding the right-hand side of Eqs. (28) in Taylor series and retaining again only the first-order terms in ϵ one can integrate the equations to obtain the following analytical expressions:

$$
\begin{aligned}
a &= a_0 + \epsilon \frac{2B_0^2 a_0^3}{\mu} \cos\beta \Big((P_{20} I_{s2} - P_{10} I_{c2}) \cos\alpha + I_{11} \sin\alpha \Big) \\
P_1 &= P_{10} + \epsilon \frac{B_0^4 a_0^2}{\mu} \Big(\cos\beta \,(-I_{c2} \cos\alpha + (I_{13} P_{10} + I_{s3} + I_{s2}) \sin\alpha) \\
&\quad - P_{20}\,(Q_{10} I_{c3} - Q_{20} I_{s3}) \sin\beta \Big) \\
P_2 &= P_{20} + \epsilon \frac{B_0^4 a_0^2}{\mu} \Big(\cos\beta \,(I_{s2} \cos\alpha + (I_{13} P_{20} + I_{c3} + I_{c2}) \sin\alpha) \\
&\quad + P_{10}\,(Q_{10} I_{c3} - Q_{20} I_{s3}) \sin\beta \Big) \\
Q_1 &= Q_{10} + \epsilon \frac{B_0^4 a_0^2}{\mu} (1 + Q_1^2 + Q_2^2) I_{s3} \sin\beta \\
Q_2 &= Q_{20} + \epsilon \frac{B_0^4 a_0^2}{\mu} (1 + Q_1^2 + Q_2^2) I_{c3} \sin\beta
\end{aligned}
\tag{32}
$$

with integrals

$$
I_{cn} = \int_{L_0}^{L} \frac{\cos L}{\Phi_0^n(L)} \mathrm{d}L \quad I_{sn} = \int_{L_0}^{L} \frac{\sin L}{\Phi_0^n(L)} \mathrm{d}L \quad I_{1n} = \int_{L_0}^{L} \frac{1}{\Phi_0^n(L)} \mathrm{d}L
$$

The integration was over the true longitude L, therefore, the time needs to be recovered by expanding the time equation

$$
\frac{\mathrm{d}t}{\mathrm{d}L} = \sqrt{\frac{a^3}{\mu} \frac{B^3}{\Phi^2(L)}} = H(a, P_1, P_2, L) \tag{33}
$$

into Taylor series to give

$$
\frac{\mathrm{d}t_0}{\mathrm{d}L} + \epsilon \frac{\mathrm{d}t_1}{\mathrm{d}L} = H(a_0, P_{10}, P_{20}, L_0) + \epsilon \left(\frac{\mathrm{d}H}{\mathrm{d}a} a_1 + \frac{\mathrm{d}H}{\mathrm{d}P_1} P_{11} + \frac{\mathrm{d}H}{\mathrm{d}P_2} P_{21} \right) \tag{34}
$$

Note that in Eq. (33) the contribution of the control acceleration was neglected. After integration, the time equations become

$$t_0 = t_{00} + \int_{L_0}^{L} \sqrt{\frac{a_0^3}{\mu} \frac{B_0^3}{\Phi_0^2(L)}} \mathrm{d}L \tag{35}$$

$$t_1 = 3\sqrt{\frac{a_0^7}{\mu^3}} B_0^5 \cos\beta(\cos\alpha I_{t1} + \sin\alpha I_{t2}) \tag{36}$$

with integrals

$$I_{t1} = \int_{L_0}^{L} \frac{1}{\Phi_0^2(L)} \left(\frac{1}{\Phi_0(L)} - \frac{1}{\Phi_0(L_0)} \right) \mathrm{d}L \qquad I_{t2} = \int_{L_0}^{L} \frac{I_{11}(L)}{\Phi_0^2(L)} \mathrm{d}L \tag{37}$$

The solutions mentioned previous were derived assuming an acceleration that was constant in the *r-t-h* reference frame. Although these formulas are simple and all the integrals find an analytical expression, they are not very convenient if the acceleration is in the tangential direction and the orbit is eccentric. However, equivalent formulas can be derived also for an inertially fixed and tangential acceleration (see [34]). Here we present only the case of a tangential acceleration because this is proven to be close to optimal if the deflection starts more than one asteroid period before the expected impact.

The accuracy of the method is proportional to the magnitude of the acceleration ϵ and to the magnitude of the local gravity field. Figure 7 shows the error in the calculation of time for a typical range of control accelerations and distances from the Sun. Although the error is very small it accumulates over several revolutions requiring an error management strategy. One method to control the propagation of the error, called *rectification*, is to update the initial conditions every n orbits by simply taking the value for $E(L)$ computed at the nth orbit as the new reference condition E_0 for the following n orbits (see [34]). In the case of a slow-push deflection the number of revolutions around the Sun is normally between 10 and 20 (see Sec. III), therefore, it is possible to get sufficiently good solutions even without rectification.

III. DEFLECTION TECHNOLOGIES

All the different deflection strategies that exist in the literature can be classified in two main categories depending on whether the deflection action is modelled with an impulsive change of velocity or with an extended thrust arc:

1. Impulsive methods (e.g., kinetic impactor and nuclear blast)
2. Slow push deflection methods (e.g., gravity tug, solar collector, low thrust tug, etc.)

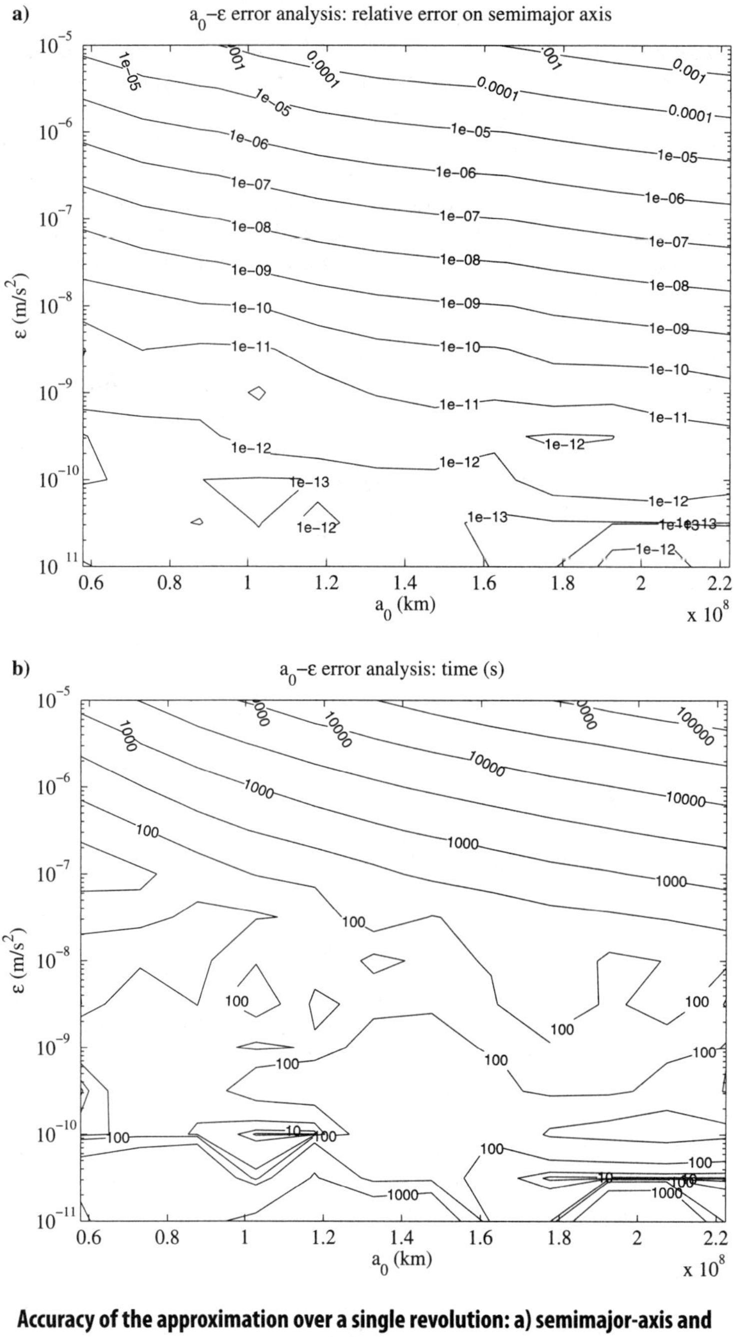

Fig. 7 Accuracy of the approximation over a single revolution: a) semimajor-axis and b) time of flight.

This classification is not representative of the physical interaction between the deflecting spacecraft and the deflected object. As it will be explained in the remainder of this section, the nuclear blast should be better grouped together with solar and laser ablation as the thrust comes from the ablation of the asteroid material whereas the kinetic impactor could be grouped with photon or ion beaming as the thrust comes from the inelastic impact of high velocity objects with the surface of the asteroid. The distinction between impulsive and slow push is more of an operational nature and helps in the definition of a particular mission and in the analysis of the outcome of the deflection. For this reason we follow this classification approach and we leave the other classification method to a future publication.

In the previous section, we have seen how a general deflection action perturbs the trajectory of an asteroid. This section is dedicated to how this action can be physically achieved with the different technologies. The intention is not to be exhaustive but to discuss some representative methods and their technological implications. The strategies reviewed here are the kinetic impactor, the nuclear interceptor, the gravity tug, the laser ablation, the solar collector, and the light tug. Although the first two fall under the the impulsive category, all the others will require a longer action time and can thus be characterized as slow-push methods.

A. KINETIC IMPACTOR

Perhaps one of the simplest way to transfer energy between a spacecraft and a target asteroid is to use the kinetic energy contained in the relative motion between the two bodies. Owing to this apparent simplicity, the kinetic impactor is, at the time this chapter is written, the most mature method for asteroid deflection and the only one that, to date, has been partially validated [35]. The impulsive change of velocity of the asteroid $\delta\boldsymbol{v}$ is commonly modelled with the conservation of the linear momentum after an inelastic collision:

$$\delta\boldsymbol{v} = \beta \frac{m_d}{m_{\mathrm{A}} + m_d} \Delta\boldsymbol{v}_{s/c} \tag{38}$$

where β represents a momentum enhancement factor that takes into account the fraction of mass ejected during the impact and $\Delta\boldsymbol{v}_{s/c}$ is the velocity of the spacecraft relative to the asteroid, m_d is the mass of the spacecraft and m_{A} the one of the asteroid. Despite the apparent simplicity of this equation, getting an accurate value for the momentum enhancement factor is critical in order to assess the efficiency of the kinetic impactor method and defining an accurate value for this parameter is still a matter of active research. High fidelity hydrodynamics codes are used to assess the importance of the contribution from the ejecta as a function of the asteroid composition and impact angle [36].

Because $\delta\boldsymbol{v}_{\mathrm{A}}$ in Eq. (38) depends on the relative velocity between impactor and asteroid, the effectiveness of the kinetic impactor depends on the trajectory of the impactor and thus on the launch date and on the transfer strategy. One can show this dependency by finding an estimation of the energy required to achieve a given $\Delta v_{s/c}$. Let us restrict to the case in which the orbits of the Earth, asteroid and impactor are all in the same plane and the asteroid is impacting the Earth at an apsidal point. In this case the impact between the spacecraft and the asteroid should preferably occur at the perihelion as demonstrated in Sec. III.A. Assuming an impact at the perihelion r_p of the orbit of an NEO with semimajor axis a_{NEO}, we have

$$\begin{aligned} \Delta v_{s/c} &= \sqrt{\frac{2\mu}{r_p} - \frac{\mu}{a_{\text{NEO}}}} - \sqrt{\frac{2\mu}{r_p} - \frac{\mu}{a_{\text{NEO}} + \delta a}} \quad \text{for } \delta a > 0 \\ \Delta v_{s/c} &= \sqrt{\frac{2\mu}{r_p} - \frac{\mu}{a_{\text{NEO}} + \delta a}} - \sqrt{\frac{2\mu}{r_p} - \frac{\mu}{a_{\text{NEO}}}} \quad \text{for } \delta a < 0 \end{aligned} \tag{39}$$

where the semimajor axis of the orbit of the impactor is $a_{\text{NEO}} + \delta a$. The Δv required to transfer from the Earth to an orbit with semimajor axis $a_{\text{NEO}} + \delta a$ is

$$\Delta v_T = \sqrt{\frac{2\mu}{r_{\text{E}}} - \frac{\mu}{r_{\text{E}} + \delta a + \delta a_{\text{N}}}} - \sqrt{\frac{\mu}{r_{\text{E}}}} \tag{40}$$

where $a_{\text{NEO}} = r_{\text{E}} + \delta a_{\text{N}}$. By combining Eqs. (39) and (40) one can get the dependency of the departure Δv as a function of the desired impact Δv:

$$\Delta v_{s/c} = \Delta v_{\text{T}} - \sqrt{\frac{2\mu}{r_{\text{E}}} - \frac{\mu}{r_{\text{E}} + \delta a_{\text{N}}}} + \sqrt{\frac{\mu}{r_{\text{E}}}} \tag{41}$$

From Eq. (41) one can see that for high δa_{N} the kinetic impactor requires no Δv_{T} to achieve good $\Delta v_{s/c}$ but as δa_{N} tends to 0 the variation of velocity required to achieve the desired deflection increases significantly. This very simple calculation leads to two key considerations

1. Although the kinetic impactor seems to be the strategy of choice for deep crossers (NEOs with high semimajor axis and high eccentricity), it might not be ideal for shallow crossers (NEOs with semimajor axis close to 1 AU and low eccentricity). As can be seen in Table 2, despite representing a lower percentage of the NEA population, shallow crossers are almost as likely to impact the Earth as deep crossers due to their more frequent encounters.
2. The energy required to increase the $\Delta v_{s/c}$ needs to be provided by a propulsion system. A high I_{sp} propulsion system becomes necessary in the case of shallow crossers. The Δv that needs to be delivered by the propulsion system is proportional to the desired impact Δv.

TABLE 2 DIFFERENT TYPES OF ASTEROIDS

	NEAs, %	PHAs, %	Impactors, %
$Q < 1.05$ AU	1	1	11
$Q > 0.95$ AU	8	22	38
Deep crossers	61	77	53
Low Inc ($i < 5$)	6	25	38

These considerations do not take into account the phasing between the asteroid and the Earth. If the phasing was included, then the launch date would play an important role as the achievable $\Delta v_{s/c}$ and the energy required for the transfer would depend on the launch date as well [28]. Once the launch date and transfer energy are defined, the primary challenge in the practical implementation of this deflection strategy is certainly the targeting of the asteroid at high velocity. Equation (38) would suggest that the challenge is simply not to miss the target. This alone would make hitting the asteroid with solar or electric sails more difficult than with conventional propulsion methods due to the limited control authority and long reaction time. Increasing $\Delta v_{s/c}$ further exacerbates the problem. However, the enhancement factor does depend on the relative motion between impactor and surface, therefore, the angle of impact as well as the rotation motion of the asteroid play an additional role. Therefore, the challenge becomes to hit the target in the right spot with the right angle, avoiding shallow elevation angles especially in the case the surface material is brittle or easily removed. The hypothesis of inelastic impact in this case would need to be reassessed.

Another concern with the kinetic impactor is the level of energy involved during the collision. As explained by Sanchez et al. [37], in some cases, the minimum level of energy required to deviate an asteroid by a distance that ensures a successful deflection may rise above the theoretical catastrophic fragmentation limit. The catastrophic fragmentation limit is defined as the specific energy required to completely fracture an asteroid (see Fig. 8).

Although this minimum level of collisional energy is in general strongly dependent on the lead time, or time prior to the impact with the Earth at which the deflection maneuver is applied, the predicted specific energy required to completely fracture an asteroid can be inferred from experimental work in impact fragmentation [41] and numerical modeling [38]. Figure 8 shows the critical specific energy Q^* for asteroids ranging from 40 m to 1 km diameter, computed by using the scaling laws provided by the aforementioned authors. The upper fragmentation limit at 1000 J/kg is above any of the specific energies Q^* expected from the scaling laws in Fig. 8, including those of basalt strength from Ryan and Melosh [38] and nonporous rocky bodies from Housen and Holsapple [39]. Even in some cases, this upper limit is more than one order of magnitude above the

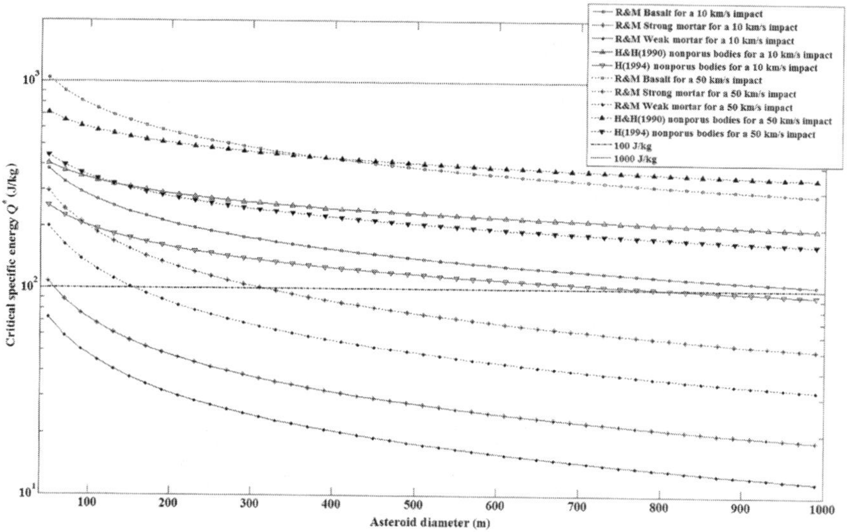

Fig. 8 Critical specific energy Q^* for barely catastrophically disrupting asteroids with a diameter ranging from 40 m to 1 km, calculated using the work of Ryan and Melosh (R&M) [38], Housen and Holsapple (H&H) [39], and Holsapple (H) [40].

predicted Q^*. Hence, for the studied range of diameters, the limit at 1000 J/kg is here considered as a fated catastrophic fragmentation. On the other hand, the lower fragmentation limit at 100 J/kg is an approximate mean energy level among all the critical energies Q^* predicted by Fig. 8, and more importantly, the 100 J/kg limit is, in general, above the four predicted Q^* using the mortar strengths from Ryan and Melosh [38]. Hence, the 100 J/kg limit may be considered as a reasonable fragmentation limit. As a consequence, an asteroid that underwent an impulsive deflection attempt may fragment in a number of pieces having different mass and velocity. If f_r is the fragmentation ratio, defined as

$$f_r = \frac{m_{\max}}{M_a} \tag{42}$$

where $m_{\max}$ is the mass of the largest fragment and M_a the initial mass of the asteroid, then a catastrophic fragmentation is defined as a fragmentation where $f_r < 0.5$. The number of fragments and their size distribution can be described with fairly good approximation by a power law [41], whereas the distribution of velocity of the fragments produced by the catastrophic break-up is described by a Gaussian function with a standard velocity deviation varying with fragment size. Figure 9 illustrates the damage ratio computed statistically between different fragmentation scenarios and the unperturbed case. The figures says that in the case of

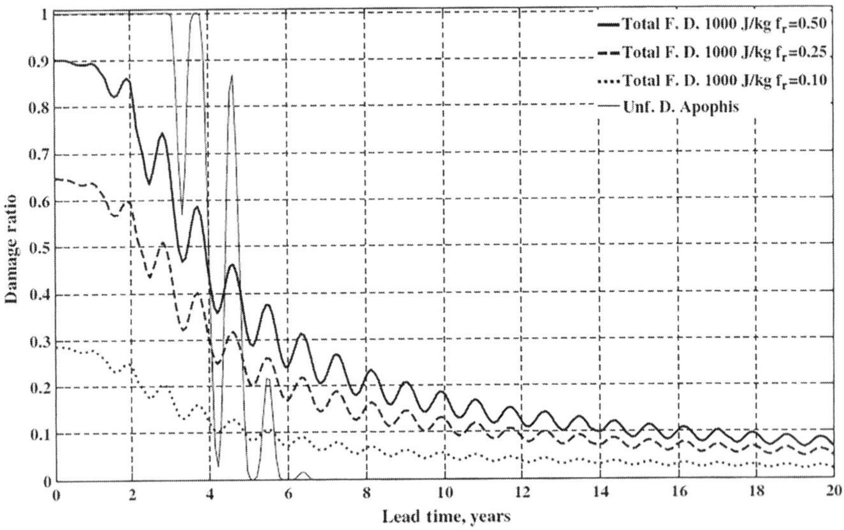

Fig. 9 Total fragmentation damage ratio of Apophis: fragmented case $f_r = 0.50$ (black solid line), $f_r = 0.25$ (black dashed line), $f_r = 0.10$ (black dotted line) and unshattered case (thin black line).

a fragmentation the probability that a large fragment (large enough to cause damage) impacts the Earth never goes to zero even if the deflection is applied far in advance of the predicted impact (for more details the interested reader can refer to Ref. [37]).

1. SMART CLOUD

A way around both the fragmentation and targeting issues would be to distribute the collisional energy by impacting the asteroid with a cloud of particle having a total mass equivalent to a single spacecraft. Technically, this could be done by releasing a swarm of small particle or even smaller spacecraft before the interception [42]. Conceptually, this approach offers the following advantages:

1. Unlike the kinetic impactor, the cloud would provide a distributed impact, more like a drag, hence eliminating the risk of fragmentation
2. The relative velocity between the cloud and the asteroid can be very high, up to 50 km/s for highly eccentric asteroids.
3. Unlike the kinetic impactor there is no need to pin-point the asteroid as an extended area is covered by the cloud.

TABLE 3 ENERGY FRACTION F_I OVER ALL THE PRODUCTS OF A NUCLEAR EXPLOSION

Source	1–x-ray	2–Neutrons	3–Gamma rays	4–Debris	5–Others
Fission	0.7	0.01	0.02	0.2	0.07
Fusion	0.55	0.2	0.01	0.2	0.04

B. NUCLEAR INTERCEPTOR

Nuclear devices are able to carry the highest energy density among all of the deviation methods currently available, which make them a very attractive option among the possible mitigation strategies.

The energy released during the explosion is carried by the debris of the exploded spacecraft and by the radiation. Table 3 shows the fraction f_i (with $i \in \{1, 2, 3, 4, 5\}$) of energy associated to each of the products of the explosion for the case of a fusion and fission devices [43, 44]. The energy delivered during the explosion, Y_0, is computed from the yield-to-mass ratio and is conservatively assumed to have a value $YTW = 0.75$ ktons/kg for fusion devices and $YTW = 0.075$ ktons/kg for fission devices[†]:

$$Y_0 = YTWm_{wh} \tag{43}$$

where m_{wh} is the mass of the bomb. In this paper, no buried or surface detonation are considered due to the added difficulty of landing and digging on an asteroid and only stand-off explosions are modelled.

With reference to Fig. 10, the explosion is assumed to happen at a distance H from the surface of the asteroid, therefore, only a portion m_{debris} of the total mass

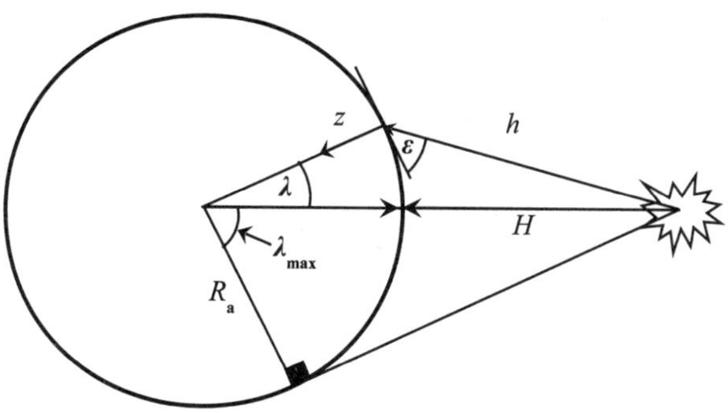

Fig. 10 Standoff configuration for the nuclear interceptor method.

[†]From data available online at http://nuclearweaponarchive.org/ (retrieved 9th Feb. 2016).

of debris m_d will hit the surface:

$$m_{\text{debris}} = S m_d \tag{44}$$

If one assumes that the exploding device sees a spherical cap with radius R_{A}, then the fraction S can be expressed as

$$S = \frac{1}{2} - \frac{\sqrt{H}}{2} \frac{\sqrt{H + 2R_{\text{A}}}}{R_{\text{A}} + H} \tag{45}$$

The ejection velocity of the debris v_{debris} is then computed from the fraction $f_4 = 0.2$ (see Table 3) of the total energy Y_0 released during the blast:

$$v_{\text{debris}} = \sqrt{\frac{2 f_4 Y_0}{m_d}} \tag{46}$$

The variation of velocity δv_{debris} due to the debris cloud only is then given by

$$\delta v_{\text{debris}} = \beta S_{sc} \frac{m_{\text{debris}} v_{\text{debris}}}{m_{\text{A}}} \tag{47}$$

where S_{sc} is a scattering factor and β the momentum enhancement factor [45], which is conservatively assumed to be equal to two.

The contribution from the radiations is derived from the Beer–Lambert law of absorption. Given a radiation with frequency ν and knowing the incident radiation energy per unit area $I_0^{\nu}(\lambda)$ and the depth z, the energy per unit area $I^{\nu}(\lambda, z)$ transmitted beyond a given depth is computed as follows:

$$I^{\nu}(\lambda, z) = \sin \epsilon(\lambda) I_0^{\nu}(\lambda) \exp\left(-\rho_{\text{A}} \kappa_{\nu} \frac{z}{\sin \epsilon(\lambda)}\right) \tag{48}$$

The incident radiation density $I_0^{\nu}(\lambda)$ is given by

$$I_0^{\nu}(\lambda) = \frac{f_i}{4\pi h^2(\lambda)} Y_0 \tag{49}$$

where the h distance is computed as

$$h = \sqrt{(H + (1 - \cos \lambda) R_{\text{A}})^2 + R_{\text{A}}^2 \sin^2 \lambda} \tag{50}$$

and ϵ is given by

$$\sin \epsilon = \frac{(R_{\text{A}} + H) \cos \lambda - R_{\text{A}}}{h} \tag{51}$$

TABLE 4 OPACITY κ_ν, OR LINEAR MASS-ABSORPTION COEFFICIENT, FOR AN ASTEROID MADE OF FORSTERITE

Radiation type	X-ray	Neutron	Gamma ray
Value	$1.5\ m^2/kg$	$0.0044\ m^2/kg$	$0.005\ m^2/kg$

The linear mass-absorption coefficient κ_ν for each type of radiation is given in Table 4 [44]. Note that quantities in Table 4 are to be considered as mean values over the range of frequencies of x-rays and Gamma-rays.

The amount of energy absorbed per unit mass at a given depth is then obtained by considering the cumulative absorption of each radiation type:

$$E(\lambda, z) = -\sum_\nu \frac{dI^\nu}{dz} = \sum_\nu \kappa_\nu I_0^\nu \exp\left(-\rho_A \kappa_\nu \frac{z}{\sin \epsilon(\lambda)}\right) \tag{52}$$

Part of this energy goes into the vaporization process of the asteroid, whereas the excess energy is converted into thermal excitation. The local average velocity of the gas molecules can then be estimated by writing a simple energy balance with E_v the total vaporization enthalpy per unit mass:

$$\bar{v}(\lambda, z) = \sqrt{2(E(\lambda, z) - E_v)} \tag{53}$$

where the assumption is that the vaporization is so fast that losses by conduction and radiation can be neglected. In Sec. C a more complete energy balance model will be presented for the case of laser and solar ablation methods. Eqation (52) allows one to define a limit depth z_{MAX} below which the vaporization process cannot continue as the energy absorbed is lower than the vaporization enthalpy. Given a certain distance H and yield Y_0, the value of z_{max} is numerically computed by finding the value of z that satisfies the relationship $E(\lambda, z) = E_v$ for each λ considered. The change in linear momentum generated by the expelled material is then expressed, for an infinitesimal volume, as

$$dP = \frac{\cos \lambda}{2} \rho_A \bar{v}(\lambda, z)\, dV \tag{54}$$

where the cosine function comes from the fact that we only retain the tangential component and the $1/2$ factor is coming from the assumption of an equiprobable scattering of the gas molecules from the ablated surface over a hemisphere. The area of a spherical cap is given by

$$S = 2\pi R_A^2(1 - \cos\lambda) \tag{55}$$

The infinitesimal volume dV is thus given by

$$dV = 2\pi R_A^2 \sin\lambda \, dz \, d\lambda \tag{56}$$

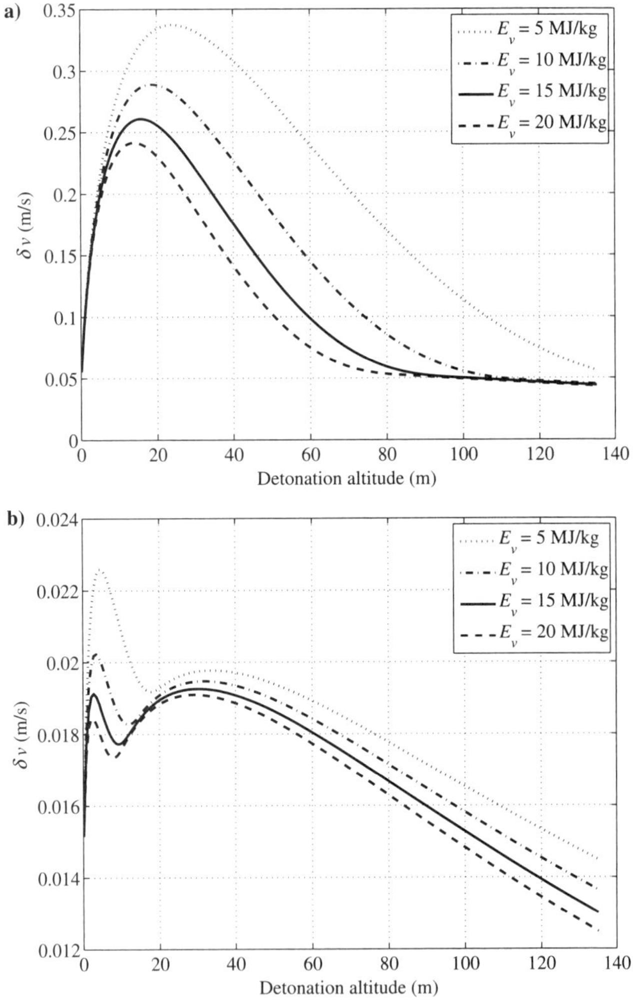

Fig. 11 Impulsive change of velocity as a function of the detonation altitude for different values of the enthalpy of vaporization: a) 600 kg fusion device and b) 600 kg fission device.

TABLE 5 ORBITAL AND PHYSICAL PROPERTIES OF TEST ASTEROID

Element		Measured Value
Semimajor axis	a_0	0.9224 AU
Eccentricity	e_0	0.1912
Period	T_0	323.5969 days
Mean motion	n_0	1.2876×10^{-5} deg/s
Mass	m_A	2.7×10^{10} kg
Gravitational constant	μ_a	1.801599×10^{-9} km^3/s^2
Physical dimensions	a_l, b_l, c_l	196 m, 112 m, 112 m
Rotational velocity	w_a	3.3×10^{-3} deg/s
Total vaporisation enthalpy	E_v	15 MJ/kg
Density	ρ_a	2650 kg/m^3

Integrating relation (54) and dividing by the mass of the asteroid eventually allows one to express the change of velocity $\delta v_{\text{radiation}}$ due to the radiations:

$$\delta v_{\text{radiation}} = \frac{\pi R_A^2}{M_A} \int_0^{\lambda_{\max}} \int_0^{z_{\max}(\lambda)} \rho_A \bar{v}(\lambda, z)\, dz \sin\lambda \cos\lambda\, d\lambda \tag{57}$$

Figure 11a shows the total $\delta v = \delta v_{\text{radiation}} + \delta v_{\text{debris}}$ imparted to an asteroid, with mass and density reported in Table 5, assuming a fusion device of 600 kg at different altitudes of detonation and for different values of the enthalpy of vaporization, whereas Fig. 11b shows the δv imparted to the same asteroid by a fission device of equal mass at different altitudes of detonation and for different values of the enthalpy of vaporization.

1. ELONGATED ASTEROID

In this section, we consider the case of an elongated asteroid with an elongation factor e_l, that is, an ellipsoidal shape with semimajor axis $a_I = e_l^{2/3} R_A$ and semiminor axes $b_I = c_I = (R_A/e^{1/3})$. The mean radius is still identical to the one used in the spherical case previously derived, so that the elongated and the spherical asteroids considered have an identical volume. Considering as a worst case scenario the configuration where the bomb is detonated along the longer side, the distance $h(\lambda)$ is now given as

$$h = \sqrt{(H + (1 - \cos\lambda) e_l^{2/3} R_A)^2 + \left(\frac{R_A}{e_l^{1/3}}\right)^2 \sin^2\lambda} \tag{58}$$

We need now to distinguish between λ, the angle in elliptical coordinates corresponding to the concentric circle or radius a_I and $\tilde{\lambda}$, the angle between the normal to the ellipsoidal surface and the horizontal direction. They can be related through the following formula:

$$\cos\tilde{\lambda} = \frac{\cos\lambda}{\sqrt{1 + (e_l^2 - 1)\sin^2\lambda}} \tag{59}$$

The value of $\sin\epsilon$ is obtained by computing the scalar product between the direction normal to the ellipsoidal surface $\mathbf{n}$ and the direction $-\mathbf{h}$, which gives

$$\sin\epsilon = \frac{(e_l^{2/3}R_\mathrm{A} + H/e_l^{2/3}R_\mathrm{A})\cos\lambda - 1}{(e_l^{1/3}h/R_\mathrm{A})\sqrt{(\cos^2\lambda/e_l^2) + \sin^2\lambda}} \tag{60}$$

Last but not the least, the infinitesimal volume is now expressed as

$$\mathrm{d}V = 2\pi e_l^{1/3}R_\mathrm{A}^2\frac{\sin^2\lambda}{\sin\tilde{\lambda}}\mathrm{d}z\,\mathrm{d}\lambda \tag{61}$$

Keeping a constant detonation altitude of 17 m, Fig. 12 shows how the δv produced compares to the spherical case, considering again a 600 kg fusion device.

2. NUCLEAR CYCLER MISSION CONCEPT AND DESIGN

The key idea underneath the nuclear cycler concept is to incrementally change the velocity of the asteroid by releasing and detonating a series of relatively

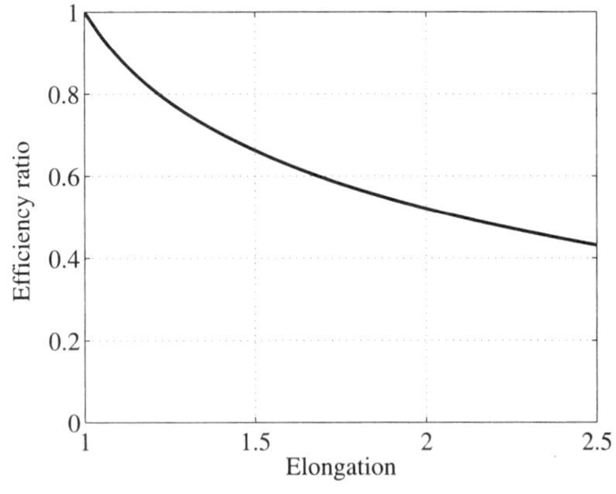

Fig. 12 Performance comparison between the spherical case and a cigar-shaped asteroid as a function of the elongation.

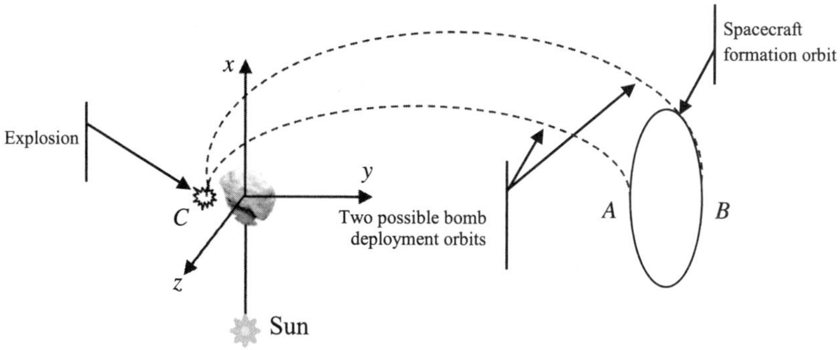

Fig. 13 Sketch (not to scale) of the nuclear cycler concept.

small nuclear bombs from a vantage point at a safety distance from the asteroid. Figure 13 shows a possible configuration with a carrier-spacecraft flying in formation with the asteroid on a periodic orbit at a distance from the asteroid and releasing two bombs at two different times. The detonation occurs on the far side of the asteroid, with respect to the spacecraft so that the asteroid is shielding the spacecraft from radiations and debris. In this particular configuration, the orbit of the carrier and the one of the bomb are timed in such a way that by the time the bomb goes from point A to point C, the carrier has moved from point A to point B and by the time the bomb goes from B to C, the carrier has moved from B back to A, closing the cycle. A new cycle can now begin. In addition, the data from the previous explosions can be collected and analyzed to control the altitude and timing of the subsequent explosions or to control the direction of the resulting $\delta \mathbf{v}$. In the remainder of this paper we will analyze only the special configuration in which point A corresponds to the perihelion of the orbit of the asteroid and point B the aphelion. In this case two bombs are released every revolution of the asteroid around the Sun.

The trajectory of the bomb, including the effect of the gravity of the asteroid, can be calculated with a shooting method by back-propagating the trajectory of the bomb from the desired detonation point to the carrier's orbit. The motion of the bombs and of the carrier spacecraft are governed by the following system of proximal motion dynamic equations:

$$
\begin{aligned}
\ddot{x} &= -\ddot{r} + 2\dot{\theta}\dot{y} + \dot{\theta}^2 + \ddot{\theta}y - \frac{\mu_{\text{sun}}(r+x)}{r_{sc}^3} - \frac{\mu_{\text{A}}}{\delta r^3}x + \frac{\partial U_{20+22}}{\partial x} \\
\ddot{y} &= -2\dot{\theta}(\dot{r}+\dot{x}) - \ddot{\theta}(r+x) + \dot{\theta}^2 y - \frac{\mu_{\text{sun}}}{r_{sc}^3}y - \frac{\mu_{\text{A}}}{\delta r^3}y + \frac{\partial U_{20+22}}{\partial y} \\
\ddot{z} &= -\frac{\mu_{\text{sun}}}{r_{sc}^3}z - \frac{\mu_{\text{A}}}{\delta r^3}z + \frac{\partial U_{20+22}}{\partial z}
\end{aligned}
\tag{62}
$$

with

$$
\begin{aligned}
\ddot{\theta} &= -\frac{2\dot{r}r\dot{\theta}}{r^2} \\
\ddot{r} &= \dot{\theta}^2 r - \frac{\mu_{\text{sun}}}{r^2}
\end{aligned}
\tag{63}
$$

where μ_{sun} and μ_{A} are the gravity constants of the Sun and the asteroid, respectively. As a first approximation, the asteroid shape can be assumed to be an ellipsoid with $a_{\text{I}} \geq b_{\text{I}} = c_{\text{I}}$ the three semi-axes along the three body-fixed orthogonal axes. The value of the semi-axes is given in Table 5. Note that the total expelled mass is assumed to be negligible compared to the mass of the asteroid, thus we used the conservative assumption that the mass of the asteroid remains unchanged. Furthermore, in the reduced model presented in this chapter, the spinning rate and axis are assumed to remain constant.

The axis c_{I} is assumed to be aligned with the vector of angular momentum, which corresponds to the z-axis of the asteroid Hill frame $\mathcal{A}$ (see Fig. 1b). The gravity field of the asteroid is expressed as the sum of a spherical field $(\mu_{\text{A}}/\delta r^2)$ plus a second-degree and second-order field [46, 47]:

$$
U_{20+22} = \frac{\mu_{\text{A}}}{\delta r^3}\left[C_{20}\left(1 - \frac{3}{2}\cos^2\varphi\right) + 3C_{22}\cos^2\varphi\cos 2\alpha\right]
\tag{64}
$$

where φ is the elevation over the x–y plane and the harmonic coefficients C_{20} and C_{22} are a function of the semi-axes,

$$
\begin{aligned}
C_{20} &= -\frac{1}{10}(2c_{\text{I}}^2 - a_{\text{I}}^2 - b_{\text{I}}^2) \\
C_{22} &= \frac{1}{20}(a_{\text{I}}^2 - b_{\text{I}}^2)
\end{aligned}
\tag{65}
$$

and α is defined as

$$
\alpha = \arctan\left(\frac{y}{x}\right) + w_{\text{A}} t
$$

The initial conditions for the backward integration, in the case the bomb is released from point B, are

$$
\begin{aligned}
x(\theta = 0) &= 0 \\
y(\theta = 0) &= y_C \\
z(\theta = 0) &= 0 \\
\dot{x}(\theta = 0) &= \delta v_C \cos\gamma \\
\dot{y}(\theta = 0) &= \delta v_C \sin\gamma \\
\dot{z}(\theta = 0) &= 0
\end{aligned}
\tag{66}
$$

The modulus of the velocity δv_C was varied from $v_{ex} + \epsilon_{ex1}$ to $v_{ex} + \epsilon_{ex2}$, with $v_{ex} = \sqrt{(2\mu_A/y_c)}$ at distance y_C from the center of the asteroid, whereas the γ angle was constrained to be in the interval $[0.6\,\pi/2]$. The trajectory was then propagated backward for a true anomaly $\Delta\theta = \pi$ using an adaptive Runge–Kutta, Dormand–Prince scheme implemented in the MATLAB© function *ode*45, with absolute and relative tolerance equal to 1*e*-6 and 1*e*-7, respectively. If the bomb is released from point B, the arrival conditions need to satisfy the constraints:

$$x(\theta = \pi)^2 + (y(\theta = \pi) - y_B)^2 + z(\theta = \pi)^2 = 0 \tag{67}$$

This analysis, however, assumes that for each value of δv_C a particular spacecraft formation orbit is targeted, rather than targeting always the same orbit. Therefore, the MATLAB function *fmincon* was used to find the optimal value of γ that satisfies the relaxed soft constraint $x(\theta = \pi)^2 + z(\theta = \pi)^2 = 0$ for different values of δv_C. As it is shown in Figs. 14 this yields a continuous and compact set of trajectories. The minimum distance from the asteroid, however, is limited by the need to have the bomb completing the transfer in half a revolution of the asteroid around the Sun. For low values of δv_C the gravity of the asteroid accelerates the turning of the velocity of the bomb leading to a crossing of the y-axes in less than half a period. The value of δv_C in this analysis, therefore, was set to quickly clear the Hill's sphere of influence of the asteroid. Figures 14 show a set of optimized trajectories superimposed to the theoretical ones obtained using the linear proximity equations. Figure 14a shows the case in which $y_C = -0.1287$ km, $\epsilon_{ex1} = -1.44917e - 5$ km/s and $\epsilon_{ex2} = -1.42417e - 5$ km/s whereas Fig. 14b shows the case in which $y_C = -0.210$ km, $\epsilon_{ex1} = 8.1178e - 6$ km/s, and $\epsilon_{ex2} = 8.3678e - 6$ km/s. These two cases correspond to an optimal explosion along the semiminor and semimajor axes of the asteroid, respectively. Note that, according to Fig. 11, a variation of 81.3 m in the detonation altitude significantly decreases the effectiveness of the explosion. On the other hand, it is possible to find trajectories from point A or point B to any point C in the interval $[-0.1287, -0.210]$ hence it is possible to target the optimal detonation distance for each explosion. Figure 15 shows a close up of Fig. 14a around point C. One can see that the bomb would approach the asteroid almost head on, although δv_C is so small that was neglected in the calculation of the deflection δv. In this analysis the assumption is that the bomb has a guidance, navigation, and control system that can compensate for any additional effect, such as light pressure.

The propellent cost to deploy each bomb can be evaluated by computing the velocity difference at points A and B of the trajectory of the bomb and the trajectory of the spacecraft. The magnitude of the deployment Δv maneuver, at the beginning of the deflection cycle, is reported in Fig. 16 for point A and B and for the two detonation altitudes. This translates into a mass fraction of maximum 5.546e-5 per bomb assuming a cold gas engine with an $I_{sp} = 68s$. Finally, the formation orbit of the carrier designed with the linear proximity equations (3) were reoptimized to keep into account the full dynamics. A small $\Delta \mathbf{v}$ correction

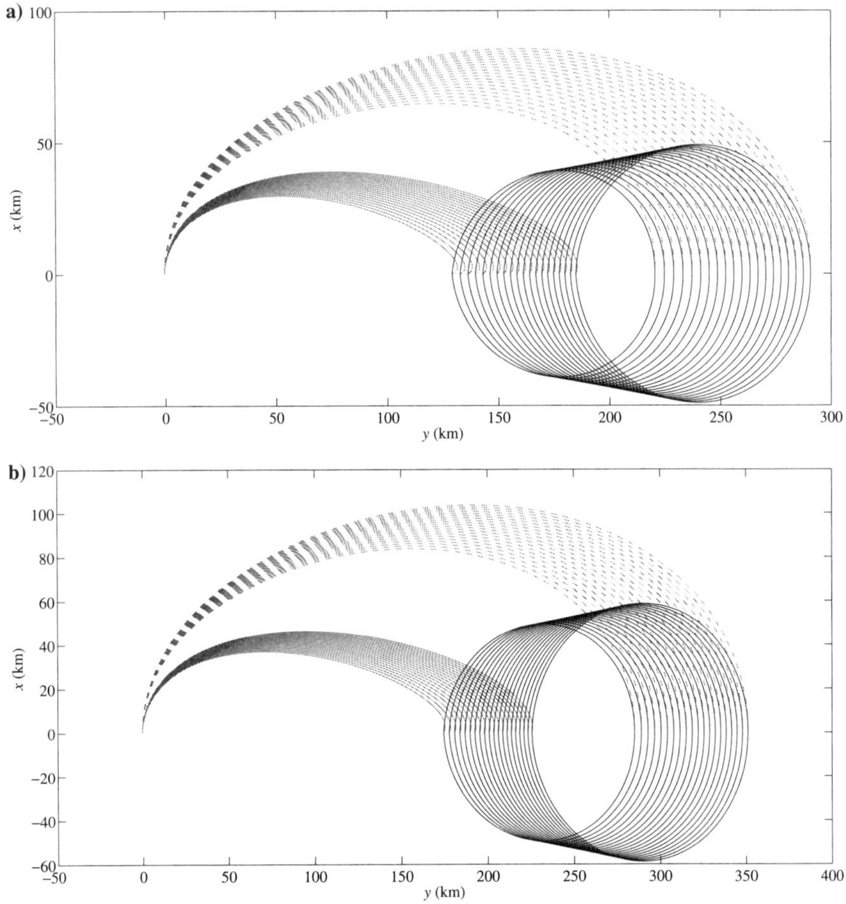

Fig. 14 Families of formation and deployment trajectories for spacecraft and bombs: a) detonation point at 210 m and b) detonation point at 128 m.

maneuver was inserted at A and B to match the A-to-B trajectory with the B-to-A trajectory such that the A and B point remain the initial conditions of the bomb deployment trajectories and the periodicity of the formation orbit is maintained. The optimized correction maneuvers, however, are of negligible size, even lower than the deployment maneuvers, and are not reported here.

At every explosion the velocity of the asteroid is modified and so are its orbital elements. As a consequence, the carrier needs to maneuver to maintain its relative orbital motion with respect to the asteroid. For the strategy presented in this paper, there is no out-of-plane component of the deflection, therefore, the spacecraft needs only an in-plane $\Delta \boldsymbol{v}_c$ correction. The component of this correction

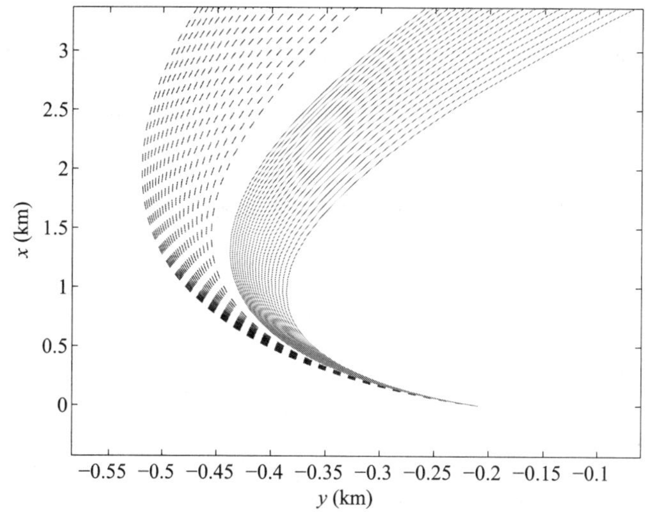

Fig. 15 Close-up of the families of deployment trajectories for a detonation point at 210 m.

along the tangential direction has to compensate for the variation of the semimajor axis of the asteroid and, therefore, is of the same magnitude of δv_t in Eq. (1). Given that the explosion occurs at the apsidal points then $\delta v_t = \delta v$. Therefore, it is here assumed that the spacecraft compensates only for a variation in a and e with a single maneuver equal to δv at each explosion. This does not restore completely the exact initial proximal motion but guarantees anyway to maintain the

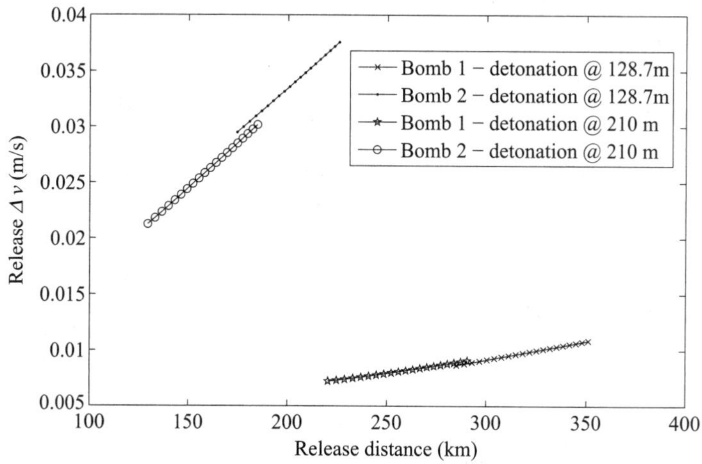

Fig. 16 Total Δv at the point of release of the bomb from the spacecraft.

formation with the asteroid. Furthermore, the result in Figs. 14 demonstrates that one can generate a continuous set of formation orbits, at different distances from the asteroid, reaching the detonation point with different values of δv_C, hence at every explosion a new bomb deployment trajectory can be defined provided that the spacecraft maintains formation.

3. COMPARISON BETWEEN A SINGLE DETONATOR AND THE NUCLEAR CYCLER

The nuclear cycler method has been applied to the case of an Apophis-like asteroid considering two different warning times of 6 years and 3 years, respectively. The warning time is here defined as the time from the first explosion to the expected impact of the undeviated asteroid with the Earth. Relevant properties of this asteroid can be found in Table 5. The initial inclination, right ascension, argument of the pericenter and mean anomaly were set so that the asteroid impacts the Earth on 13140 MJD.

An interesting first result is obtained by computing the total δv produced by either a single or a fractionated detonation for the same total mass of the bombs.

The results of our model, in Fig. 17, indicate that a fractionated explosion may be better than a single explosion for the same total mass. The explanation of this result is in the dependency of the δv on the view angle λ in Eq. (57) and the penetration depth z_{max}. Figure 18 shows the optimal detonation altitude as a function of the spacecraft mass for a single interceptor. One can see that the

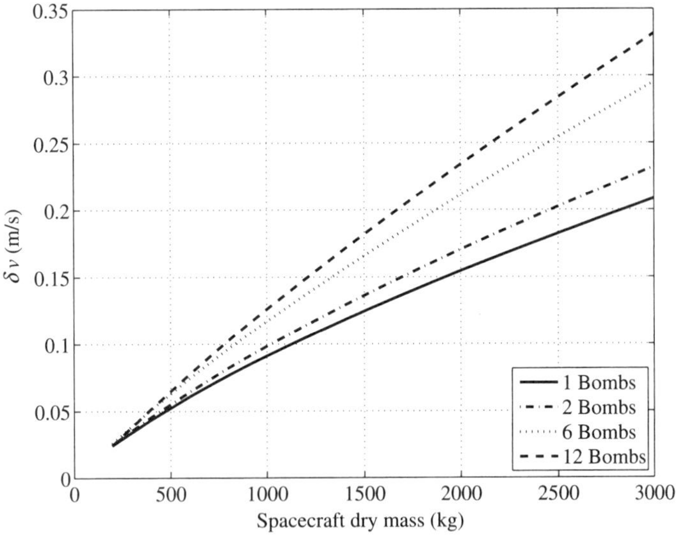

Fig. 17 Total δv as a function of the dry mass of the spacecraft for different numbers of explosions.

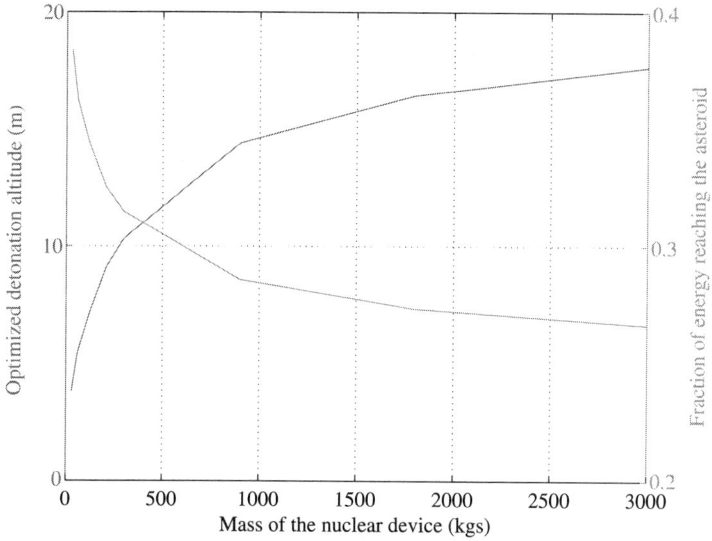

Fig. 18 Optimal altitude of detonation and fraction of the total energy reaching the asteroid for different sizes of the nuclear device.

optimal altitude is indeed lower for a smaller bomb, therefore, the fraction of the total released energy actually reaching the asteroid is bigger in this case.

However, once the deflection parameter b is used as performance indicator, one can see, in Figs. 19, that the single interceptor method outperforms the cycler one, thanks to the fact that the whole velocity variation is delivered at the very beginning of the first cycle and thus its effect propagates for a longer period. The comparison is done by considering identical dry masses of the spacecraft with the cumulative mass of the nuclear bombs representing 30% of the total dry mass of the spacecraft in both cases. For the single interceptor method, the whole mass of the spacecraft contributes to the ejecta, whereas only the mass of the bombs contributes to ejecta for the cycler method. Last but not the least, in both cases, the detonation occurs at the optimal altitude. As one would expect, Figs. 19 show that a longer warning time is beneficial in term of deflection distance. Also, a warning time of only 3 years constrains the maximum number of explosion to 6 for the nuclear cycler method if explosions occur only at the apsidal points. Another interesting result is obtained by normalizing the value of the b parameter obtained for the case of a fractionated detonation with the result of the single interceptor method. The results in Figs. 20 indicate that the mass efficiency (the ratio between b parameter and mass of the spacecraft) can be as low as 40% for small spacecraft and reduces to 75% for larger spacecraft. Note the interesting overtaking between the curves for 2 and 6 explosions for the 6 years lead time case when a mass larger than 1 ton is considered.

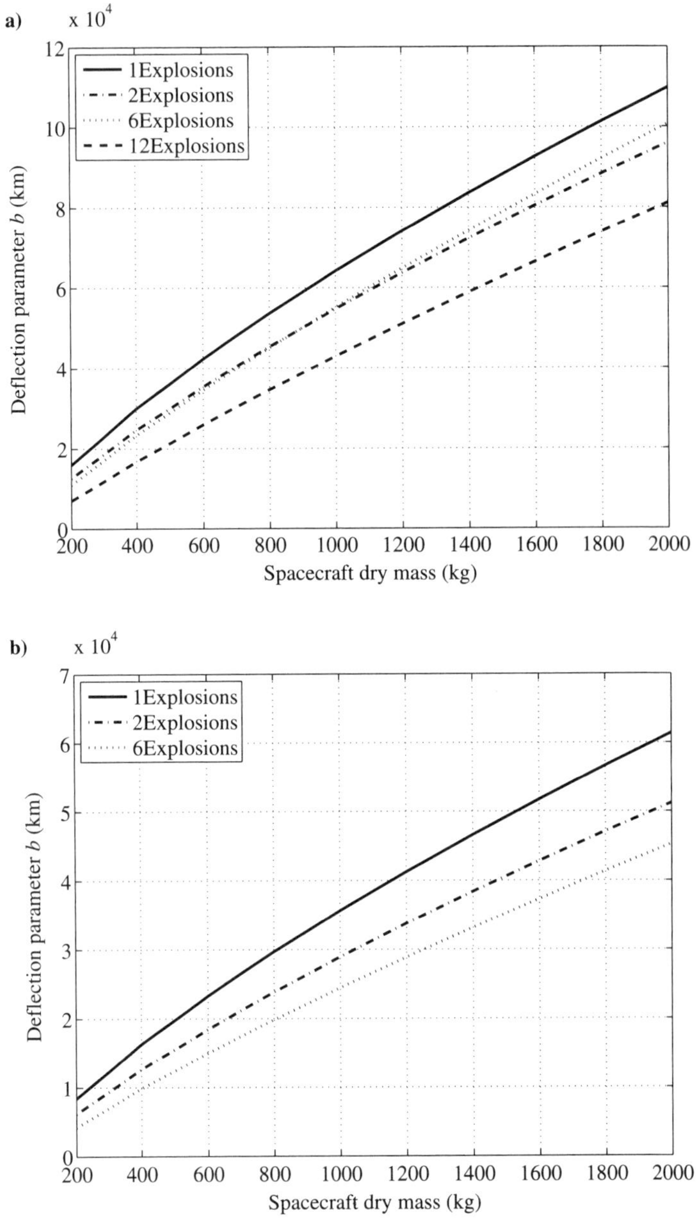

Fig. 19 Impact parameter for a varying number of explosions: a) 6 years' lead time and b) 3 years' lead time.

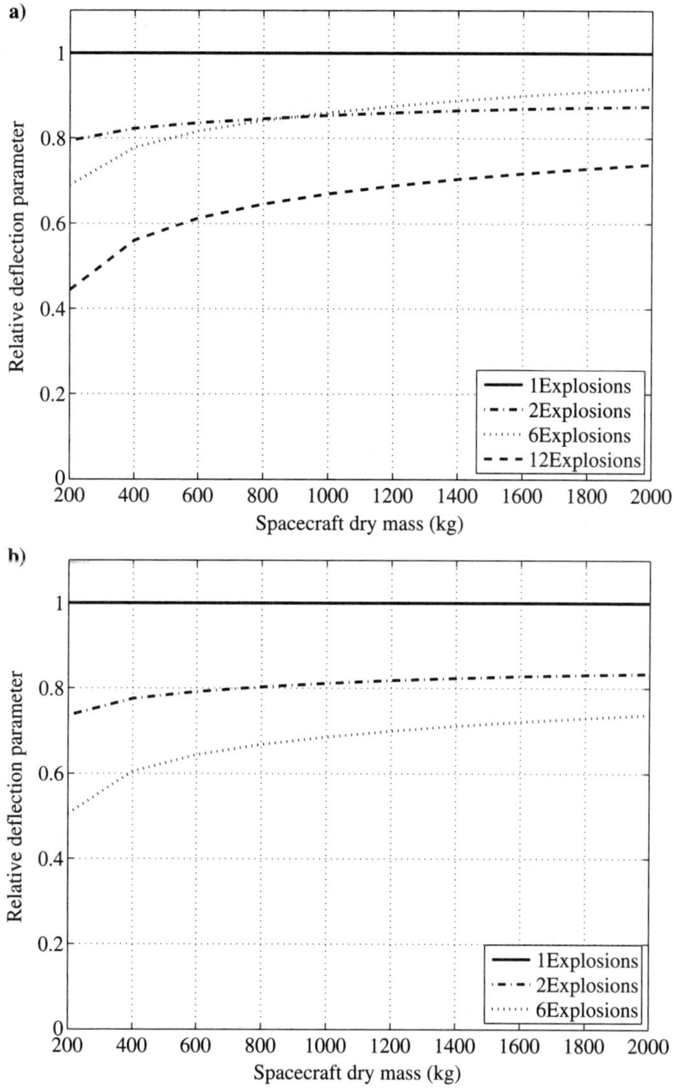

Fig. 20 Efficiency of the nuclear cycler method compared to the single interceptor method: a) 6 years' lead time and b) 3 years' lead time.

4. GRAVITY TUG

The idea is to use the mutual attraction between the spacecraft and the asteroid to progressively change the velocity of the asteroid. In the more traditional configuration, the spacecraft is placed at a distance d from the asteroid and two thruster mounted in a slanted configuration, to avoid thrust impingement, would balance the gravity attraction. By doing so the net result is a constant acceleration on a compound spacecraft-asteroid, see Fig. 21. For the position of the spacecraft to be fixed, the gravity force must equate the net thrust F_{hover}:

$$F_{\text{hover}} = 2T_S \cos\left(\arcsin\left(\frac{R_a}{d}\right) + \phi\right) \tag{68}$$

$$F_g = \frac{GM_a m_S(t)}{d^2} \tag{69}$$

$$F_{\text{hover}} = F_g \tag{70}$$

where T_S is the thrust of a single engine in the slanted configuration, R_a is the radius of the asteroid, M_a its mass, G the gravity constant and $m_S(t)$ is the mass of the spacecraft at time t. The tugging acceleration is simply

$$a_{\text{gtug}}(t) = \frac{Gm_S(t)}{d^2} \tag{71}$$

If the engines are assumed to be always on and the initial mass of the spacecraft is m_i, the mass of the spacecraft at time t can be expressed as

$$m_S(t) = m_i \exp\left(-\frac{GM_a(t - t_0)}{d^2 \cos\left(\arcsin\left(R_a/d\right) + \phi\right) I_{sp} g_0}\right) \tag{72}$$

where I_{sp} is the specific impulse of the engine and g_0 the gravity acceleration on the surface of the Earth and t_0 is the beginning of the deflection action. Figure 22 shows the comparison between the slanted gravity tug method and a low-thrust strategy overtime.

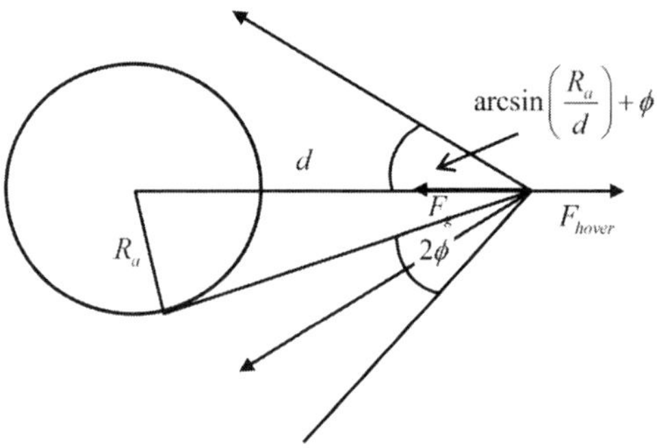

Fig. 21 Detailed sketch of the gravity tug approach with slanted engines.

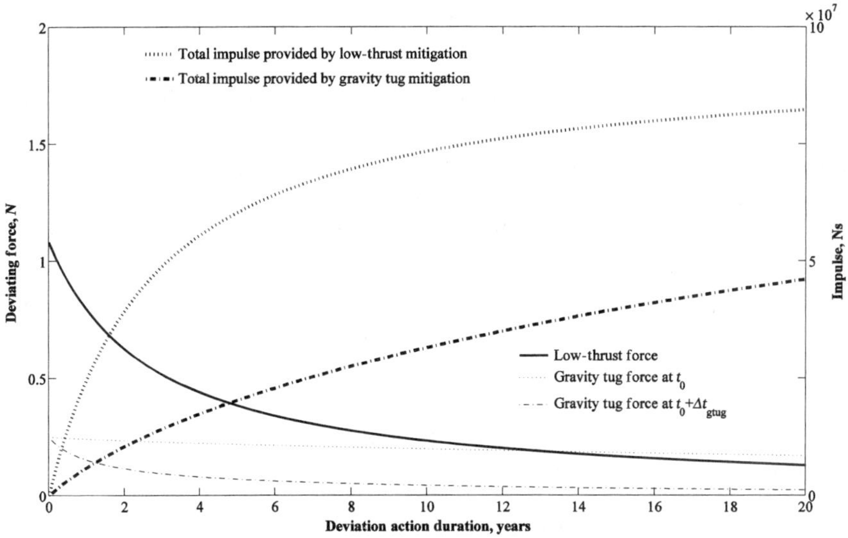

Fig. 22 Comparison of the low-thrust and gravity tug options.

5. GRAVITY TUG ON HALO ORBITS

In Refs. [25, 48, 49] a variant of the original gravity tractor concept was proposed to remove the need for a slanted configuration and to use the thrust more efficiently. The idea is to place the tractor on a displaced halo orbit that is artificially maintained by a constant thrust. If the plume of gas of the engine generating the thrust is not impinging the asteroid, the net result is traction on the asteroid in the direction of the thrust vector. Compared to the slanted configuration, the halo configuration requires only one engine but the thrust delivered by the engine has to be lower or the distance from the asteroid has to be shorter. In fact the traction for the halo configuration is given by

$$a_{\text{gtug}}^{H} = \frac{Gm_H(t)\cos\psi}{d^2} \tag{73}$$

where ψ is the angle between the thrust direction and the asteroid-spacecraft position vector. Therefore, the mass of the tractor would need to be increased by a factor $1/\cos\psi$ to maintain the same thrust level or the deflection time should be extended to get the same miss distance with a reduced deflection action. An extended thrust time implies, of course, more propellant. To get an estimation of the mass of propellant required for the halo configuration in comparison to the slanted configuration, one can consider the simple case in which the distance of the slanted configuration from the center of the asteroid is kept constant. In this case, the mass of the spacecraft is given by Eq. (72) and the tugging force is

$$F_g = \frac{GM_a m_i e^{-(GM_s t)/(I_{sp} g_0 d^2 \cos\alpha)}}{d^2} = 2T_S \cos\alpha = T_H \tag{74}$$

where $\alpha = \arcsin\left(R_a/d\right) + \phi$ and the further assumption is that the thrust of the halo configuration follows the same time law of the tugging force given by the slated configuration. As a result the variation of mass of the halo configuration is

$$\dot{m}_H = -I_{sp} g_0 \frac{GM_a m_i e^{-(cGM_a t)/(d^2 \cos\alpha)}}{d^2} = -\frac{cGM_a m_i}{d^2} e^{-At} \tag{75}$$

with $c = I_{sp} g_0$ and

$$A = \frac{cGM_a}{d^2 \cos\alpha} \tag{76}$$

The mass of the halo spacecraft as a function of time results to be

$$m_H = \frac{I_{sp} g_0 GM_a m_i}{d^2} \frac{e^{-At}}{A} = m_i \cos\alpha e^{-At} \tag{77}$$

If the initial mass of the halo and slanted configurations are the same, then:

$$m_H = m_S \cos\alpha \tag{78}$$

Given that both configurations generate the same tugging force, the total time to achieve a given deflection is the same and, therefore, the mass of the halo configuration is simply $\cos\alpha$ lower than the mass of the slanted configuration.

C. SLOW-PUSH ABLATION METHODS

Ablation-based methods make use of the asteroid material to generate the deflection action. The asteroid itself, therefore, is the source of the propellent required to produce the required thrust. Although the nuclear interceptor is based on the ablation of the surface of the asteroid, in this section we examine other two techniques that use high intensity lasers or focused sunlight to ablate the asteroid. The resulting thrust is generally low and these methods fall under the slow-push category.

1. THRUST MODEL

The mass flow rate per unit area $\dot{\mu}$ generated during ablation can be derived from a simple energy balance between the input energy and the different energy losses during the ablation process:

$$\left(E_v + \frac{\bar{v}^2}{2} + C_p(T_s - T_0) + C_v(T_s - T_0)\right)\dot{\mu} = P_i - q_{\text{rad}} - q_{\text{cond}} \tag{79}$$

where E_v is the total enthalpy of melting and vaporization, T_s the temperature at which the ablation process occurs, T_0 the initial temperature, C_p and C_v the specific heats at constant pressure and constant volume, respectively. The term $\bar{v}^2/2$ accounts for the energy that goes into increasing the kinetic energy of the gas produced during the vaporization process. Note that the terms on the left of Eq. (79) represent a conservative approximation. In fact, part of the energy would go into heating up the solid phase to the melting temperature T_m and part into heating up the molten phase to the vaporization temperature T_s. Instead, the increment from T_0 to T_s is used to account for possible unmodeled components. The radiation loss q_{rad} per unit area is computed from the black-body radiation law:

$$q_{\mathrm{rad}} = \sigma_{sb}\epsilon T_s^4 \tag{80}$$

with σ_{sb} the Stefan–Boltzmann constant and ϵ the emissivity of the surface. The conduction loss can be computed with the 1-D model:

$$\chi_{\mathrm{A}}\frac{\partial^2 T}{\partial x^2} = \frac{\partial T}{\partial t} \tag{81}$$

where χ_{A} is the thermal diffusivity of the material given by

$$\chi_{\mathrm{A}} = \frac{k_{\mathrm{A}}}{\rho_{\mathrm{A}} c_{\mathrm{A}}}$$

The boundary conditions are

$$-k_{\mathrm{A}}\frac{\partial T}{\partial x} + \epsilon\sigma_{sb}T^4 = P_{\mathrm{in}} \tag{82a}$$

$$T(0,x) = T_0;\quad T(t,L) = T_0 \tag{82b}$$

The initial surface temperature T_0 is 278 K, which is assumed here to be constant from the surface down to the core of the asteroid. Note that Eq. (82) does not include any vaporization effect that would slow down the heating process. If one assumes that steady-state conditions are not reached during ablation, that is, the speed of ablation is slower that the thermal diffusivity, then an approximated analytical solution for the conduction loss can be derived in the form:

$$q_{\mathrm{cond}} = (T_s - T_0)\sqrt{\frac{C_v\rho_a k_a}{\pi t}} \tag{83}$$

where the thermal conductivity is given by

$$k_a = k_{a0}\left(\frac{298}{T_s}\right)^{0.5} \tag{84}$$

The ablation temperature is also related to the local pressure through the Clausius–Clapeyron formula

$$\ln \frac{p_s}{p_{\text{ref}}} = \frac{E_v}{R}\left(\frac{1}{T_{\text{ref}}} - \frac{1}{T_s}\right) \tag{85}$$

and to the mass flow through the Hertz–Knudsen equation

$$\dot{\mu} = (1-\kappa)p_s\left(\frac{1}{2\pi R_s T_s}\right)^{1/2} \tag{86}$$

with κ representing the fraction of molecules that recondense, which increases with the local pressure. In practice, the change in thrust due to the recondensation is limited. Over a wide range of κ value the maximum variation in thrust is only 4% [50]. The temperature T_s, pressure p_s, and mass flow simultaneously satisfy Eqs. (79), (85), and (86). R_s is the specific gas constant. R_s can be expressed as a function of the molecular mass M_a and the Universal gas constant, $R = 8.3144\,\text{J/K/Mol}$ where $R_S = R/M_a$. Once the mass flow rate per unit area $\dot{\mu}$ is available, the mass flow rate is given by the area integral

$$\frac{\mathrm{d}m_{\text{exp}}}{\mathrm{d}t} = 2v_{\text{rot}}\int_{y_0}^{y_{\max}}\int_{t_{\text{in}}}^{t_{\text{out}}}\frac{1}{H}\left(P_{\text{in}} - q_{\text{rad}} - q_{\text{cond}}\right)\,\mathrm{d}t\,\mathrm{d}y \tag{87}$$

where $[t_{\text{in}}, t_{\text{out}}]$ is the time for which a point under the spot light is illuminated, $[y_0, y_{\max}]$ are the vertical limits of the illuminated surface area (i.e., orthogonal to the direction of rotation of the surface of the asteroid), $H = \left(E_v + (\bar{v}^2/2) + C_p(T_s - T_0) + C_v(T_s - T_0)\right)$ is an augmented enthalpy of ablation, v_{rot} is the linear velocity of a point as it travels under the illuminated spot area.

The thrust delivered by the ablation process can be estimated as the product of the mass flow rate and the ejection speed of the gas and debris v_{ex}:

$$F = \lambda_s \dot{m} v_{ex} \tag{88}$$

with

$$v_{ex} = \sqrt{\frac{8k_b T_s}{\pi M_a}} \tag{89}$$

k_b the Boltzman constant, M_a the molar mass of the vapor and λ_s a scattering factor that accounts for the fact that the vapor does not expand in a single direction. The scattering factor is here computed as the integral of the trigonometric part of Eq. (91). Previous work has shown that the sublimation temperature of a range of Mg-Fe and Si-Fe oxides can vary from 3175–3800 K [51]. A lower sublimation temperature may also be caused by the transparency of pure minerals [52]. Olivine is a terrestrial rock and can, therefore, be classified as an inhomogeneous mixture of Fe_2Si0_4, $CaAl_2Si0_4$, $CaMg_2Si_20_8$ and $CaFeSi_20_8$, $KaAlSi_30_8$ and $NaAlSi_30_8$ [51, 53]. These additional compounds represent an impurity within the olivine sample and critically will have a different sublimation temperature. This variation will affect the mass flow rate of the ablated ejecta and cause an increase in the local vapor pressure. It will vary on a case-by-case basis and is element and molecule dependent. It is, therefore, assumed that the reference temperature at 1 atm is $T_{ref} = 3800\,K$. The experimental results showed that for an olivine sample, the ablated material will dissociate into diatomic oxides. This has a prevalence of SiO and its molar mass is considered to be $M_a = 0.06\,kg/mol$. Determined through the scanning electron microscope, the experimental results also showed the incoherent sublimation of the target material. This occurs when the Fe/Mg ratio of the pre-ablated and deposited ejecta is different [52, 54, 55]. This variation is caused by the chemical breakdown—evaporation, condensation, recombination, and decomposition—of the target material [56]. It is achieved by the elemental and isotopic fragmentation of the magnesium and silicon elements and the release of secondary chemical species. This includes, but is not limited to, the ablation of MgO, FeO, SiO, SiO_2, and O_2. Furthermore, as the surface material is ablated away, new underlying material is brought forward. This resulted in the surface enrichment of carbon, sodium, aluminum, chlorine, potassium, and calcium. Around the ablation rim, a relatively large layer of semi-material is created. This is caused by the thermal propagation of the laser beam and results in the recrystallization of part of the original source material. This has the potential to extend the ablation area. See Table 6 for a summary of the physical constants used in this chapter.

TABLE 6 MEASURED AND ASSUMED PARAMETERS OF THE OLIVINE ASTEROID MATERIAL

Parameter	Value
Bulk density of the Olivine sample ρ_a	3500 kg/m^3
Complete sublimation enthalpy of Olivine E_v	14.5×10^6 J/kg
Black body emissivity ϵ	0.97
Temperature of the target before sublimation T_0	273 K
Solid heat capacity C_v	1361 J/kgK
Thermal conductivity k_{A0}	4.51 W/mK
Gas heat capacity C_p	1350 J/kgK
Molar mass M_a	0.06 kg/mol

2. AVAILABLE POWER AND INCIDENT POWER

We restrict our discussion here to the case of solar-pumped lasers and solar concentrators. Both methods collect the sunlight with a concentrating mirror but the laser option needs to convert the light source in electrical energy through solar cells and the power system on-board the spacecraft, the solar concentrators directly reflect the sunlight onto the asteroid. In the laser system, the energy loss due to the conversion process is counterbalanced by the higher focusing capabilities and the consequent higher power density at the ablation point. For both methods, the input power per unit area can be expressed as

$$P_{\text{in}} = \tau_c \sigma_{\text{sys}} C_r \alpha_{\text{A}} S_0 \left(\frac{r_{\text{AU}}}{r_{\text{A}}} \right)^2 \tag{90}$$

where α_{A} is the absorptivity of the surface material, τ_c is a degradation factor due to contamination, $S_0 = 1367\ \text{W/m}^2$ is the solar flux at 1 AU, r_{A} the Sun-asteroid distance, σ_{sys} is the system efficiency (which depends on the method used), and C_r is the concentration ratio (the ratio between the power density from the Sun on the mirror surface, and that of the spot area on the asteroid).

3. CONTAMINATION MODEL

For solar pumped lasers and solar concentrators, the contamination of the mirror surfaces due to the debris plume can be modeled based on the work by Kahle et al. [57] The study is based on a number of initial assumptions regarding the expansion of the plume and sublimation process. The first assumption holds that the sublimation process is comparable to the generation of tails in comets. The asteroid is assumed to contain a reservoir of material underneath the surface, with the gas expanding both outward as expected, and inward through a throat into vacuum within the asteroid itself. This assumption holds true, for example,for a loose rubble-pile asteroid model. The second assumption is that the plume expansion is similar to the expansion of gas of a rocket engine outside the nozzle. The density of the expelled gas ρ_{exp} is computed analytically,

$$\rho_{\text{exp}}(r_{s/sc}, \varphi) = j_c \frac{\dot{m}_{\text{exp}}}{\bar{v} A_{\text{spot}}} \left(\frac{d_{\text{spot}}}{2\delta r + d_{\text{spot}}} \right)^2 (\cos\Theta)^{2/(k-1)} \tag{91}$$

where $r_{s/sc}$ is the distance from the spot on the surface of the asteroid to the spacecraft, and $\Theta = \pi\varphi/2\varphi_{\max}$ where φ is the angle between the spot-spacecraft vector and the y axis of the Hill reference frame. The jet constant j_c was set to 0.345, the maximum expansion angle $\varphi_{\max} = 130.45$ deg, and adiabatic index $k = 1.4$, based on the values for diatomic particles [58].

The position vector $\mathbf{r}_{s/sc}$ from the spot to the spacecraft is defined as

$$\mathbf{r}_{s/sc} = \begin{bmatrix} x + r_{\text{ell}} \sin w_{\text{A}} t \cos(w_{\text{A}} t + \theta_{v_{\text{A}}}) - r_{\text{ell}} \cos w_{\text{A}} t \sin(w_{\text{A}} t + \theta_{v_{\text{A}}}) \\ y - r_{\text{ell}} \cos w_{\text{A}} t \cos(w_{\text{A}} t + \theta_{v_{\text{A}}}) - r_{\text{ell}} \sin w_{\text{A}} t \sin(w_{\text{A}} t + \theta_{v_{\text{A}}}) \\ z \end{bmatrix} \tag{92}$$

where the radius of the ellipse is given by

$$r_{\text{ell}} = \frac{a_{\text{I}} b_{\text{I}}}{\sqrt{\left(b_{\text{I}} \cos(w_{\text{A}} t + \theta_{v_{\text{A}}})\right)^2 + \left(a_{\text{I}} \sin(w_{\text{A}} t + \theta_{v_{\text{A}}})\right)^2}} \tag{93}$$

and, with reference to Fig. 1b, the position of the spacecraft with respect to the center of the asteroid is $\delta\mathbf{r} = [x, y, z]^T$. We assume here that the asteroid is spinning around the z axis with a rotational velocity w_{A}. The direction of the velocity of the asteroid in the heliocentric reference frame projected onto the Hill reference frame $\mathcal{A}$ is $\theta_{v_{\text{A}}}$. In other words, in order to have a deflection thrust aligned with the velocity of the asteroid, the spot is assumed to be at an elevation angle over the y axis equal to $\theta_{v_{\text{A}}}$.

The third assumption made is that all the particles impacting the surface of the mirror condense and stick to the mirror. The exhaust velocity is constant, therefore, the thrust depends only on the mass flow. A higher thrust results in a higher mass flow and thus in a faster contamination. Following the approach used to compute the contamination of surfaces due to out-gassing, a view factor ψ_{vf} was added equal to the angle between the normal to the mirror and the incident flow of gas. The resulting variation of the thickness of the material condensing on the mirror can be computed with

$$\frac{dh_{\text{cnd}}}{dt} = \frac{2\,\overline{v}\,\rho_{\text{exp}}}{\rho_{\text{layer}}} \cos \psi_{vf} \tag{94}$$

The average debris velocity $\overline{v}$ is multiplied by a factor of 2 to account for the expansion of the gas in a vacuum. The layer density ρ_{layer} is to set to 1 g/cm^3. The power density on the asteroid surface is decreased based on the contamination of the mirrors. A degradation factor τ_c is applied to the power beamed to the asteroid surface, based on the Lambert–Beer–Bouguer law [57]

$$\tau_c = \exp^{-2\eta h_{\text{cnd}}} \tag{95}$$

where $\eta = 10^4$/m is taken as the absorption coefficient for forsterite from the experiment reported in Ref. [59]. Equation (94) can be numerically integrated, along with Gauss equations, for the period of ablation to account for the degradation of the input power in Eq. (90).

4. TUGGING EFFECT

If the laser or the solar concentrator is carried by a spacecraft flying in formation with the asteroid at a distance δr, a tugging effect is possible if the gravity of the asteroid has to be compensated by a low-thrust system and the plume does not impinge the asteroid. The maximum tugging acceleration experienced by the asteroid, in the direction opposite to the thrust coming from the ablation process, is given by Eq. (71). As an example, if we assume a concentration ratio $C_r = 100$ at aphelion of the orbit of Apophis, with a 62 m diameter mirror at a distance of 750 m from the surface of the asteroid the tug acceleration is about $a_{\text{tug},a} = 2.52 \times 10^{-13}\,\text{m/s}^2$ and increase to $a_{\text{tug},p} = 4.83 \times 10^{-13}\,\text{m/s}^2$ at perihelion. This is several orders of magnitude lower than the thrust produced by the ablation process and can be considered only a simple perturbation.

D. ENHANCED YARKOVSKY EFFECT

If the power density is not high enough to ablate the surface material, it may be sufficient to heat up the surface creating a low thrust by means of the Yarkovsky effect combined with light pressure (photon beaming). The assumption in this case is that vaporization is small and produces a negligible effect. Given the emissivity ϵ of the asteroid, the force due to the light projected onto the asteroid is

$$\begin{aligned} F_{\text{light}} &= 2(1 - \varsigma_{\text{A}})\, n_{sc}\, \Lambda_{\text{light}}\, \sigma_{\text{sys}}\, C_r A_{\text{spot}}\, S_{\text{srp}} \left(\frac{r_{\text{AU}}}{r_{sc}}\right)^2 \\ &\quad + n_{sc}\, \varsigma_{\text{A}}\, \sigma_{\text{sys}}\, C_r A_{\text{spot}}\, S_{\text{srp}} \left(\frac{r_{\text{AU}}}{r_{sc}}\right)^2 \\ &= n_{sc}\, \sigma_{\text{sys}}\, A_{\text{M}}\, S_{\text{srp}} \left(\frac{r_{\text{AU}}}{r_{sc}}\right)^2 \left(2\Lambda_{\text{light}}(1 - \varsigma_{\text{A}}) + \varsigma_{\text{A}}\right) \end{aligned} \tag{96}$$

where $S_{\text{srp}} = 4.562 \times 10^{-6}\,\text{N/m}$ is the solar pressure at 1 AU, A_{spot} is the 2-D illuminated surface area on the asteroid, A_{M} is the effective surface area of the primary mirror that is perpendicular to the Sun, $C_r = A_{\text{M}}/A_{\text{spot}}$ is the concentration ratio, $\Lambda_{\text{light}} = 4/(3\pi)$ is the scattering factor and $n_{s/c}$ is the number of illuminated spot areas, $\alpha_{\text{A}} = 1 - \varsigma_{\text{A}} = 0.8$. The first component is the reflected light and the second component is the absorbed light. The scattering factor is the average of the reflected force assuming that the reflection can have any direction angle α_{L} between $[0, \pi]$. Therefore, the scattering factor here is defined as $\Lambda_{\text{light}} = (1/\pi) \int_0^{\pi} \cos^3 \alpha_{\text{L}}\, d\alpha_{\text{L}}$. For an asteroid surface temperature T_{A}, the emission of photons will add a force component F_{ir} [60]

$$F_{ir} = n_{\text{spot}} \Lambda \frac{\epsilon \sigma T_{\text{A}}^4}{c} A_{\text{spot}} \tag{97}$$

where c is the speed of light. The temperature T_A can be calculated solving Eqs. (81) and (82).

E. DISAGGREGATED COLLABORATIVE ABLATION PROCESS

From Eq. (90) one can understand that the power density at the spot depends on two critical system parameters: the concentration ratio C_r and the system efficiency σ_{sys}. If solar concentrators are used, the system efficiency depends solely on the reflectivity of the mirror and one can assume $\sigma_{sys} = 0.9$ at the beginning of life. On the other hand, the concentration ratio depends on the size of the mirror and on the distance from the spot as source, the Sun, has a finite size. This implies that high-power densities can be achieved only with large mirrors or very short distances. The latter accelerates the contamination of the surface and the mirror whereas the former brings along a lot of structural and control complications. An alternative, proposed in Ref. [29] is to use a formation of mirrors (mirror bees) and superimpose the illuminated areas to achieve the desired intensity. If $n_{s/c}$ spacecraft superimpose their beams then P_{in} in Eq. (90) is multiplied by $n_{s/c}$, alternatively, if the beams are not superimposed the mass flow rate in Eq. (87) is multiplied by $n_{s/c}$.

If solar-pumped lasers are used, the concentration ratio does not depend on the size of the primary mirror, on the other hand the system efficiency is significantly lower with values $\sigma_{sys} = 0.21$ or lower, if one assumes an optimistic solar cell efficiency of 0.4, a laser wall-plug efficiency of 0.6 and a power system efficiency of 0.9. As a consequence a considerable amount of power needs to be rejected, in the form of heat, at different stages. Furthermore, lower power lasers at low temperature normally achieve better efficiencies. This implies that smaller systems are preferable. As for the case of the solar concentrators in Ref. [24] it was proposed to use a swarm of solar-pumped lasers (laser bees) coherently combining their beams.

In both cases, the swarm would fly in formation with the asteroid on a set of periodic orbits as in Fig. 23a that minimize plume impingement. A further option for the case of the laser formation, proposed in Ref. [24], is to maintain an artificially displaced periodic orbit below the y-axis so that only the back of the mirror (see conceptual spacecraft configuration in Fig. 23b) would be exposed to the plume.

Figures 24 show the achievable impact parameter for a swarm of lasers, powered by either a 5 m in diameter or a 10 m diameter mirror, deflecting the asteroid Apophis. The warning time is the time from the start of the deflection to the expected time of the impact. Figures 25 show the achievable impact parameter for a swarm of solar concentrators 62 m in diameter, deflecting the asteroid Apophis. The warning time is, once again, the time from the start of the deflection to the expected time of the impact. Figure 24a shows the result with ablation of the surface whereas Figs. 24b without ablation and only the effect of light

a)

b)

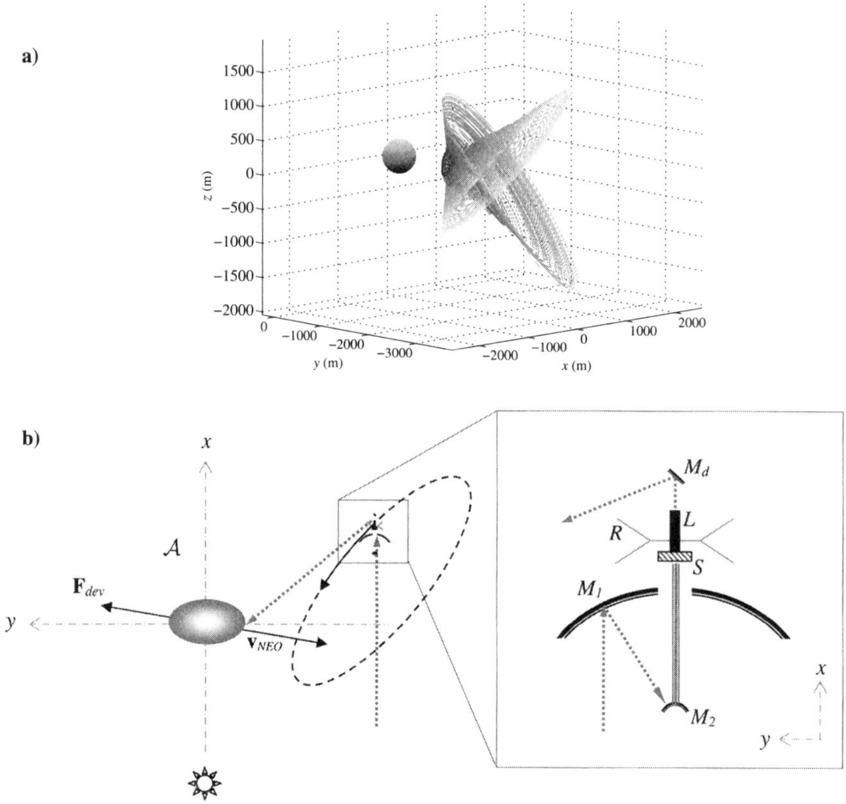

Fig. 23 a) Formation orbits with minimum distance of 500 m and b) illustration of the spacecraft and laser system, showing the two parabolic mirrors (M_1, M_2), the directional steering mirror (M_d), the solar arrays (S) which pump the laser (L), and the radiators (R).

pressure and enhanced Yarkovsky (photon beaming). As one can see the performance of the ablation process is poor, mainly due to contamination that, de facto, stops the ablation process prematurely. The combined effect of light pressure and Yarkovsky although small is still not negligible.

IV. COMPARISON OF DEFLECTION METHODS

In this section, we propose a simple way to compare different methods on an objective basis. First we compare laser ablation against methods based on electric propulsion and the slow-push methods against the kinetic impactor.

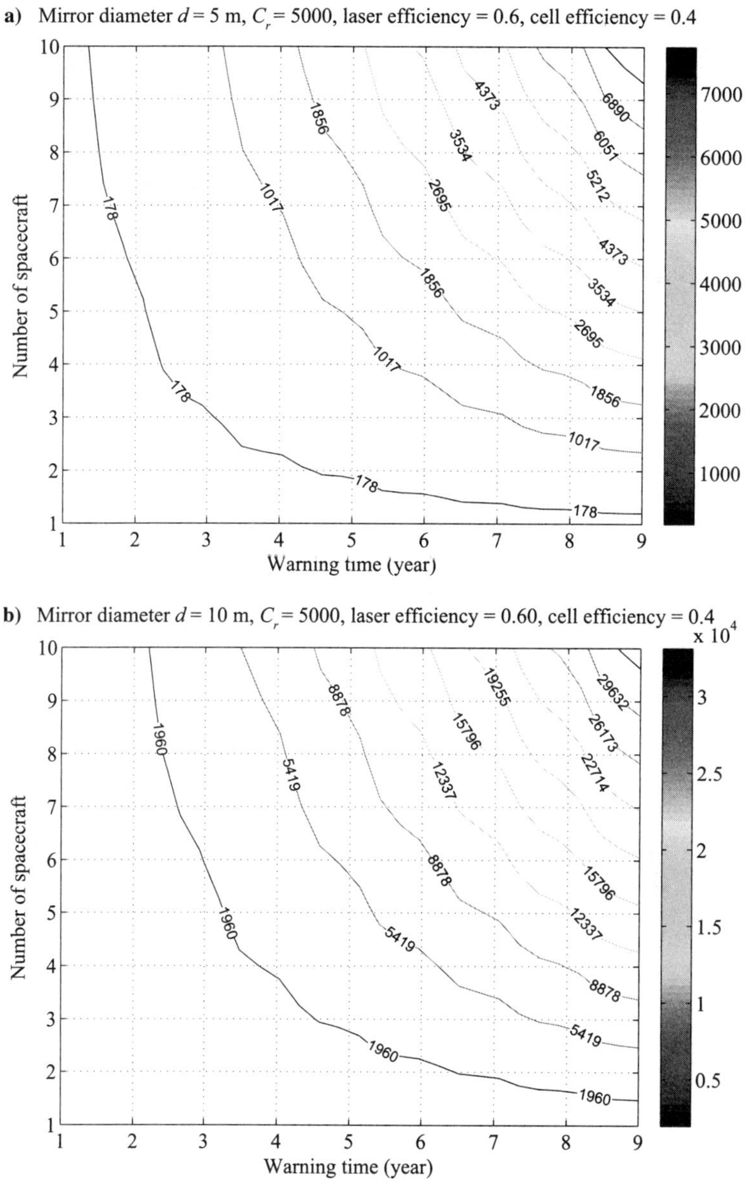

Fig. 24 Impact parameter achieved when deflecting the asteroid Apophis with a swarm of solar pumped lasers: a) primary mirror diameter of 5 m and b) primary mirror diameter 10 m.

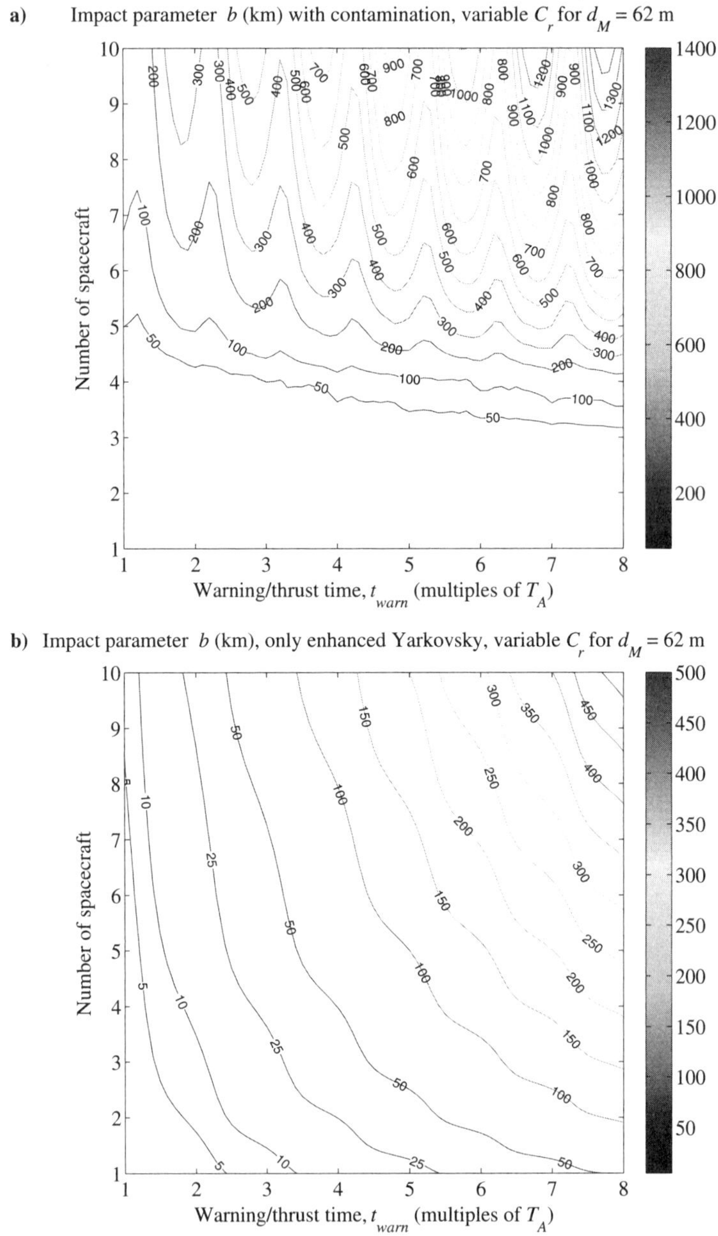

Fig. 25 Impact parameter achieved when deflecting the asteroid Apophis with a swarm of solar concentrators carrying a 62 m in diameter mirror: a) ablation and contamination and b) only Yarkovsky and light pressure.

A. LASER ABLATION VS METHODS BASED ON LOW-THRUST ENGINES

It is now interesting to compare the laser system to other contactless deflection methods based on electric propulsion. Rather than comparing laser ablation against specific technologies like ion beaming of gravitational tractoring, the analysis in this section considers a generic electric propulsion system that employs two identical electric engines operating at the same time and providing the best possible momentum transfer, that is, the asteroid is subject to the full thrust delivered by one of the engines. Two metrics are used in the comparison: the momentum coupling that relates the achievable thrust to the power installed on the spacecraft and the deflection system mass that relates the achievable deflection Δv to the mass of the system required to produce that Δv. Both the laser ablation and the electric propulsion systems are assumed to have the same power system and the same configuration with deployable solar arrays. From the thrust delivered by the sublimation process and the input power P_L to the laser system one can derive the momentum coupling coefficient

$$C_m = \frac{F_{\text{sub}}}{P_L} \tag{98}$$

This definition is slightly different than the one found in Phipps et al. [61] as the interest here is to size the power system onboard the spacecraft. The power input to the laser is

$$P_L = \eta_P \eta_S \frac{P_{1\,\text{AU}} A_{SA}}{R_{AU}^2} \tag{99}$$

where A_{SA} is the area of the array, R_{AU} is the distance from the Sun in Astronomical Units (AU), $P_{1\,\text{AU}}$ is the power per square meter at 1 AU, η_p is the efficiency of the power system and η_S the efficiency of the solar arrays.

To better understand the impact of C_m on the size of the laser system, one can take, for comparison, some existing electric propulsion systems. The momentum coupling as a function of the input power to the deflection system is represented in Fig. 26 for a laser wall-plug efficiency of 55% and three different electric propulsion systems: the RIT-22, with $I_{sp} = 4500$ s and a thrust of 150 mN at 5000 W of input power, the RIT-10 with $I_{sp} = 3325$ s at a nominal thrust of 15 mN and input power of 460 W, and the PPS1350G with a thrust of 67 mN at 1190 W and an $I_{sp} = 1540$ s. The material characteristics are in Table 6, the rotation speed of the asteroid is $\omega_A = 500$ rotations per day, 10% of the incident laser light is absorbed by the vapor and only 80% of the remaining light is absorbed by the surface material. Note that the assumption here is that the efficiency of ion engines does not change with the power level. This is not generally true as for very low propulsion level the efficiency currently drops to 40% or less. The momentum coupling of ion engines is, therefore, expected to be lower for low-thrust levels. As one can see the momentum coupling of a laser system is comparable to the one of an electric propulsion system using two engines.

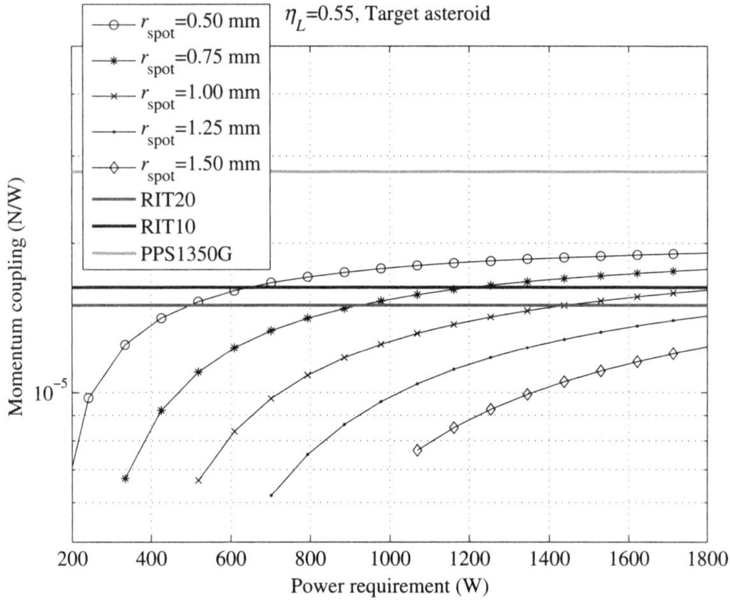

Fig. 26 Momentum coupling for different spot sizes (laser efficiency 55%).

The momentum coupling, however, is not sufficient to define the total mass of the system. If the interest is in a given Δv delivered by the deflection method, the deflection system mass can be defined as the portion of the mass of the spacecraft that is required to obtain that Δv. Following this definition, the mass of the laser system $m_{\rm LS}$ can be estimated as

$$m_{\rm LS} = \alpha_P P_L + \rho_R(1 - \eta_L)\frac{P_L}{\sigma\varepsilon_R T_R^4} + m_L \tag{100}$$

where m_L is the mass of the laser itself plus the optics, $\rho_R = 0.5\,\mathrm{kg/m^2}$, is the mass of the radiators, ϵ_R their emissivity, T_R their operating temperature, α_P is the specific mass, or mass per unit power, of the power system, P_L is the input power to the laser, η_L is the efficiency of the laser that is assumed to be 55% in this analysis. The specific mass of the power system, which derived again from system level considerations, is 40 kg/kW and the mass of the laser is 10 kg/kW. This assumes that the heat sink is included in the thermal control system, that is, the radiators. The temperature of the radiators is assumed to operate at the temperature at which the diodes of the laser need to function. This is 283 K and the emissivity was taken to be 0.8. The input power P_L is the one defined in Eq. (99).

The deflection system mass of the laser system can be compared to the deflection system mass of an electric propulsion (EP) system that produces the same Δv.

The mass of the EP system should include the mass of two engines (whether it is an ion beaming system or a gravity tractor), the mass of the related power system, the mass of the radiators, the mass of propellant, and the mass of the tanks. The mass of the radiator and power system is computed using the same figures and assumptions used for the laser system except for the efficiency of the engine that is always equal to 60%. The mass of the propellant is simply, $m_p = 2F_{\mathrm{EP}}\Delta t_{\mathrm{thrust}}/I_{sp}g_0$, the mass of the EP system is, therefore,

$$m_{\mathrm{EP}} = 2.2\frac{F_{\mathrm{EP}}}{g_0 I_{sp}}\Delta t_{\mathrm{thrust}} + \alpha_P P_{\mathrm{EP}} + \rho_R(1-\eta_{\mathrm{EP}})\frac{P_{\mathrm{EP}}}{\sigma\varepsilon_R T_R^4} + 2m_e \tag{101}$$

where m_e is the mass of a single engine, I_{sp} is the specific impulse of the engine and η_{EP} is the efficiency of the EP system that is assumed to be 60% in this analysis. The assumptions here are that the mass of the tanks is only 10% of the mass of the propellant. Furthermore, the transfer of momentum is considered to be equal to one, which corresponds to an ideal ion beaming system. The specific impulse in this comparison is taken to be 3500 s. The mass of the engine is assumed to be 5 kg/kW. This excludes electronics and the PDCU, which are assumed to be included in the power system. Note that the temperature of the radiators T_R for the laser system is lower than for the EP system. The assumption is that the laser needs to operate at 283 K whereas the EP system can operate at 374 K. Figure 27a shows the mass of the laser system (LS), assuming a spot size of 1mm and a distance of 250 m, versus the mass of the asteroid. The optics does not change with the power as the distance and spot size are constant, therefore, one can add a fixed mass equal to 100 kg corresponding to a 160 mm optics to the mass of the deflection system. Figure 27b shows the mass of the EP system with the same input power of the laser system and delivering the same Δv. To have a correct and fair comparison of the two concepts it is assumed that the thrust of the EP system follows the same profile of the thrust of the LS system and is on for the same length of time. The difference in mass is mainly due to the propellant consumption and associated tanks.

Figures 27 demonstrate that for the same installed power and for the same resulting Δv the laser system is more advantageous. On the other hand, an increase in the efficiency of ion engines could potentially allow a faster deflection with the same overall system mass. At the same time the laser system needs to cope with the contamination that can be higher than the model predicts and with the actual energy efficiency of the ablation process that can be lower than expected if a 3-D heat diffusion model is considered. This last point is the subject of current investigation.

B. DEFLECTION MASS COEFFICIENT AND OPTIMAL DEFLECTION TIME

One can now derive a simple estimation of the mass of the laser system as a function of the Δv imparted by the deflection system on the asteroid. Assuming a constant momentum coupling, a constant thrust and Eq. (98), the mass of the

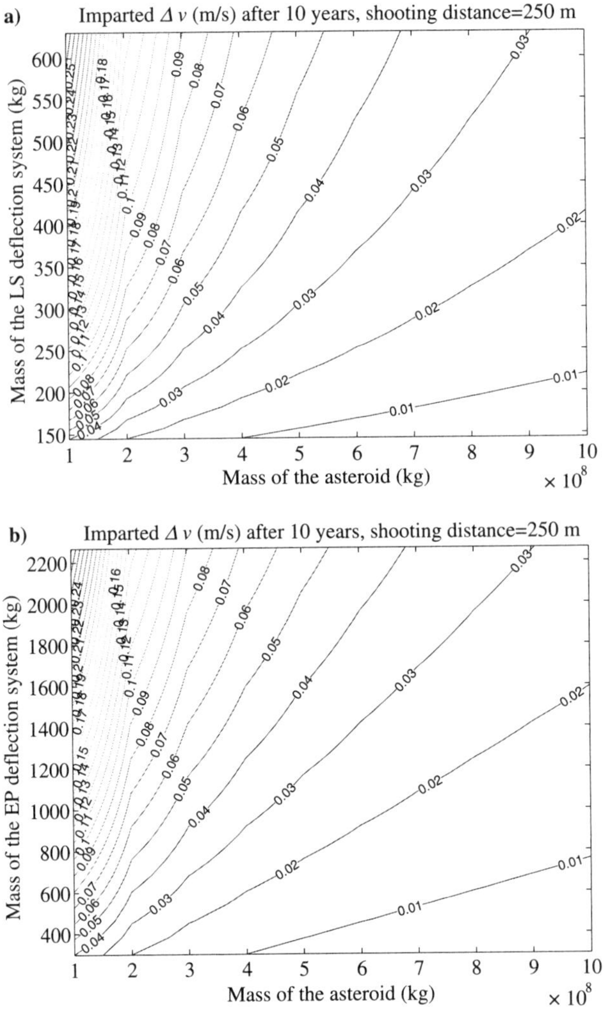

Fig. 27 Imparted Δv as a function of asteroid mass and mass of the deflection system: a) laser ablation and b) ion propulsion.

laser system as a function of the imparted Δv and mass of the asteroid can be written as

$$m_{\text{LS}} = \alpha_P \frac{F_{\text{sub}}}{C_m} + \frac{F_{\text{sub}}}{C_m} \frac{\rho_R(1-\eta_L)}{\sigma \varepsilon_R T_R^4} + \mu_L \frac{F_{\text{sub}}}{C_m} \tag{102}$$

One can now assume that $F_{\text{sub}} = m_{\text{A}}\Delta v/\Delta t$ from which the mass of the laser system becomes

$$m_{\text{LS}} = \left[\alpha_P + \alpha_{SP}e^{\nu_L \Delta t} + \frac{\rho_R(1-\eta_L)}{\sigma \varepsilon_R T_R^4} + \mu_L\right]\frac{\Delta v m_{\text{A}}}{\Delta t C_m} \tag{103}$$

where it was assumed that the mass of laser and optics is proportional to the power delivered to the asteroid through a factor $\mu_L = m_L/P_L$. This is true for the laser but only partially true for the optics as the optics does not depend on the power and remains almost constant over a wide range of powers. The effect of the contamination is introduced through the factor $\alpha_{SP}e^{\nu_L \Delta t}$ where a constant rate of contamination ν_L is assumed given that the thrust is constant. The exponential degradation affects only the solar arrays and, therefore, it is assumed that the solar arrays with specific mass α_{SP} are oversized to compensate for the reduced power at the end of life. Given the mass of the laser system in Eq. (103), the deflection mass coefficient can now be defined as

$$\eta_{\text{LS}} = \left[\alpha_P + \alpha_{SP}e^{\nu_L \Delta t} + \frac{\rho_R(1-\eta_L)}{\sigma \varepsilon_R T_R^4} + \mu_L\right]\frac{m_{\text{A}}}{\Delta t C_m} \tag{104}$$

A similar derivation can also be considered for an ion-propulsion system. Note that in this case the exponential increase of the arrays applies only in the case of the back spattering effect of ion-beaming. If the power input to the engine P_{EP} is defined as

$$P_{\text{EP}} = \frac{F_{\text{EP}} I_{sp} g_0}{2\eta_{\text{EP}}} \tag{105}$$

then the mass of the EP system can be recasted as

$$m_{\text{EP}} = \left[\frac{2.2}{g_0 I_{sp}} + \left(\alpha_P + \alpha_{SP}e^{\nu_{\text{EP}} \Delta t} + 2\mu_e\right)\frac{I_{sp} g_0}{2\eta_{\text{EP}}\Delta t} + \frac{\rho_R(1-\eta_{\text{EP}})I_{sp} g_0}{\Delta t 2\eta_{\text{EP}}\sigma \varepsilon_R T_R^4}\right]\Delta v m_{\text{A}} \tag{106}$$

where μ_e is the mass per unit power of the engines. The deflection mass coefficient of the EP system is then defined as

$$\eta_{\text{EP}} = \left[\frac{2.2}{g_0 I_{sp}} + \left(\alpha_P + \alpha_{SP}e^{\nu_{\text{EP}} \Delta t} + 2\mu_e\right)\frac{I_{sp} g_0}{2\eta_{\text{EP}}\Delta t} + \frac{\rho_R(1-\eta_{\text{EP}})I_{sp} g_0}{\Delta t 2\eta_{\text{EP}}\sigma \varepsilon_R T_R^4}\right] m_{\text{A}} \tag{107}$$

From these two expressions it is clear that the mass of the laser system is diminishing with time for the same Δv and mass of the asteroid if no contamination is considered. Hence long deflection times with small power levels are to be preferred. If contamination is included the rate of degradation is the driving parameter to decide which system is optimal. In the case of ion-engine systems, a similar consideration applies to all terms in Eq. (106) except for the first term in the square brackets. That term does not change with time and establishes a direct proportionality between mass of the system and mass of the asteroid. It is interesting to note that the efficiency of the ablation process does not affect the mass of the laser, which is instead dependent on the momentum coupling. On the contrary, the mass of the EP system is a function of the efficiency of the engines. This is a distinctive feature of laser ablation compared to EP deflection techniques. Figure 28 shows the estimated Δv and deflection system mass from Eq. (103) assuming a constant $\nu_L = 1.145 \times 10^{-9}\,\mathrm{s}^{-1}$. The values of Δv are a good approximation of the outcome of the simulations in Fig. 27a. Note that in the case of Fig. 28 the thrust is constant and the mass of the array is augmented to maintain the minimum required input power at the end of life, thus Figs. 27a and 28a cannot display an exact match. Figure 28b shows the mass of the laser system (103) divided by the mass of the asteroid. The same trend is displayed by the mass of the EP system (106). In the case of no contamination the deflection mass coefficient of laser ablation goes to zero with time whereas it remains proportional to the mass of the asteroid in the case of ion propulsion. In the case of contamination one can compute the optimal deflection time for both deflection systems by taking the first derivative of the mass coefficient with respect to time:

$$\frac{\mathrm{d}\eta_{\mathrm{LS}}}{\mathrm{d}\Delta t} = \frac{\nu_L \alpha_{SP} e^{\nu_L \Delta t} m_{\mathrm{A}}}{\Delta t C_m} - \left[\alpha_P + \alpha_{SP} e^{\nu_L \Delta t} + \frac{\rho_R(1-\eta_L)}{\sigma \varepsilon_R T_R^4} + \mu_L\right] \frac{m_{\mathrm{A}}}{\Delta t^2 C_m} = 0 \tag{108}$$

then by collecting the term $(m_{\mathrm{A}}/\Delta t^2 C_m)$ one can get:

$$\frac{\mathrm{d}\eta_{\mathrm{LS}}}{\mathrm{d}\Delta t} = (\nu_L \Delta t - 1)\,\alpha_{SP} e^{\nu_L \Delta t} - \left[\alpha_P + \frac{\rho_R(1-\eta_L)}{\sigma \varepsilon_R T_R^4} + \mu_L\right] = 0 \tag{109}$$

The same derivation applies to the case of the mass of the EP system:

$$\begin{aligned}\frac{\mathrm{d}\eta_{\mathrm{EP}}}{\mathrm{d}\Delta t} &= \Bigg[\left(\nu_{\mathrm{EP}} \alpha_{SP} e^{\nu_{\mathrm{EP}} \Delta t}\right) \frac{I_{sp} g_0}{\eta_{\mathrm{EP}} \Delta t} \\ &\quad - \left(\alpha_P + \alpha_{SP} e^{\nu_{\mathrm{EP}} \Delta t} + 2\mu_e\right) \frac{I_{sp} g_0}{\eta_{\mathrm{EP}} \Delta t^2} - \frac{\rho_R(1-\eta_{\mathrm{EP}}) I_{sp} g_0}{\Delta t^2 \eta_{\mathrm{EP}} \sigma \varepsilon_R T_R^4}\Bigg] m_{\mathrm{A}} = 0\end{aligned} \tag{110}$$

$$\frac{\mathrm{d}\eta_{\mathrm{EP}}}{\mathrm{d}\Delta t} = (\nu_{\mathrm{EP}} \Delta t - 1)\,\alpha_{SP} e^{\nu_{\mathrm{EP}} \Delta t} - \left(\alpha_P + \frac{\rho_R(1-\eta_{\mathrm{EP}})}{\sigma \varepsilon_R T_R^4} + 2\mu_e\right) = 0 \tag{111}$$

Equations (109) and (111) are saying that the deflection time that provides the minimum system mass for the same Δv is independent of the mass of the asteroid. It depends instead on the contamination rate, on the power conversion efficiency of the system and on the mass per unit power of engine and power system. If the

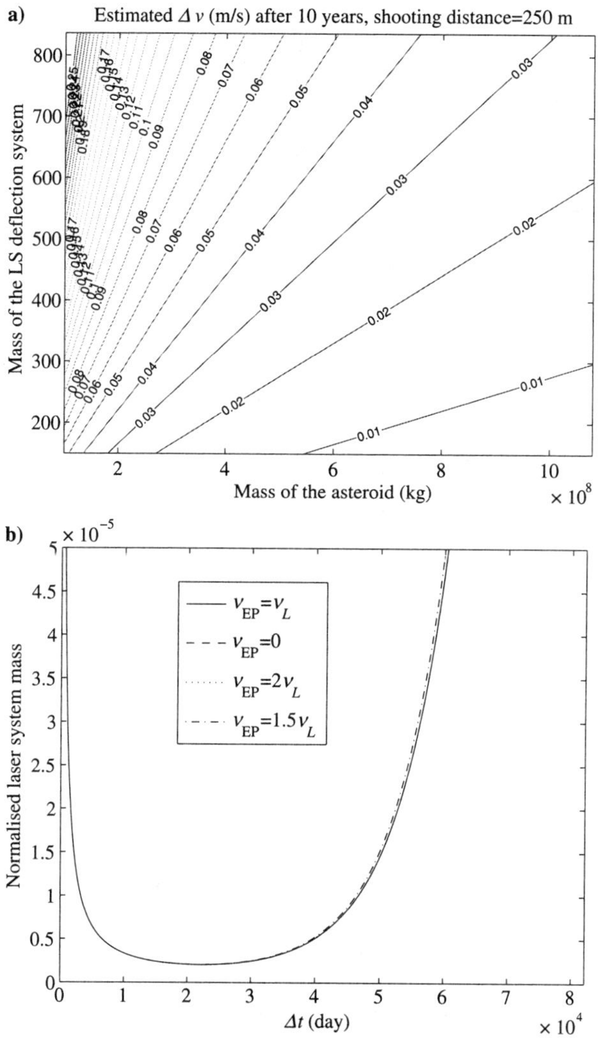

Fig. 28 a) Estimated mass of the laser system and b) normalized mass of a low-thrust system with contamination.

contamination rate, power conversion efficiency, and specific mass of the system were the same, laser ablation and ion propulsion would have the same optimal deflection time.

One can now analyse the behavior of the ratio between mass of the EP system and mass of the laser system for different values of the contamination factor ν_{EP}

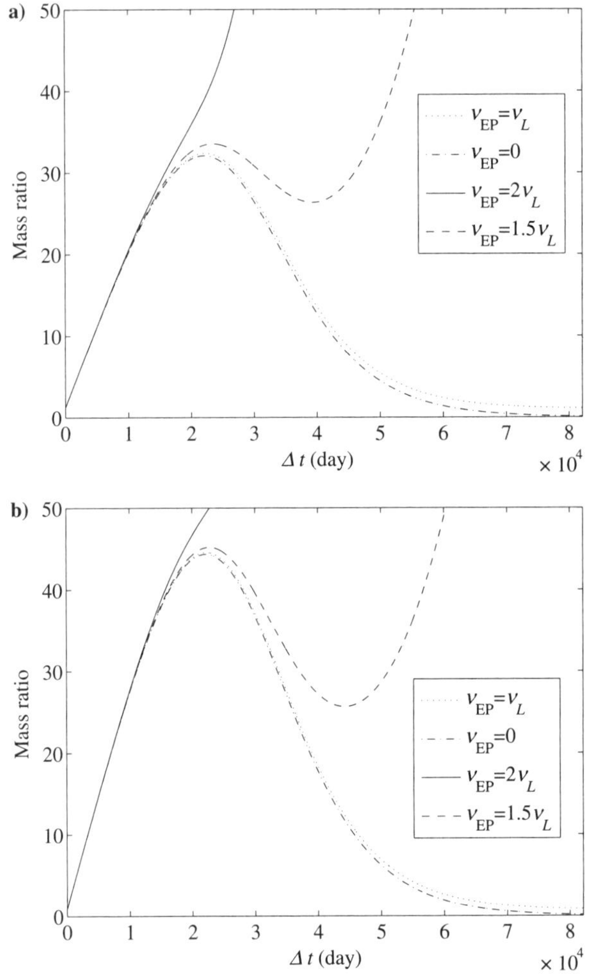

Fig. 29 Ratio between mass of the EP system and mass of the laser system: a) momentum coupling $C_m = \eta_{EP}/(I_{sp}g_0)$ and b) momentum coupling $C_m < \eta_{EP}/(I_{sp}g_0)$.

and different values of the momentum coupling C_m:

$$\frac{m_{\text{EP}}}{m_{\text{LS}}} = \frac{2.2(\Delta t/g_0 I_{sp}) + \left[\alpha_P + \alpha_{SP} e^{\nu_{\text{EP}}\Delta t} + (\rho_R(1-\eta_{\text{EP}})/\sigma \varepsilon_R T_R^4) + 2\mu_e\right](I_{sp}g_0/\eta_{\text{EP}})}{\left[\alpha_P + \alpha_{SP} e^{\nu_L \Delta t} + (\rho_R(1-\eta_L)/\sigma \varepsilon_R T_R^4) + \mu_L\right](1/C_m)} \tag{112}$$

Figures 29 show the $m_{\text{EP}}/m_{\text{LS}}$ defined in Eq. (112) as function of time for two different values of the momentum coupling. In both cases if the contamination factors for the LS and EP system are equal, then the mass ratio converges asymptotically to 1, on the other hand if $\nu_{EP} < \nu_{LS}$ the mass ratio converges to 0. In both cases the ratio reaches a maximum point that defines the time at which the laser system is performing at the best compared to the EP system. For all values of ν_{EP} there is a first turning time t_1^{EP} at which the mass ratio becomes greater than 1. Then, for $\nu_{EP} < \nu_{LS}$, there is a second turning time t_2^{EP} at which the mass ratio becomes lower than 1. For all deflection times t such that $t_1^{EP} < t < t_2^{EP}$ the laser system has a lower deflection system mass than the EP system.

C. COMPARISON BETWEEN SLOW-PUSH AND IMPULSIVE METHODS

One can now use the same concept of system mass ratio to compare a slow-push method against an impulsive one. This section will be limited to the kinetic impactor as some considerations extend to all purely impulsive approaches. From Sec. III. A one can write the variation of velocity of an asteroid due to a kinetic impact as

$$\Delta v = \frac{m_d}{m_A}(\Delta v_T - q) \tag{113}$$

where q is

$$q = \sqrt{\frac{2\mu}{r_E} - \frac{\mu}{r_E + \delta a_N}} - \sqrt{\frac{\mu}{r_E}} \tag{114}$$

The mass of the spacecraft at impact m_d can be related to mass at launch m_{si} through

$$\Delta v_T = I_{sp} g_0 \ln \frac{m_{si}}{m_d} \tag{115}$$

from which

$$\Delta v = \frac{m_d}{m_A}\left(I_{sp} g_0 \ln \frac{m_{si}}{m_d} - q\right) \tag{116}$$

If the mass fraction is $\epsilon = m_d/m_{si}$ then the mass at launch can be written as

$$m_{si} = \frac{\Delta v m_A}{I_{sp} g_0 \epsilon \ln \frac{1}{\epsilon} - \epsilon q} \tag{117}$$

Now one can take the minimum with respect to the mass ratio by finding the zeros of the derivative of m_{si} with respect to ϵ which leads to the optimal mass ratio:

$$\epsilon = e^{-\frac{q}{I_{sp} g_0} - 1} \tag{118}$$

Figures 30 show what we can call the mass momentum coupling (MMC) of the kinetic impactor or ratio between the launch mass and the delivered change in momentum $m_{si}/(m_A \Delta v)$. The figures show the MMC for a shallow crosses (dotted

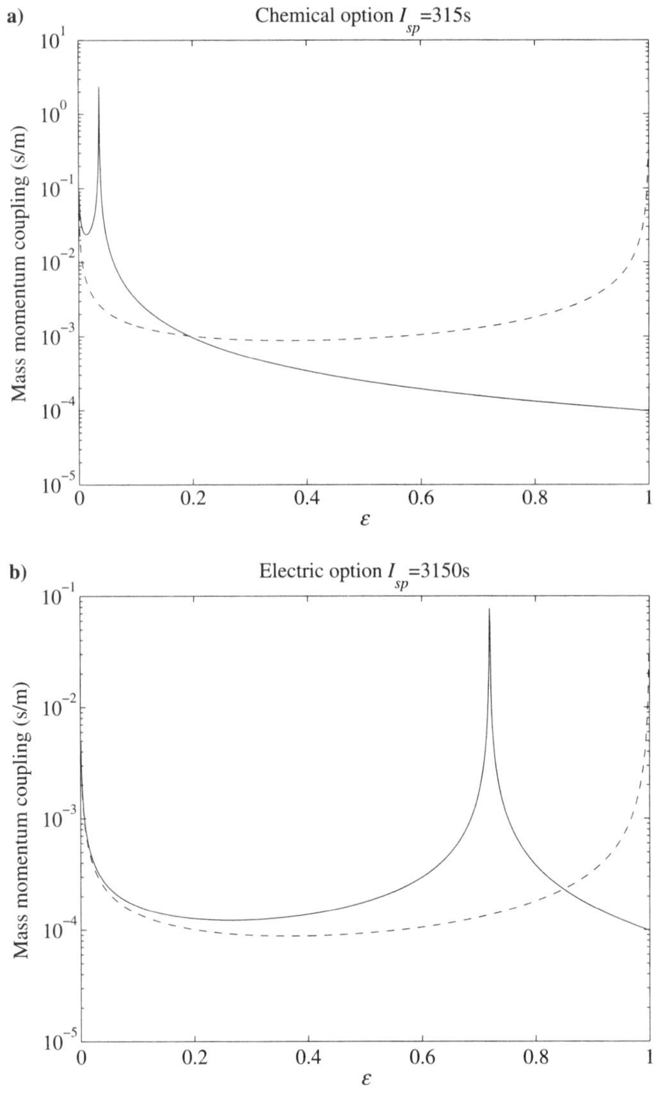

Fig. 30 Mass momentum coupling vs mass fraction for a simple kinetic impactor.

line) and a deep crosser (solid line) for two different I_{sp}'s. In all cases, the first minimum from the left to the right of the range of ϵ is given by Eq. (118) but for deep crossers the figures demonstrate that the optimal mass ratio is actually 1, the second minimum, which suggests no transfer. One can now rewrite the mass of low-thrust tug (ion beaming or similar) at launch $m_{\mathrm{EP}i}$ as a function of the total Δv delivered by the kinetic impactor

$$m_{\mathrm{EP}i} = \left(\frac{2.2\epsilon(-I_{sp}g_0 \ln \epsilon - q)}{g_0 I_{sp}} m_{si} + A \right) e^{\frac{q}{I_{sp}g_0}} \tag{119}$$

with

$$A = (\alpha_P + 2\mu_e) \frac{F I_{sp} g_0}{2\eta_{\mathrm{EP}}} + \frac{F \rho_{\mathbb{R}}(1 - \eta_{\mathrm{EP}}) I_{sp} g_0}{2\eta_{\mathrm{EP}} \sigma \varepsilon_R T_{\mathbb{R}}^4} \tag{120}$$

Equation (119) shows that in this simple model the mass of an equivalent low-thrust system is directly proportional to the mass of a kinetic impactor at launch. The constant of proportionality depends on the mass fraction impacting the asteroid and the I_{sp} required to deliver that mass. The mass of the low-thrust tug is amplified by the transfer cost $e^{q/(I_{sp}g_0)}$. For shallow crossers, $q = 0$, the amplification is 1 and the difference in mass between a low-thrust tug and an impactor depends only on the system mass A and the ratio between the I_{sp} of the impactor and the one of the low-thrust tug. For deep crossers, the amplification is >1 and q enters in the proportionality between $m_{\mathrm{EP}i}$ and m_{si}. For this latter case, therefore, a kinetic impactor is to be preferred whereas in the former case the low-thrust tug might be preferable. This results is in-line with what can be found in Ref. [3]. The reader should be warned that, this simplified analysis does not include the effect of the orbit geometry on the deflection or the phasing between asteroid and impactor. As shown in Refs. [24, 28] these two factor play a significant role on the end value to the miss distance.

V. CONCLUSION

This chapter has reviewed a number of basic techniques to estimate the miss distance achievable with either impulsive or slow push deflection methods. Furthermore, some simple models were provided to calculate the deflection action for some representative impulsive and slow-push techniques. The techniques to calculate the miss distance are of general validity, however, the reader needs to keep in mind that the model used to calculate the deflection action are subject to improvement and further development to fully account for the physics of the interaction with the asteroid and the technology installed on board the spacecraft.

ACKNOWLEDGMENTS

The author would like to acknowledge the students who worked with Massimiliano Vasile on the papers from which part of the results in this chapter were taken. In particular the authors would like to acknowledge the work of Christie Maddock, Federico Zuiani, Alison Gibbings, Camilla Colombo, and Joan-Pau Sanchez Cuartielles.

REFERENCES

[1] Stokes, G. H., Yeomans, D. K., Bottke, W. F., Jewitt, D., Chesley, S. R., Kelso, T. S., Evans, J. B., McMillan, R. S., Gold, R. E., Spahr, T. B., Harris, A. W., and Worden, S. P., "Study to Determine the Feasibility of Extending the Search for Near-Earth Objects to Smaller Limiting Diameters," Near-Earth Object Science Definition Team, NASA, Aug. 2003.

[2] Toon, O. B., Zahnle, K., Morrison, D., Turco, R. P., and Covey, C., "Environmental Perturbations Caused by the Impact of Asteroids and Comets," *Reviews of Geophysics*, Vol. 35, 1997, pp. 41–78.

[3] Sanchez Cuartielles, J. P., Colombo, C., Vasile, M., and Radice, G., "Multi-Criteria Comparison Among Several Mitigation Strategies for Dangerous Near Earth Objects," *Journal of Guidance, Control, and Dynamics*, Vol. 32, No. 1, Jan.–Feb. 2009, pp. 121–142.

[4] Conway, B., "Near-Optimal Deflection of Earth-Approaching Asteroids," *Journal of Guidance, Control, and Dynamics*, Vol. 24, No. 5, 2001, pp. 1035–1037.

[5] Vasile, M., "Robust Optimization of Trajectory Intercepting Dangerous Neo," *AAS/AIAA Astrodynamic Specialist Conference*, Monterey, California, 5–8 Aug. 2002.

[6] Carusi, A., Valsecchi, G. B., D'abramo, G., and Bottini, A., "Deflecting NEOs in Route of Collision with the Earth," *Icarus*, Vol. 159, No. 2, 2002, pp. 417–422.

[7] Conway, B. A., "Optimal Low-Thrust Interception of Earth-Crossing Asteroids," *Journal of Guidance, Control, and Dynamics*, Vol. 20, 1997, pp. 995–1002.

[8] McInnes, C., "Deflection of Near-Earth Asteroids by Kinetics Energy Impacts from Retrograde Orbits," *Planetary and Space Science*, Vol. 52, 2003, pp. 587–590.

[9] Gibbings, A., and Vasile, M., "A Smart Cloud Approach to Asteroid Deflection," *62nd International Astronautical Congress*, Cape Town, South Africa, 2011.

[10] Walker, R., Izzo, D., De Negueruela, C., Ayre, M., Summerer, L., and Vasile, M., "Concepts for Near Earth Asteroid Deflection Using Spacecraft with Advanced Nuclear and Solar Electric Propulsion System," *Journal of the British Interplanetary Society*, Vol. 58, No. 7/8, 2005, pp. 268–278.

[11] Lu, E. T., and Love, S. G., "Gravitational Tractor for Towing Asteroids," *Nature*, Vol. 438, 2005, pp. 177–178.

[12] Phipps, C. R., "Laser Deflection of NEO's," *NASA Near Earth Object: Interception Workshop*, Albuquerque, New Mexico, 1992.

[13] Phipps, C. R., "Laser Deflection of Near-Earth Asteroids and Comet Nuclei," *Proceedings of International Conference on Lasers 96, STS Press*, 1997, pp. 580–587.

[14] Park, S.-Y., and Mazanek, D. D., "Deflection of Earth-Crossing Asteroids/Comets Using Rendezvous Spacecraft and Laser Ablation," *Journal of Astronautical Sciences*, Vol. 53, No. 1, Jan.–Mar. 2005, pp. 21–37.

[15] Yoo, S.-M., Songa, Y.-J., Park, S.-Y., and Choi, K.-H., "Spacecraft Formation Flying for Earth-Crossing Object Deflections Using a Power Limited Laser Ablating," *Advances in Space Research*, Vol. 43, No. 12, 2009, pp. 1873–1889.

[16] Melosh, H. J., and Nemchinov, I. V., "Solar Asteroid Diversion," *Nature*, Vol. 366, 1993, pp. 21–22.

[17] Vasile, M., and Maddock, C., "On the Deflection of Asteroids with Mirrors," *Celestial Mechanics and Dynamical Astronomy*, Vol. 107, No. 1–2, 2010, pp. 265–284.

[18] Bombardelli, C., and Pelez, J., "Ion Beam Shepherd for Asteroid Deflection," *Journal of Guidance, Control, and Dynamics*, Vol. 34, No. 4, 2011, pp. 1270–1272.

[19] Varghese, S., "Electro Magnetic Tractor," *ICAMAME 2015: International Conference on Aerospace, Mechanical, Automotive and Materials Engineering*, Vol. 2, No. 10, 2014, p. 134.

[20] Olds, J., Charania, A., Graham, M., and Wallace, J., "The League of Extraordinary Machines: A Rapid and Scalable Approach to Planetary Defense Against Asteroid Impactors," Technical Report, NASA Institute for Advanced Concepts, CP-NIAC 02-02, Atlanta, USA, April 2004.

[21] Bombardelli, C., "Artificial Spin-Up and Fragmentation of Subkilometre Asteroids," *Acta Astronautica*, Vol. 65, No. 7, 2009, 1162–1167.

[22] Vasile, M., "A Multi-Mirror Solution for the Deflection of Dangerous NEOs," *Communications in Nonlinear Science and Numerical Simulation*, Vol. 14, No. 12, 2009, pp. 4139–4152.

[23] Maddock, C., and Vasile, M., "Design of Optimal Spacecraft-Asteorid Formations Through a Hybrid Global Optimization Approach," *Journal of Intelligent Computing and Cybernetics*, Vol. 1, No. 2, 2008, pp. 239–268.

[24] Vasile, M., and Maddock, C., "Design of a Formation of Solar Pumped Lasers for Asteroid Deflection," *Advances in Space Research*, Vol. 50, No. 7, Oct. 2012, pp. 891–905.

[25] Wie, B., "Dynamics and Control of Gravity Tractor Spacecraft for Asteroid Deflection," *Journal of Guidance, Control, and Dynamics*, Vol. 31, No. 5, 2008, pp. 1413–1423.

[26] Spitale, J. N., "Asteroid Hazard Mitigation Using the Yarkovsky Effect," *Science*, Vol. 296, 2002, p. 77.

[27] Schaub, H., and Junkins, J. L., *Analytical Mechanics of Space Systems*, AIAA Education Series, AIAA, Reston, VA, 2003.

[28] Vasile, M., and Colombo, C., "Optimal Impact Strategies for Asteroid Deflection," *Journal of Guidance, Control, and Dynamics*, Vol. 31, No. 4, July 2008, pp. 858–872.

[29] Vasile, M., and Maddock, C., "On the Deflection of Asteroids with Mirrors," *Celestial Mechanics and Dynamical Astronomy*, Vol. 107, No. 1, May 2010, pp. 265–284.

[30] Colombo, C., Vasile, M., and Radice, G., "Semi-Analytical Solution for the Optimal Low-Thrust Deflection of Near-Earth Objects," *Journal of Guidance, Control, and Dynamics*, Vol. 32, No. 3, May–June 2009, pp. 796–809.

[31] Zuiani, F., Vasile, M., and Gibbings, A., "Evidence-Based Robust Design of Deflection Actions for Near Earth Objects," *Celestial Mechanics and Dynamical Astronomy*, Vol. 114, No. 1–2, 2012, pp. 107–136.

[32] Zuiani, F., and Vasile, M., "Extension of Finite Perturbative Elements for Multi-Revolution Low-Thrust Transfer Optimisation," *63rd International Astronautical Congress*, IAC, Naples, Sept. 2012.

[33] Zuiani, F., Vasile, M., and Gibbings, A., "Evidence-Based Robust Design of Deflection Actions for Near Earth Objects Celestial Mechanics and Dynamical

Astronomy," *Celestial Mechanics and Dynamical Astronomy*, Vol. 114, No. 1–2, 2012, pp. 107–136.

[34] Zuiani, F., and Vasile, M., "Extended Analytical Formulas for the Perturbed Keplerian Motion Under a Constant Control Acceleration," *Celestial Mechanics and Dynamical Astronomy*, Vol. 121, No. 3, 2015, 275–300.

[35] Hampton, D., Baer, J., Huisjen, M., Varner, C., Delamere, A., Wellnitz, D., A'Hearn, M., and Klaasen, K., "An Overview of the Instrument Suite for the Deep Impact Mission," *Space Science Reviews*, Vol. 117, No. 1–2, 2005, pp. 43–93.

[36] Housen, K. R., and Holsapple, K.A., "Experimental Measurement of Momentum Transfer in Hypervelocity Collisions," *46th Lunar and Planetary Science Conference*, Number 2894, Texas, USA, March 2015.

[37] Sanchez Cuartielles, J. P., Colombo, C., Vasile, M., and Radice, G., "Multi-Criteria Comparison Among Several Mitigation Strategies for Dangerous Near Earth Objects," *Journal of Guidance, Control, and Dynamics*, Vol. 32, 2009, pp. 121–142.

[38] Ryan, E. V., and Melosh, H. J., "Impact Fragmentation: From the Laboratory to Asteroids," *Icarus*, Vol. 133, April 1998, pp. 1–24.

[39] Housen, K. R., and Holsapple, K. A., "On the Fragmentation of Asteroids and Planetary Satellites," *Icarus*, Vol. 84, 1990, pp. 226–253.

[40] Holsapple, K. A., "Catastrophic Disruptions and Cratering of Solar System Bodies: A Review and New Results," *Planetary and Space Science*, Vol. 42, May 1994, pp. 1067–1078.

[41] Davis, D. R., and Ryan, E. V., "On Collisional Disruption: Experimental Results and Scaling Laws," *Icarus*, Vol. 83, 1990, p. 182.

[42] Gibbings, A., and Vasile, M., "A Smart Cloud Approach to Asteroid Deflection," *62nd International Astronautical Congress*, Cape Town, 2011.

[43] Glasstone, S., and Dolan, P. J., *The Effects of Nuclear Weapons*, 3rd ed., Princeton University, U.S. Government Printing Office, US Department of Defense, 2007.

[44] Hammerling, P., and Remo, J. L., "NEO Interaction with Nuclear Radiation," *Acta Astronautica*, Vol. 36, No. 6, 1995, pp. 337–346.

[45] Tedeschi, W. J., Remo, J. L., Schulze, J. F., and Young, R. P., "Experimental Hypervelocity Impact Effects on Simulated Planetesimal Materials," *International Journal of Impact Engineering*, Vol. 17, 1995, pp. 837–848.

[46] Hu, W., and Scheeres, D. J., "Spacecraft Motion About Slowly Rotating Asteroids," *Journal of Guidance, Control, and Dynamics*, Vol. 25, No. 4, July–Aug. 2002, pp. 765–775.

[47] Rossi, A., Marzari, F., and Farinella, P., "Orbital Evolution Around Irregular Bodies," *Earth, Planets, Space*, Vol. 51, 1999, pp. 1173–1180.

[48] McInnes, C., "Near Earth Object Orbit Modication Using Gravitational Coupling," *Journal of Guidance, Control, and Dynamics*, Vol. 30, No. 3, 2007, pp. 870–873.

[49] Yamaguchi, K., and Yamakawa, H., "Orbital Deflection of Potentially Hazardous Asteroids Using a Coulomb Force Attractor," *Transactions of the Japan Society for Aeronautical and Space Sciences, Aerospace Technology Japan*, Vol. 12, No. 29, 2014, pp. Tr_27-Tr_33.

[50] Vasile, M., Gibbings, A., Watson, I., and Hopkins, J.-M., "Improved Laser Ablation Model for Asteroid Deflection," *Acta Astronautica*, Vol. 103, 2014, pp. 382–394.

[51] Ahrens, T. J., and O'Keefe, J. D., "Shock Melting and Vaporization of Lunar Rocks and Minerals," *The Moon*, Vol. 4, Nos. 1–2, 1972, pp. 214–249.

[52] Nagahara, H., Kushiro, I., and Mysen, B. O., "Evaporation of Olivine: Low Pressure Phase Relations of the Olivine System and Its Implications for the Origin of Chondritic Components in the Solar Nebula," *Geochimica et Cosmochimica Acta*, Vol. 58, 1994, pp. 1951–1963.

[53] Farquhar, J., and Rumble, D., "Comparison of Oxygen Isotope Data Obtained by Laser Fluorination of Olivine with krf Excimer Laser and CO_2 Laser," *Geochimica et Cosmochimica Acta*, Vol. 62, No. 18, 1998, pp. 3141–3149.

[54] Nagahara, H., Kushiro, I., Mori, H., and Mysen, B. O., "Experimental Vaporisation and Condensation of Olivine Solid Solution," *Nature*, Vol. 331, 1988, pp. 516–518.

[55] Nagahara, H., Kita, N. T., Ozawa, K., and Morishita, Y. "Condensation of Major Elements During Chondrule Formation and Its Implications to the Origin of Chondrules," *Nature*, Vol. 331, 2009, pp. 516–518.

[56] Hashimoto, A., "Evaporation Kinetics of Forsterite and Implications for the Early Solar Nebula," *Nature*, Vol. 347, 1990, pp. 53–55.

[57] Kahle, R., Kührt, E., Hahn, G., and Knollenberg, J., "Physical Limits of Solar Collectors in Deflecting Earth-Threatening Asteroids," *Aerospace Science and Technology*, Vol. 10, 2006, pp. 253–263.

[58] Legge, H., and Boettcher, R. D., "Modelling Control Thruster Plume Flow and Impingement," *Rarefied Gas Dynamics*, Vol. 2, 1985, pp. 983–992.

[59] Gibbings, A., Vasile, M., Watson, I., Hopkins, J.-M., and Burns, D., "Experimental Analysis of Laser Ablated Plumes for Asteroid Deflection and Exploitation," *Acta Astronautica*, Vol. 90, No. 1, 2013, pp. 85–97. NEO Planetary Defense: From Threat to Action—Selected papers from the *2011 IAA Planetary Defense Conference*, Bucharest.

[60] Brož, M., "Yarkovsky Effect and the Dynamics of the Solar System," Ph.D. Thesis, Faculty of Mathematics and Physics, Astronomical Institute, Charles University, Prague, Czech Republic, 2006.

[61] Phipps, C. R., "An Alternate Treatment of the Vapor-Plasma Transition," *International Journal of Aerospace Innovation*, Vol. 3, No. 1, 2011, pp. 45–50.

INDEX

Note: Page numbers followed by *f* or **t** (indicating figures or tables).

SUPPLEMENTAL MATERIALS

A complete listing of titles in the Progress in Astronautics and Aeronautics series is available from AIAA's electronic library, Aerospace Research Central (ARC) at arc.aiaa.org. Visit ARC frequently to stay abreast of product changes, corrections, special offers, and new publications.

AIAA is committed to devoting resources to the education of both practicing and future aerospace professionals. In 1996, the AIAA Foundation was founded. Its programs enhance scientific literacy and advance the arts and sciences of aerospace. For more information, please visit www.aiaafoundation.org.